Enantioselective Organocatalyzed Reactions I

Rainer Mahrwald
Editor

Enantioselective Organocatalyzed Reactions I

Enantioselective Oxidation, Reduction, Functionalization and Desymmetrization

Editor
Prof. Dr. Rainer Mahrwald
Institut für Chemie der
Humboldt-Universität zu Berlin
Brook-Taylor-Str. 2
12 489 Berlin
Germany
rainer.mahrwald@chemie.hu-berlin.de

ISBN 978-90-481-3864-7 e-ISBN 978-90-481-3865-4
DOI 10.1007/978-90-481-3865-4
Springer Dordrecht Heidelberg London New York

Library of Congress Control Number: 2011933564

Printed on acid-free paper

Springer is part of Springer Science+Business Media (www.springer.com)

Foreword

It is part of the human experience to wonder and marvel at the beauty of life in the world around us. An astronomer might look up in awe and question the size, shape and age of the universe. A biologist might pay keen attention to the anatomical features unique to a given organism and see the beauty in the organism's adaptability. The organic chemist, however, tends to look deeper and ponder the actual molecules, focusing on the unique mechanisms that create a wide array of complex molecules that in turn conspire to create life itself. This is as true today as it was more than a century ago when the great organic chemist Emil Fischer stated "If we wish to catch up with Nature, we shall need to use the same methods as she does, and I can foresee a time in which physiological chemistry will not only make greater use of natural enzymes, but will actually resort to creating synthetic ones."[1] Fischer foresaw that if we could understand how Nature's enzymes catalyze reactions, we could create our own synthetic catalysts. Indeed, it was in recreating Nature's aldolase enzymes that we were led to re-examine the chemistry of Hajos and Parrish in a new light. Through experimentation, we realized that the simple amino acid proline could recapitulate the 'complex' chemistry of an aldolase enzyme thereby providing a stunningly simple solution to the direct asymmetric aldol, Michael, Mannich and other reactions. Indeed, catalytic activity of amino acids, particularly in enamine and iminium chemistry, is not restricted to the amino acid proline but rather is a feature that most, if not all, amino acids have in common.

A decade has now passed since the studies of my laboratory and those of David MacMillan's refocused the considerable attention of the community on the profound potential of small organic molecules to catalyze asymmetric reactions. In this time, the scope of organocatalysis has enlarged considerably with respect both to the type of reactions catalyzed (aldol, cycloaddition, redox, asymmetric assembly and domino reactions, conjugate addition reactions, etc.) and the mechanisms used

[1] Fischer E: Synthesen in der Purin- und Zuckergruppe. In Les Prix Nobel en 1902. Edited by Cleve PT, Hasselberg C-B, Morner K-A-H: P-A Norstedt & Fils; 1905.

to affect catalysis (enamine, iminium, hydrogen bonding, Bronsted or Lewis acid/base chemistry, SOMO activation, etc.). With each new reaction and mechanistic lever applied, "the veil behind which Nature has so carefully concealed her secrets is being lifted."[1] Indeed, there is much to be gained in envisioning an enzyme as an organic flask that affects catalysis through the side chains of amino acids, their intervening amide linkages, organic cofactors, and the flask itself.

Although the notion of organocatalysis has been with us since the very beginnings of organic chemistry, in the last decade organocatalysis has created a sea change with respect to our abilities both to synthesize molecules asymmetrically and to understand how these molecules were synthesized in a prebiotic world before enzymes themselves existed. We speculate that the first carbohydrates might have been first synthesized via amino acid catalysis and can envision a route to homochirality that might have been exploited to create life. Thus organocatalysis might have been the key chemistry available in the prebiotic word and might have provided the homochiral building blocks that allowed life to form. Indeed, biosynthetic reactions occurring in organisms today that are catalyzed by small organic molecules might explain our inability to find protein enzymes for certain reactions. The secret of life now feels a bit closer. Organocatalysis has also changed our notions concerning what is possible in catalytic asymmetric synthesis. We are not so much concerned with creating a single stereogenic center now or a single bond connection, but with creating arrays of 2, 3, 4, 5, or more stereocenters and bond connections under organocatalysis in a single pot with excellent control of enantio- and diastereoselectivity. Such reports are now becoming commonplace whereas a decade ago they would have been greeted as major triumphs. This has led many to suggest that we are in the golden age of organocatalysis, that the creative stage of this endeavor has somehow now passed. I believe we are just at the beginning of this endeavor and that much fascinating and unexpected chemistry lies ahead of us.

The explosive growth of asymmetric organocatalysis has been driven by a new-found appreciation for reactivity among small organic molecules, but we have barely begun to explore this space. How big is this space and what might we find? It is easier to answer the first part of this question. It is vast indeed. If we consider calculations performed to define the space of small molecules of molecular weight of less than 500 Da that consist of only C, H, N, O, P, S, Cl, and Br and are stable at room temperature and to oxygen and water, the number is unimaginable – more than 10^{63} molecules.[2] With this in mind, I believe that the organocatalysts known today represent islands of reactivity or catalytic potential in a near infinite sea. The discovery of new islands of catalytic potential will rely on the scientific method, chemical intuition, and luck but they are out there to be discovered and they promise new and ever more stunning catalytic syntheses of complex molecules and ever greater understanding of how atoms born in stars eventually conspired to create life. As Emil Fischer told us more than a century ago, progress towards this goal will "not

[2] W.C. Guida et al. (1996) Med. Res. Rev. 16, 3–50.

so much be determined by brilliant achievements of individual workers, but rather by the planned collaboration of many observers."[1] And so, the future of organocatalysis is vast and bright and will continue to benefit from the community of chemists that join this endeavor. Stunning reactions and reactivities await us.

The Scripps Research Institute
La Jolla, CA

Carlos F. Barbas, III

Preface

Without any doubt organocatalysis belongs to the most exciting and innovative chapters of organic chemistry today.

Since this chapter was first opened systematically 10 years ago, a plethora of methods and catalysts have been developed to solve problems of organic chemistry. More and more these methodologies have been applied in total syntheses of natural products. This is what this two-volume book set wants to demonstrate – the full power of organocatalysis.

Asymmetric C-C bond formation processes form the subject of the second book while functionalization, catalysts and general aspects of organocatalysis are covered in the first one. Overlappings cannot be entirely avoided by such an approach. However, often these overlappings are desirable and valuable in order to illustrate a methodology by different views as this is true for asymmetric hydrogenation, enantioselective conjugate hydride addition, oxidation or transfer hydrogenation catalyzed by chiral primary amines.

It took great pleasure in working together with a team of leading experts in this field. I have to thank them for organizing these overviews.

Also, I thank the team at Springer UK to help us to publish these results.

Berlin
Spring 2011

Rainer Mahrwald

Contents

Contributors

Kristin Brand Leibniz Institute of Plant Biochemistry, Weinberg 3, D-06120 Halle (Saale), Germany, Kristin.Brand@ipb-halle.de

Martin C. Nin Brauer Leibniz Institute of Plant Biochemistry, Weinberg 3, D-06120 Halle (Saale), Germany, MartinClaudio.NinBrauer@ipb-halle.de

Jean-François Brière Institut de Recherche en Chimie Organique Fine (IRCOF) associé au CNRS (UMR COBRA-6014 & FR 3038), INSA-Université de Rouen, Rue Tesnière, Mont-Saint-Aignan F-76130, France, jean-francois.briere@insa-rouen.fr

Benjamin R. Buckley Department of Chemistry, Loughborough University, Loughborough, Leicestershire LE11 5EE, UK, b.r.buckley@lboro.ac.uk

Liujuan Chen Beijing National Laboratory for Molecular Sciences (BNLMS), CAS Key Laboratory of Molecular Recognition and Function, Institute of Chemistry, the Chinese Academy of Sciences, Beijing 100190, P.R. China, chenliujuan@iccas.ac.cn

Gabriela Guillena Instituto de Síntesis Orgánica (ISO) and Departamento de Química Orgánica, Facultad de Ciencias, Universidad de Alicante, Apdo. 99, E-03080 Alicante, Spain, gabriela.guillena@ua.es

Vincent Levacher Institut de Recherche en Chimie Organique Fine (IRCOF) associé au CNRS (UMR COBRA-6014 & FR 3038), INSA-Université de Rouen, Rue Tesnière, Mont-Saint-Aignan F-76130, France, vincent.levacher@insa-rouen.fr

Sanzhong Luo Beijing National Laboratory for Molecular Sciences (BNLMS), CAS Key Laboratory of Molecular Recognition and Function, Institute of Chemistry, the Chinese Academy of Sciences, Beijing 100190, P.R. China, luosz@iccas.ac.cn

Tommaso Marcelli Dipartimento di Chimica, Materiali ed Ingegneria Chimica "Giulio Natta", Politecnico di Milano, Via Mancinelli, 7, 20131 Milan, Italy, tommaso.marcelli@chem.polimi.it

Stephen P. Neary Department of Chemistry, Loughborough University, Loughborough, Leicestershire LE11 5EE, UK, s.p.neary@lboro.ac.uk

Sylvain Oudeyer Institut de Recherche en Chimie Organique Fine (IRCOF) associé au CNRS (UMR COBRA-6014 & FR 3038), INSA-Université de Rouen, Rue Tesnière, Mont-Saint-Aignan F-76130, France, sylvain.oudeyer@insa-rouen.fr

Petri M. Pihko Department of Chemistry, University of Jyväskylä, P.O. Box 35, FI-40014 Jyväskylä, Finland, Petri.Pihko@jyu.fi

Thomas Poisson Institut de Recherche en Chimie Organique Fine (IRCOF) associé au CNRS (UMR COBRA-6014 & FR 3038), INSA-Université de Rouen, Rue Tesnière, Mont-Saint-Aignan F-76130, France, thomas.poisson@insa-rouen.fr

Hasibur Rahaman Department of Chemistry, University of Jyväskylä, P.O. Box 35, FI-40014 Jyväskylä, Finland, hasibur.h.rahaman@jyu.fi

Diego J. Ramón Instituto de Síntesis Orgánica (ISO) and Departamento de Química Orgánica, Facultad de Ciencias, Universidad de Alicante, Apdo. 99, E-03080 Alicante, Spain, djramon@ua.es

Ludger A. Wessjohann Leibniz Institute of Plant Biochemistry, Weinberg 3, D-06120 Halle (Saale), Germany, wessjohann@ipb-halle.de

Chapter 1
Organocatalysed Asymmetric Oxidation Reactions

Benjamin R. Buckley and Stephen P. Neary

Abstract This review highlights the progress of highly enantioselective organocatalytic systems for asymmetric oxidation. For example, since the first reported enantioselective dioxirane catalysts developed independently by Curci and Marples, several systems have been developed which give greater than 90% ee for epoxidation. Dioxiranes have also been used for a range of other oxidation reactions including C–H oxidation. Oxaziridines have been widely used for the enantioselective oxidation of sulfides, alkenes and they have recently been reported, for the first time, to catalyse the epoxidation and C–H hydroxylation of a range of substrates. Oxaziridinium salts, the quaternised analogues of oxaziridines are prepared *in situ* from the corresponding iminium ions and ees of up to 97% have been achieved. The epoxidation of α,β-unsaturated ketones has been successfully achieved using asymmetric phase transfer catalysis and *in situ* amine/enamine catalysis.

1.1 Introduction

From both an economic and environmental viewpoint there is an urgent need for more atom efficient systems that employ environmentally benign oxidants. Coupled with the central goal for green chemistry, i.e. reduction or elimination of hazards from chemical products and processes, the constraints in this area are even greater. Organocatalysis, that is catalysis by organic molecules in the absence of metals, is currently attracting considerable attention. In comparison with transition metal catalysed processes organic based catalysts have several advantages in not only ease

B.R. Buckley (✉) • S.P. Neary
Department of Chemistry, Loughborough University, Loughborough, Leicestershire LE11 5EE, UK
e-mail: b.r.buckley@lboro.ac.uk; s.p.neary@lboro.ac.uk

R. Mahrwald (ed.), *Enantioselective Organocatalyzed Reactions I: Enantioselective Oxidation, Reduction, Functionalization and Desymmetrization*,
DOI 10.1007/978-90-481-3865-4_1,

of product purification, greater ease of handling, and reduced toxicity, but also in their inherent stability to air and moisture, and they are often cheaper than their transition metal counterparts. Hence they are extremely desirable as they satisfy the required criteria from the view point of green chemistry.

Selective oxidation is a common requirement for wide range of processes and is particularly problematic in areas where an enantioselective reaction is required, as is common in the production of pharmaceutical and agrochemical actives and intermediates. Oxidation processes in general are of great importance in organic chemistry, both in industry and in academic laboratories, and epoxidation, hydroxylation, and oxidative cleavage of alkenes are among the most common transformations.

1.2 Epoxidation

Epoxides, or oxiranes as they are otherwise known, are reactive heterocycles and are unlike most other cyclic ethers, which are generally used as unreactive organic solvents. The reason for their reactivity can be attributed to the ring strain associated with their structure, where the bond angles are close to 60° rather than ~109° that is expected for tetrahedral carbon or oxygen atoms. Thus, ring opening of the epoxide functionality is the driving force behind many reactions. Epoxides readily undergo reactions in which the three membered ring is opened by nucleophiles, and this can take place under neutral, basic or acidic conditions, giving the corresponding alcohols (Scheme 1.1).

Scheme 1.1 Nucleophilic ring opening of the epoxide functionality

This high degree of reactivity has established the epoxide moiety as an important molecule not only in organic synthesis but also in biological systems. Epoxides are found in a large variety of natural products and as key intermediates in the synthesis of several biologically active compounds. The usefulness of epoxides as key intermediates in organic synthesis is well documented. There are plethoras of elegant synthesis using this moiety. Of the many synthetically useful examples chiral epoxides have been successfully used to give enantiomerically enriched 1,2-aminoalcohols [1], by attack of the epoxide with either ammonia or sodium azide; 1,2 diamines [2], by ring opening with a secondary amine and subsequent aziridinium formation followed by attack of a primary amine; and aziridines [3], through reaction of the epoxide with sodium azide followed by treatment with triphenylphosphine.

1.2.1 Achiral Synthesis of Epoxides

The most widely used method for the production of epoxides is through the oxidation of alkenes using organic peracids and was first discovered in 1909 by Prileshaev [4]. The oxidants required for this type of reaction are generally made from hydrogen peroxide. These are usually perbenzoic acid, its substituted derivatives or, mainly used in industry, aliphatic peracids.

Of the percarboxylic acids, commercially available *m*-chloroperbenzoic acid (*m*-CPBA) **1** is the most favoured and inert solvents such as dichloromethane, chloroform and benzene are commonly used in the reaction media (Scheme 1.2).

Scheme 1.2 Epoxidation of alkenes mediated by *m*-CPBA

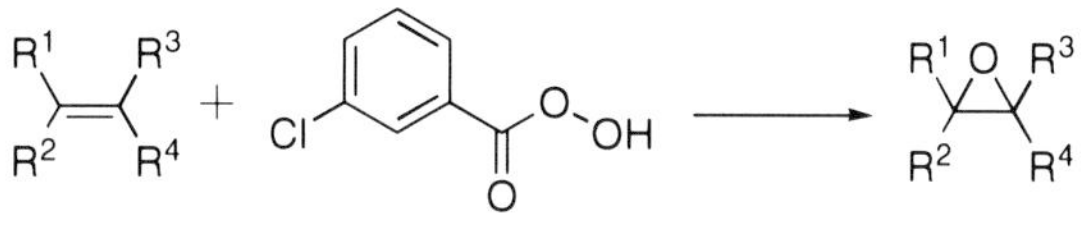

Peracid oxidation is an electrophilic process in which the driving force of the reaction is provided by the electron donating nature of the alkene double bond and the electron accepting nature of the peracid group. The transition state of this type of reaction was originally believed to be planar or "butterfly" [5]. However, recent calculations favour the spiro transition state (Scheme 1.3) [6]. The transfer of oxygen is concerted and hence the reactions are stereospecific.

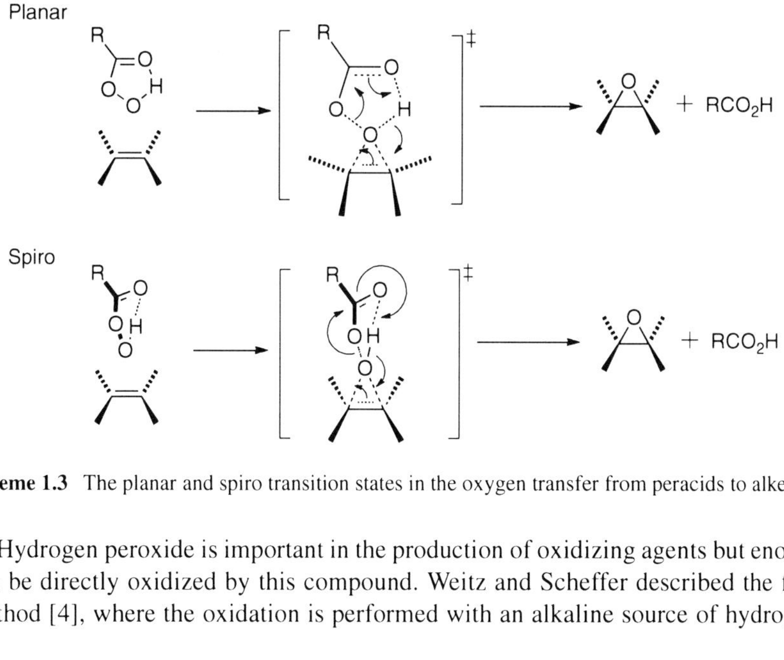

Scheme 1.3 The planar and spiro transition states in the oxygen transfer from peracids to alkenes

Hydrogen peroxide is important in the production of oxidizing agents but enones can be directly oxidized by this compound. Weitz and Scheffer described the first method [4], where the oxidation is performed with an alkaline source of hydrogen peroxide. A reversible attack by the nucleophilic hydroperoxy anion **2**, in a conjugated enone system, followed by ring closure and hydroxide elimination from the intermediate enolate anion **3**, affords the epoxide (Scheme 1.4).

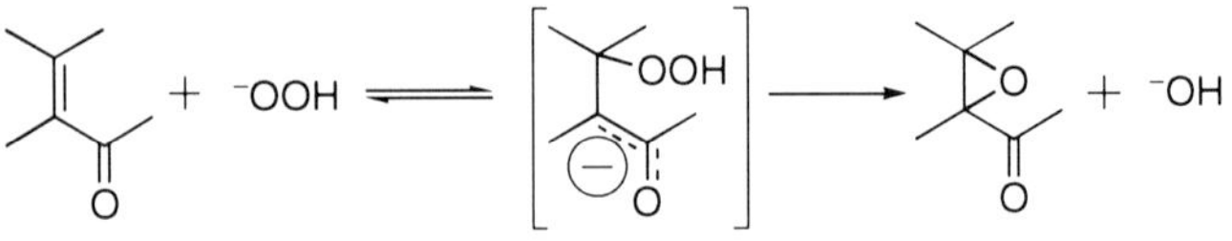

Scheme 1.4 The Weitz and Scheffer hydrogen peroxide mediated epoxidation

Good yields are generally observed for this type of reaction; however, it is non-stereospecific, due to the extended life time of the intermediate enolate anion, which has time to rotate around the alpha and beta carbon-carbon bond. Oxides of transition metals are often used as catalysts with hydrogen peroxide to afford the epoxides of electron poor alkenes.

Payne has also reported that alkenes present in an alkaline solution of hydrogen peroxide can be epoxidized. The presence of a nitrile is required and alkenes are epoxidized in good to excellent yields [7, 8]. It is thought that the reaction proceeds through a peroxyimidic acid **4**, which is the addition adduct resulting from nucleophilic attack of hydroperoxide on the nitrile carbon atom, but this intermediate has never been isolated (Scheme 1.5). In the absence of substrate, the hydroperoxyimine (or peroxyimidic acid) disproportionates to give the corresponding primary amide and molecular oxygen, and this process constitutes the Radziszewski reaction [7]. In the presence of an olefin the reactive intermediate is intercepted, and epoxidation occurs, with the primary amide (hydrated nitrile), being the only side product of the reaction. The transition state for the oxygen transfer step presumably resembles that described for the epoxidation of alkenes by peracids but no definite proof exists. The reaction is stereospecific and therefore retention of alkene geometry is observed.

Scheme 1.5 Formation of Peroxyimidic acid, the reactive intermediate in the Payne epoxidation of alkenes

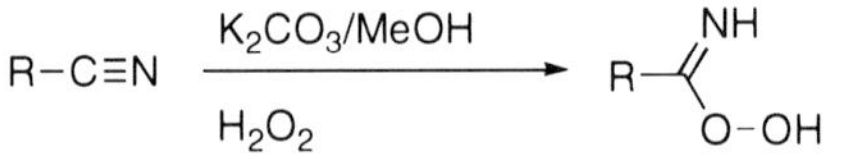

R = Alkyl, aryl

1.2.2 *Catalytic Asymmetric Epoxidation Methods*

Non-racemic chiral peracids are of limited value for asymmetric epoxidation [9]. An asymmetric version of the related Payne procedure, mediated by a nitrile and hydrogen peroxide, has provided high ees [10]. 2-Cyanoheptahelicene **5** was reported to epoxidise *trans*-stilbene and α-methylstyrene under Payne conditions, with high enantiocontrol. The use of helicenes in asymmetric epoxidation, however, has not been elaborated, presumably due to the difficulty of synthesis of helicenes in enantiomerically pure form (Scheme 1.6) [11].

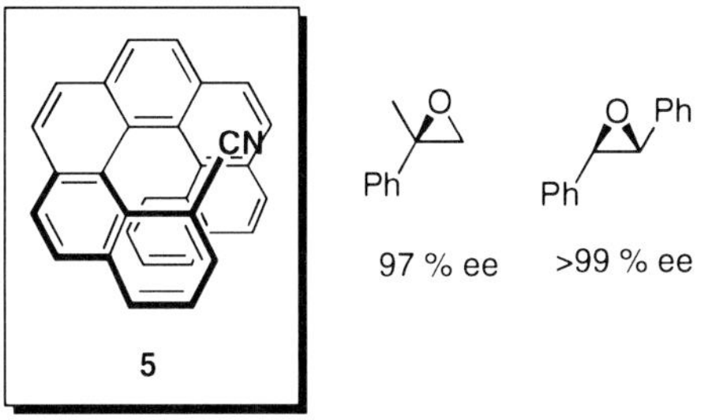

CN
5
Ph
97 % ee
O Ph
Ph
>99 % ee

Scheme 1.6 2-Cyanoheptahelicene as a chiral mediator in the classical Payne epoxidation

1.2.2.1 The Julià Epoxidation of α,β-Unsaturated Ketones

Synthetic polypeptides were found, by Julià, to epoxidize α,β-unsaturated ketones with high enantioselectivity [12]. The Julià process can be easily performed at 0°C, and using a triphasic system comprising of toluene, water and polyalanine in the presence of alkaline hydrogen peroxide, chalcone oxide was produced in 97% ee (Scheme 1.7). The Julià process has become the method of choice for the epoxidation of *trans*-1,3-diarylenones. However, this methodology is extremely substrate-specific, and enones with enolisable α-protons are usually poor substrates.

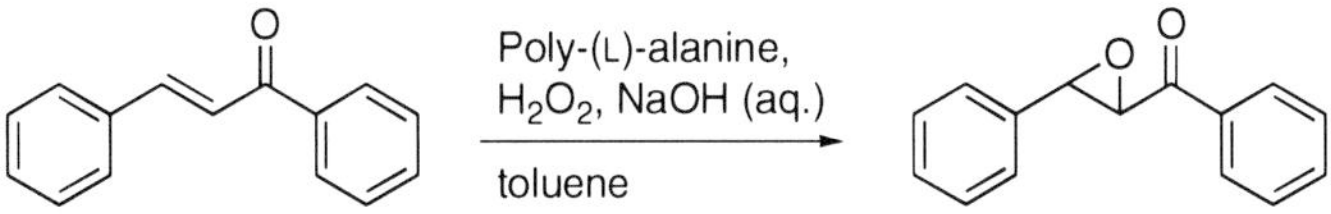

Poly-(L)-alanine,
H_2O_2, NaOH (aq.)
toluene

Scheme 1.7 The Julià catalytic asymmetric epoxidation of chalcone

Roberts has shown that the asymmetric epoxidation of chalcone can be catalysed by polyamino acid derivatives under non-aqueous conditions [13]. This improved reaction involves the use of a urea-hydrogen peroxide complex in THF, in the presence of an organic base (DBU) and immobilized poly-(L)-leucine. Under these conditions, the reaction of chalcone derivatives and related substrates provided the corresponding epoxides in 70–99% yield and 83–95% ee within 30 min. Several substrates with enolisable enones have also been epoxidized successfully [14].

The asymmetric epoxidation of the chalcone type of substrate has also been accomplished using other types of chiral catalysts [15]. Wynberg was the first to use chiral ammonium salts, and obtained chalcone oxide with 55% ee using alkaline hydrogen peroxide as the stoichiometric oxidant and a quinine-derived quaternary ammonium salt as the chiral phase transfer catalyst [16]. More recently, Lygo

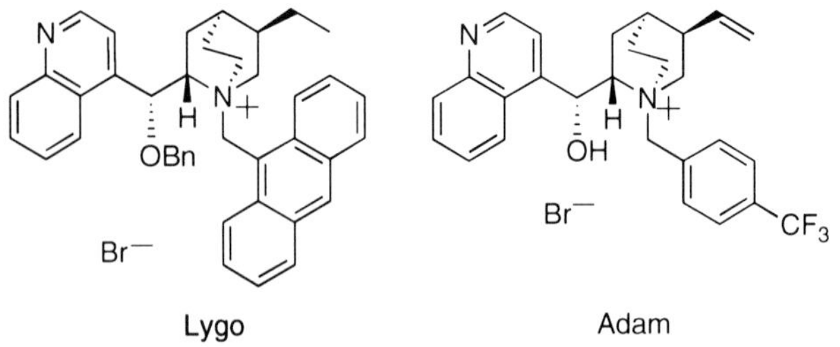

Fig. 1.1 Lygo's and Adam's chinchona-derived quaternary ammonium salts for catalytic asymmetric epoxidation

re-examined the effect of the structure of the chinchona derived ammonium salt on the enantioselectivity of the process (Fig. 1.1) [17]. Optimization led to superior catalysts for the asymmetric epoxidation of 1,3-diarylenones and some substrates with aliphatic substituents. The enantioselectivities for the substrates tested ranged between 69% and 89% ee. Further modification by Adam has produced ees of up to 98% [18]. For further advances in this area please see Chap. 5.

1.2.2.2 Dioxirane-Based Systems for Asymmetric Epoxidation

Dioxirane-based systems for asymmetric epoxidation have received much attention over the past decade, and they have emerged as one of the most effective methods for producing enantiomerically enriched oxiranes. Their reactivity stems from the ring strain associated with the three membered ring and the relatively weak O-O bond. They are readily attacked even by poor nucleophiles such as olefins. The general method to produce a dioxirane is through the use of a ketone and a stoichiometric oxidant (usually Oxone) in either acetonitrile or dimethoxymethane (Scheme 1.8).

Scheme 1.8 Generation of dioxiranes using Oxone as stoichiometric oxidant in the epoxidation of alkenes

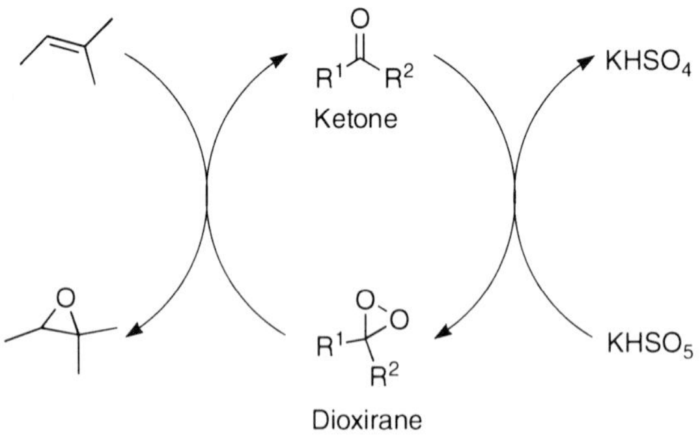

Initially several research groups (Curci and Marples) studied chiral ketones as catalysts for asymmetric epoxidation, but until recently enantioselectivities with these systems were only up to *ca.* 20% ee (Fig. 1.2) [19–21].

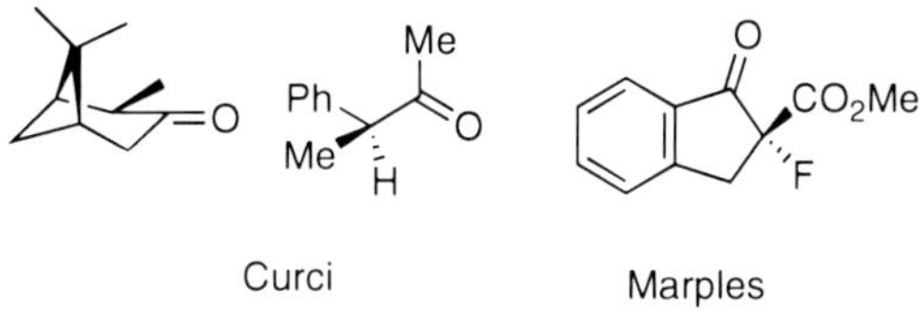

Fig. 1.2 Initial chiral ketones screened by Curci and Marples for dioxirane-based asymmetric epoxidation

Highly Enantioselective Systems Developed by Yang

In 1996 Yang first reported the asymmetric epoxidation of olefins mediated by a C_2 symmetric dioxirane generated from the corresponding ketone [22]. The chiral ketone was derived from BINAP, and exhibited enantioselectivities typically between 5% and 50% ee under stoichiometric conditions, and 87% ee in the epoxidation of *trans*-4,4′- diphenylstilbene. With modification of the original C2 symmetric ketone and development of the reaction to run catalytically, Yang was able to increase the enantioselectivity of the process [23, 24]. With very hindered alkenes, such as *trans*-4,4′-di*tert*-butylstilbene, enantioselectivities of up to 95% ee were achieved (Scheme 1.9).

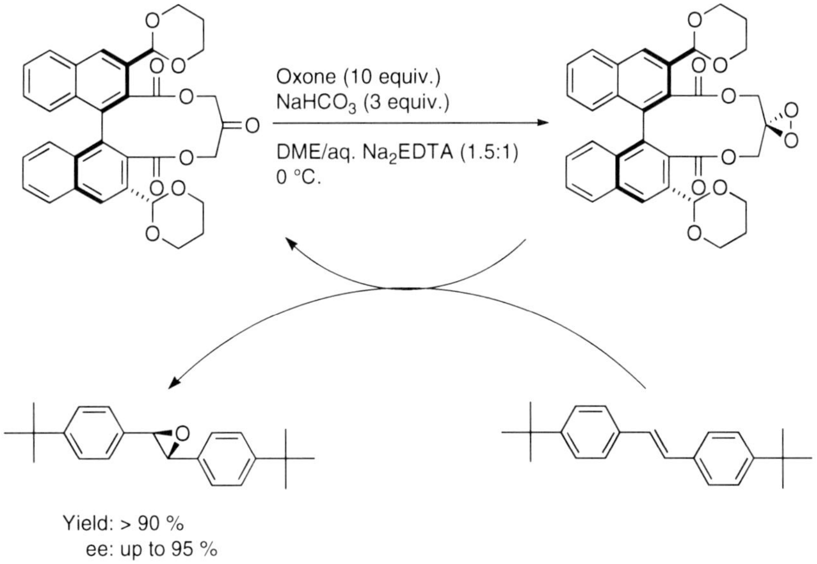

Scheme 1.9 Yang's chiral ketone derived dioxirane based on binaphtyl-2.2'-dicarboxylic acid

Denmark's Dioxiranes for Asymmetric Epoxidation

After establishing oxo ammonium salts (Fig. 1.3) as useful mediators in the catalytic epoxidation of several alkenes [25, 26], Denmark developed two chiral systems which imparted moderate enantioselectivities (Fig. 1.3), catalyst **6**, for example, producing 1-phenylcyclohexene oxide with 58% ee.

Fig. 1.3 Denmark's Oxo Ammonium Salts

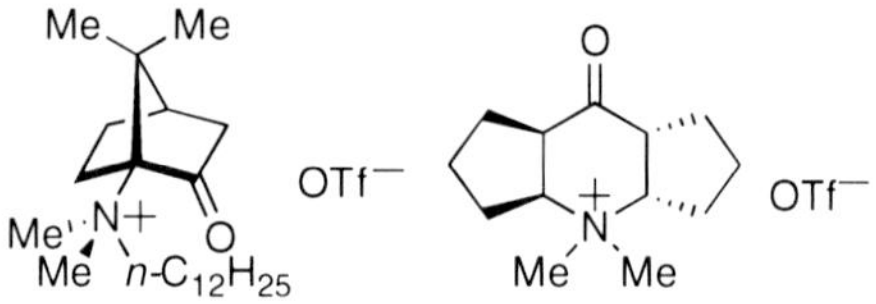

Me Me
OTf⁻
Me N+ O
Me n-$C_{12}H_{25}$
O
OTf⁻
N
Me Me
6

Fig. 1.4 Denmark's aliphatic Oxo Bis(ammonium) salts

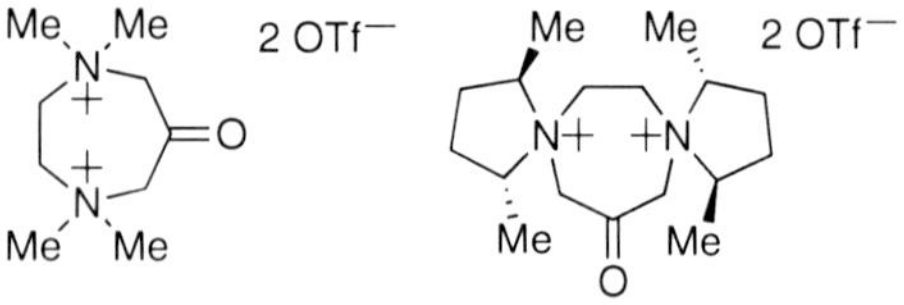

Me Me
2 OTf⁻
N
O
N
Me Me
7

Me Me
2 OTf⁻
N+ +N
Me Me
O
8

Furthering this work, Denmark produced a range of oxo bis(ammonium) salts to enhance the electrophilicity of the carbonyl carbon and it was hoped this would eliminate the Baeyer-Villiger reaction. The first two attempts resulted in no formation of epoxide in the oxidation reaction. Oxone decomposition studies revealed that the catalysts were rapidly destroyed by oxone.

It was believed that aliphatic bis(ammonium) ketones would not suffer the oxidative instability displayed by the aromatic counterparts (Fig. 1.4). Indeed ketone **7** when using 10 mol% epoxidized a variety of olefins in good yield, but the chiral variants only gave poor ees, for example, *ca* 10% ee for *trans*-β-methylstyrene (using catalyst **8**).

Fluorine-containing ketones have proven to be one of the most successful types of catalyst for dioxirane-mediated epoxidation [27]. Denmark has shown that good to excellent enantioselectivities can be achieved with catalyst **9** for *trans*-olefins (Scheme 1.10). However, catalyst loadings are high (30 mol%) [28].

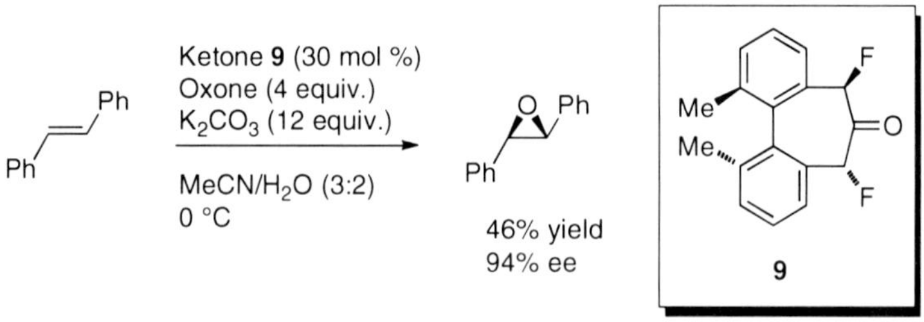

Ph
Ph
Ketone **9** (30 mol %)
Oxone (4 equiv.)
K_2CO_3 (12 equiv.)
$MeCN/H_2O$ (3:2)
0 °C
O Ph
Ph
46% yield
94% ee
Me
Me
F
O
F
9

Scheme 1.10 Denmark's most successful system containing a fluoroketone

Armstrong's α-fluoroketone as a Precursor in Dioxirane Mediated Epoxidation

Several other attempts at producing highly enantioselective dioxirane-based catalysts for asymmetric epoxidation have been reported [29, 30]; of these Armstrong's tropinone-derived α-fluoroketone was found to be a good mediator (Scheme 1.11)

[31]. The ketone exhibited enantioselectivities in the range of 64–83% ee for various substrates. Armstrong argued that the α-fluoro group exerts a stabilizing and directing effect by interacting with the olefinic proton in the substrate during the transition state of oxygen transfer.

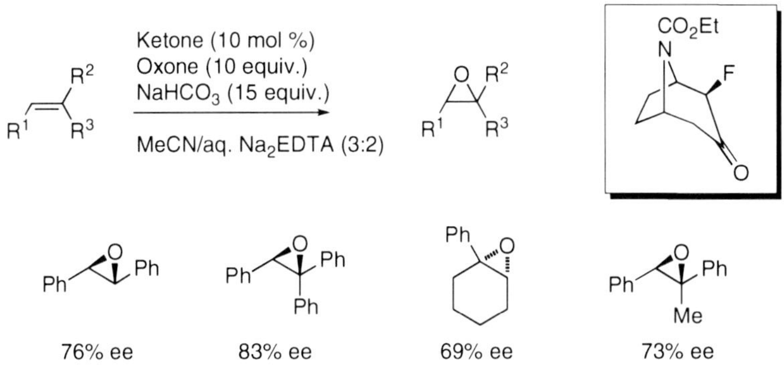

Ketone (10 mol %)
Oxone (10 equiv.)
$NaHCO_3$ (15 equiv.)
MeCN/aq. Na_2EDTA (3:2)

76% ee 83% ee 69% ee 73% ee

Scheme 1.11 Armstrong's tropinone-derived α-fluoroketone gave ees of up to 83%

Shi's Highly Enantioselective Dioxirane Epoxidation Mediated by a Fructose-derived Dioxirane

A breakthrough in dioxirane-mediated epoxidation was achieved by Shi in the late 1990s. He reported excellent ees for a wide variety of substrates using the fructose-derived ketone **10** [32]. This catalyst is easily prepared in two steps, and typical enantioselectivities range from 80% to 95% ee. However, the chiral ketone decomposes under the reaction conditions (pH 7–8), presumably through Baeyer-Villiger oxidation, and initially a large excess of the mediator had to be used (3 equivalents, with respect to the substrate) (Scheme 1.12).

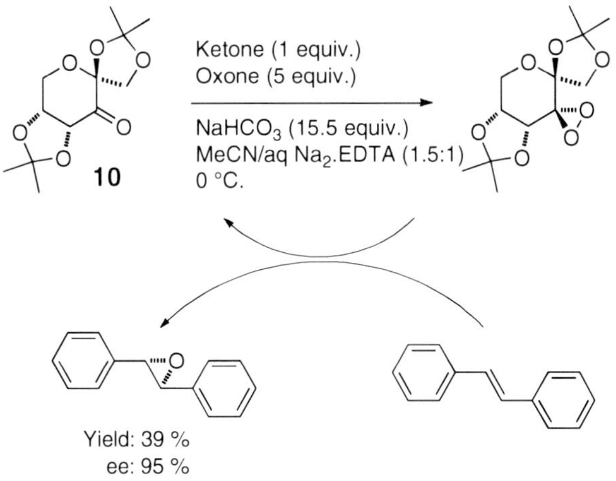

Ketone (1 equiv.)
Oxone (5 equiv.)
$NaHCO_3$ (15.5 equiv.)
MeCN/aq Na_2.EDTA (1.5:1)
0 °C.

Yield: 39 %
ee: 95 %

Scheme 1.12 Shi's fructose derived ketone for asymmetric epoxidation

After experimentation it was found that Baeyer-Villiger oxidation could be suppressed and the amount of catalyst could be reduced to 20 mol% if the reaction was carried out between pH 10 and 11 (Scheme 1.13) [33, 34]. Yields were increased (65–95%) and the catalytic system resulted in slightly higher ees (91–97% ee). The synthetic utility of this system has been widely explored with the successful asymmetric epoxidation of various hydroxyalkenes (90–94% ee) [35], enol ethers and enol esters (80–91% ee) [36], enynes (90–97% ee) [37], vinylsilanes (84–94% ee) [38], cis-alkenes (84–97% ee)[39, 40], terminal alkenes (30–94% ee) [41], and mono-epoxidation of conjugated dienes (90–97% ee) [42].

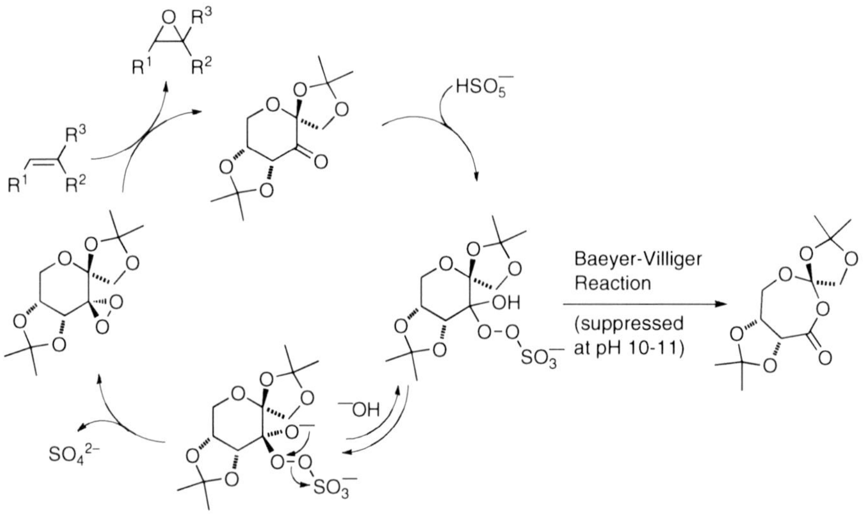

Baeyer-Villiger Reaction
(suppressed at pH 10-11)

Scheme 1.13 Shi's proposed catalytic cycle for dioxirane mediated epoxidation

Shi has even expanded this work to the kinetic resolution of racemic cyclic olefins [43]. Successful resolution of cyclic olefins with the chiral centre at the allylic position has been achieved (Scheme 1.14).

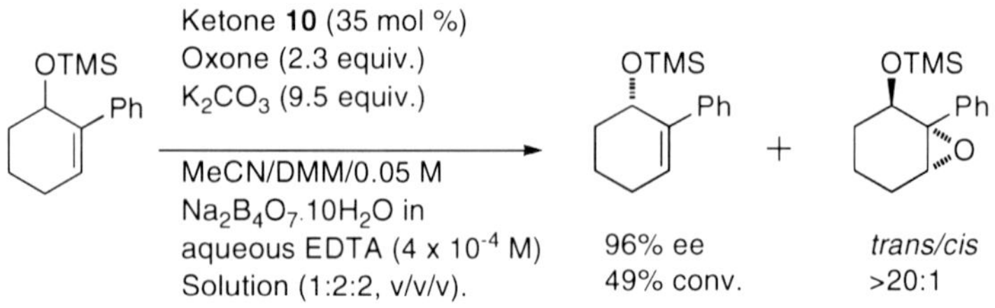

Ketone **10** (35 mol %)
Oxone (2.3 equiv.)
K_2CO_3 (9.5 equiv.)
MeCN/DMM/0.05 M
$Na_2B_4O_7 \cdot 10H_2O$ in
aqueous EDTA (4 x 10^{-4} M)
Solution (1:2:2, v/v/v).

96% ee
49% conv.

trans/cis
>20:1

Scheme 1.14 Shi's Kinetic resolution of cyclic olefins

Shi has also recently reported asymmetric epoxidation mediated by alkaline hydrogen peroxide [44, 45]. High yields and ees were obtained under these reaction conditions with up to 95% ee for 1-phenylcyclohexene oxide using the original fructose derived catalyst **10**. Peroxyimidic acid **11** is postulated to be the active oxidant (Scheme 1.15).

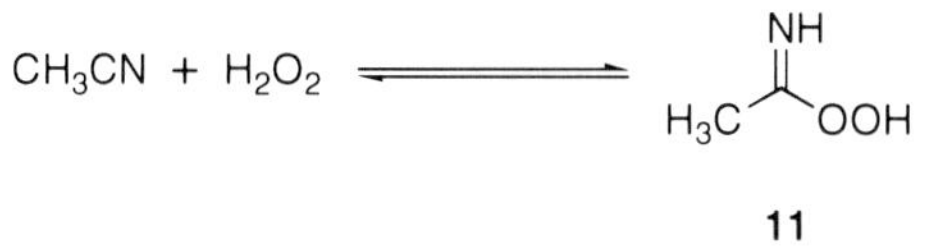

$CH_3CN + H_2O_2$ ⇌ H_3C–C(=NH)–OOH

11

Scheme 1.15 Postulated formation of peroxyimidic acid in the asymmetric epoxidation reaction using hydrogen peroxide as primary oxidant

In one of his most recent publications, highly efficient asymmetric epoxidation of α,β-unsaturated esters has been achieved using a modified fructose catalyst [46]. It was found that the original catalyst **10** was not effective toward α,β-unsaturated esters due to its decomposition under the reaction conditions by Baeyer-Villiger oxidation. Replacement of the fused ketal in **10** with more electron withdrawing groups (acetates) produced a active and highly enantioselective catalyst (**12**, Scheme 1.16).

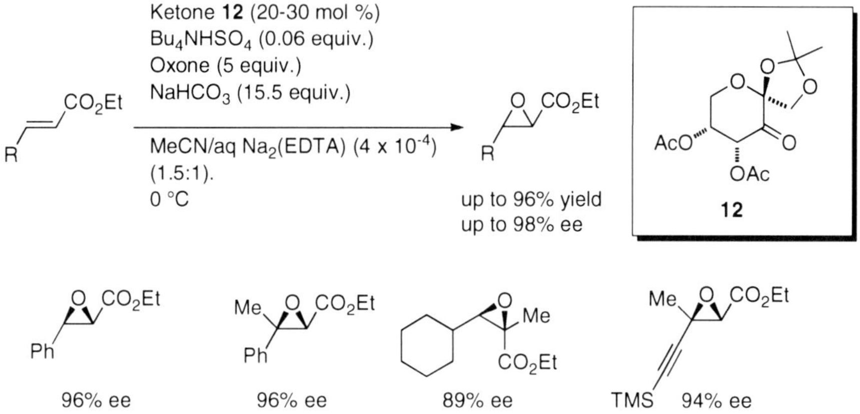

Ketone **12** (20-30 mol %)
Bu_4NHSO_4 (0.06 equiv.)
Oxone (5 equiv.)
$NaHCO_3$ (15.5 equiv.)
MeCN/aq $Na_2(EDTA)$ (4×10^{-4})
(1.5:1).
0 °C

up to 96% yield
up to 98% ee

AcO, O, O, O, O, OAc, **12**

Ph: 96% ee; Me, Ph: 96% ee; cyclohexyl, Me: 89% ee; Me, TMS: 94% ee

Scheme 1.16 Shi's fructose-derived catalyst for α,β-unsaturated esters

More recently Shi has reported several highly enantioselective epoxidation systems based on the oxazolidinone catalysts **13** and **14**, modified from the original fructose-derived ketone **10** [47]. Highly chemo- and enantioselective epoxidation of *cis*-enynes was achieved using oxone as the stoichiometric oxidant (Scheme 1.17). In a separate report ketone **14** has been used in the epoxidation of a variety of substrates using hydrogen peroxide as the stoichiometric oxidant, for example the

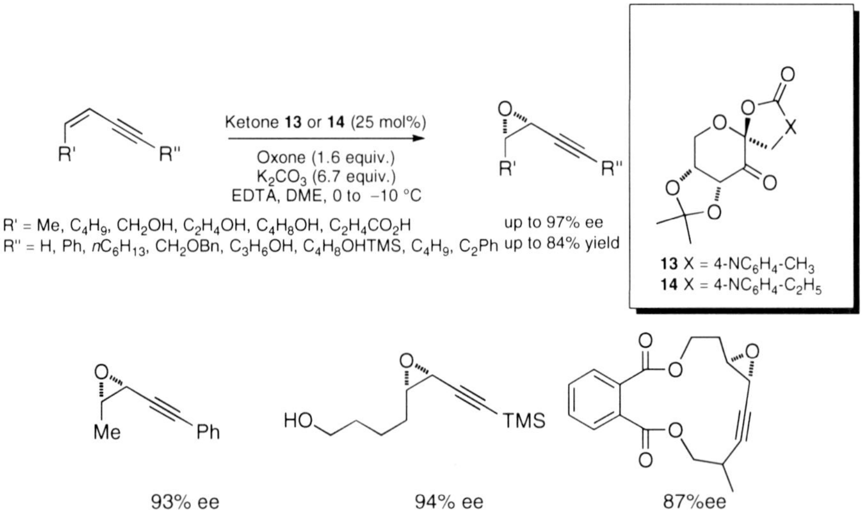

Ketone **13** or **14** (25 mol%)
Oxone (1.6 equiv.)
K_2CO_3 (6.7 equiv.)
EDTA, DME, 0 to −10 °C

R' = Me, C_4H_9, CH_2OH, C_2H_4OH, C_4H_8OH, $C_2H_4CO_2H$ up to 97% ee
R'' = H, Ph, nC_6H_{13}, CH_2OBn, C_3H_6OH, C_4H_8OHTMS, C_4H_9, C_2Ph up to 84% yield

13 X = 4-NC_6H_4-CH_3
14 X = 4-NC_6H_4-C_2H_5

93% ee 94% ee 87%ee

Scheme 1.17 Shi's highly chemo- and enantioselective epoxidation of *cis*-enynes

alkene in Scheme 1.18 [48]. Ager and co-workers have also reported the first large scale application of this technology, with over 100 kg of the lactone **15** produced using ketone **10** (Scheme 1.19) [49].

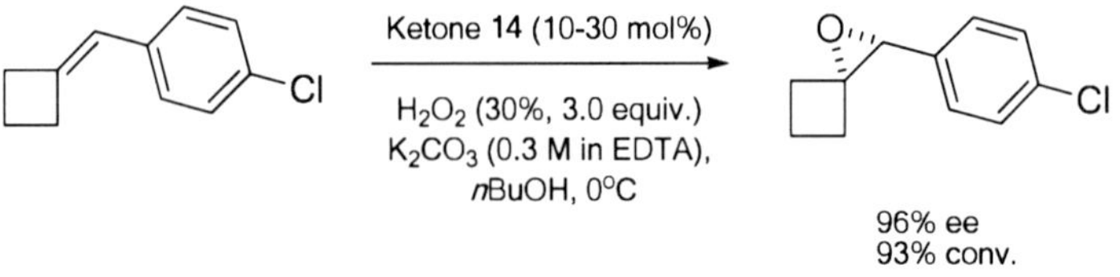

Ketone 14 (10-30 mol%)
H_2O_2 (30%, 3.0 equiv.)
K_2CO_3 (0.3 M in EDTA),
*n*BuOH, 0°C

96% ee
93% conv.

Scheme 1.18 Epoxidation of a variety of substrates using hydrogen peroxide as the stoichiometric oxidant

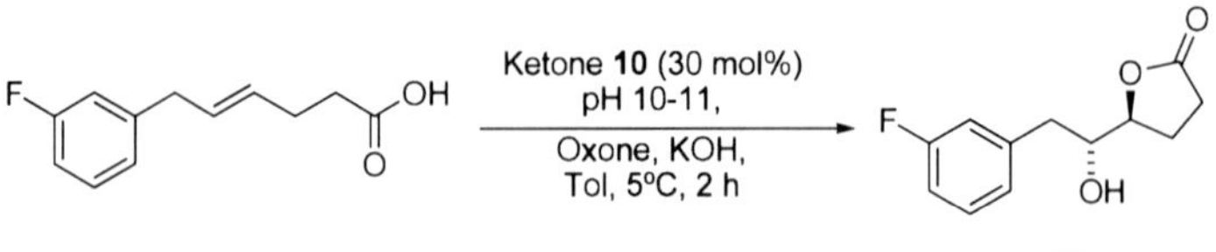

Ketone **10** (30 mol%)
pH 10-11,
Oxone, KOH,
Tol, 5°C, 2 h

15

Scheme 1.19 The first large scale application of the Shi epoxidation

The success of Shi's catalyst is believed to stem from a well ordered transition state, and results favour the spiro transition state (Fig. 1.5). The favoured transition state for *trans* or trisubstituted olefins is shown mediated by catalyst **10b** and the

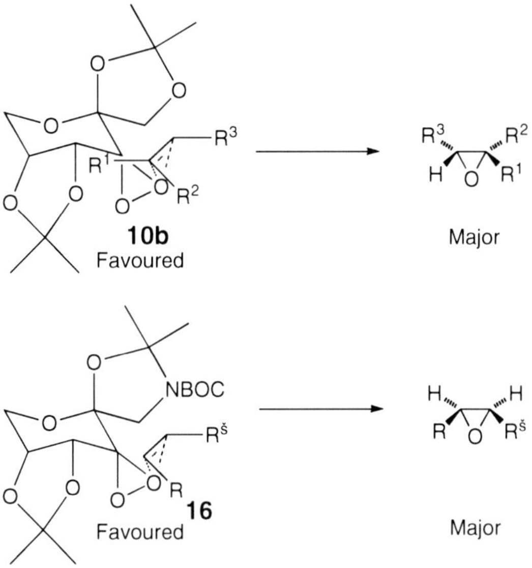

Fig. 1.5 Shi's proposed transition states for dioxirane mediated epoxidation

favoured transition state for *cis* disubstituted olefins is shown, mediated by the BOC protected catalyst **16**.

A recent application in the synthesis of (–)-glabrescol, by Corey, illustrates the scope of dioxirane-mediated asymmetric epoxidation (Scheme 1.20) [50].

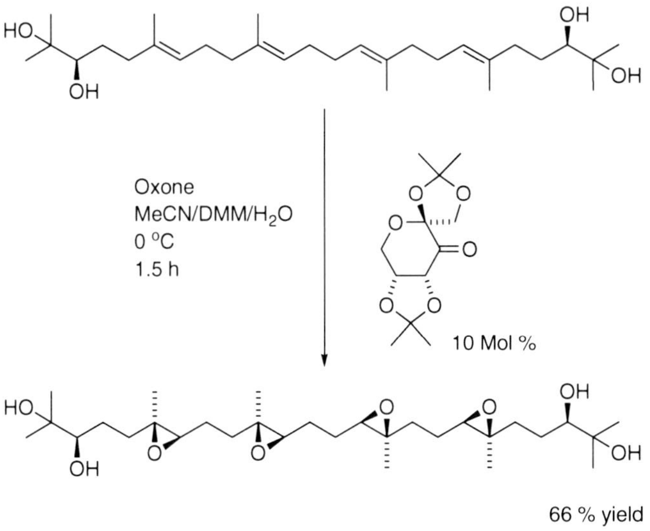

Scheme 1.20 Corey's Synthesis of (–)-glabrescol using Shi's fructose derived ketone

1.2.2.3 Oxaziridines

Oxaziridines, the nitrogen analogues of dioxiranes, also act as oxygen transfer agents. Major accomplishments in this area have been pioneered by Davis. It was shown that oxaziridines can epoxidize simple alkenes, but the reaction is significantly slower (3–12 h at 60°C) than for dioxiranes [51]. Oxaziridines are generally accessed by oxidation of the corresponding imines, which are produced upon oxygen transfer to substrates [52, 53]. This is a stoichiometric method, and at present no catalytic procedure for the epoxidation reaction mediated by oxaziridines has been described. However, this method appears attractive for sensitive substrates because epoxidations can take place under neutral conditions with no additional reagents. The oxygen transfer occurs with retention of configuration of the original olefin geometry and is therefore believed to be a concerted process.

Davis has described many examples of asymmetric epoxidation mediated by oxaziridines. The first chiral oxaziridine reported was derived from bromocamphor **17**, but the observed enantioselectivities were low, the best being 40% ee for 1-phenylcyclohexene oxide (Scheme 1.21) [54].

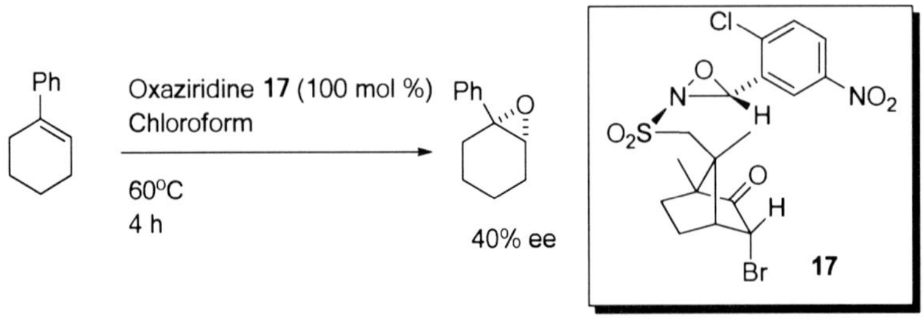

Scheme 1.21 Davis's first chiral oxaziridine for asymmetric epoxidation

Increased enantioselectivities were observed when oxaziridine **18** was employed, and ees of greater than 90% for *trans*-α-methylstilbene oxide were observed. However, reaction times were extremely long (2 weeks at 60°C) (Scheme 1.22) [55].

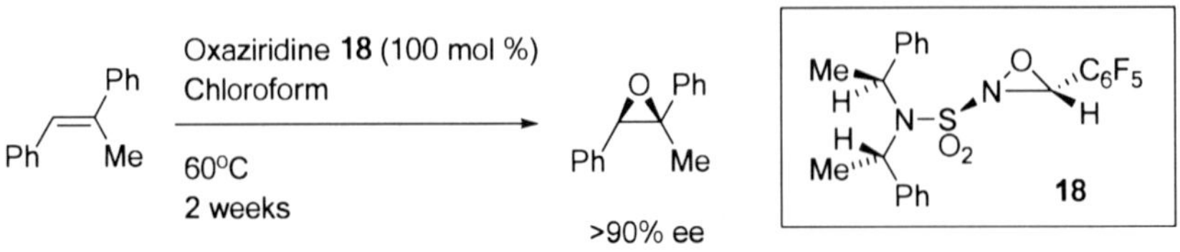

Scheme 1.22 Davis's highly enantioselective oxaziridine for asymmetric epoxidation

A large proportion of Davis's work has been involved in the elucidation of the transition state employed in the transfer of the oxygen from the oxaziridine to the olefin substrate. Davis favoured the planar transition state and was at the time supported by theoretical calculations; however, more recent calculations favour the spiro transition state [6]. Davis has also described the asymmetric oxidation of enolate anions by chiral oxaziridines, which led to α-hydroxyketones with enantioselectivities of up to 95% ee [56, 57]. Silyl enol ethers have also been reported to give epoxides when treated with oxaziridines, but the instability of these compounds is too great to allow isolation [37, 58, 59]. To date, only Davis has reported successful isolation of α-silyloxy epoxides [60].

The benzoxathiazine dioxide **19** has recently been reported, for the first time, to catalyse the epoxidation and C–H hydroxylation of a range of substrates (Scheme 1.23) [61].

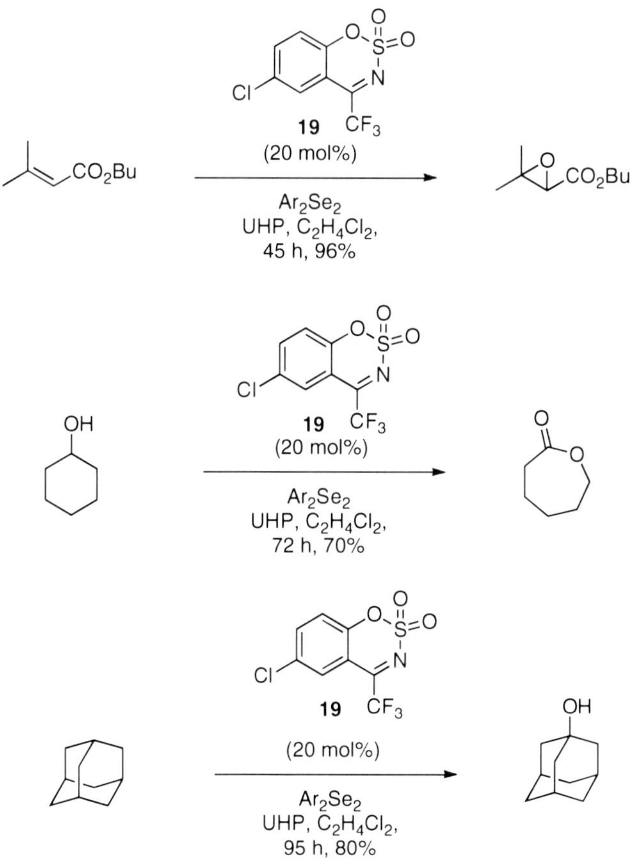

CO_2Bu → **19** (20 mol%), Ar_2Se_2, UHP, $C_2H_4Cl_2$, 45 h, 96% → CO_2Bu

OH → **19** (20 mol%), Ar_2Se_2, UHP, $C_2H_4Cl_2$, 72 h, 70% → O, O

19 (20 mol%), Ar_2Se_2, UHP, $C_2H_4Cl_2$, 95 h, 80% → OH

Cl, O, S=O, O, N, CF_3

Scheme 1.23 Oxaziridine mediated oxidation using **19** and urea hydrogen peroxide (UHP)

1.2.2.4 Oxaziridinium/Iminium Salt Systems for Asymmetric Epoxidation

Oxaziridinium salts are the quarternized analogues of oxaziridines, and as a result of being more electrophilic, transfer oxygen efficiently to nucleophilic substrates.

Initial Studies by Lusinchi

The first oxaziridinium salt, described by Lusinchi in 1976 [62–64], was based on a steroidal pyrrolinic skeleton. Through peracid oxidation of the steroidal imine and quaternization using methylfluorosulfonate it was shown that an oxaziridinium species can be formed (Scheme 1.24). This new species was rather unstable, and upon decomposition reverted to an iminium salt, which could be directly prepared from the imine. However, it was not until some 11 years later that the potential of this type of system to transfer oxygen was realized [65, 66].

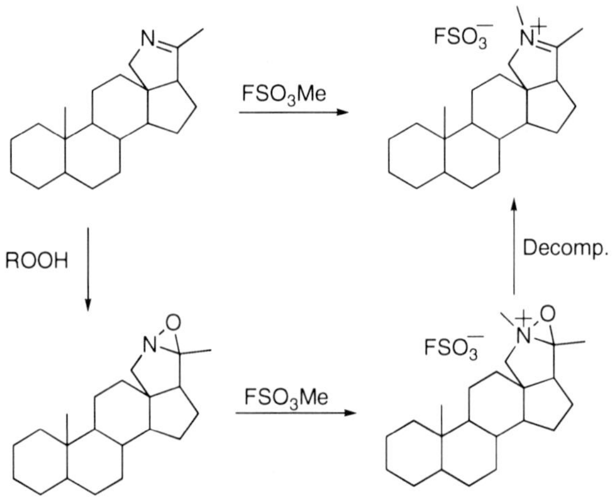

FSO_3Me / ROOH / Decomp. / FSO_3^-

Scheme 1.24 The first example of an oxaziridinium salt developed by Lusinchi

Using an oxaziridinium salt derived from dihydroisoquinoline, Lusinchi was able to transfer the oxygen to several simple alkenes in good yield (Scheme 1.25) [66, 67]. Following this work, the first enantiomerically pure oxaziridinium salt was prepared [68]. Quaternization of a chiral oxaziridine **20**, derived from (1*S*,2*R*)-(+)-norephedrine, produced the oxaziridinium salt **21** (Scheme 1.26).

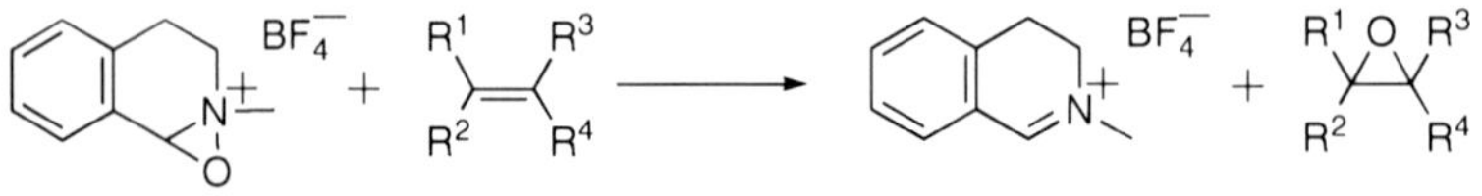

BF_4^- + $R^1R^2C{=}CR^3R^4$ → BF_4^- + epoxide (R^1, R^2, R^3, R^4)

Scheme 1.25 Oxygen transfer to alkenes mediated by a oxaziridinium salt derived from tetrahydroisoquinoline

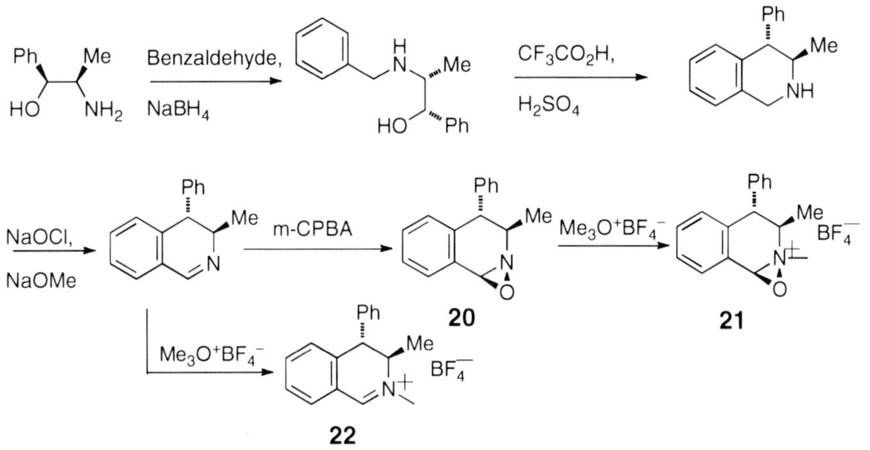

Scheme 1.26 The first enantiomerically pure oxaziridinium salt derived from (1S,2R)-(+)-norephedrine

This oxaziridinium salt was also able to transfer oxygen to olefins, and induced moderate enantiocontrol, epoxidizing *trans*-stilbene with 33% ee. With the side product of the reaction being an iminium salt **22**, there was potential to develop this chemistry catalytically. If this iminium salt could be re-oxidized to the oxaziridinium *in situ*, catalytic transfer of oxygen to alkenes could be achieved (Scheme 1.27).

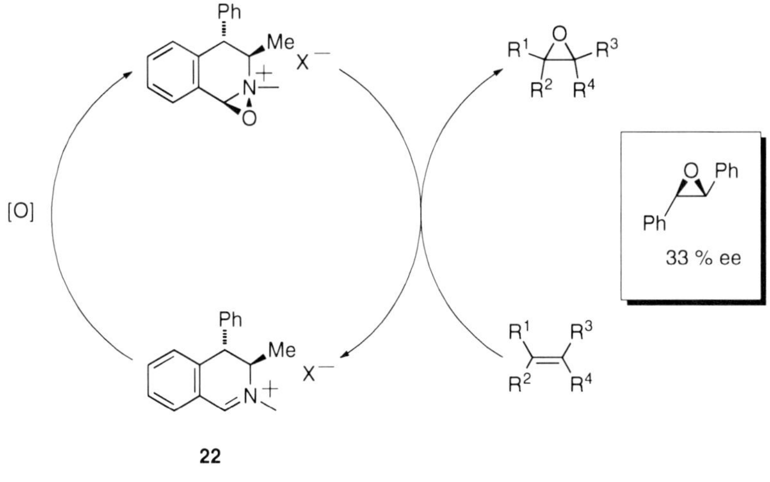

Scheme 1.27 Catalytic cycle for epoxidations mediated by oxaziridinium salts

Lusinchi was able to develop this catalytic system using Oxone, similar to that developed for dioxiranes; however, less pH control is required as there is no competitive Baeyer-Villiger oxidation. The enantiomerically pure iminium salt **22** is

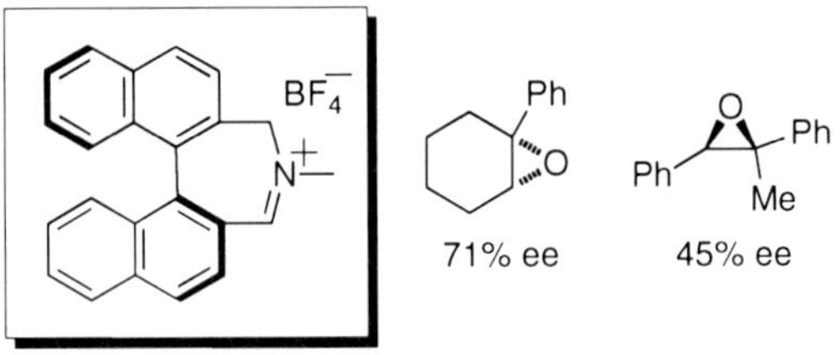

Fig. 1.6 Aggarwal's Binapthylene based iminium salt catalyst

thus able to epoxidize *trans*-stilbene catalytically (20 mol%) with the same degree of selectivity as the stoichiometric oxaziridinium salt (33% ee). Lusinchi has also shown that oxaziridinium salts are capable of transferring oxygen to other nucleophilic substrates such as sulfides to form sulfoxides [69], amines to form nitrones, and imines to form oxaziridines [70].

Lusinchi's group has reported the only X-ray determination of an oxaziridinium salt [68, 71]. Its geometry is similar to that of the parent oxaziridine, and the N-O bond length of 1.468 Å in the oxaziridinium salt is shortened compared with the mean bond length of 1.508 Å observed for oxaziridines. It is also interesting to note that the oxaziridine ring is perpendicular to the isoquinoline ring.

Aggarwal and Wang's Binapthylene Based Iminium Salt

Since Lusinchi's early work, several groups have identified an interest in oxaziridinium/iminium salt chemistry. Aggarwal and Wang produced a cyclic binaphthalene-derived iminium salt, which was shown to be effective in the epoxidation of 1-phenylcyclohexene giving 71% ee [72]. However, this catalyst appeared somewhat substrate dependent, the best ee for other olefins tested being only 45% (Fig. 1.6).

Armstrong's Acyclic Iminium salts

Armstrong has shown that even acyclic iminium salts can mediate epoxidation by Oxone [73]; however, enantiomeric excesses are low [74]. By condensing *N*-trimethylsilylpyrrolidine **23** with a range of aromatic aldehydes in the presence of trimethylsilyltriflate, Armstrong was able to produce a range of substituted exocyclic iminium salts (Scheme 1.28).

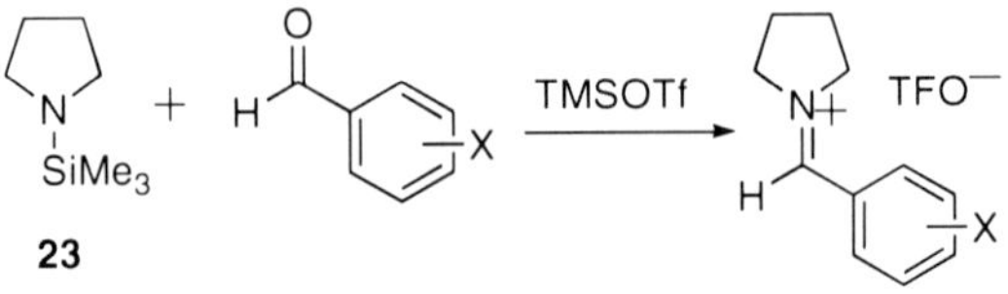

Scheme 1.28 Armstrong's synthesis of exocyclic iminium salts

Fig. 1.7 Armstrong's most successful chiral exocyclic iminium salt catalyst

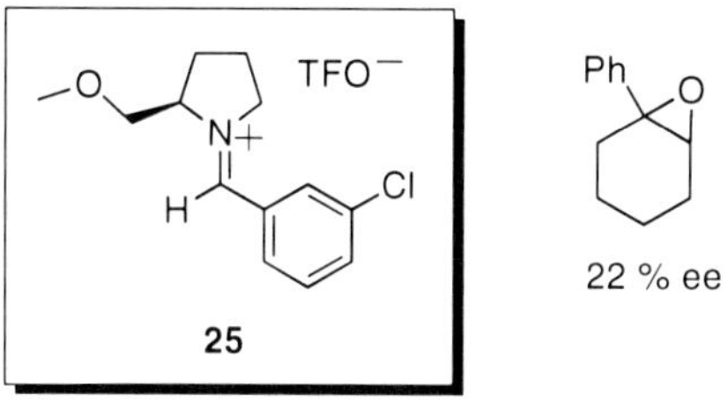

It was found that only those compounds with electron withdrawing groups present on the aromatic ring were active mediators. Catalytic reactions were carried out with the *ortho*-Cl **24** derivative, giving good conversion to epoxide (Scheme 1.29).

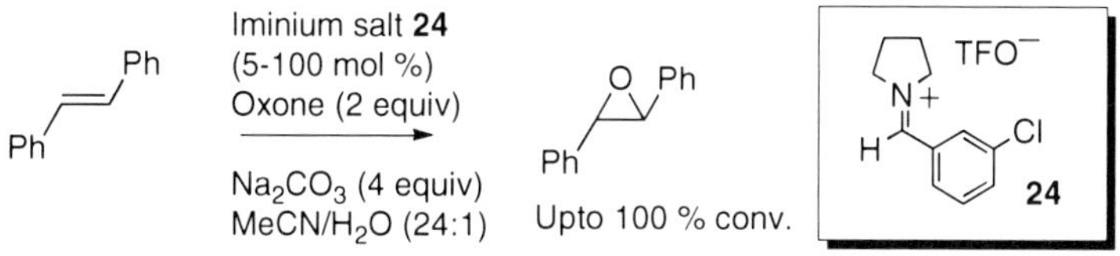

Scheme 1.29 Epoxidation of *trans*-stilbene with Armstrong's catalyst **24**

Despite many attempts at producing chiral variants of this catalyst system, Armstrong was unable to gain significant ees, catalyst **25** giving only 22% ee for 1-phenylcyclohexene (Fig. 1.7).

Komatsu and Wong's Exocyclic Iminium Salts

Two other groups have shown that exocyclic iminium salts can be useful mediators in asymmetric epoxidation. Komatsu has developed a system based on ketiminium salts [75], through the condensation of aliphatic cyclic amines and ketones. A chiral variant was also produced, derived from prolinol and cyclohexanone, which gave 70% yield and 39% ee for cinnamyl alcohol (Scheme 1.30).

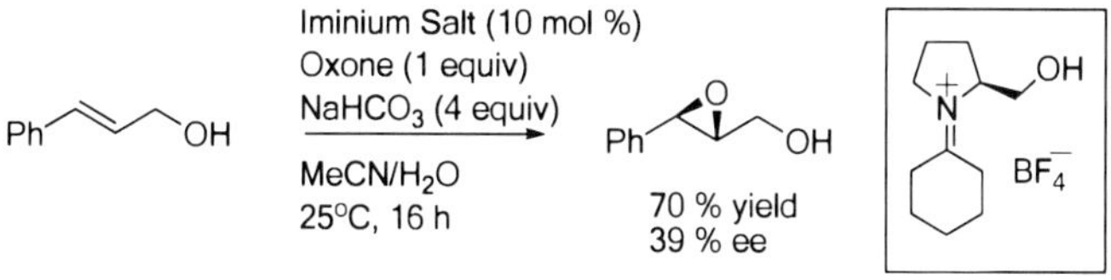

Scheme 1.30 Komatsu's ketiminium salt mediated epoxidation

Moderate ees have been achieved by Yang using another exocyclic iminium salt system [76]. However, these salts are not isolated and are generated *in situ*, thus obviating the difficulties inherent in the preparation and isolation of unstable iminium salts. A major drawback to this type of system is the catalyst loadings; for an efficient rate of reaction up to 50 mol% of iminium salt is generally required. Nevertheless, ees of up to 65% have been achieved. A range of amines and aldehydes were screened. A novel proline based amine and ß-branched butyraldehyde were found to be the best precursors (Scheme 1.31).

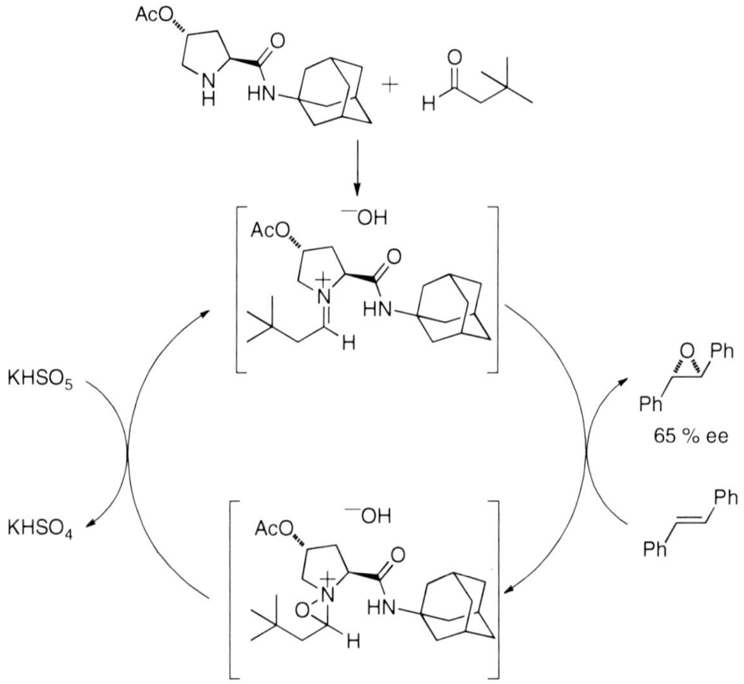

Scheme 1.31 Wong's *in situ* based iminium salt epoxidation system

Armstrong's Intramolecular Epoxidation

Armstrong has also published an *in situ* epoxidation, mediated by an intramolecular oxaziridinium salt, which gave good regioselectivity (Scheme 1.32) [77].

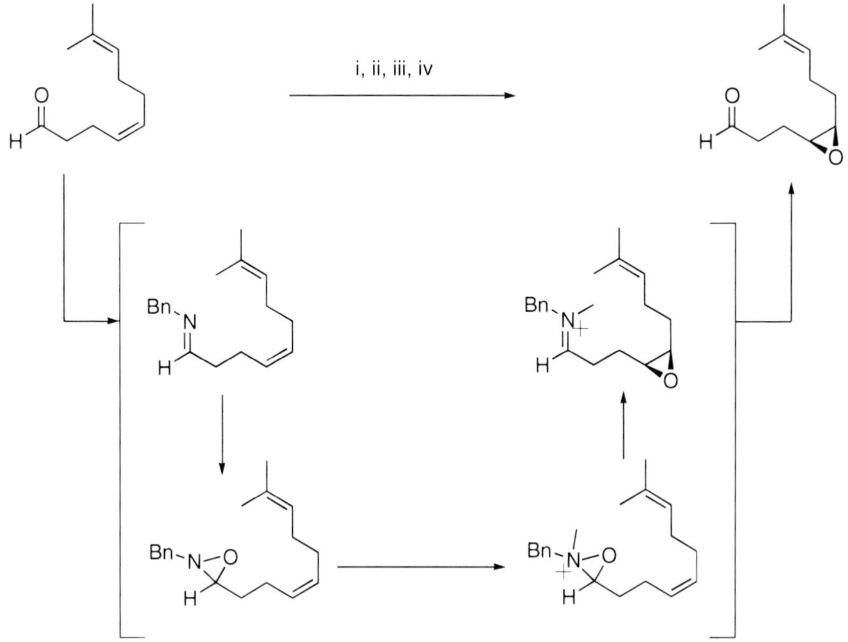

Scheme 1.32 Armstrong's intramolecular epoxidation. *Reagents and Conditions: i*: $BnNH_2$, 4 Å mol. sieves, CH_2Cl_2; *ii*: Oxone, $NaHCO_3$, $MeCN/H_2O$; *iii*: MeOTf, CH_2Cl_2; *iv*: $NaHCO_3$ (aq.)

With modification of this procedure, using an oxaziridine, Armstrong was able to demonstrate the synthetic utility of this method by introducing a chiral amine to afford enantiomerically enriched products, and greater than 98% ee was obtained for **26**, which is a terminal epoxide (Scheme 1.33) [78]. However, a loss of selectivity was observed when chain lengths between the aldehyde and alkene exceeded 3 atoms.

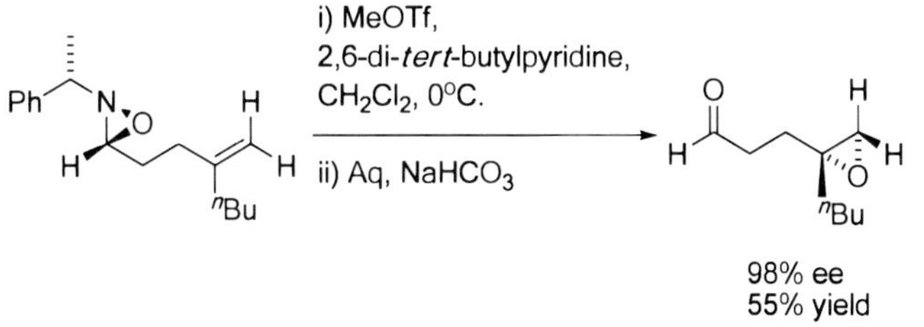

Scheme 1.33 Armstrong's intramolecular oxaridinium mediated epoxidation

Bohé's Improved Achiral Catalyst

More recently Bohé, a former co-worker with Lusinchi, has reported an improved achiral catalyst which prevents some common side reactions observed in iminium salt-mediated epoxidation [79]. Two factors are known to reduce the catalytic efficiency of the epoxidation process; hydrolysis of the iminium salt directly by the reaction medium generally only affects the acyclic systems. However, loss of active oxygen from the intermediate oxaziridinium species, through a reaction which does not regenerate the imin can occur to all systems containing protons α to the nitrogen atom. This is an irreversible base-catalysed isomerization (Scheme 1.34).

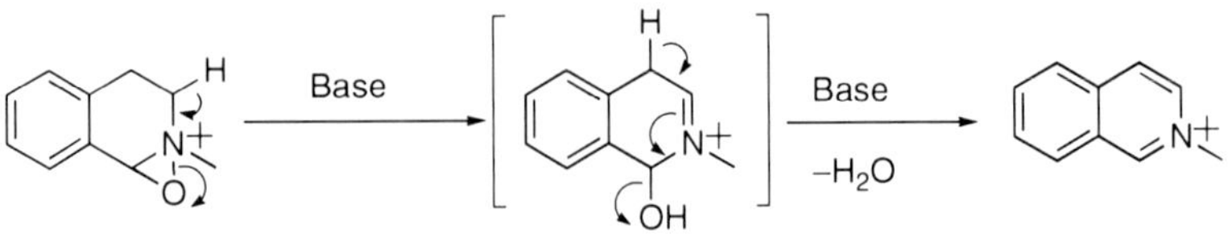

Base

Base

$-H_2O$

Scheme 1.34 Bohé's proposed irreversible base-catalysed isomerization

A dramatic increase in catalyst efficiency is observed when the 3,3-disubstituted-dihydroisoquinolinium salt **28** is used in place of catalyst **27**, thus eliminating the base catalysed isomerization (Scheme 1.35).

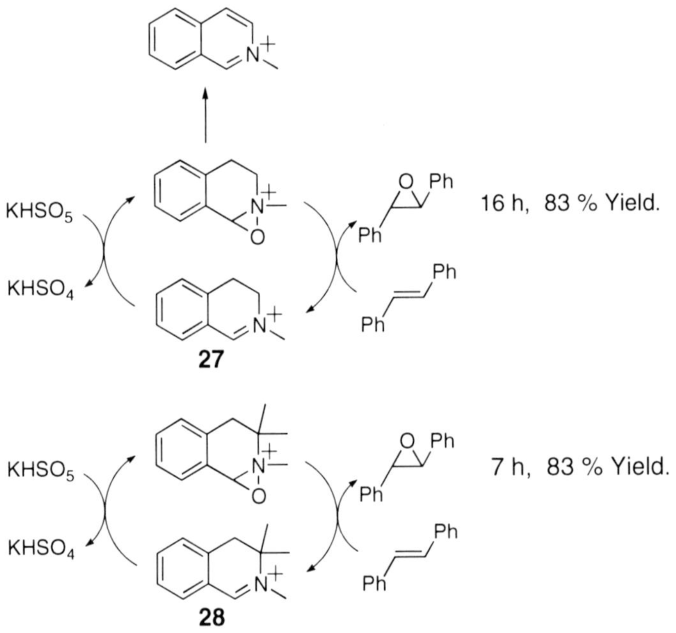

$KHSO_5$ $KHSO_4$ 16 h, 83 % Yield. **27**

$KHSO_5$ $KHSO_4$ 7 h, 83 % Yield. **28**

Scheme 1.35 Bohé's improved achiral catalyst for catalytic epoxidation

Page Group Findings

In the search for a new and highly enantioselective system for iminium salt mediated catalytic asymmetric epoxidation, several parameters and catalyst substructures were examined. The method of oxidation, using Oxone, was established after some experimentation and a possible catalytic cycle for an oxaziridinium ion as the oxidative intermediate is depicted in Scheme 1.36 [80, 81]. The first stage is the formation of an

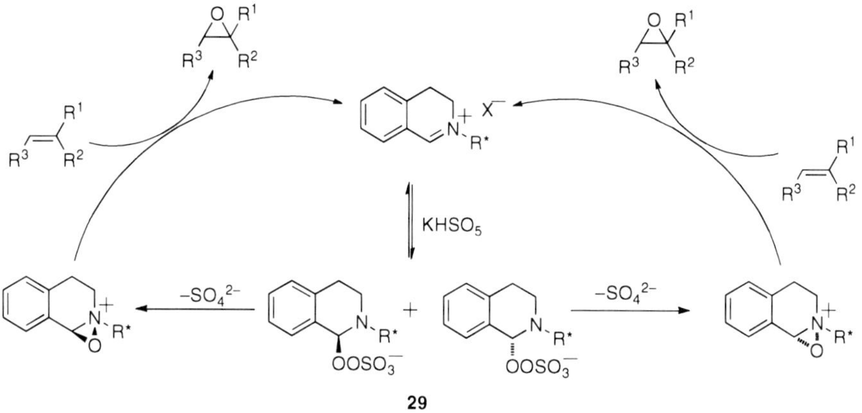

Scheme 1.36 Catalytic cycle for the oxaziridinium ion as an oxidative intermediate in the epoxidation reaction

initial adduct **29**, uncharged at nitrogen, formed by (probably reversible) nucleophilic attack of the oxidant on the iminium salt. This is followed by irreversible expulsion of sulfate to give the oxaziridinium ion, which is believed to be the rate-determining step under the reaction conditions. Oxygen may then be transferred to a substrate in a subsequent step, the rate of which would not be expected to have any great solvent dependence. An interesting but complicating feature of these processes is that it is not one but two diastereoisomeric oxaziridinium salts which may be formed, by attack of oxidant at the *si* or *re* face of the iminium species. Each may deliver the oxygen atom to either of the prochiral faces of the alkene substrate with a different degree of enantiocontrol, and they may be in competition for the alkene substrate.

An ideal method for testing a wide variety of substructures was developed through the condensation of enantiomerically pure chiral primary amines with 2-(2-bromoethyl)benzaldehyde **30** as shown in Scheme 1.37.

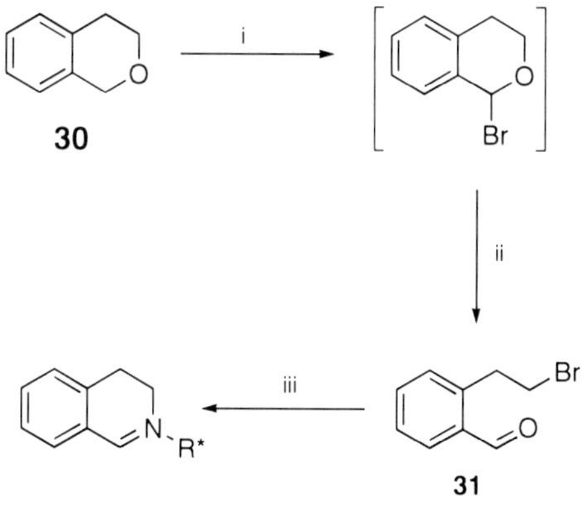

Scheme 1.37 The 2-(2-bromoethyl) benzaldehyde method for forming dihydroisoquinolinium salts. *Reagents and conditions: i*: Br_2, CCl_4, 1 h; *ii*: HBr (conc), Δ, 10 min; *iii*: (**a**) R^*NH_2, EtOH, 0°C-r.t., 12 h, (**b**) $NaBPh_4$, MeCN, 5 min

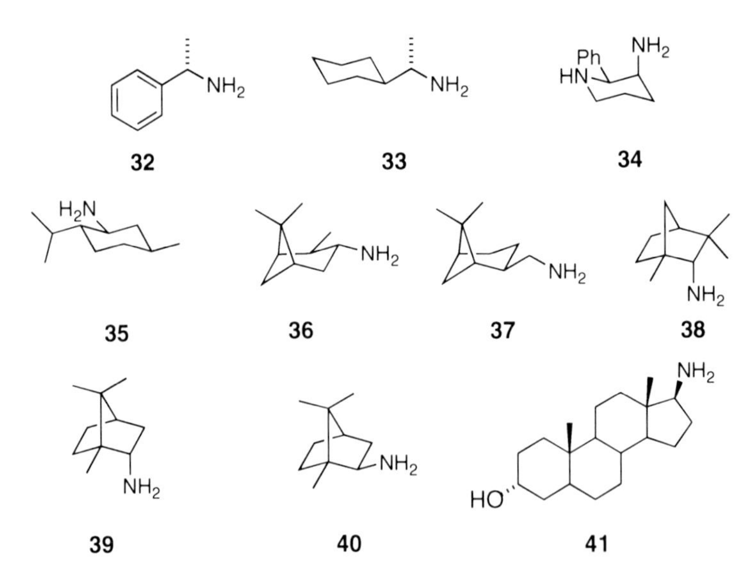

Fig. 1.8 Initial primary amines condensed with 2-(2-bromoethyl)benzaldehyde **31** to form iminium salts

The iminium salts prepared by this method have the advantage that they are extremely easy to prepare on any scale and that the structural variation available is large, because the chirality is resident in the amine component. Treatment of isochroman **30** with bromine in carbon tetrachloride under reflux for 1 h followed by exposure to concentrated hydrobromic acid provides 2-(2-bromoethyl)benzaldehyde **31** in 65% yield [82]. Primary amines condense smoothly with this material to furnish the corresponding dihydroisoquinolinium bromides. These organic salts are generally oils, and the inherent difficulties in purification by conventional methods, necessitated a change in counterion. Addition of sodium tetraphenylborate, at the end of the reaction, in the minimum amount of acetonitrile induces rapid formation of the corresponding tetraphenylborate salts as crystalline solids. Overall yields of catalyst are generally between 30% and 80%, limited in part as a consequence of a side reaction, elimination of hydrogen bromide from the bromoethyl moiety of the precursor. No chromatography is necessary at any point in this sequence.

Very hindered amines give inferior yields of iminium salt, typically 25–30%, presumably due to an increased tendency to act as bases rather than nucleophiles, as evidenced by the increased levels of 2-vinylbenzaldehyde and derived imine.

A range of structurally different chiral primary amines was converted to their corresponding iminium tetraphenylborate salts (Fig. 1.8) and tested in the asymmetric epoxidation of the standard test substrate 1-phenylcyclohexene, using oxone (4 equivalents) as the stoichiometric oxidant, sodium carbonate (8 equivalents) as base, in acetonitrile/water (2:1) at 0°C (Table 1.1).

With the first two entries in Table 1.1, using the structurally simplest amines, no asymmetric induction was observed, and it became clear that a conformationally more defined and rigid system was required to impart reasonable enantioselectivities.

Table 1.1 Epoxidation of 1-phenylcyclohexene with dihydroisoquinolinium tetraphenylborate salts derived from chiral primary amines[a]

Entry	Amine precursor	Catalyst load (mol %)	Yield (%)	ee (%)	Configuration
1	**32**	5.0	54	0	–
2	**33**	5.0	70	0	–
3	**34**	1.0	39	25	(−)-(1S,2S)
4	**35**	0.5	63	19	(+)-(1R,2R)
5	**36**	0.5	68	27	(+)-(1R,2R)
6	**37**	0.5	66	12	(−)-(1S,2S)
7	**38**	0.5	45	32	(+)-(1R,2R)
8	**39**	0.5	60	18	(+)-(1R,2R)
9	**40**	0.5	58	8	(+)-(1R,2R)
10	**41**	0.5	47	14	(−)-(1S,2S)

[a]Conditions: 4 equiv. Oxone; 8 equiv. Na_2CO_3; 0°C; H_2O/MeCN 1:2; reactions monitored by TLC

Both the camphor and menthyl based systems gave low ees, although these are two of the more common systems upon which chiral auxiliaries have been based. The fenchyl derivative is the most selective under these reaction conditions. However, the *N*-(isopinylcampheyl) dihydroisoquinolinium salt, which is considerably less sterically hindered than the fenchyl, is almost as selective, giving a better yield and increased rate of reaction.

Using this methodology a range of iminium salt catalysts were prepared [83]. (1*S*,2*S*)-5-Amino-2,2-dimethyl-4-phenyl-1,3-dioxane **42** reacted smoothly with 2-(2-bromoethyl)benzaldehyde **31** under the usual conditions to furnish the corresponding dihydroisoquinolinium tetraphenylborate salt **43** in greater than 75% yield (Scheme 1.38). This iminium salt was tested in the catalytic asymmetric epoxidation of several alkenes at 0°C, and a comparison of the results with those obtained using the catalysts in Table 1.1 indicated that catalyst **43** in general induces much higher enantioselectivity in asymmetric epoxidation (up to 60% ee, Scheme 1.39).

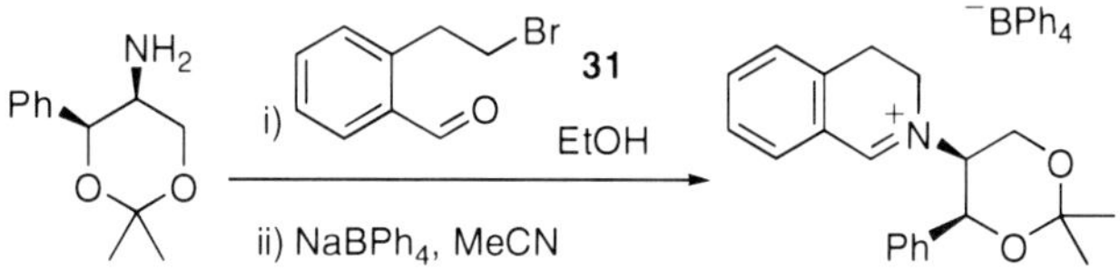

Scheme 1.38 The amino acetal catalyst derived from (1*S*,2*S*)-5-amino-2,2-dimethyl-4-phenyl-1,3-dioxane

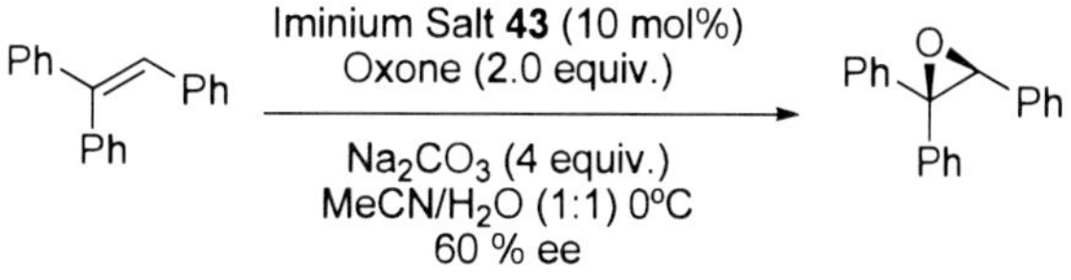

Scheme 1.39 Epoxidation of triphenylethylene with the iminium salt **43**

In their ongoing efforts to develop new and more selective catalysts based on iminium salts a new family of catalyst was produced, in which the dihydroisoquinolinium moiety has been replaced by a biphenyl structure fused to a sevenmembered azepinium salt [84]. A similar system was also developed by Aggarwal but with axial chirality, achiral at the nitrogen [72]; the system gave some good results, although the enantioselectivity of the catalyst was dependent upon the substitution pattern of the alkene.

Initially, a new iminium salt catalyst was prepared; the 1,3-dioxane catalyst **44**, derived from amine **42**, and also gave up to 60% ee in the epoxidation of 1-phenylcyclohexene, interestingly this series of catalysts were far more reactive than the related dihydroisoquinolinium catalysts, for example epoxidation using catalyst **44** at 5 mol% loading was complete after only 3 min (Scheme 1.40) compared to the 60 min required for catalyst **43**.

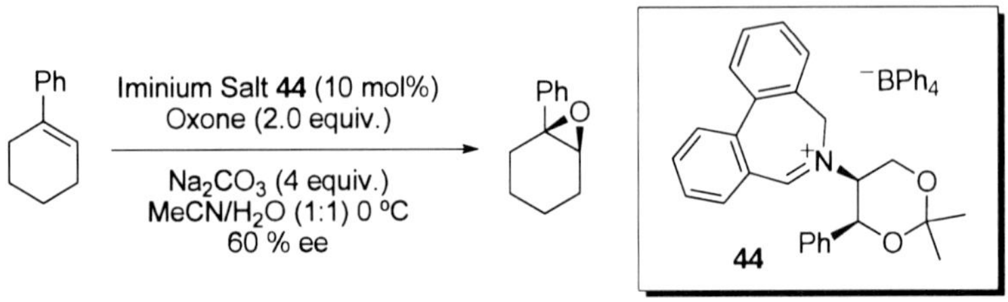

Ph
Iminium Salt **44** (10 mol%)
Oxone (2.0 equiv.)
Na_2CO_3 (4 equiv.)
$MeCN/H_2O$ (1:1) 0 °C
60 % ee
Ph O
$^{-}BPh_4$ N Ph O O **44**

Scheme 1.40 Epoxidation of 1-phenylcyclohexene with the biphenyl iminium salt

Lacour has also shown that the related bridged biphenyl azepine derivatives (such as **45**) can afford epoxides with good enantioselectivity (Scheme 1.41) [85].

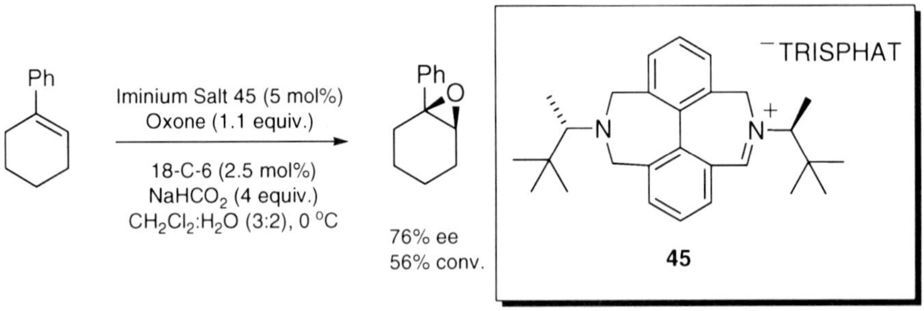

Ph
Iminium Salt 45 (5 mol%)
Oxone (1.1 equiv.)
18-C-6 (2.5 mol%)
$NaHCO_2$ (4 equiv.)
CH_2Cl_2:H_2O (3:2), 0 °C
Ph O
76% ee
56% conv.
$^{-}$TRISPHAT N N + **45**

Scheme 1.41 Lacour's biphenyl azepine catalyst

After various iterations and modifications the binaphthylene based catalyst **46** was reported and ees of up to 95% were observed for the epoxidation of 1-phenyldihydronaphthalene (Scheme 1.42) [86].

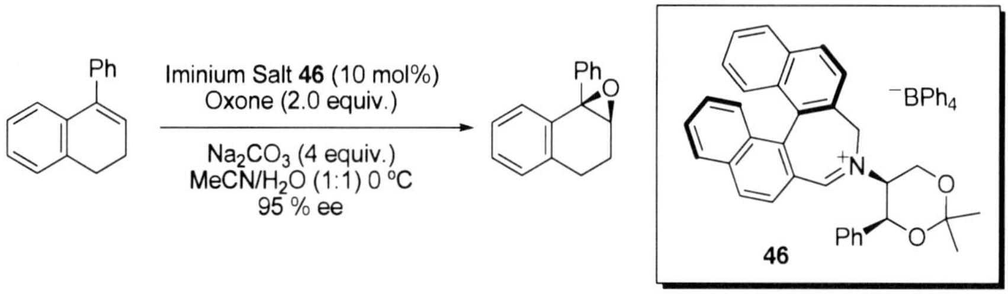

Scheme 1.42 The first highly enantioselective iminium salt catalyst used in the epoxidation of 1-phenyldihydronaphthalene

Rather than modifying the catalyst systems Page and co-workers have reported several new oxidation systems for oxaziridinium salt mediated epoxidation. Conditions employing hydrogen peroxide as the stoichiometric oxidant and sodium carbonate to generate percarbonate were reported to give ees of up to 56% [87]. In a subsequent report the group also describe an electrochemical batch system for the generation of percarbonate and persulfate; addition of an iminium salt and alkene affords the corresponding epoxides in up to 64% ee [88]. The latter system is particularly attractive as the stoichiometric oxidant can be prepared directly without transport or storage issues.

The best results so far have been with the tetraphenylphosphonium monoperoxysulfate system [89]. The standard conditions employed in epoxidation reactions catalysed by iminium salts involve the use of oxone as stoichiometric oxidant, a base (2 molar equivalents of Na_2CO_3 per equivalent of oxone) and water/acetonitrile as solvent mixture (Scheme 1.43): the presence of water is essential for oxone solubility. Under the reaction conditions, there are separate aqueous and organic phases; it is possible that the catalyst acts as a phase transfer agent in these reactions.

Scheme 1.43 General iminium salt epoxidation conditions when using Oxone

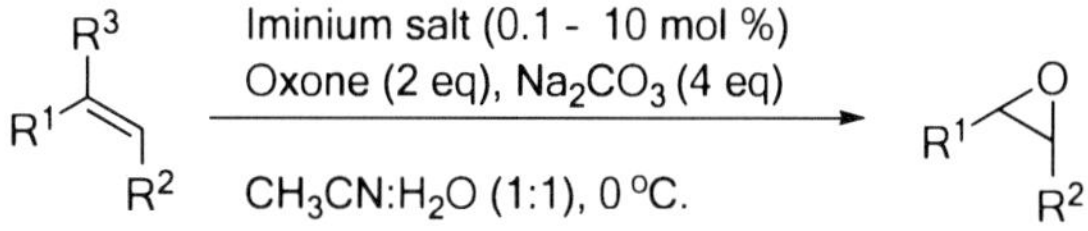

The principal limitation to this system is the restricted range of temperatures in which the epoxidation can be performed (0°C to room temperature). The upper limit is determined by the oxone, which decomposes relatively quickly in the basic medium at room temperature. The lower limit is determined by the use of the aqueous medium; the normal ratio of the water and acetonitrile solvents used is 1:1, and this mixture freezes at around −8°C. Therefore the development of the TPPP system (tetraphenylphosphonium monoperoxybisulfate) has allowed, for the first time, oxaziridinium salt mediated epoxidation to occur at sub zero temperatures [90]. This has resulted in the highly enantioselective epoxidation of *cis*-alkenes using catalyst **48** (Scheme 1.44) and the asymmetric total syntheses of levcromakalim **49** [91], (−)-(3'*S*)-lomatin **50**, and (+)-(3'*S*,4'*R*)-*trans*-khellactone **51** (Fig. 1.9) [92].

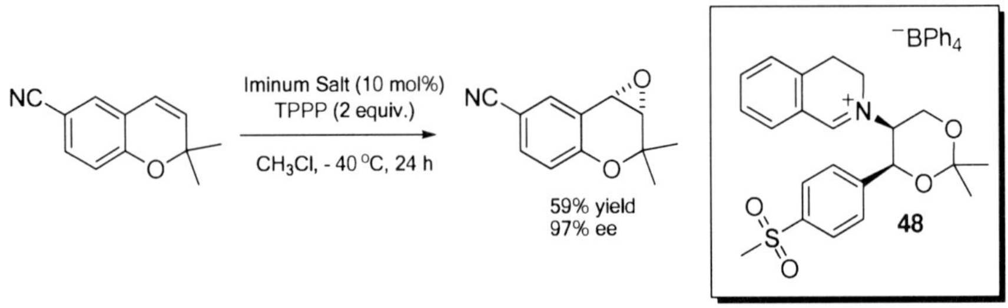

Iminum Salt (10 mol%)
TPPP (2 equiv.)
CH_3Cl, - 40 °C, 24 h
59% yield
97% ee
48

Scheme 1.44 The highly enantioselective epoxidation of *cis*-alkenes using TPPP

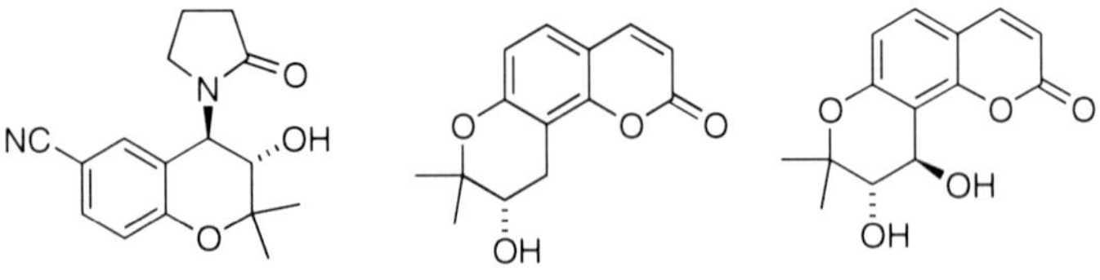

49 50 51

Fig. 1.9 Natural products prepared using TPPP and iminum salt 48

Epoxidation of α,β-unsaturated aldehydes has recently been independently reported by both Jørgensen (Scheme 1.45) [93] and MacMillan (Scheme 1.46) [94] with excellent levels of enantiomeric excess. Lattanzi has reported an extension to his group's original findings on epoxidation of chalcones using the β-amino alcohol **52** as catalyst (Scheme 1.47) [95]. High ees for a range of chalcones were observed when using this amino alcohol catalyst. Subsequent modification of the amino alcohol structure proved fruitless, with ees in some cases being dramatically decreased.

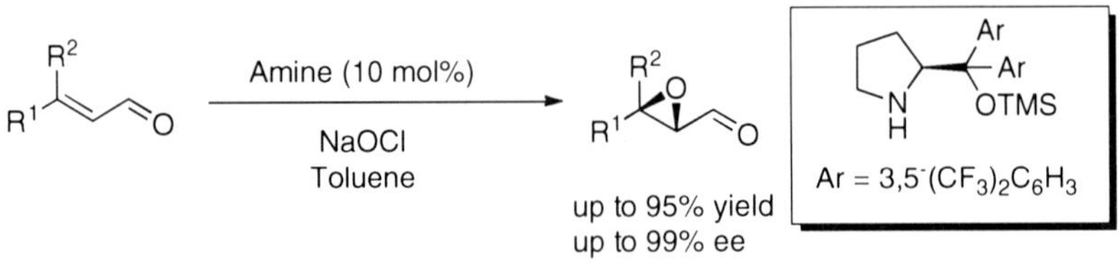

Amine (10 mol%)
NaOCl
Toluene
up to 95% yield
up to 99% ee
Ar = 3,5-$(CF_3)_2C_6H_3$

Scheme 1.45 Jørgensen's amine catalysed epoxidation reaction

Amine (20 mol%)
NsNIPh
CH2Cl2-AcOH, –30 °C
up to 95% yield
up to 99% ee
.$HClO_4$

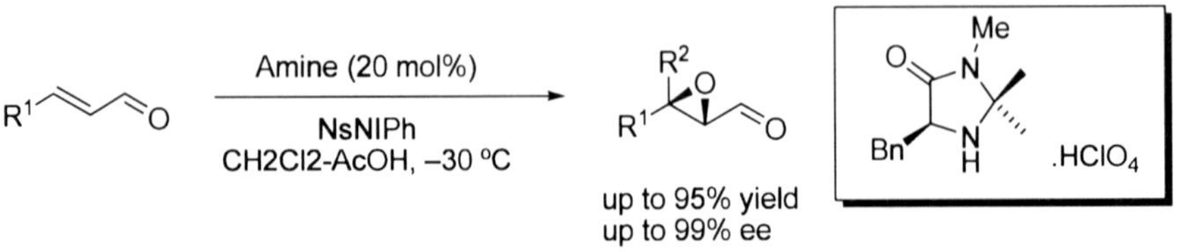

Scheme 1.46 MacMillan's amine catalysed epoxidation reaction

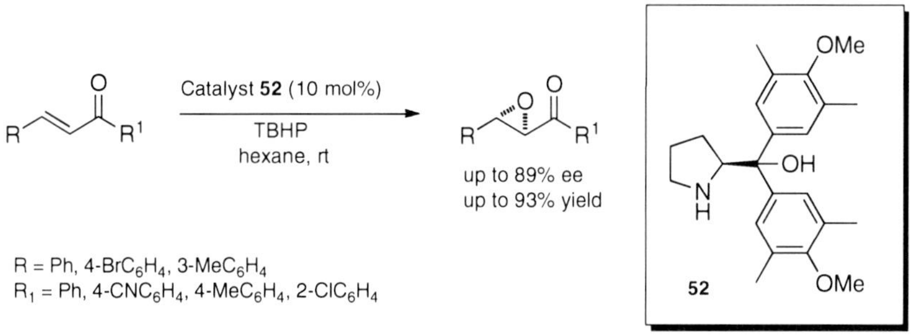

Catalyst **52** (10 mol%)
TBHP
hexane, rt

up to 89% ee
up to 93% yield

R = Ph, 4-BrC_6H_4, 3-MeC_6H_4
R_1 = Ph, 4-CNC_6H_4, 4-MeC_6H_4, 2-ClC_6H_4

Scheme 1.47 Lattanzi's amine epoxidation of α,β-unsaturated ketones

1.2.2.5 Amine Catalysis

In 2000, Aggarwal reported the use of simple amines as epoxidation catalysts using Oxone as the stoichiometric oxidant (Scheme 1.48) [96]. Yang and co-workers have also reported the effect of substituents on amine catalysed oxidation. It was found that fluorine atoms give the highest catalytic efficiency with 100% conversion, 87% yield and ees of up to 50% (Scheme 1.49). These results were then improved by repeating the reactions at lower temperatures of 0°C to −20°C to afford up to 61% ee [97].

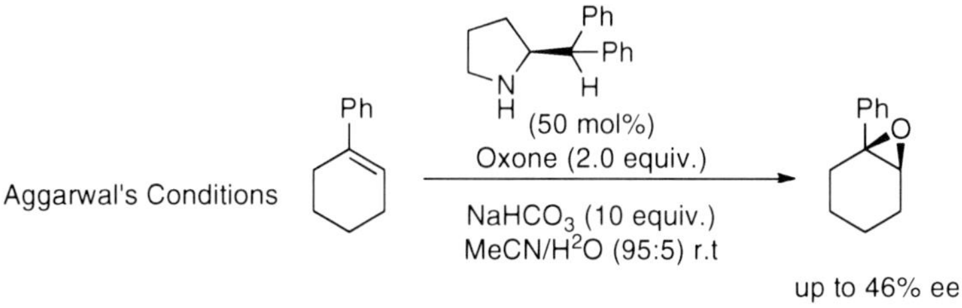

Aggarwal's Conditions

(50 mol%)
Oxone (2.0 equiv.)
$NaHCO_3$ (10 equiv.)
MeCN/H^2O (95:5) r.t

up to 46% ee

Scheme 1.48 Aggarwal's amine catalysed epoxidation of 1-phenylcyclohexene

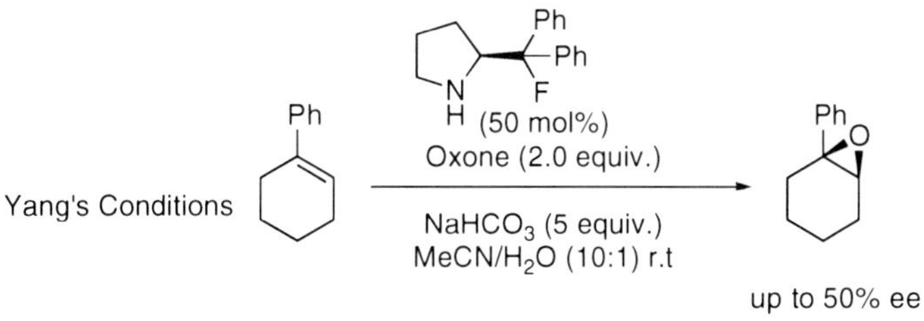

Yang's Conditions

(50 mol%)
Oxone (2.0 equiv.)
$NaHCO_3$ (5 equiv.)
MeCN/H_2O (10:1) r.t

up to 50% ee

Scheme 1.49 Yang's fluorinated amine catalysed epoxidation process

In concluding this work, Yang found that under slightly acidic reaction conditions, the fluorinated amine could be protonated *in-situ*, which removes the need to pre-form the ammonium salts that are essential for epoxidation. These findings agree

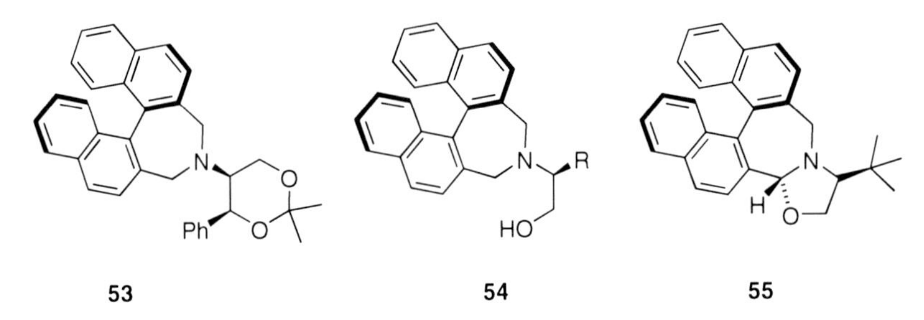

Fig. 1.10 Chiral amines for asymmetric epoxidation

with those of Aggarwal, in that the amines role in the reaction is to act as a phase transfer catalyst as well as a Oxone activator [98].

Recently, Lacour and Page have shown that the amine analogues of some of the previously reported iminium salt catalysts also catalyse the epoxidation of unfunctionalized alkenes, for example amine **53** (Fig. 1.10) [99]. Page and co-workers have also shown that amino alcohol derived binaphthyl amines such as **54** catalyse the epoxidation of 1-phenylcyclohexene with ees of up to 81% [100]. Interestingly when **55** was used in the epoxidation reaction the corresponding oxazolidinone was isolated suggesting that the reactions proceed through a iminum/oxaziridinium intermediate.

1.2.2.6 Miscellaneous

Deng and co-workers have reported the highly enantioselective peroxidation of α,β-unsaturated ketones [101]. The nature of the stoichiometric oxidant and the temperature of the reaction played a significant role in determining which product was formed. When using *tert*-butyl hydroperoxide at room temperature, the hydroperoxide addition product was the major constituent, but when using dimethylphenyl hydroperoxide as the stoichiometric oxidant and carrying the reaction out at 0°C, the epoxide was the major product.

List and co-workers have also reported on asymmetric epoxidation, however, their first report was restricted to cyclic α,β-unsaturated ketones [102]. Excellent levels of enantiomeric excess were observed over a range of substrates and catalysts. In a later publication List also described independent work on the highly enantioselective peroxidation of α,β-unsaturated ketones using catalyst **56** (Scheme 1.50) [103]. Interestingly the stoichiometric oxidant employed, hydrogen peroxide, affords the peroxyhemiketal product, which can either be reduced to form the hydroxyl group or treated with base to give the epoxide (Table 1.2). Excellent ees for the peroxyhemiketal, epoxide or hydroxyl products were observed over a range of substrate types. In a separate publication List also reports the highly enantioselective epoxidation of enals using chiral phosphoric acid salts using counteranion-directed catalysis [104].

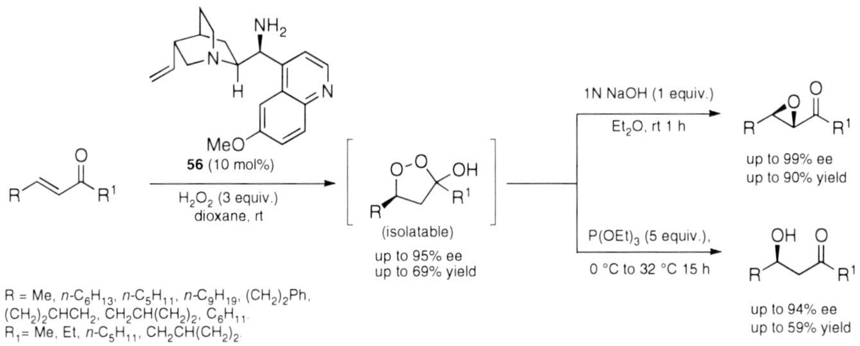

NH_2 H N MeO N
56 (10 mol%)
H_2O_2 (3 equiv.)
dioxane, rt
(isolatable)
up to 95% ee
up to 69% yield
1N NaOH (1 equiv.)
Et_2O, rt 1 h
up to 99% ee
up to 90% yield
$P(OEt)_3$ (5 equiv.),
0 °C to 32 °C 15 h
up to 94% ee
up to 59% yield

R = Me, n-C_6H_{13}, n-C_5H_{11}, n-C_9H_{19}, $(CH_2)_2Ph$, $(CH_2)_2CHCH_2$, $CH_2CH(CH_2)_2$, C_6H_{11}.
R_1 = Me, Et, n-C_5H_{11}, $CH_2CH(CH_2)_2$.

Scheme 1.50 Asymmetric peroxidation

Table 1.2 List's asymmetric peroxyhemiketal formation and transformation to epoxides or alcohols

Entry	Alkene	Reaction conditions	Product	Yield (%)	ee (%)
1	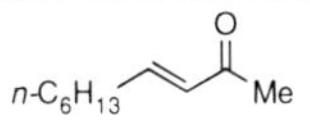 n-C_6H_{13} … Me	[A]	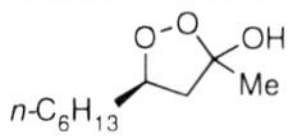 n-C_6H_{13} … OH, Me	65	95
2	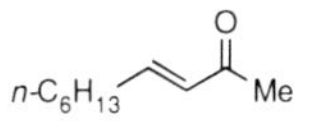 n-C_6H_{13} … Me	[B]	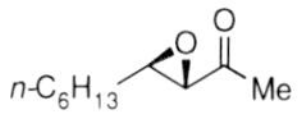 n-C_6H_{13} … Me	72	97
3	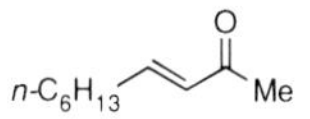 n-C_6H_{13} … Me	[C]	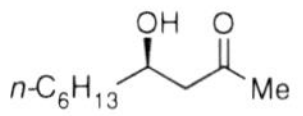 n-C_6H_{13} … OH … Me	59	94
4	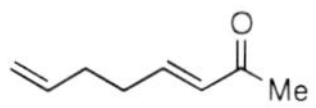 … Me	[A]	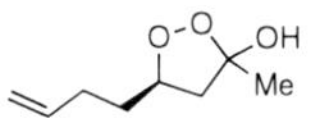 … OH, Me	69	94
5	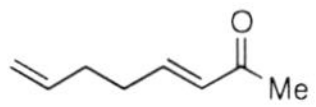 … Me	[B]	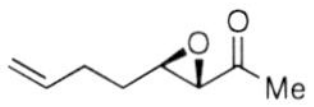 … Me	76	97
6	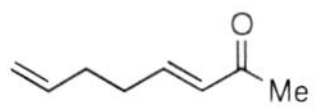 … Me	[C]	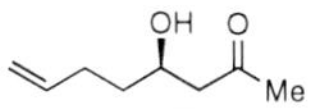 … OH … Me	55	92

[A] **56** (10 mol%), H_2O_2 (3.0 equiv.), dioxane, 32°C, 36–48 h
[B] (a) **56** (10 mol%), H_2O_2 (3.0 equiv.), dioxane, 32°C, 12–48 h; (b) 1 N NaOH (1.0 equiv.), Et_2O, r.t. 1 h
[C] (a) **56** (10 mol%), H_2O_2 (3.0 equiv.), dioxane, 32°C, 12–48 h; (b) $P(OEt)_3$ (5.0 equiv.) 0–32°C, 15 h

1.3 Asymmetric Sulfur Oxidation Methods

1.3.1 Introduction

Andersen reported the preparation of the first enantiopure sulfoxide in the 1960s [105]. This was achieved by nucleophilic displacement of a leaving group from a diastereomerically pure sulfinate ester. Despite obtaining high yields of enantiopure sulfoxides, the preparation was difficult and there was a limited availability of substrates. However, following on from this initial work, Ruano and Senanayake both independently used chiral auxillaries that could undergo two consecutive nucleophilic displacements to give an enantiopure sulfoxide [106, 107].

One of the most widely used oxidation methods for the synthesis of chiral sulfoxides is the titanium-based Kagan oxidation method, first reported in 1980. This method has even been used to synthesise the drug esomeprazole on a large scale [108]. Bolm has also reported a robust metal mediated oxidation process based on vanadium, which involved the *in-situ* formation of a catalyst from vanadyl acetylacetonate and a Schiff base [109]. The oxygen source for this method was hydrogen peroxide, the reaction is not moisture sensitive as was the case for that of Kagan, so could be carried out in an open reaction vessel. This gave high enantioselectivities of the sulfoxide products.

1.3.2 Organocatalytic Methods

1.3.2.1 Achiral Syntheses

In 2009, Russo and Lattanzi carried the out highly chemoselective oxidation of sulfides, where catalyst loading was as low as 1 mol% of *N,N′*-Bis [(3,5-bistrifluoromethyl) phenyl]thiourea using *tert*-butyl hydroperoxide (TBHP) as the oxidant at room temperature in dichloromethane (Scheme 1.51) [110].

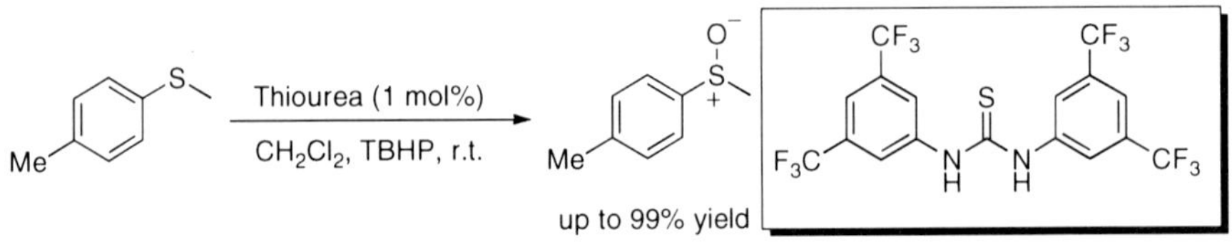

Thiourea (1 mol%)

CH_2Cl_2, TBHP, r.t.

up to 99% yield

Scheme 1.51 Russo and Lattanzi's oxidation of sulfides using hydrogen bonding catalysis

This particular thiourea used by Russo for sulfoxidation, gives a catalytic turnover that competes well with transition metal complexes generally used for sulfoxidation reactions. The effectiveness of the TBHP activation could be rationalised according to a double hydrogen-bonding interaction of the thiourea with the proximal oxygen of TBHP, which should enhance the elctrophilic character of the distal oxygen attacked by a sulfide. Formation of the TBHP and thiourea complex (Fig. 1.11) was confirmed by ^{1}H NMR spectroscopic analysis, where the chemical shift of H_a moves downfield from 7.88 to 7.91 ppm, and the proton in the TBHP also shifts downfield from 7.14 to 7.42 ppm.

In 2009, Habibi showed that sulfoxidation reactions are possible with sodium perborate or sodium percarbonate and silica sulfuric acid in the presence of KBr, under mild heterogeneous conditions with moderate to good yields [111].

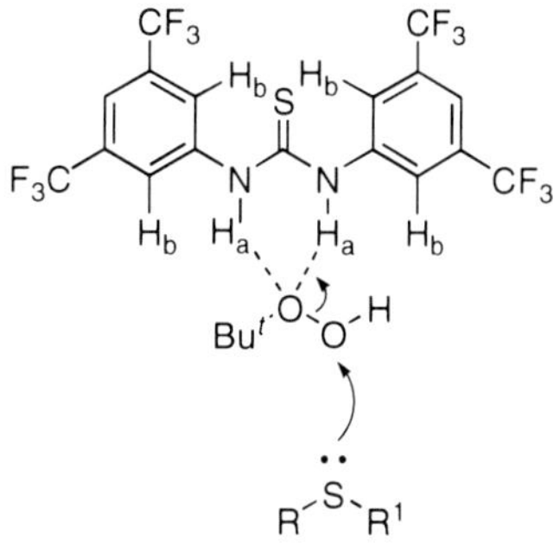

Fig. 1.11 Transition state model for the thiourea mediated oxidation process

Recently it has been shown that the bromonium ion (Br^+) can effectively be applied to the oxidation of different types of organic compounds [112]. Following on from this Habibi tried to introduce a new catalytic media, based on the *in situ* generation of Br^+ using sodium percarbonate and/or sodium perborate and catalytic amounts of bromide in the presence of an activator for the effective oxidation of sulfides to sulfoxides.

Firstly, the group carried out a study that determined the best activator to be silica sulfuric acid giving 100% yield for the oxidation of benzyl phenyl sulfide to the sulfoxide, in dichloromethane at room temperature using sodium perborate in the presence of catalytic amounts of KBr. Without KBr as the activator then these reactions do not work.

1.3.2.2 Asymmetric Syntheses

Metal free asymmetric sulfur oxidation has been reported using oxaziridines and hydroperoxides. To achieve oxygen transfer, oxaziridines have been developed that have electron withdrawing substituents on the nitrogen atom, or on both the nitrogen and carbon atom of the three-membered ring.

In the early 1980s, Davis carried out a sulfoxidation reaction using chiral 2-sulfonyloxaziridines, and found that asymmetric oxidation of unfunctionalized substrates such as sulfides and disulfides could be achieved [113, 114]. This was made possible through incorporation of a rigid environment around the active site of the oxidation reaction, such as in 2-sulfonyloxaziridines. From studies of asymmetric oxidations using these chiral oxidising agents, factors important in controlling absolute configuration of the product appear to be steric in nature. The difference in the group size effect in the oxaziridine and substrate play an important role in determining the absolute configuration of the product. Davis also showed that as the group size difference increased, so did the enantioselectivity.

In 1987, Davis reported the asymmetric oxidation of non-functionalized sulfides using the 2-sulfamyloxaziridine diastereoisomers **57** and **58** (Fig. 1.12) with ees ranging between 21% and 45% for the corresponding sulfoxides [115].

Fig. 1.12 Davis oxaziridines for the oxidation of sulfides

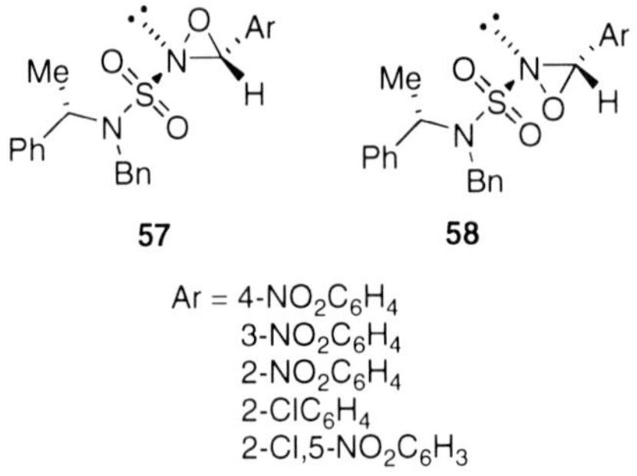

In 1988 Davis reported the synthesis and properties of (camphorylsulfonyl) oxaziridine, achieving a 77% yield, starting from the inexpensive camphorsulfonic acid. The oxaziridine reported was the first optically active *N*-sulfonyloxaziridine to be obtained as a single isomer on oxidation of the sulfonimine (Scheme 1.52) [116].

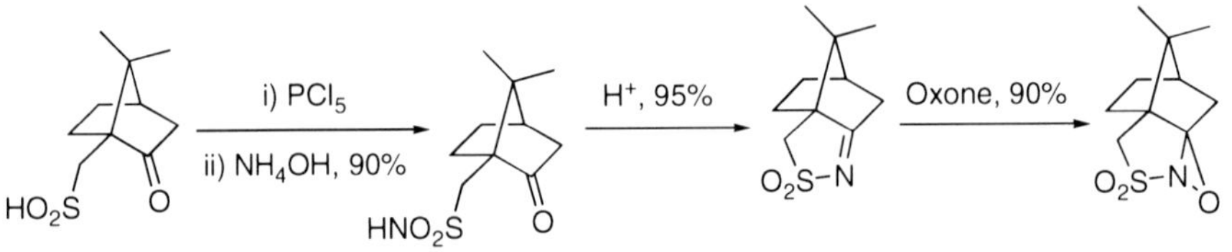

Scheme 1.52 The synthesis of the first camphorylsulfonyl oxaziridine

In 1988, Boyd's phosphinoyloxaziridines were used for the oxidation of thioethers [117, 118], and in 1988, Lusinchi reported the oxidation of weakly basic and nucleophilic thioether substrates, by oxaziridines. This can be performed if the oxygen transfer reaction is promoted by an acid [119].

In 1994, the oxidation of sulfides by perfluoro-*cis*-2,3-dialkyloxaziridines, was reported by DesMarteau [120]. On using either of the fluorine containing oxaziridines (**59** or **60**) high yields were obtained of the sulfoxide (90–97%, Scheme 1.53).

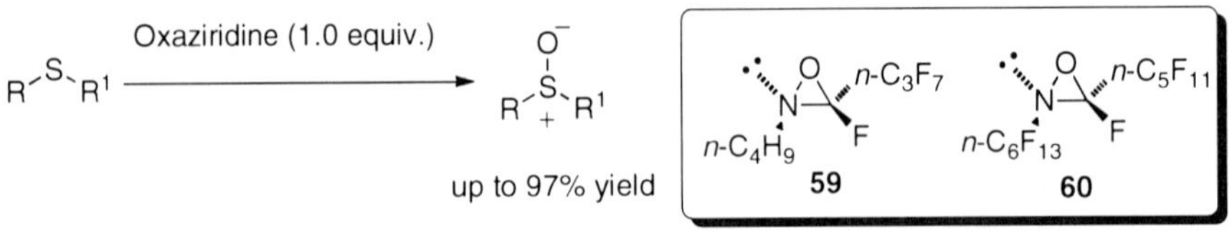

R^1	R^2	Oxaziridine	Yield (%)
Bn	Me	60	95
Ph	Me	59	97
Ph	CH_2CO_2H	59	95
allyl	Me	59	90

Scheme 1.53 DesMarteau's oxaziridine mediated sulfoxidation

DesMarteau has also reported a number of useful syntheses (Scheme 1.54) using **59**. The hydrochlorides of promazine, chloropromazine, and promethazine, are three typical neuroleptic drugs commonly employed in human therapy. On treating the hydrochlorides with an equimolar amount of oxaziridine, in trifluoroethanol, the corresponding sulfinyl products were obtained in 90–94% yields.

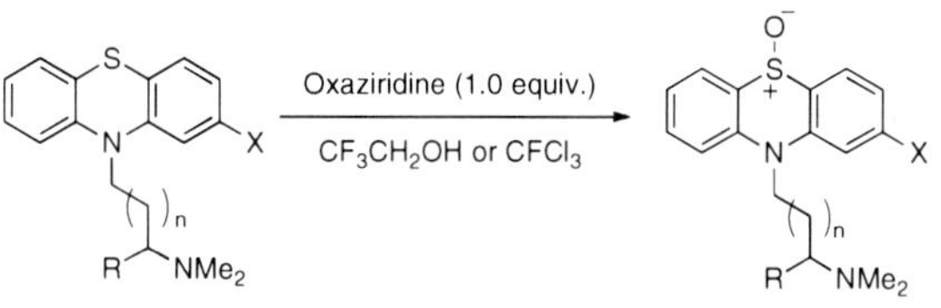

Oxaziridine (1.0 equiv.)
CF_3CH_2OH or $CFCl_3$

	R	X	n
Promazine.HCl	H	H	1
Chloropromazine.HCl	H	Cl	1
Promethazine.HCl	CH_3	H	0

Scheme 1.54 Oxidation of promazine, chloropromazine, and promethazine using oxaziridines

In 1995, Page reported asymmetric sulfoxidation using [(3,3-methoxycamphoryl) sulfonyl] oxaziridine **61**, and the corresponding imine **62**, as well as the dichloro derivative **63** and **64** (Fig. 1.13) [121]. **64** has been shown to be the most selective in previous studies, and can be used to prepare aryl alkyl sulfoxides in greater than 95% ee.

Page and Bethell have also developed a catalytic process [122, 123]. The most successful system is derived from camphor sulfonyl imine **62**, giving enantiomerically enriched sulfides with ees of greater than 98% (Scheme 1.55) [123]. More recently, 3-substituted-1,2-benzisothiazole-1,1-dioxides have been employed [124, 125].

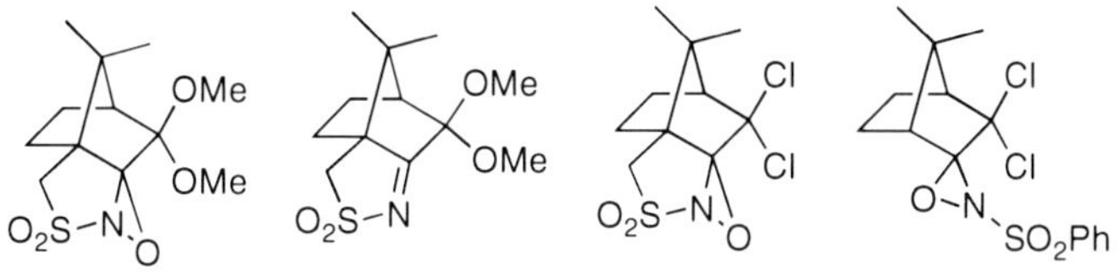

61 **62** **63** **64**

Fig. 1.13 Page's camphor derived imines and oxaziridines

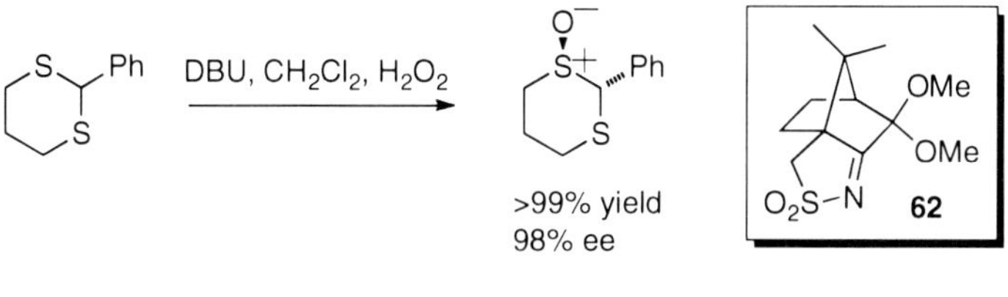

DBU, CH_2Cl_2, H_2O_2

>99% yield
98% ee

Scheme 1.55 Page and Bethell's highly enantioselective sulfur oxidation protocol

In 1998, Bohé and Lusinchi reported the oxygen atom transfer from a chiral *N*-alkyl oxaziridine promoted by acid, for the asymmetric oxidation of sulfides. The chiral oxaziridine **20**, from the corresponding dihydroisoquinoline by *m*-CPBA oxidation was used to see if chiral *N*-alkyl oxaziridines would perform in the presence of acid [69].

The results for the oxidation of *para*-tolylmethylsulfide using the oxaziridine **20**, using either TFA or methanesulfonic acid (MsOH) to promote oxygen transfer is shown below in Scheme 1.56. Both reactions were performed at room temperature, with 0.2 mmol of substrate and a slight excess of acid. Using either TFA or MsOH made almost no difference to the enantioselectivity of the reaction (42% and 44% respectively). It can also be noted that the reaction takes much longer in TFA (24 h), in comparison to MsOH (1 min).

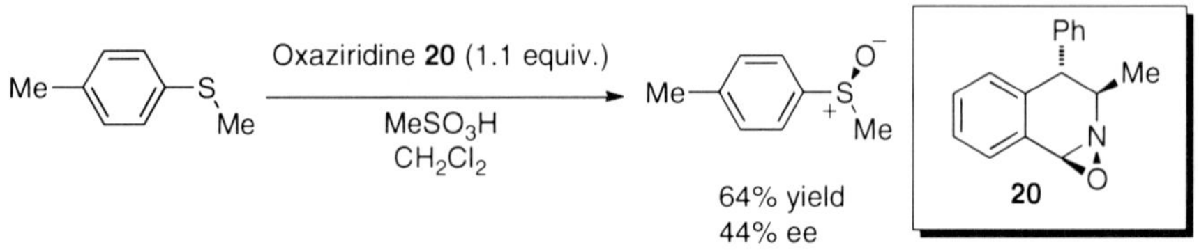

Oxaziridine **20** (1.1 equiv.), $MeSO_3H$, CH_2Cl_2; 64% yield, 44% ee

Scheme 1.56 Sulfur oxidation using the oxaziridine **20**

The proposed mechanism for this reaction (Scheme 1.57), shows that an equilibrium is established between oxaziridine **65**, and the two protonated forms **66** and **67**. The oxaziridinium **67** is theoretically the most populated, owing to the greater basicity of the nitrogen, and is thus able to transfer its oxygen to the highly nucleophilic sulfide (DMS), resulting in the imine **68**. If DMS is not present, or any other sulfide, then the O-protonated form **66** will result in the corresponding nitrone **69**.

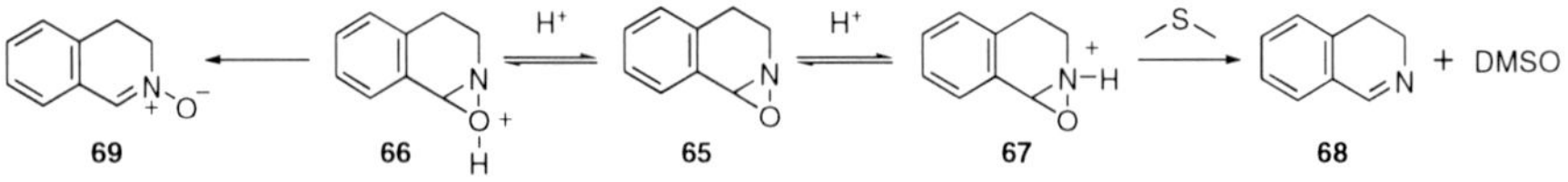

69, 66, 65, 67, 68, + DMSO

Scheme 1.57 Proposed mechanism for reactions involving oxaziridine **20** (substituents on the tetrahydroisoquinoline ring removed for clarity)

The report also stated that oxygen transfer from electrophilic reagents to sulfides are thought to be S_N2 displacements, and can be rationalised in terms of two transition states, planar and spiro.

In a planar transition state, both electron pairs on the sulfur are in the plane of the electrophilic oxygen-containing functional group. Whereas in the spiro, the plane containing the two electron pairs on the sulfur is perpendicular to the plane of the electrophilic oxygen-containing functional group. Theoretical studies on the hypothetical oxidation of hydrogen sulfide by oxaziridines have shown that there are only slight energy differences between both geometries, suggesting that the asymmetric inductions seen are probably due to steric interactions from both the transition states [126].

These results show that sulfides can be oxidized to the corresponding sulfoxides without over-oxidation to sulfones. The study of solvent effect showed that the three-component system is governed by subtle acid–base equilibria, which favours the oxygen transfer from the oxaziridine to the sulfide.

In 2004, Rozwadowska reported the use of non-racemic 3,4-dihydroisoquinolinium salts for the oxidation of sulfides with ees of up to 42% [127]. In 2007, Bohé reported a new oxaziridinium salt **70** for enantioselective oxidation of sulfides, with up to 99% ee and good yields (Scheme 1.58) [128]. Bohé, also stated that no sulfone was present in this particular reaction either, according to ^{1}H NMR spectroscopy. On lowering the reaction temperature, ees were improved with values up to 99% and a slight increase in yield to 88%.

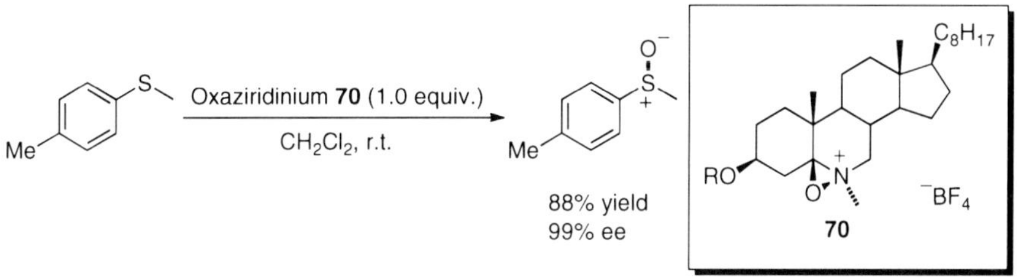

Scheme 1.58 Bohé's steroidal oxaziridinium salt for stoichiometric sulfoxidation

1.4 Asymmetric Phosphine Oxidation Methods

Chiral nonracemic phosphorus compounds are widely used in catalytic asymmetric synthesis, both as ligands in metal-based processes [129] and as organocatalysts [130]. However, although a very large number of such ligands have been tested, the majority contain either an asymmetric centre or axis in the backbone (C-stereogenic, e.g., BINAP and DuPHOS) instead of on the phosphorus atom (P-stereogenic, e.g., DiPAMP). Although P-stereogenic ligands have proven to be effective, relatively few have been studied because they are difficult to prepare [131]. Original approaches were based on kinetic resolution and crystallization from unequal mixtures of diastereomers, while more recent strategies include desymmetrization, enzymatic resolution, and catalytic asymmetric synthesis. Gilheaney has reported a stoichiometric asymmetric oxidation approach towards phosphines under Appel conditions and ees of up to 98% have been observed for the synthesis of a bidentate system (Scheme 1.59) [132]. This area of organocatalytic research is currently in its infancy and a successful catalytic asymmetric approach has yet to be reported.

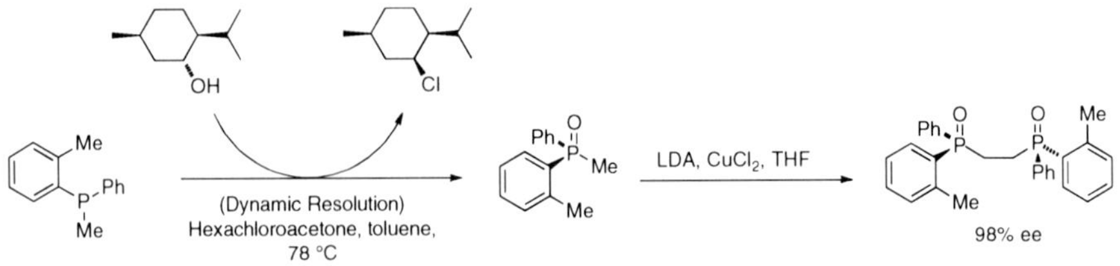

Scheme 1.59 Gilheaney's stoichiometric asymmetric oxidation approach towards phosphines under Appel conditions

References

1. Benedetti F, Berti F, Norbedo S (1998) Tetrahedron Lett 39:7971
2. O'Brien P, de Sousa SE, Poumellec P (1998) J Chem Soc Perkin Trans 1:1483
3. (a) Zwanenburg B, Legters J, Thijis L (1989) Tetrahedron Lett 30:4881; (b) Zwanenburg B, Grijzen Y, Thijis L, Gentilucci L (1995) Tetrahedron Lett 36:4665
4. Bartok M, Lang KL (1985) In: Hassner A (ed.) Small ring heterocycles, vol 42, Part 3, Wiley Interscience, Weinheim
5. Plesnicar B (1983) In: Patai S (ed) The chemistry of peroxides, Wiley, New York, p 521
6. (a) Houk KN, Liu J, DeMello NC, Condroski KR (1997) J Am Chem Soc 119:10147; (b) Washington I, Houk KN (2000) J Am Chem Soc 122:2948
7. Payne GB, Denning PH, Williams PH (1961) J Org Chem 26:659
8. Payne GB (1962) Tetrahedron 18:763
9. (a) Henbest HB (1965) Chem Soc Spec Publ 19:83; (b) Ewins RC, Henbest HB, Mckervey MA (1967) J Chem Soc Chem Commun 1085; (c) Pirkle WH, Rinaldi PL (1977) J Org Chem 42:2080; (d) Montanari FJ (1969) J Chem Soc Chem Commun 135; (e) Bowman RM, Collins JF, Grundon MF (1973) J Chem Soc Perkin Trans 1:626; (f) Bowman R M, Collins JF, Grundon MF (1967) J Chem Soc Chem Commun 1131; (g) Morrison JD, Mosher HS (1971) Asymmetric Organic Reactions, American Chemical Society, Washington DC, p 336
10. Hassine B, Gorsane M, Geerts-Evrard F, Pecher J, Martin RH, Castelet D (1986) Bull Soc Chim Belg 95:547
11. Hassine B, Gorsane M, Pecher J, Martin RH (1986) Bull Soc Chim Belg 95:557
12. Juliá S, Masana J, Vega JC (1980) Angew Chem Int Ed 19:929
13. (a) Bentley PA, Bergeron S, Cappi MW, Hibbs DE, Hursthouse MB, Nugent TC, Pulido R, Roberts SM, Wu LE (1997) J Chem Soc Chem Commun 739; (b) Allen JV, Cappi MW, Kary PD, Roberts SM, Williams NM, Wu LE (1997) J Chem Soc Perkin Trans 1:3297; (c) Adger BM, Barkley JV, Bergeron S, Cappi MW, Flowerdew BE, Jackson MP, McCague R, Nugent TC, Roberts SM (1997) J Chem Soc Perkin Trans 1:3501
14. (a) Gilmore AT, Roberts SM, Hursthouse MB, Abdul-Malik, KM (1998) Tetrahedron Lett 39:3315; (b) Allen JV, Bergeron S, Griffiths MJ, Mukherjee S, Roberts SM, Williamson NM, Wu LE (1998) J Chem. Soc Perkin Trans 1:3171
15. (a) Mazaleyrat JP (1983) Tetrahedron Lett 24:1243; (b) Baba N, Oda J, Kawaguchi M (1986) Agric Biol Chem 50:3113; (c) Shi M, Masaki Y (1994) J Chem Res (S) 250; (d) Shi M, Kazuta K, Satoh Y, Masaki Y (1994) Chem Pharm Bull 42:2625
16. (a) Helder R, Hummelen JC, Laane RWPM, Wiering JS, Wynberg H (1976) Tetrahedron Lett 17:1831; (b) Wynberg H, Gerijdanus B (1978) J Chem Soc Chem Commun 427; (c) Marsman B, Wynberg H (1979) J Org Chem 44:2312
17. Lygo B, Wainwright PG (1998) Tetrahedron Lett 39:1599

18. (a) Adam W, Bheema Roa P, Degen H-G, Levai A, Patonay T, Saha-Möller CR (2002) J Org Chem 67:259; (b) Adam W, Bheema Roa P, Degen H-G, Saha-Möller C R (2001) Tetrahedron: Asymm 12:121
19. Curci R, Fiorentino M, Serio MR (1984) J Chem Soc Chem Commun 155
20. Curci R, D'Accolti L, Fiorentino M, Rosa A (1995) Tetrahedron Lett 36:2437
21. Brown DS, Marples BA, Smith P, Walton L (1995) Tetrahedron 51:3587
22. Yang D, Yip Y-C, Tang M-W, Wong M-K, Zheng J-H, Cheung K-K (1996) J Am Chem Soc 118:491
23. Yang D, Wang XC, Wong M-K, Yip Y-C, Tang M-W (1996) J Am Chem Soc 118:11311
24. Yang D, Wong M-K, Yip Y-C, Wang XC, Tang M-W, Zheng J-H, Cheung K-K (1998) J Am Chem Soc 120:5943
25. Denmark SE, Forbs DC, Hays DS, DePue JS, Wilde RG (1995) J Org Chem 60:1391
26. Denmark SE, Wu Z (1999) Synlett 847
27. Denmark SE, Wu Z, Crudden C, M, Matsuhashi H (1997) J Org Chem 62:8288
28. Denmark SE, Matsuhashi H (2002) J Org Chem 67:3479
29. Song CE, Kim YH, Lee KC, Lee S-G, Jin BW (1997) Tetrahedron 53:2921
30. Adam W, Zaho C-G (1997) Tetrahedron: Asymm 8:3995
31. Armstrong A, Hayter BR (1998) Chem Commun 621
32. Tu Y, Wang Z-X, Shi Y (1996) J Am Chem Soc 118:9806
33. Wang Z-X, Tu Y, Frohn M, Shi Y (1997) J Org Chem 62:2328
34. Frohn M, Shi Y (1997) J Org Chem 62:2328
35. Wang Z-X, Shi Y (1997) J Org Chem 62:8622
36. Wang Z-X, Shi Y (1998) J Org Chem 63:3099
37. Zhu Y, Tu Y, Yu H, Shi Y (1998) Tetrahedron Lett 39:7819
38. Warren JD, Shi Y (1999) J Org Chem 64:7675
39. Tian H, She X, Shu L, Yu H, Shi Y (2000) J Am Chem Soc 122:2435
40. Tian H, She X, Yu H, Shu L, Shi Y (2002) J Org Chem 67:2435
41. Tian H, She X, Xu J, Shi Y (2001) Org Lett 3:1929
42. Cao GA, Wang Z-X, Tu Y, Shi Y (1998) Tetrahedron Lett 39:4425
43. Frohn M, Zhou X, Zhang J-R, Tang Y, Shi Y (1999) J Am Chem Soc 121:7718
44. Shu L, Shi Y (1999) Tetrahedron Lett 40:8721
45. Shu L, Shi Y (2000) J Org Chem 65:8807
46. Wu X-Y, She X, Shi Y (2002) J Am Chem Soc 124:8792
47. Burke CP, Shi Y (2007) J Org Chem 72:4093
48. Burke CP, Shu L, Shi Y (2007) J Org Chem 72:6320
49. Ager DJ, Anderson K, Oblinger E, Shi Y, VanderRoest J (2007) Org Process Res Dev 11:44
50. (a) Xiong Z, Corey EJ (2000) J Am Chem Soc 122:4831; (b) Xiong Z, Corey EJ (2000) J Am Chem Soc 122:9328
51. Davis FA, Abdul-Malik NF, Awad SB, Haracal ME (1981) Tetrahedron Lett 22:917
52. Davis FA, Towson JC, Weismiller MC, Lal S, Caroll PJ (1988) J Am Chem Soc 110:8477
53. Page PCB, Heer JP, Bethell D, Lund A, Collington EW, Andrews DM (1997) J Org Chem 62:6093
54. Davis FA, Haracal ME, Awad SB (1983) J Am Chem Soc 105:3123
55. Davis FA, Przeslawski RM (1991) In: Abstracts of papers of 201st national meeting of the American Chemical Society, Atlanta, American Chemical Society, Washinghton, DC, 1991, ORGN 0105
56. Davis FA, Haque MS (1986) J Org Chem 51:4083
57. Davis FA, Vishwakarma LC, Billmers JM (1984) J Org Chem 49:3241
58. Brook AG, MaCrae OM (1974) J Organomet Chem 77:C19
59. Adam W, Fell RT, Saha-Möller CR, Zhao C-G (1998) Tetrahedron: Asymm 9:397
60. Davis FA, Sheppard AC (1987) J Org Chem 52:954
61. Brodsky BH, Du Bois J (2005) J Am Chem Soc 127:15391
62. Milliet P, Picot A, Lusinchi X (1976) Tetrahedron Lett 17:1573

63. Milliet P, Picot A, Lusinchi X (1976) Tetrahedron Lett 17:1577
64. Milliet P, Picot A, Lusinchi X (1981) Tetrahedron 24:4201
65. Hanquet G, Lusinchi X, Milliet P (1987) Tetrahedron Lett 28:6061
66. Hanquet G, Lusinchi X, Milliet P (1988) Tetrahedron Lett 29:3941
67. Lusinchi X, Hanquet G (1997) Tetrahedron 53:13727
68. Hanquet G, Lusinchi X, Milliet P (1993) Tetrahedron Lett 34:7271
69. Bohé L, Lusinchi M, Lusinchi X (1999) Tetrahedron 55:155
70. Hanquet G, Lusinchi X (1994) Tetrahedron 50:12185
71. Chiaroni A, Hanquet G, Lusinchi M, Riche C (1993) Acta Crystallogr Sect C 51:2047
72. Aggarwal VK, Wang MF (1996) J Chem Soc Chem Commun 191
73. Armstrong A, Ahmed G, Garnett I, Goacolou K (1997) Synlett 1075
74. Armstrong A, Ahmed G, Garnett I, Goacolou K, Wailes JS (1999) Tetrahedron 55:2341
75. Minakata S, Takemiya A, Nakamura K, Ryu I, Komatsu M (2000) Synlett 1810
76. Wong M-K, Ho L-M, Zheng Y-S, Ho C-Y, Yang D (2001) Org Lett 3:2587
77. Armstrong A, Draffan AG (1998) Synlett 646
78. Armstrong A, Draffan AG (1999) Tetrahedron Lett 40:4453
79. Bohé L, Kammoun M (2002) Tetrahedron Lett 43:803
80. Page PCB, Rassias GA, Bethell D, Schilling MB (1998) J Org Chem 63:2774
81. Page PCB, Rassias GA, Barros D, Bethell D, Schilling MB (2000) J Chem Soc Perkin Trans 1:3325
82. Rieche A, Schmitz E (1956) Chem Ber 89:1254
83. Page PCB, Rassias GA, Barros D, Ardakani A, Buckley B, Bethell D, Smith TAD, Slawin AMZ (2001) J Org Chem 66:6926
84. Page PCB, Rassias GA, Barros D, Ardakani A, Bethell D, Merifield E (2002) Synlett 580
85. Vachon J, Rentsch S, Martinez A, Marsol C, Lacour J (2007) Org Biomol Chem 5:501
86. Page PCB, Buckley BR, Blacker AJ (2004) Org Lett 6:1543
87. Page PCB, Parker P, Rassias GA, Buckley BR, Bethell D (2008) Adv Synth Catal 350:1867
88. Page PCB, Marken F, Williamson C, Chan Y, Buckley BR, Bethell D (2008) Adv Synth Catal 350:1149
89. Page PCB, Barros D, Buckley BR, Ardakani A, Marples BA (2004) J Org Chem 69:3595
90. Page PCB, Barros D, Buckley BR, Blacker AJ, Marples BA (2005) Tetrahedron: Asymm 16:3488
91. Page PCB, Buckley BR, Heaney H, Blacker AJ (2005) Org Lett 7:375
92. Page PCB, Appleby LF, Day D, Chan Y, Buckley BR, Allin SM, McKenzie MJ (2009) Org Lett 11:1991
93. Maringo M, Franzén J, Poulsen TB, Zhuang W, Jørgensen KA (2005) J Am Chem Soc 127:6964
94. Lee S, MacMillan DWC (2006) Tetrahedron 62:11413
95. Russo A, Lattanzi A (2008) Eur J Org Chem 2767
96. (a) Aggarwal VK, Lopin C, Sandrinelli FJ (2003) J Am Chem Soc 125:7596; (b) Adamo MFA, Aggarwal VK, Sage MA (2000) J Am Chem Soc 122:8317
97. Ho C-Y, Chen Y-C, Wong M-K, Yang D (2005) J Org Chem 70:898
98. For a mini review see: Armstrong A (2004) Angew Chem Int Ed 43:1460
99. Goncalves MH, Martinez A, Grass S, Page PCB, Lacour J (2006) Tetrahedron Lett 47:5297
100. Page PCB, Farah MM, Buckley BR, Blacker AJ, Lacour J (2008) Synlett 1381
101. Lu X, Liu Y, Sun B, Cindric B, Deng L (2008) J Am Chem Soc 130:8134
102. Wang X, Reisinger CM, List B (2008) J Am Chem Soc 130:6070
103. Reisinger CM, Wang X, List B (2008) Angew Chem Int Ed 47:1119
104. Wang X, List B (2008) Angew Chem Int Ed 47:8112
105. Andersen KK, Goffield W, Papankolaou NE, Foley JW, Perkins RI (1964) J Am Chem Soc 86:5637
106. Ruano JLC, Aranda MT, Zarzuelo MM (2003) Org Lett 5:75

107. Zhengzu H, Krishnamurthy D, Grover P, Wilkinson HS, Fang QC, Su XP, Lu ZH, Magiera D, Senanayake CH (2003) Angew Chem Int Ed 42:2032
108. Kagan HB, Pitchen P (1984) Tetrahedron Lett 25:1049
109. Bolm C, Bienewald F (1995) Angew Chem Int Ed 34:2640
110. Russo A, Lattanzi A (2009) Adv Synth Catal 351:521
111. Habibi D, Zolfigol MA, Safaiee M, Shamsian A, Ghorbani-Choghamarani A (2009) Catal Commun 10:1257
112. (a) Damavandi JA, Karami B, Zolfigol MA (2002) Synlett 933; (b) Zolfigol MA, Amani K, Ghorbani-Choghamarani A, Hajjami M, Ayazi-Nasrabadi R, Jafari S (2009) Catal Commun 9:1739
113. Davis FA, Billmers J (1983) J Org Chem 48:2672
114. Davis FA, Jenkins RH, Awad SB, Stringer OD, Watson WH, Galley J (1982) J Am Chem Soc 104:5412
115. Davis FA, McCauley JP, Chattopachyay S, Harakal ME, Towson JC, Watson WH, Tavabaiepour J (1987) J Am Chem Soc 109:3370
116. Davis FA, Jewson JC, Weismiller MC, Lal S, Caroll PC (1988) J Am Chem Soc 110:8477
117. Boyd DR, Malone JF, McGuckin MR, Jennings WB, Rutherford M, Saket BM (1994) J Chem Soc Perkin Trans II 1145
118. Jennings WB, Kochanewycz MJ, Lovely CJ, Boyd DR (1994) J Chem Soc Chem Commun 2569
119. Hanquet G, Lusinchi X, Milliet P (1988) Tetrahedron Lett 29:2817
120. Desmarteau DD, Petrov VA, Montanari V, Preynolato M, Resnati G (1994) J Org Chem 59:5762
121. Page PCB, Bethell D, Heer JP, Collington EW, Andrews DM (1995) Synlett 773
122. (a) Page PCB, Graham AE, Bethell D, Park BK (1993) Synth Commun 23:1507; (b) Page PCB, Bethell D, Heer JP, Collington EW, Andrews DM (1994) Tetrahedron Lett 35:9629; (c) Page PCB, Heer JP, Bethell D, Collington EW, Andrews D (1995) Synlett 773
123. Andrews DM, Bethell D, Collington EW, Heer JP, Page PCB (1995) Tetrahedron: Asymm 6:2911
124. Page PCB, Bethell D, Stocks PA, Heer JP, Graham AE, Vahedi H, Healy M, Collington EW, Andrews DM (1997) Synlett 1355
125. Page PCB, Bethell D, Vahedi H (2000) J Org Chem 65:6756
126. Bach RD, Coddens BA, McDouall JW, Shegal BS (1990) J Org Chem 55:3325
127. Gluszynska A, Ma kowska I, Rozwadowska MD, Sienniak W (2004) Tetrahedron: Asymm 15:2499
128. Del Rio RE, Wang B, Achab S, Bohé L (2007) Org Lett 9:2265
129. (a) Ojima I (ed.) (2000) Catalytic asymmetric synthesis, 2nd edn. Chaps. 1–3. Wiley VCH, New York; (b) Zhang X (ed.) (2004) Tetrahedron:Asymm 15:2099; (c) Crepy KVL, Imamoto T (2003) Top Curr Chem 1
130. (a) Seayad J, List B (2005) Org Biomol Chem 3:719; (b) Connon SJ (2006) Angew Chem Int Ed 45:3909
131. (a) Pietrusiewicz KM, Zablocka M (1994) Chem Rev 94:1375; (b) Grabulosa A, Granell J, Muller G (2007) Coord Chem Rev 251:25
132. Bergin E, O'Connor CT, Robinson SB, McGarrigle EM, O'Mahony CP, Gilheany DG (2007) J Am Chem Soc 129:9566

Chapter 2
Asymmetric Transfer Hydrogenations Using Hantzsch Esters

Tommaso Marcelli

Abstract 1,4-Dihydropyridines are extensively used as reducing agents in asymmetric organocatalytic protocols. In particular, readily available Hantzsch esters are competent hydrogen equivalents for iminium ion-, Brønsted acid- and hydrogen bonding-catalyzed reactions. These methodologies give often unsurpassed degrees of stereocontrol in the reduction of α,β-unsaturated carbonyl compounds, cyclic/acyclic imines and activated olefins. In addition, dihydropyridine-mediated reductions can be easily implemented in asymmetric cascade processes using one or more catalysts. The versatility of these catalytic protocols has been demonstrated by their application to the synthesis of a variety of biologically active compounds.

2.1 Introduction

The stereocontrolled addition of hydrogen to an unsaturated substrate is probably the asymmetric catalytic transformation which has been studied in more detail both in industry and in academia [1]. Countless examples of transition metal-catalyzed enantioselective reductions can be found in the chemical literature, in many cases with virtually perfect degree of stereocontrol. While hydrogen gas remains the most popular reducing agent for asymmetric hydrogenations, the use of small molecules able to formally transfer two atoms of hydrogen to a substrate (transfer hydrogenation) has become a well-established strategy in view of its advantages in terms of safety and operational simplicity [2]. Initially confined to transition metal catalysis (especially ruthenium complexes), asymmetric transfer hydrogenation can nowadays be performed with different substrates classes using chiral organic molecules

T. Marcelli (✉)
Dipartimento di Chimica, Materiali ed Ingegneria Chimica "Giulio Natta",
Politecnico di Milano, Via Mancinelli, 7, 20131 Milan, Italy
e-mail: tommaso.marcelli@chem.polimi.it

R. Mahrwald (ed.), *Enantioselective Organocatalyzed Reactions I: Enantioselective Oxidation, Reduction, Functionalization and Desymmetrization*,
DOI 10.1007/978-90-481-3865-4_2,

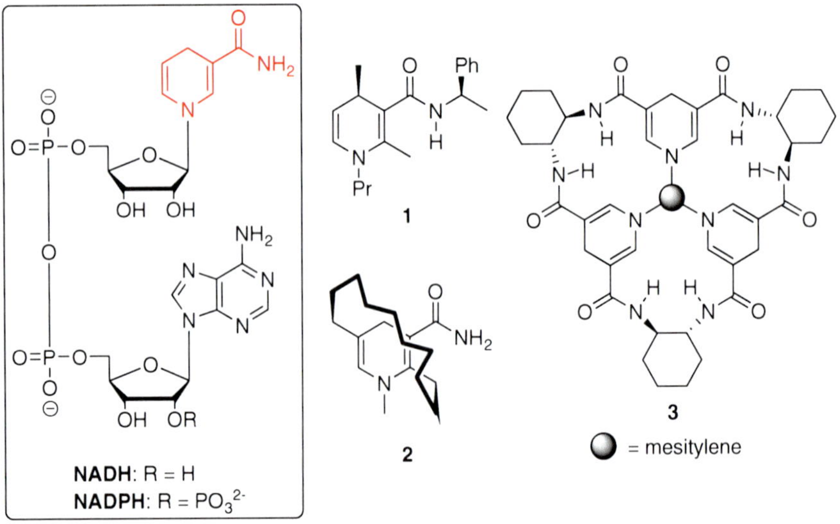

Fig. 2.1 NAD(P)H and chiral N-alkyl nicotinamides

as catalysts. In this respect, nearly the entirety of these protocols makes use of dihydropyridines as hydrogen donors. This Chapter gives an overview of organocatalyzed asymmetric transfer hydrogenations, highlighting their applications to the synthesis of biologically relevant molecules. For the sake of completeness, non-asymmetric catalytic reductions using dihydropyridines are also briefly discussed.

2.1.1 NADH and NADPH Mimics

Nicotinamide adenine dinucleotide (NADH) and nicotinamide adenine dinucleotide phosphate (NADPH) are ubiquitous cofactors mediating an impressive variety of redox processes in living organisms (Fig. 2.1) [3]. These compounds owe their unique reactivity to the dihydropyridine core of the nicotinamide subunit which can (formally) lose a hydride at the C_4 position with concomitant aromatization of the heterocycle yielding the oxidized form of the coenzyme, NAD(P)$^+$. Throughout the second half of the twentieth century, considerable efforts have been spent in the development of small-molecule NAD(P)H mimics, mainly to gain insights on mechanistic aspects of redox biological processes [4–6]. In particular, a major controversy addressed using synthetic nicotinamides involved whether the oxidation of NAD(P)H takes place as a single-step hydride transfer or as a sequential electron-proton-electron transfer. In this respect, depending on the type of process examined, there are experimental data in support of both hypotheses [7].

The reactivity of synthetic nicotinamides has been studied with several substrates, often in the presence of Lewis acids. Among the remarkable variety of

NAD(P)H mimics which have been synthesized, various chiral nicotinamides were tested for their ability to deliver a hydride to unsaturated substrates in an enantioselective fashion; representative examples are given in Fig. 2.1.

Compound **1** contains two stereocenters, one of them on the C_4 position which is destroyed upon hydride transfer [8]. Bicyclic amide **2** features an alkyl chain bridging C_2 and C_5 of the dihydropyridine core to mimic the lipophilic region of the active site of L-lactate dehydrogenase [9]. Finally, in trimeric compound **3** the chirality comes from 1,2-*trans*-cyclohexyldiamine, a very popular building block for asymmetric (organo)catalysts [10]. These compounds have all been shown to reduce α-ketoesters in a highly enantioselective fashion in the presence of magnesium perchlorate.

2.1.2 Hantzsch Esters

1,4-Dihydropyridines find important applications in medicine (mainly as calcium channel blockers) and their chemistry has been thoroughly investigated over the last decades [11–14]. The first synthesis of dihydropyridines **4** was reported by Arthur Rudolf Hantzsch over a century ago and involved the condensation between ammonia, an aldehyde and two molecules of a β-ketoester (Scheme 2.1) [15]. Use of a modified procedure allowed synthesis of unsymmetrical dihydropyridines starting from two different β-ketoesters [16]. Compounds **4** are known as Hantzsch esters and have been extensively used as pyridine precursors in the synthesis of heterocyclic compounds.

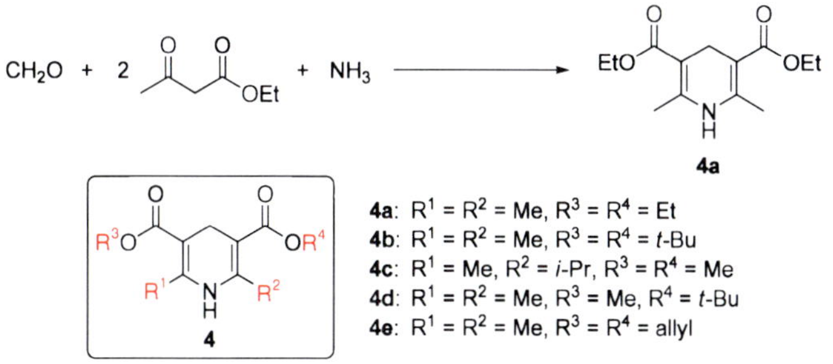

4a: $R^1 = R^2 = Me$, $R^3 = R^4 = Et$
4b: $R^1 = R^2 = Me$, $R^3 = R^4 = t\text{-Bu}$
4c: $R^1 = Me$, $R^2 = i\text{-Pr}$, $R^3 = R^4 = Me$
4d: $R^1 = R^2 = Me$, $R^3 = Me$, $R^4 = t\text{-Bu}$
4e: $R^1 = R^2 = Me$, $R^3 = R^4 = \text{allyl}$

Scheme 2.1 Hantzsch synthesis of dihydropyridines and common Hantzsch esters used in asymmetric organocatalytic transfer hydrogenations

The hydride donor ability of Hantzsch ester **4a** has been experimentally determined to be slightly higher than that of N-benzylnicotinamide, both being comparable to the borane-triethylamine complex [17]. Together with trichlorosilane [18], Hantzsch esters **4** are the reagents of choice to effect asymmetric reductions using chiral organocatalysts.

2.1.3 Organocatalysis

Within the recent wave of interest in organocatalysis, several research groups tackled the challenging task of developing transition metal-free asymmetric hydrogenations, mainly using Hantzsch esters as the hydrogen source. In little more than 5 years, an impressive variety of organocatalytic protocols for the asymmetric reduction of different functionalities (such as activated olefins and imines) has been reported [19–21]. In general, three main types of organocatalytic transfer hydrogenation can be identified, depending on the substrate and the type of catalysts (Scheme 2.2).

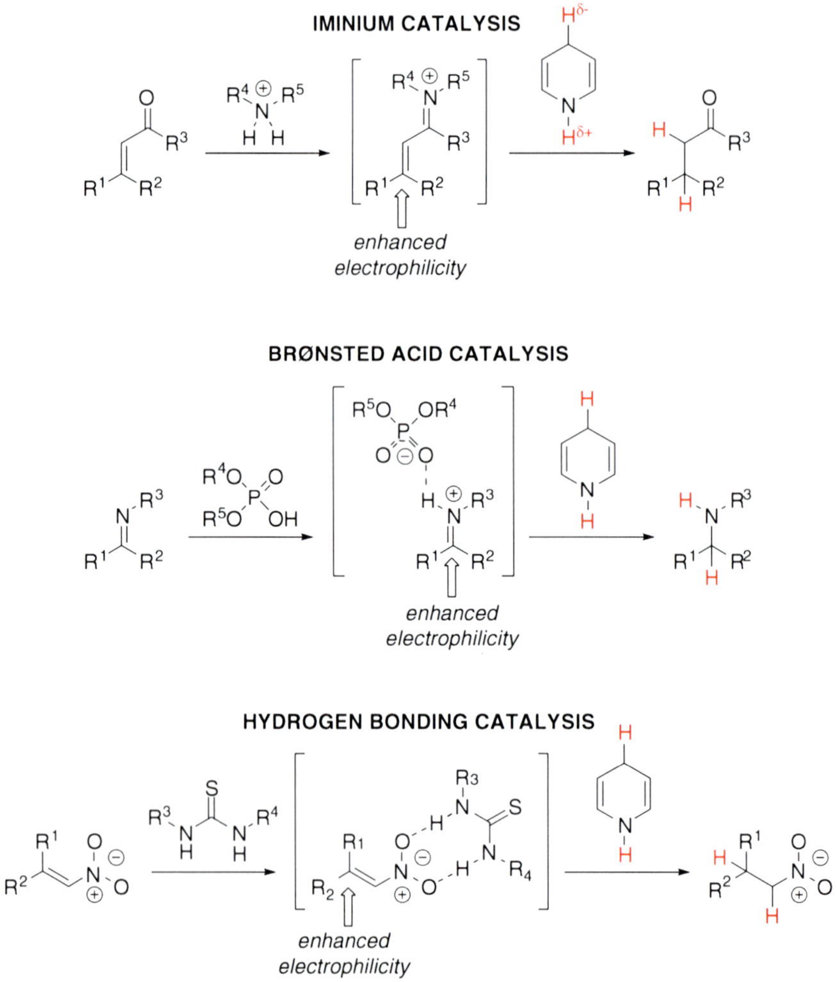

IMINIUM CATALYSIS

enhanced electrophilicity

BRØNSTED ACID CATALYSIS

enhanced electrophilicity

HYDROGEN BONDING CATALYSIS

enhanced electrophilicity

Scheme 2.2 Classes of organocatalyzed transfer hydrogenations using dihydropyridines

For all these reactions, there is general consensus that the transfer hydrogenation takes place *via* hydride transfer rather than electron/proton/electron donation. These three classes of reactions are very different and are individually discussed in the following sections. Nevertheless, the role of the catalyst is common to all these processes and is namely to increase of electrophilicity of a carbon atom by either: (1) formation of a covalent intermediate with the substrate [22], (2) explicit protonation of an electronegative atom [23], or (3) complexation by hydrogen bonding [24]. In addition to that, catalysts occasionally also interact with the Hantzsch esters, increasing their reactivity and positioning them in a geometric arrangement suitable for the subsequent hydride transfer.

The following pages give an overview of the various types of organocatalytic asymmetric transfer hydrogenations which have been realized so far, mostly yielding valuable chiral building blocks with very high enantioselectivities. In addition to that, the well-known versatility of iminium and Brønsted acid catalysis has been often exploited to couple an enantioselective reduction step with other organocatalytic transformations in cascade processes giving access to highly enantioenriched, drug-like compounds [25]. The examples described below are divided according to the type of mechanism; cascade reactions with more than one catalyst are included in the section corresponding to the mode of substrate activation for the hydride transfer step.

2.2 Iminium Ion Catalysis

2.2.1 Achiral Catalysts

The capability of dialkylammonium salts to react with α,β-unsaturated aldehydes generating an iminium ion particularly activated towards 1,4-additions was exploited to realize a highly selective conjugate reduction of enals [26]. The optimized conditions for this reaction envision the use of 5 mol% dibenzylammonium trifluoroacetate **5a** in combination with Hantzsch ester **4a**, to yield reduced aldehydes with short reaction times in the case of electron-deficient substrates. The intrinsic regioselectivity of this approach found application in the synthesis of various lepidopteran sex pheromones [27] (Scheme 2.3).

2.2.2 Chiral Catalysts

Several protocols for the asymmetric reduction of α,β-unsaturated carbonyl compounds rely on the use of acid salts of chiral imidazolidinones **6**, a very popular catalyst class for enantioselective conjugate additions [22] (Fig. 2.2).

For instance, the trifluoroacetic (TFA) salt of imidazolidinone **6a** proved very effective for the enantioselective 1,4-reduction of enals [28]. Interestingly, reaction

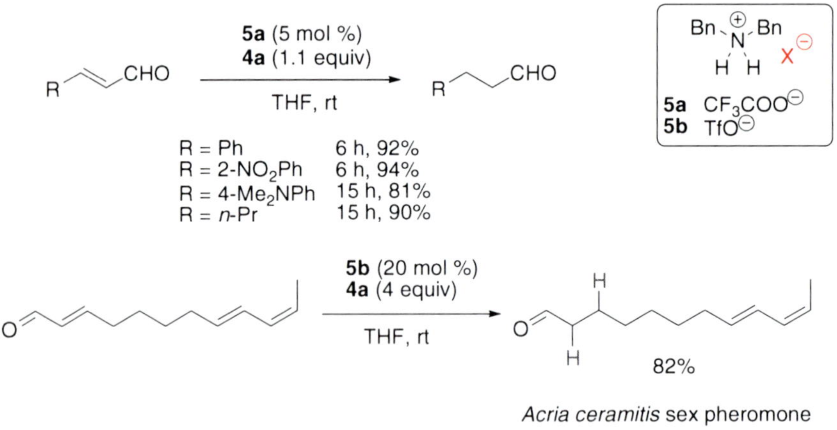

Scheme 2.3 Organocatalytic conjugate reduction of enals

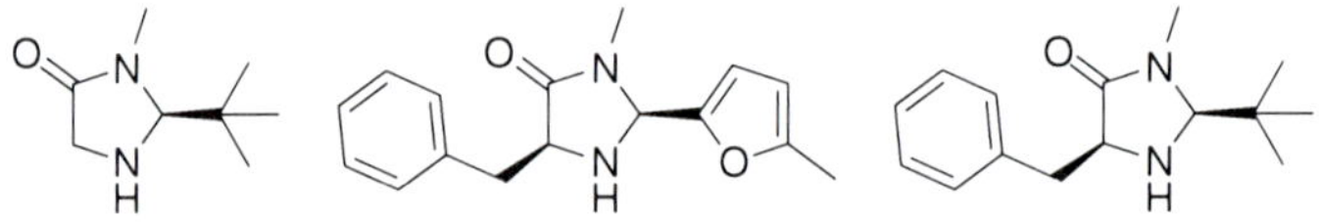

Fig. 2.2 Imidazolidinones catalysts used in asymmetric transfer hydrogenation (represented as free bases)

of the two isomers of a same olefin gives identical stereochemical outcome, in stark contrast with what observed for many metal-catalyzed hydrogenations. While this catalyst is not able to promote the conjugate reduction of cyclic enones, furyl-substituted catalyst **6b·TCA** (trichloroacetic acid) in combination with Hantzsch ester **4b** was shown to successfully reduce cyclic α,β-unsaturated ketones of different ring sizes (5–7 atoms) with good levels of asymmetric induction [29]. For enals, it has been proposed that the catalyst reacts with the substrate leading to the predominant formation of an *E,E*-iminium ion which undergoes hydride attack. Likewise, for cyclic enones the *E* iminium ion leaves the bottom face of the conjugated system unshielded. High-level DFT calculations confirmed this hypothesis reproducing the experimental enantioselectivity [30]. The computational results also indicate that in the energetically most accessible arrangement of the reactants for the hydride transfer step, the Hantzsch ester and the iminium ion lie in an *anti* geometry, hence with the dihydropyridine N-H pointing away from the catalyst (Scheme 2.4).

These protocols for conjugate reductions were exploited in the synthesis of various biologically relevant targets. For instance, it was shown that α,β-unsaturated aldehydes containing either a thiazole or an oxazole as β-substituent are competent substrates for the transfer hydrogenation even though the reduced products are obtained with slightly lower enantiomeric excesses. The resulting chiral heterocyclic

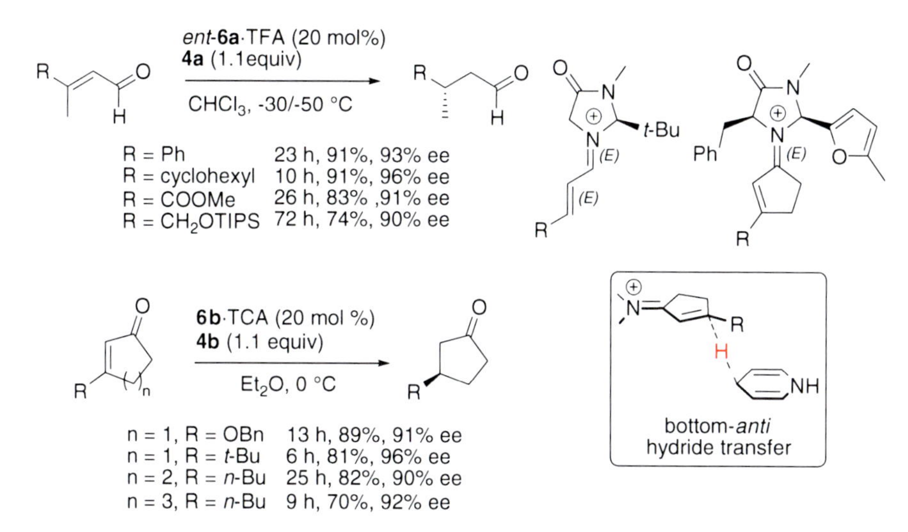

Scheme 2.4 Enantioselective conjugate reduction of α,β-unsaturated carbonyl compounds

derivatives represent a rather common motif in biological compounds and this methodology was applied in the synthesis of the C7–C14 fragment of Ulupualide A [31]. The imidazolidinone-catalyzed conjugate reduction was also employed in a total synthesis of (+)-neopeltolide A, a highly cytotoxic marine macrolide [32]. Enantioselective reduction of a cyclic enone containing an ester group as β-substituent provided a chiral building block which was further elaborated to a spliceosome inhibitor designed starting from an overlay of the structures of FR901464 and pladienolide, two natural bioactive compounds presumably sharing the same mode of action [33] (Fig. 2.3).

In asymmetric counteranion-directed catalysis (ACDC), chiral acids are used to generate the active form of the catalyst (in the abovementioned cases, an imidazolidinonium cation). This strategy, which has been applied to various organocatalytic transformations, proved very successful in the case of conjugate reductions using Hantzsch esters [34]. In particular, salts obtained from bulky chiral phosphoric acids and α-amino esters or achiral secondary amines have been shown to be powerful catalysts for the transfer hydrogenation of problematic substrates (Scheme 2.5). For instance, catalyst **7a** can be used to obtain high degrees of stereocontrol with in the reduction of nonhindered aliphatic enals, whereas imidazolidinone-based catalysts give unsatisfactory results [35]. The levels of enantioselectivities obtained using this catalyst remain high even at elevated temperatures, as demonstrated in the synthesis of the fragrance (*S*)-Florhydral® [36]. Valine-derived catalyst **7b** gave excellent results in the reduction of cyclic enones and, remarkably, promoted the transfer hydrogenation of acyclic α,β-unsaturated ketones, although with somewhat lower enantioselectivities [37].

Highly enantioselective reductions in aqueous media were realized using peptide catalysts covalently attached to polyethylene glycol grafted on polystyrene resin. Catalyst **8** gave the best results among various oligopeptides screened for the

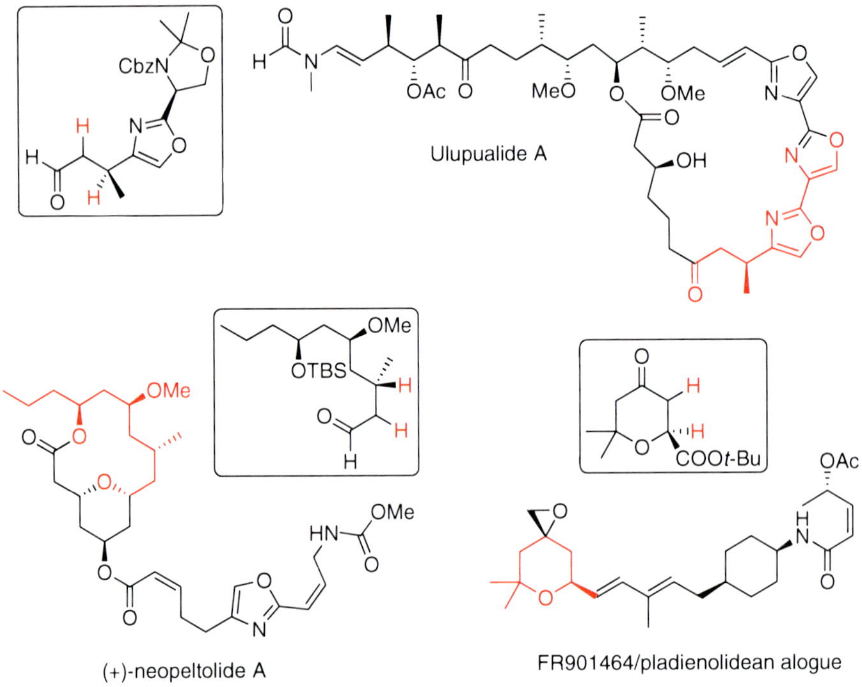

Fig. 2.3 Applications of the asymmetric conjugate reduction to the synthesis of bioactive compounds

reduction of β-methyl cinnamaldehydes (Scheme 2.6) [38]. Key elements in the catalyst design included the presence of a 2-aminoisobutyrric acid / D-proline (Aib – d-Pro) motif, which is known to induce a β-turn in organic solvents, and a rather long hydrophobic polyleucine chain (25.4 residues on average). This catalyst gave good results with several α,β-unsaturated aldehydes although it failed in promoting the reaction of sterically hindered substrates. IR studies on the solution structure of the catalyst indicated that the polyleucine chain folds to a stable α helix and confirmed the presence of a β-turn [39].

2.2.3 Cascade Processes

Upon addition of a hydride to the α,β-unsaturated iminium intermediate, the resulting nucleophilic enamine can be trapped with a suitable electrophile present in the reaction mixture (tandem iminium-enamine catalysis) [40]. While in the abovementioned cases the enamine intermediate is quenched with a proton, some methodologies exploit the reactivity of enamines to couple conjugate reduction to the formation of a new C-C bond. For instance, α,β-unsaturated aldehydes containing a second Michael acceptor in their structure undergo reductive cyclization processes resulting in the formation of a cyclopentane ring with two contiguous stereocenters

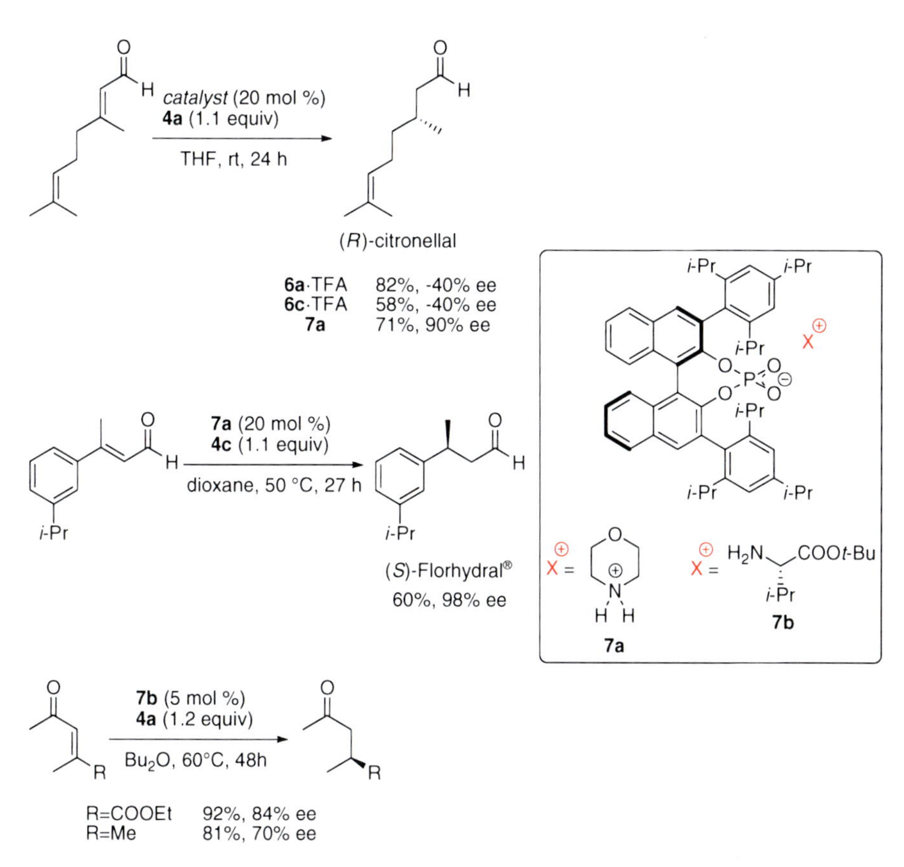

Scheme 2.5 Counteranion-directed enantioselective conjugate reduction of α,β-unsaturated carbonyl compounds

8·TFA (20 mol %)
4a (1.1 equiv)
THF/H_2O 1:2, rt, 48 h

Pro – D-Pro– Aib– Trp – Trp – (Leu)25.4

8

R = Ph 75%, 90% ee
R = 4-MeC_6H_4 76%, 95% ee
R = 3-ClC_6H_4 69%, 93% ee
R = 2-ClC_6H_4 <1%

Scheme 2.6 Oligopeptide-catalyzed enantioselective reduction of enals

(Scheme 2.7) [41]. The same concept has also been applied to intermolecular reactions such as a reductive Mannich reaction between enals and α-imino esters using Hantzsch esters as the hydrogen source [42]. In this case, however, the electrophile (α-imino ester) is added after the reduction of the starting material is complete and both reaction medium and temperature are adjusted to optimize yield and diastereoselectivity in the Mannich step.

Scheme 2.7 Asymmetric reductive Michael cyclization and schematic representation of the tandem iminium-enamine catalysis mechanism

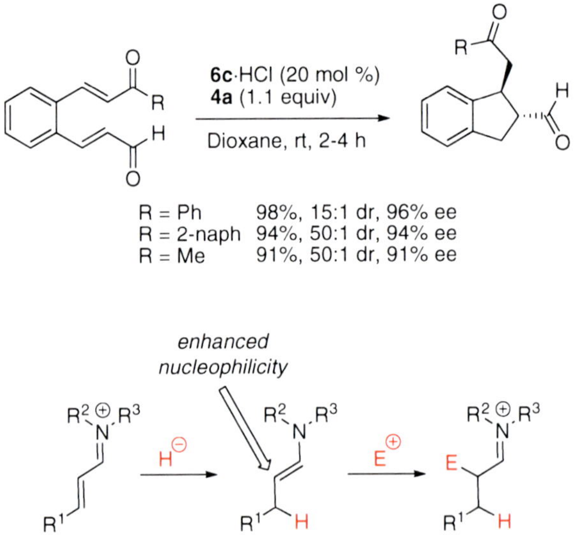

6c·HCl (20 mol %)
4a (1.1 equiv)
Dioxane, rt, 2-4 h

R = Ph 98%, 15:1 dr, 96% ee
R = 2-naph 94%, 50:1 dr, 94% ee
R = Me 91%, 50:1 dr, 91% ee

enhanced nucleophilicity

Hanztsch esters have also been used as hydrogen donors for organocatalytic cascade reductive alkylations of carbonyl compounds. In this approach, a Knoevenagel condensation between a ketone and an activated methylene compound is followed by a transfer hydrogenation [43, 44]. While these reactions are catalyzed by protonated cyclic secondary amines (such as proline and derivatives), from a mechanistical point of view they share little resemblance to the previously discussed examples. In more detail, the role of the catalyst is to promote the alkylation step and the resulting intermediate is activated enough to undergo spontaneous hydride transfer (Scheme 2.8). Although in most cases the enantiomeric excess of the products was not determined, this mechanistic hypothesis implies the substantial absence of stereocontrol in the hydride transfer step. Nevertheless, this tandem

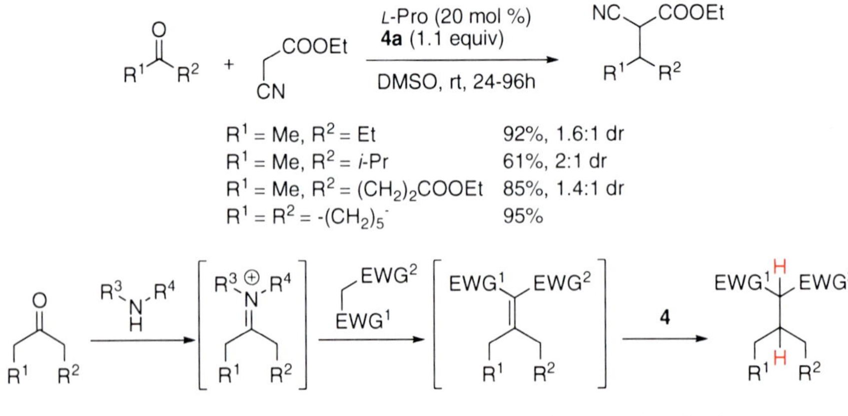

L-Pro (20 mol %)
4a (1.1 equiv)
DMSO, rt, 24-96h

R^1 = Me, R^2 = Et 92%, 1.6:1 dr
R^1 = Me, R^2 = *i*-Pr 61%, 2:1 dr
R^1 = Me, R^2 = $(CH_2)_2COOEt$ 85%, 1.4:1 dr
R^1 = R^2 = $-(CH_2)_5-$ 95%

Scheme 2.8 Tandem Knoevenagel/transfer hydrogenation and proposed mechanism

Knoevenagel / transfer hydrogenation has been coupled to other organocatalytic reactions (often in one-pot procedures) to synthesize libraries of enantioenriched chiral compounds, either by means of chiral starting materials [45] or by using well-developed enantioselective reactions before or after the reductive alkylation step [46, 47].

2.3 Brønsted Acid Catalysis

2.3.1 *Achiral Catalysts*

Early reports on the use of Hantzsch esters as reducing agents for carbon-nitrogen double bonds described the screening of a variety of Brønsted acids as catalysts. Interestingly, several of them were found to promote the transfer hydrogenation of N-*para*-methoxyphenyl (PMP) ketimines [48] and quinolines [49]. In both cases, however, the best results were obtained using diphenyl phosphate **9** as catalyst (Scheme 2.9). Quinolines react with two equivalents of Hantzsch ester yielding tetrahydroquinolines at remarkably low catalyst loadings. This reaction is presumed to proceed via an initial conjugate reduction followed by isomerization to the 3,4-dihydropyridine and subsequent 1,2-reduction. In alternative to this catalytic approach, a methodology making use of acyl chlorides in stoichiometric amount for quinoline activation (hence yielding N-acyl derivatives) has also been developed [50].

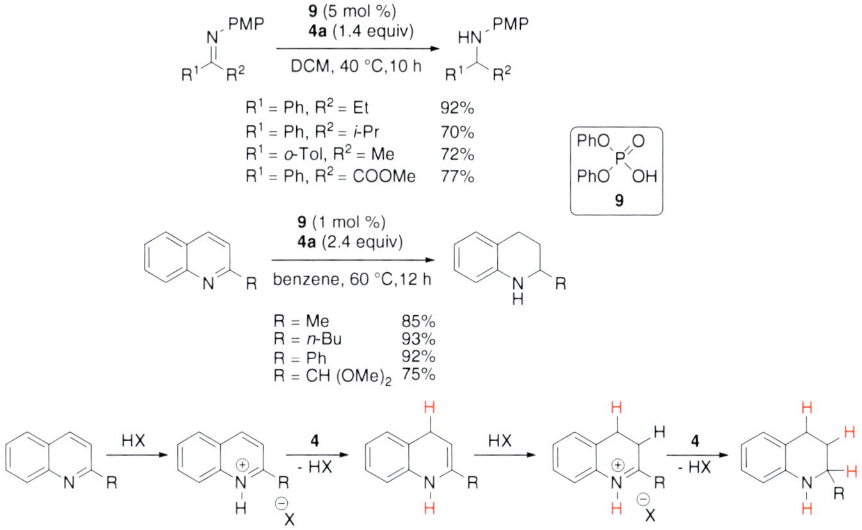

9 (5 mol %), **4a** (1.4 equiv), DCM, 40 °C, 10 h

R^1 = Ph, R^2 = Et	92%
R^1 = Ph, R^2 = *i*-Pr	70%
R^1 = *o*-Tol, R^2 = Me	72%
R^1 = Ph, R^2 = COOMe	77%

9 (1 mol %), **4a** (2.4 equiv), benzene, 60 °C, 12 h

R = Me	85%
R = *n*-Bu	93%
R = Ph	92%
R = CH $(OMe)_2$	75%

Scheme 2.9 Brønsted acid-catalyzed reduction of N-aryl ketimines and 2-substituted quinolines. PMP = *p*-methoxyphenyl

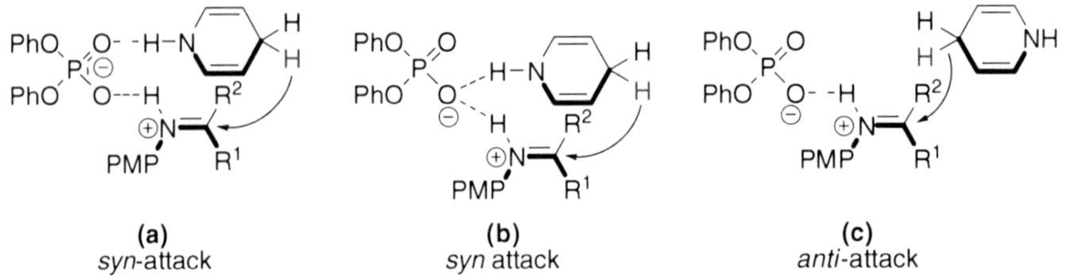

(a) *syn*-attack (b) *syn* attack (c) *anti*-attack

Fig. 2.4 Mechanistic possibilities for the phosphoric acid-catalyzed hydrogenation of N-PMP imines

Computational studies on the mechanism of the transfer hydrogenation of N-aryl imines using an achiral phosphoric acid catalyst identified different possible modes of substrate(s) activation [51, 52] (Fig. 2.4). In more detail, upon protonation of the imine, the phosphate group can (a) engage in two hydrogen bonds with the substrate and the Hantzsch ester using both oxygens; (b) interact with the iminium and the dihydropyridine N-H with the same oxygen; (c) interact exclusively with the N-aryl iminium ion. The calculations indicate that mode (a) is by far preferred over the other two which are not energetically accessible. Moreover, calculations on the reaction of *E*- and *Z*-iminium ions showed that both isomers have similar energy once they are complexed by the phosphate anion and they are equally competent substrates for hydride transfer.

2.3.2 Chiral Catalysts

The advent of axially chiral phosphoric acids as powerful organocatalysts for nucleophilic additions to C=N bonds [53] prompted several research groups to investigate the feasibility of enantioselective transfer hydrogenations of imines (and related compounds) using Hantzsch esters. In general, bulky 2,2′-disubstituted BINOL-derived phosphoric acids **10** gave excellent results with many substrates combinations. In some cases, partially hydrogenated BINOL derivatives or C_2-symmetric catalysts based on other scaffolds gave advantages in terms of reactivity and selectivity (Fig. 2.5).

The enantioselective reduction of N-PMP ketimines using BINOL-derived catalysts **10** has been investigated by different research groups [54–56]. These studies showed that consistently high levels of chiral induction can be obtained in this reaction for structurally different imines, including challenging substrates such as 2-butanone PMP imine. Moreover, the imines can be efficiently generated *in situ* starting from the corresponding ketones and *p*-anisidine without altering the reaction outcome (Scheme 2.10). The origin of enantioselectivity for this type of reactions was investigated using different computational techniques. The calculations indicate that the bulky substituents in the 2 and 2′ positions of the catalyst control the binding geometry of substrate and Hantzsch ester. As a result, hydride transfer to the Z-iminium ion becomes energetically favoured and governs the stereochemical outcome of the reaction [51, 52].

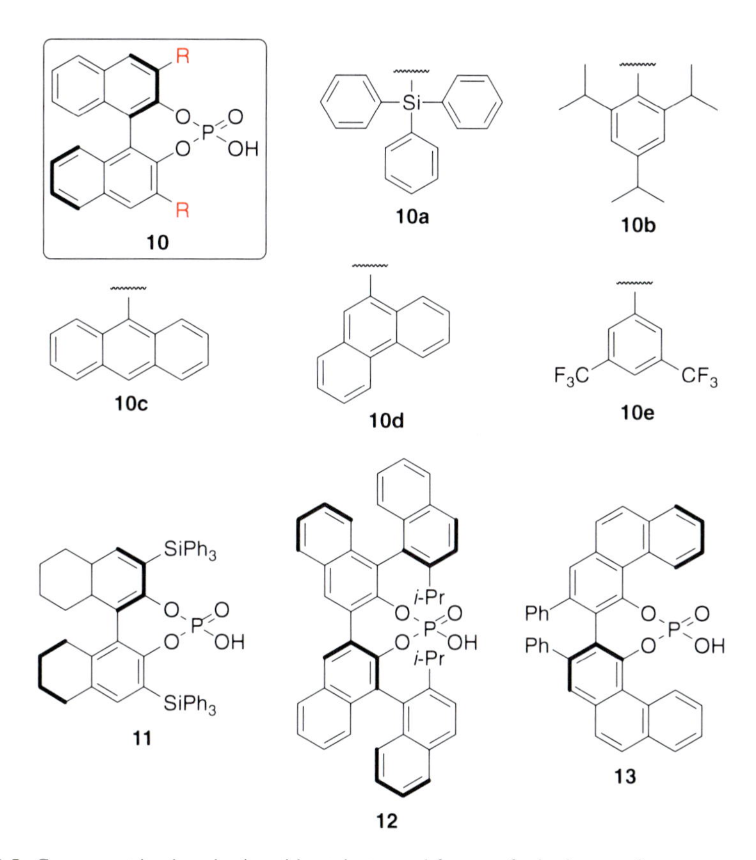

Fig. 2.5 C_2-symmetric phosphoric acid catalysts used for transfer hydrogenations

These results prompted other research groups to investigate the use of different catalysts and hydrogen donors for this type of transfer hydrogenation. A survey of different nucleotides containing Brønsted acid moieties revealed that adenosine 5′-diphosphate (ADP) can promote the reductive amination of ketones with *p*-anisidine, although with lower enantioselectivities compared to catalysts **10** [57]. Recent studies showed that benzothiazolines (such as **14**) can efficiently replace Hantzsch esters as hydrogen donors in this reaction under very similar conditions [58]. These compounds, which can be conveniently generated *in situ* from commercially available precursors, lead to occasionally higher enantiomeric excesses in the transfer hydrogenation of N-PMP ketimines. They have been recently exploited in a two-step reductive amination/aza-Michael asymmetric synthesis of tetrahydroisoquinolines and β-carbolines [59].

The ease of racemization of α-branched N-aryl aldimines in the presence of a Brønsted acid was exploited in a dynamic kinetic resolution *via* asymmetric transfer hydrogenation [60]. In this reaction, imines are formed *in situ* and hydride addition does not generate a new chiral center but takes place preferentially with one enantiomer of the substrate. As a result, (nearly) full conversion of an aldehyde to an

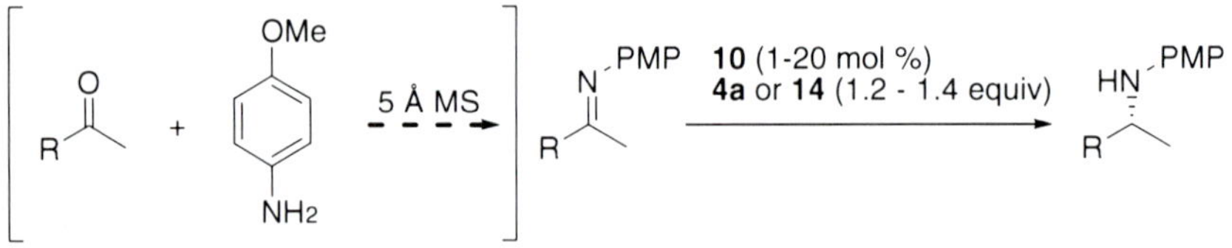

OMe

5 Å MS

N PMP

R

NH2

10 (1-20 mol %)
4a or 14 (1.2 - 1.4 equiv)

HN PMP

R

R = Ph

	preformed imine	*in situ* imine
10e (20 mol %), **4a** (1.4 equiv), benzene, 60 °C	76%, 74% ee	not reported
ent-**10b** (1 mol %), **4a** (1.4 equiv), toluene, 35 °C	96%, -88% ee	92%, -88% ee
10a (10 mol %), **4a** (1.2 equiv), benzene, 40 °C	not reported	87%, 94% ee

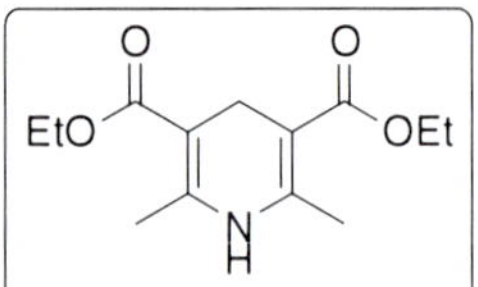

EtO OEt
4a

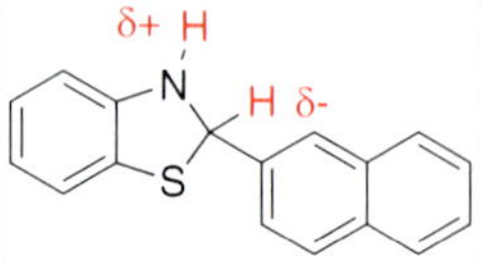

δ+ H
H δ-
N
S
14

in situ imine,
10a (10 mol %),
4a (1.2 equiv),
benzene, 40 °C

R = 2-FC_6H_4	60%, 83% ee
R = 2-naph	73%, 96% ee
R = Cy	49%, 86% ee
R = Et	71%, 83% ee

preformed imine,
10b (2 mol %),
14 (1.4 equiv),
mesitylene, 50 °C

R = Ph	90%, 98% ee
R = 2-naph	96%, 98% ee
R = *p*-Tol	93%, 98% ee
R = Cy	80%, 98% ee

Scheme 2.10 Enantioselective reduction of N-PMP imines

enantiomerically enriched β-branched amine can be achieved using chiral phosphoric acid catalysis (Scheme 2.11). DFT calculations showed that, in this case, the presence of an α-substituent renders hydride transfer to the *E*-iminium energetically favored and the stereoselectivity of the hydride addition is governed by interactions between the catalyst's bulky substituents and the PMP group of the iminium ion [61].

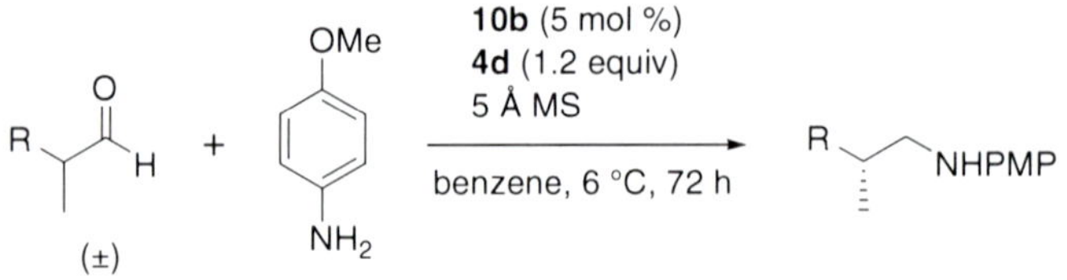

R O H
(±)

OMe
NH_2

10b (5 mol %)
4d (1.2 equiv)
5 Å MS

benzene, 6 °C, 72 h

R NHPMP

R = Ph	87%, 96% ee
R = 2-naph	96%, 96% ee
R = *t*-Bu	77%, 80% ee
R = CF_3	40%, 80% ee

Scheme 2.11 Dynamic kinetic resolution of aldehydes *via* organocatalytic transfer hydrogenation

α-Imino esters were also shown to be competent substrates for the enantioselective reduction using Hantzsch esters and phosphoric acid catalysts. In this reaction, giving access to valuable α-amino acids derivatives, VAPOL-derived catalyst **13** [62] performs somewhat better than BINOL derivatives **10** with respect to both activity and stereocontrol [63] (Scheme 2.12). Interestingly, the sense of stereoinduction for the hydride transfer was found to depend on the nature of the imine substituent, with alanine and phenylglycine derivatives precursors yielding opposite enantiomers. With respect to the nature of the hydrogen donor, benzothiazoline **14** also affords the reduction products with excellent enantiomeric excesses, sometimes higher than those obtained with Hantzsch ester **4a** [64].

ent-**10c** (10 mol %), **4a** (1.2 equiv), toluene, 60 °C, 48 h

R1 = Ph, R^2 = Et	88%, 92% ee
R1 = Ph, R^2 = *i*-Pr	87%, 97% ee
R1 = 2-naph, R^2 = *i*-Pr	93%, 98% ee
R1 = Cy, R_2 = *i*-Pr	46%, 88% ee

13 (5mol%), **4a** (1.4 equiv), toluene, 50 °C, 18-22 h

R^1 = Ph, R2 = Et	93%, 96% ee
R^1 = 4-$CF_3C_6H_4$, R^2 = Et	98%, 96% ee
R^1 = Me, R^2 = Et	88%, 99% ee
R^1 = $PhCH_2CH_2$, R^2 = Et	85%, 98% ee

ent-**10c** (1 mol %), *4a* (2.2 equiv), Et_2O, rt, 18-22h

R=Ph	58%, 94% ee
R=*p*-Tol	42%, 95% ee
R=1-naph	27%, 83% ee
R=$PhCH_2CH_2$	<5%

Scheme 2.12 Asymmetric reduction of α-imino esters

The scope of this reaction was expanded with the development of the asymmetric transfer hydrogenation of β-γ-alkynyl α-imino esters. In this reaction, both imine and alkyne undergo reduction to yield synthetically challenging *trans*-alkenyl amino acid derivatives [65]. It was shown that conjugate hydride transfer to the alkyne takes place first, as propargyl α-amino esters failed to undergo reduction under the reaction conditions.

Enamides were explored as substrates for Brønsted acid organocatalytic reductions. While this reaction formally involves hydrogenation of a C=C bond, substrate protonation yields an iminium ion which is subsequently reduced by Hantzsch ester **4a** [66] (Scheme 2.13). Optimization of the reaction conditions revealed that the loading of phosphoric acid catalyst could be dramatically decreased while retaining high degrees of stereocontrol by using acetic acid as cocatalyst to increase the concentration of iminium ion in the reaction mixture.

The previously described double hydrogenation of quinolines to tetrahydroquinolines was also investigated with axially chiral phosphoric acid catalysts. 2-Substituted quinolines were readily reduced in high enantioselectivities using

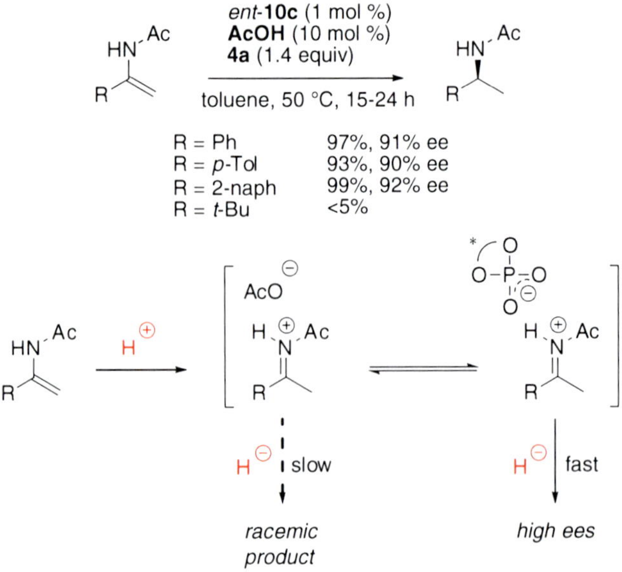

ent-**10c** (1 mol %)
AcOH (10 mol %)
4a (1.4 equiv)
toluene, 50 °C, 15-24 h

R = Ph 97%, 91% ee
R = *p*-Tol 93%, 90% ee
R = 2-naph 99%, 92% ee
R = *t*-Bu <5%

AcO
H slow
racemic product
H fast
high ees

Scheme 2.13 Enantioselective reduction of enamides using an achiral Brønsted acid cocatalyst

either BINOL-derived catalyst **10d** [67] or bulky phosphoric acid **12** [68] (Scheme 2.14). In the latter case, the catalyst loading could be substantially reduced while maintaining excellent levels of asymmetric induction. This methodology provides easy access to a variety of tetrahydroquinoline alkaloids with an alkyl

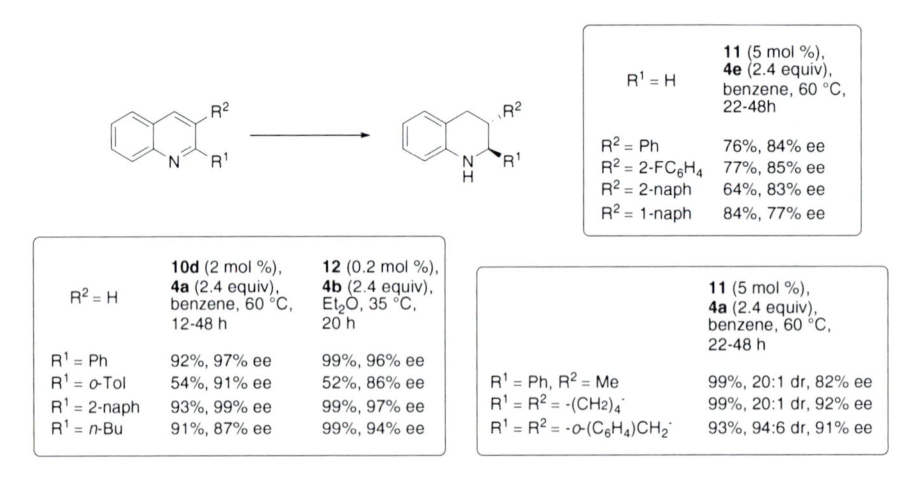

Scheme 2.14 Asymmetric transfer hydrogenation of quinolines

substituent at the 2-position. Asymmetric hydrogenation of 3-substituted quinolines was also successfully accomplished [69]. In this case, it is worth pointing out that the stereochemistry is determined by protonation of the 1,4-dihydroquinoline intermediate rather than by hydride transfer (see Scheme 2.9). Finally, 2,3-disubstituted quinolines were also shown to be competent substrates for this reaction, yielding products with two adjacent stereogenic centers in a *trans* relative arrangement [68].

This approach was later extended to the enantioselective transfer hydrogenation of other nitrogen heterocycles. Benzoxazines, benzothiazines and benzoxazinones were all reduced using remarkably low catalyst loadings with consistently high levels of asymmetric induction [70] (Scheme 2.15). 2-Substituted pyridines containing an electron-withdrawing group on C_3 could be reduced twice yielding chiral tetrahydropyridines with mostly high enantioselectivities [71]. 1,10-Phenanthrolines were also subjected to transfer hydrogenation using BINOL-derived phosphoric acid catalysts [72]. In this case, yields were generally less satisfactory. Finally, quinoxalines and quinoxalinones were shown to be competent substrates for organocatalytic transfer hydrogenation. However, significantly higher catalyst loadings were required to achieve full conversion in an acceptable reaction time [73].

2.3.3 Cascade Processes

In addition to the example discussed in section 1.2.3, the complementary character of iminium and enamine mechanisms was also exploited for the transfer hydrogenation of C=N bonds. Reaction of δ-diketones with aromatic amines in the presence of a Hantzsch ester and an acid catalyst yields prevalently *trans*-disubstitued cyclohexylamines [74] (Scheme 2.16). In this transformation, the initially formed enamine undergoes cyclization yielding an α,β-unsaturated iminium ion which in turn reacts with two molecules of Hantzsch ester to liberate the final product. The acid catalyst is crucial to maintain a high concentration of the iminium ion in the reaction mixture.

The previously described asymmetric hydrogenation of pyridines [71] was modified and implemented in a cascade sequence beginning with the reaction of an enamine with an α,β-unsaturated ketone [75]. Upon exposure to a Brønsted acid catalyst, these two reaction partners undergo a tandem Michael addition / cyclization yielding a 1,4-dihydropyridine derivative. Protonation of this intermediate provides the cationic substrate for the hydride transfer. A similar approach was used in the cascade synthesis of tetrahydroquinolines. In this case, however, generation of the substrate for transfer hydrogenation was accomplished using gold(I) catalysis. In short, propinyl-substituted anilines were cyclized to 1,4-dihydroquinolines and subsequently reduced by Hantzsch ester **4a** using catalyst **10d** [76]. The optical purities of the resulting adducts are comparable with those obtained in the Brønsted-acid catalyzed transfer hydrogenation [67] (Scheme 2.14).

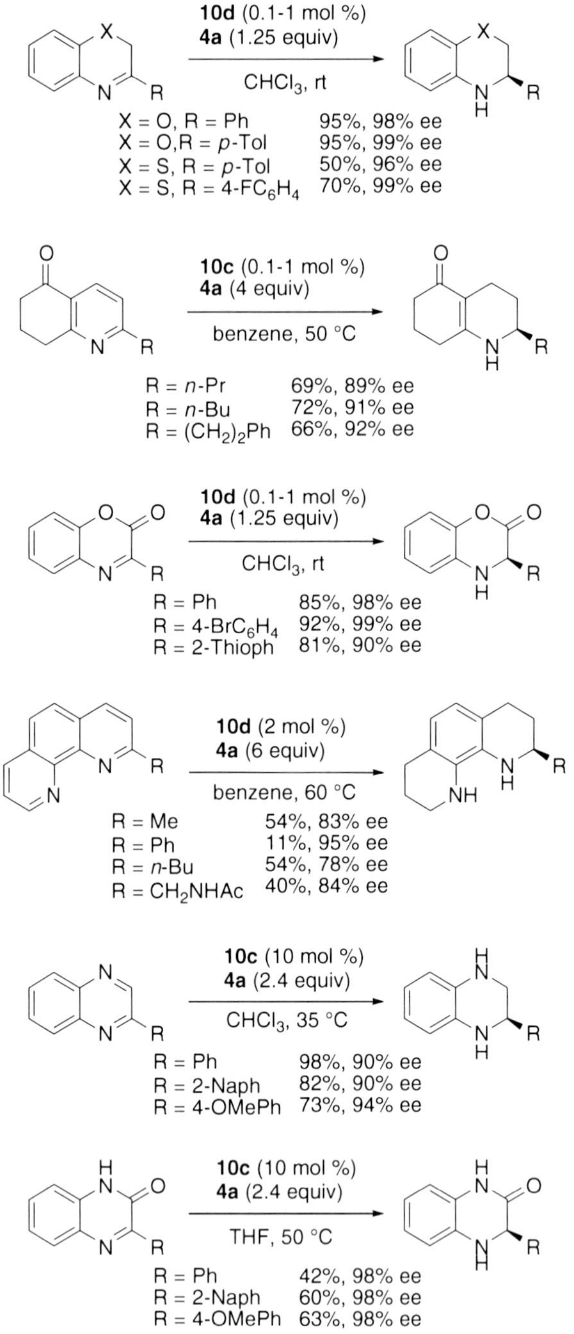

10d (0.1-1 mol %)
4a (1.25 equiv)
$CHCl_3$, rt

X = O, R = Ph 95%, 98% ee
X = O,R = p-Tol 95%, 99% ee
X = S, R = p-Tol 50%, 96% ee
X = S, R = 4-FC_6H_4 70%, 99% ee

10c (0.1-1 mol %)
4a (4 equiv)
benzene, 50 °C

R = n-Pr 69%, 89% ee
R = n-Bu 72%, 91% ee
R = $(CH_2)_2Ph$ 66%, 92% ee

10d (0.1-1 mol %)
4a (1.25 equiv)
$CHCl_3$, rt

R = Ph 85%, 98% ee
R = 4-BrC_6H_4 92%, 99% ee
R = 2-Thioph 81%, 90% ee

10d (2 mol %)
4a (6 equiv)
benzene, 60 °C

R = Me 54%, 83% ee
R = Ph 11%, 95% ee
R = n-Bu 54%, 78% ee
R = CH_2NHAc 40%, 84% ee

10c (10 mol %)
4a (2.4 equiv)
$CHCl_3$, 35 °C

R = Ph 98%, 90% ee
R = 2-Naph 82%, 90% ee
R = 4-OMePh 73%, 94% ee

10c (10 mol %)
4a (2.4 equiv)
THF, 50 °C

R = Ph 42%, 98% ee
R = 2-Naph 60%, 98% ee
R = 4-OMePh 63%, 98% ee

Scheme 2.15 Enantioselective reduction of different nitrogen heterocycles

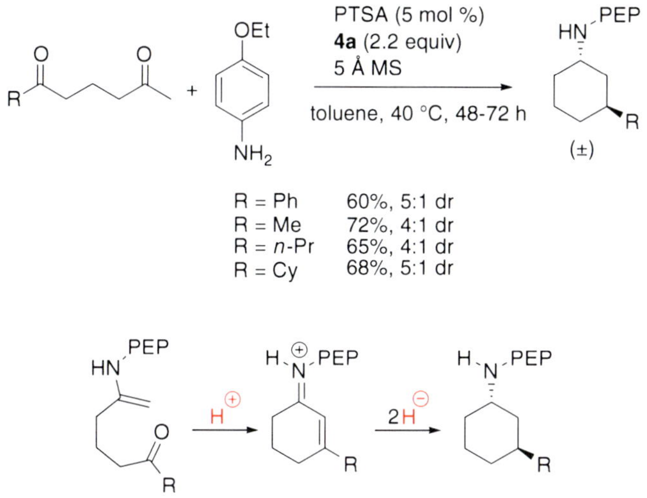
PTSA (5 mol %)
4a (2.2 equiv)
5 Å MS
toluene, 40 °C, 48-72 h
OEt
NH2
HN PEP
R
(±)
R = Ph 60%, 5:1 dr
R = Me 72%, 4:1 dr
R = n-Pr 65%, 4:1 dr
R = Cy 68%, 5:1 dr
2H

Scheme 2.16 Cascade enamine/iminium synthesis of cyclohexylamines. PEP = *p*-ethoxyphenyl; PTSA = *p*-toluenesulfonic acid

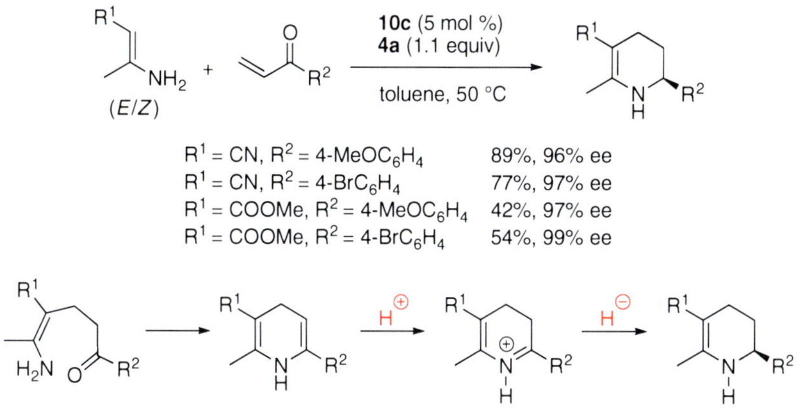
10c (5 mol %)
4a (1.1 equiv)
toluene, 50 °C
(E/Z)
R1 = CN, R2 = 4-MeOC6H4 89%, 96% ee
R1 = CN, R2 = 4-BrC6H4 77%, 97% ee
R1 = COOMe, R2 = 4-MeOC6H4 42%, 97% ee
R1 = COOMe, R2 = 4-BrC6H4 54%, 99% ee

Scheme 2.17 Cascade synthesis of chiral dihydropyridines

2.4 Hydrogen Bonding Catalysis

2.4.1 *Achiral Catalysts*

The organocatalytic activation of electrophiles through (double) hydrogen bonding has been exploited in the transfer hydrogenation of imines and nitroolefins using Hanztsch esters. Unsubstituted thiourea was reported as efficient catalyst for the one-pot reductive amination of both aldehydes [77] and ketones [78] with various aromatic amines. However, these results were heavily questioned by another research group who failed to reproduce them in their entirety. Instead, electron-poor thiourea **15** was described as an efficient promoter for the transfer hydrogenation of preformed N-PMP aldimines [79] (Scheme 2.18). The same catalyst was also successfully applied to the conjugate reduction of nitroalkenes [80]. Other substrates, such as chalcones, cyclic enones and acrolein derivatives failed to react under similar conditions.

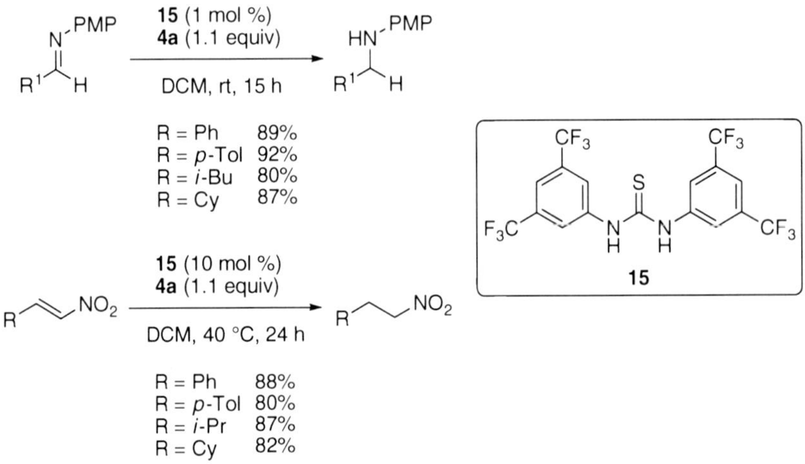

Scheme 2.18 Transfer hydrogenations using thiourea catalyst **15**

2.4.2 *Chiral Catalysts*

A highly enantioselective variant of the transfer hydrogenation of nitroolefins has been developed replacing thiourea **15** with chiral catalyst **16** [81, 82]. Using this protocol, β,β-disubstituted nitroalkenes can be reduced with high degrees of stereocontrol. Unlike in the case of α,β-unsaturated carbonyl compounds, the olefin geometry plays a key role in the stereochemical outcome of the reaction, as substrates in a 1:1 *E*/*Z* ratio yield substantially racemic product. Olefins containing an ester group as one of the β-substituents can be efficiently reduced to chiral β-nitroesters, valuable precursors for the synthesis of enantiomerically enriched β-amino acids (Scheme 2.19).

A chiral reductase mimic containing an electron-poor thiourea and an N-benzyl nicotinamide group was tested in the reduction of diketones (Scheme 2.20) [83]. So far, this compound (**17**) constitutes the only example of an organocatalyst incorporating a dihydropyridine group which can mediate the transfer hydrogenation of a substrate and be regenerated by a stoichiometric reducing agent (sodium dithionate). While the *in situ* regeneration of the dihydropyridine moiety was successfully demonstrated, the application of this catalyst to asymmetric reductions gave disappointing results, in that the products were found to undergo fast racemization under the reaction conditions.

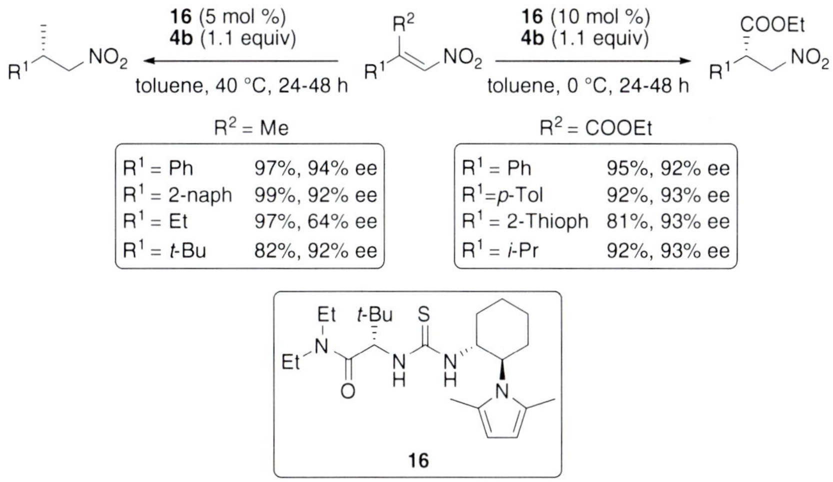

Scheme 2.19 Enantioselective conjugate reduction of nitroolefins

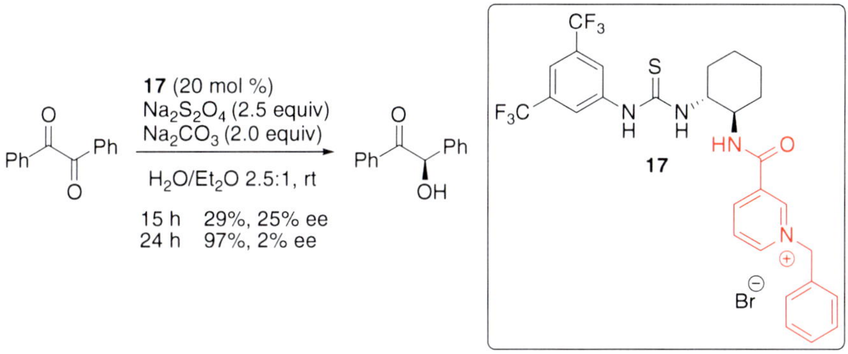

Scheme 2.20 Transfer hydrogenation using a thiourea reductase mimic

2.5 Conclusions

The examples described in this chapter unambiguously show that asymmetric organocatalytic reductions using Hantzsch esters have quickly become a powerful tool for the synthesis of chiral compounds. The typical mildness of the reaction conditions coupled with the high chemoselectivity characterizing organocatalytic reactions render some of these protocols very attractive choices for an asymmetric reduction step at a late stage of a synthesis. On the other hand, the combination of these transfer hydrogenations with other catalytic reactions, which has been successfully demonstrated in some remarkable cascade processes, offers a variety of possibilities which still have to be fully explored. It is worth noting how metal-catalyzed and organocatalyzed transfer hydrogenations are often complementary in scope and they should be regarded as two sides of the same coin. On the other hand, the atom economy of organocatalytic reductions using Hantzsch esters is poor, compound **4a** having a molecular weight of 253. For this reason, recycling of the pyridine byproduct or *in situ* regeneration of active reducing species should be further investigated in the coming years to render this approach attractive for preparative purposes. Considering the pace at which innovative discoveries in asymmetric organocatalysis are reported, it is reasonable to expect several new reductive processes exploiting the peculiar reactivity of dihydropyridines in the near future as well as greener methodologies for transfer hydrogenation making use of these intriguing compounds.

References

1. van Leeuwen PWNM (2004) Homogeneous catalysis: understanding the art. Kluwer Academic, Dordrecht
2. Gladiali S, Alberico E (2006) Chem Soc Rev 35:226
3. Frey PA, Hegeman AD (2007) Coenzymes I: organic coenzymes. In: Frey PA, Hegeman AD (eds.) Enzymatic reaction mechanisms. Oxford University Press, New York
4. Mauzerall D, Westheimer FH (1955) J Am Chem Soc 77:2261
5. de Nie-Sarink MJ, Pandit UK (1978) Tetrahedron Lett 19:1335
6. van Bergen TJ, Mulder T, Kellogg RM (1976) J Am Chem Soc 98:1960
7. Gębicki J, Marcinek A, Zielonka J (2004) Acc Chem Res 37:379
8. Ohno A, Ikeguchi M, Kimura T, Oka S (1979) J Am Chem Soc 101:7036
9. Kanomata N, Nakata T (2000) J Am Chem Soc 122:4563
10. Wang N, Zhao J (2009) Adv Synth Catal 351:3045
11. Shah A, Bariwal J, Molnár J, Kawase M, Motohashi N (2008) Top Heterocycl Chem 15:201
12. Lavilla RJ (2002) J Chem Soc Perkin Trans 1:1141
13. Natale NR (2000) Chem Innov 30(11):22
14. Stout DM, Meyers AI (1982) Chem Rev 82:223
15. Hantzsch A (1881) Chem Ber 14:1637
16. Knoevenagel E, Fries A (1898) Chem Ber 31:761
17. Richter D, Mayr H (2009) Angew Chem Int Ed 48:1958
18. Benaglia M, Guizzetti S, Pignataro L (2008) Coord Chem Rev 252:492
19. Ouellet SG, Walji AM, Macmillan DWC (2007) Acc Chem Res 40:1327
20. You S (2007) Chem Asian J 2:820
21. Connon SJ (2007) Org Biomol Chem 5:3407
22. Lelais G, MacMillan DWC (2006) Aldrichim Acta 39:79
23. Akiyama T (2007) Chem Rev 107:5744
24. Doyle AG, Jacobsen EN (2007) Chem Rev 107:5713
25. Enders D, Grondal C, Hüttl MRM (2007) Angew Chem Int Ed 46:1570
26. Yang JW, Fonseca MTH, List B (2004) Angew Chem Int Ed 43:6660
27. De Figueiredo RM, Berner R, Julis J, Liu T, Türp D, Christmann M (2007) J Org Chem 72:640
28. Ouellet SG, Tuttle JB, Macmillan DWC (2005) J Am Chem Soc 127:32
29. Tuttle JB, Ouellet SG, Macmillan DWC (2006) J Am Chem Soc 128:12662
30. Gutierrez O, Iafe RG, Houk KN (2009) Org Lett 11:4298
31. Hoffman TJ, Dash J, Rigby JH, Arseniyadis S, Cossy J (2009) Org Lett 11:2756
32. Paterson I, Miller NA (2008) Chem Commun 4708–4710
33. Lagisetti C, Pourpak A, Jiang Q, Cui X, Goronga T, Morris SW, Webb TR (2008) J Med Chem 51:6220

34. Lacour J, Moraleda D (2009) Chem Commun 7073–7089
35. Mayer S, List B (2006) Angew Chem Int Ed 45:4193
36. Stadler M, List B (2008) Synlett 597–599
37. Martin NJA, List B (2006) J Am Chem Soc 128:13368
38. Akagawa K, Akabane H, Sakamoto S, Kudo K (2008) Org Lett 10:2035
39. Akagawa K, Akabane H, Sakamoto S, Kudo K (2009) Tetrahedron: Asymm 20:461
40. List B (2006) Chem Commun 819.
41. Yang JW, Hechavarria Fonseca MT, List B (2005) J Am Chem Soc 127:15036
42. Zhao G, Córdova A (2006) Tetrahedron Lett 47:7417
43. Ramachary DB, Kishor M, Reddy GB (2006) Org Biomol Chem 4:1641
44. Ramachary DB, Kishor M, Ramakumar K (2006) Tetrahedron Lett 47:651
45. Ramachary DB, Vijayendar Reddy Y (2010) J Org Chem 75:74
46. Ramachary DB, Kishor M (2008) Org Biomol Chem 6:4176
47. Ramachary DB, Sakthidevi R (2008) Org Biomol Chem 6:2488
48. Rueping M, Azap C, Sugiono E, Theissmann T (2005) Synlett 2367.
49. Rueping M, Theissmann T, Antonchick A (2006) Synlett 2006:1071
50. Babu TH, Shanthi G, Perumal PT (2009) Tetrahedron Lett 50:2881
51. Marcelli T, Hammar P, Himo F (2008) Chem Eur J 14:8562
52. Simón L, Goodman JM (2008) J Am Chem Soc 130:8741
53. Connon SJ (2006) Angew Chem Int Ed 45:3909
54. Hoffmann S, Seayad AM, List B (2005) Angew Chem Int Ed 44:7424
55. Rueping M, Sugiono E, Azap C, Theissmann T, Bolte M (2005) Org Lett 7:3781
56. Storer R, Carrera D, Ni Y, Macmillan D (2006) J Am Chem Soc 128:84
57. Kumar A, Sharma S, Maurya RA (2010) Adv Synth Catal 352:2227
58. Zhu C, Akiyama T (2009) Org Lett 11:4180
59. Enders D, Liebich J, Raabe G (2010) Chem Eur J 16:9763
60. Hoffmann S, Nicoletti M, List B (2006) J Am Chem Soc 128:13074
61. Marcelli T, Hammar P, Himo F (2009) Adv Synth Catal 351:525
62. Li G, Liang Y, Antilla J (2007) J Am Chem Soc 129:5830
63. Kang Q, Zhao Z, You S (2007) Adv Synth Catal 349:1657
64. Zhu C, Akiyama T (2010) Adv Synth Catal 352:1846
65. Kang Q, Zhao Z, You S (2008) Org Lett 10:2031
66. Li G, Antilla JC (2009) Org Lett 11:1075
67. Rueping M, Antonchick AP, Theissmann T (2006) Angew Chem Int Ed 45:3683
68. Guo Q, Du D, Xu J (2008) Angew Chem Int Ed 47:759
69. Rueping M, Theissmann T, Raja S, Bats J (2008) Adv Synth Catal 350:1001
70. Rueping M, Antonchick AP, Theissmann T (2006) Angew Chem Int Ed 45:6751
71. Rueping M, Antonchick A (2007) Angew Chem Int Ed 46:4562
72. Metallinos C, Barrett F B, Xu S (2008) Synlett 720.
73. Rueping M, Tato F, Schoepke F (2010) Chem Eur J 16:2688
74. Zhou J, List B (2007) Synlett 2037
75. Rueping M, Antonchick A (2008) Angew Chem Int Ed 47:5836
76. Han Z, Xiao H, Chen X, Gong L (2009) J Am Chem Soc 131:9182
77. Menche D, Arikan F (2006) Synlett 841.
78. Menche D, Hassfeld J, Li J, Menche G, Ritter A, Rudolph S (2006) Org Lett 8:741
79. Zhang Z, Schreiner P (2007) Synlett 1455.
80. Zhang Z, Schreiner P (2007) Synthesis 2559
81. Martin N, Ozores L, List B (2007) J Am Chem Soc 129:8976
82. Martin NJA, Cheng X, List B (2008) J Am Chem Soc 130:13862
83. Procuranti B, Connon S J (2007) Chem Commun 1421.

Chapter 3
Organocatalyzed Enantioselective Protonation

Thomas Poisson, Sylvain Oudeyer, Jean-François Brière, and Vincent Levacher

Abstract This chapter gathers the main organocatalytic processes involving an enantioselective proton transfer. These include decarboxylation of malonates, addition of protic nucleophiles to ketenes and to α,β-unsaturated carbonyl compounds, photodeconjugation of α,β-unsaturated esters, protonation of silyl enol ethers, along with some other miscellaneous chemical transformations. The main purpose is to provide an overall view of what is being done in this field by discussing the scope and limitations, and the mechanism of each organocatalytic process. To account for the catalytic nature of the enantioselective protonation, a special emphasis will be placed on the description of the proton transfer pathway from the achiral proton source to the prochiral substrate.

3.1 Introduction

Proton transfer processes are involved in many biochemical events and often play a key role in the catalytic activity of enzymes. Of particular interest, are enantioselective proton transfer processes frequently encountered in a number of biosynthetic sequences. Over the last few years, esterase and decarboxylase enzymes have appeared as appealing biocatalysts to achieve enantioselective protonation of enol acetates [1] and enantioselective decarboxylation of malonate derivatives [2] respectively. Conceptually, enantioselective protonation provides a simple

T. Poisson • S. Oudeyer • J.-F. Brière • V. Levacher (✉)
Institut de Recherche en Chimie Organique Fine (IRCOF) associé au CNRS (UMR COBRA-6014 & FR 3038), INSA-Université de Rouen, Rue Tesnière, Mont-Saint-Aignan F-76130, France
e-mail: thomas.poisson@insa-rouen.fr; sylvain.oudeyer@insa-rouen.fr; jean-francois.briere@insa-rouen.fr; vincent.levacher@insa-rouen.fr

R. Mahrwald (ed.), *Enantioselective Organocatalyzed Reactions I: Enantioselective Oxidation, Reduction, Functionalization and Desymmetrization*, DOI 10.1007/978-90-481-3865-4_3, © Springer Science+Business Media B.V. 2011

methodological approach for generating a tertiary carbon stereocenter. The presence of this stereoelement in numerous biologically active compounds as well as in valuable chiral scaffolds prompted synthetic chemists to design efficient chemical tools (Fig. 3.1) for achieving enantioselective proton transfer [3]. The major obstacles that have hampered the development of chemical strategies in the past years are the high reactivity of the proton which makes the stereocontrol of such proton transfer tricky, the competition between C- and O-protonation of enolates and the difficulty in controlling the enolate geometry of acyclic substrates. To date, the protonation of metal-enolates has largely dominated the literature in this area. The challenge faced by synthetic chemists is to develop catalytic processes under metal-free and environmentally benign conditions. Organocatalytic approaches have emerged only recently as a useful synthetic tool, although the organocatalyzed decarboxylation of malonates was first reported as early as 1904 [4] and organocatalyzed protonation of ketenes in the 1960s [5].

This chapter covers organocatalytic processes where the enantio-determining step involves a proton transfer. Most of the organocatalytic processes outlined herein share a key step in common, *i.e.* the enantioselective protonation of an enolate or enol intermediate species obtained *in situ* from various precursors. The main organocatalytic approaches reported in the literature may be classified according to the nature of these precursors (Scheme 3.1). Special emphasis will be given to decarboxylation of malonates, addition of protic nucleophiles (NuH) to ketenes or to α,β-unsaturated carbonyl compounds. We will also focus on tautomerisation of enols formed *in situ* via photodeconjugation of α,β-unsaturated esters and on the protonation of silyl enolates. Finally, a last section will be devoted to other miscellaneous substrates.

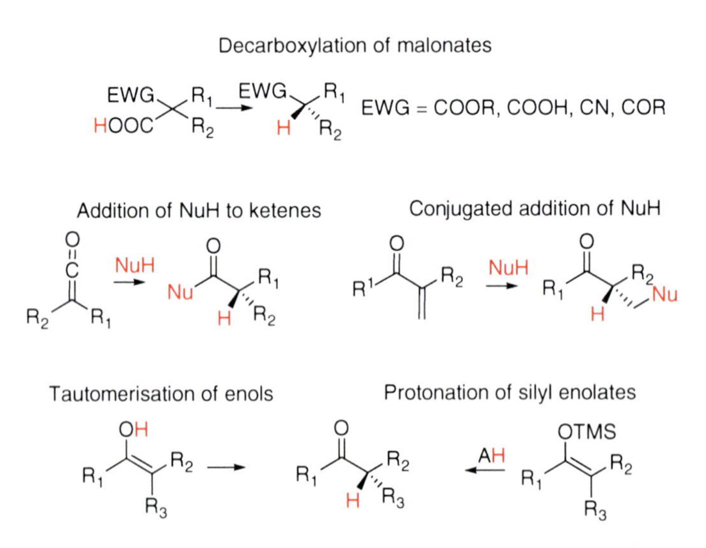

Scheme 3.1 Main strategies investigated in organocatalyzed enantioselective protonation

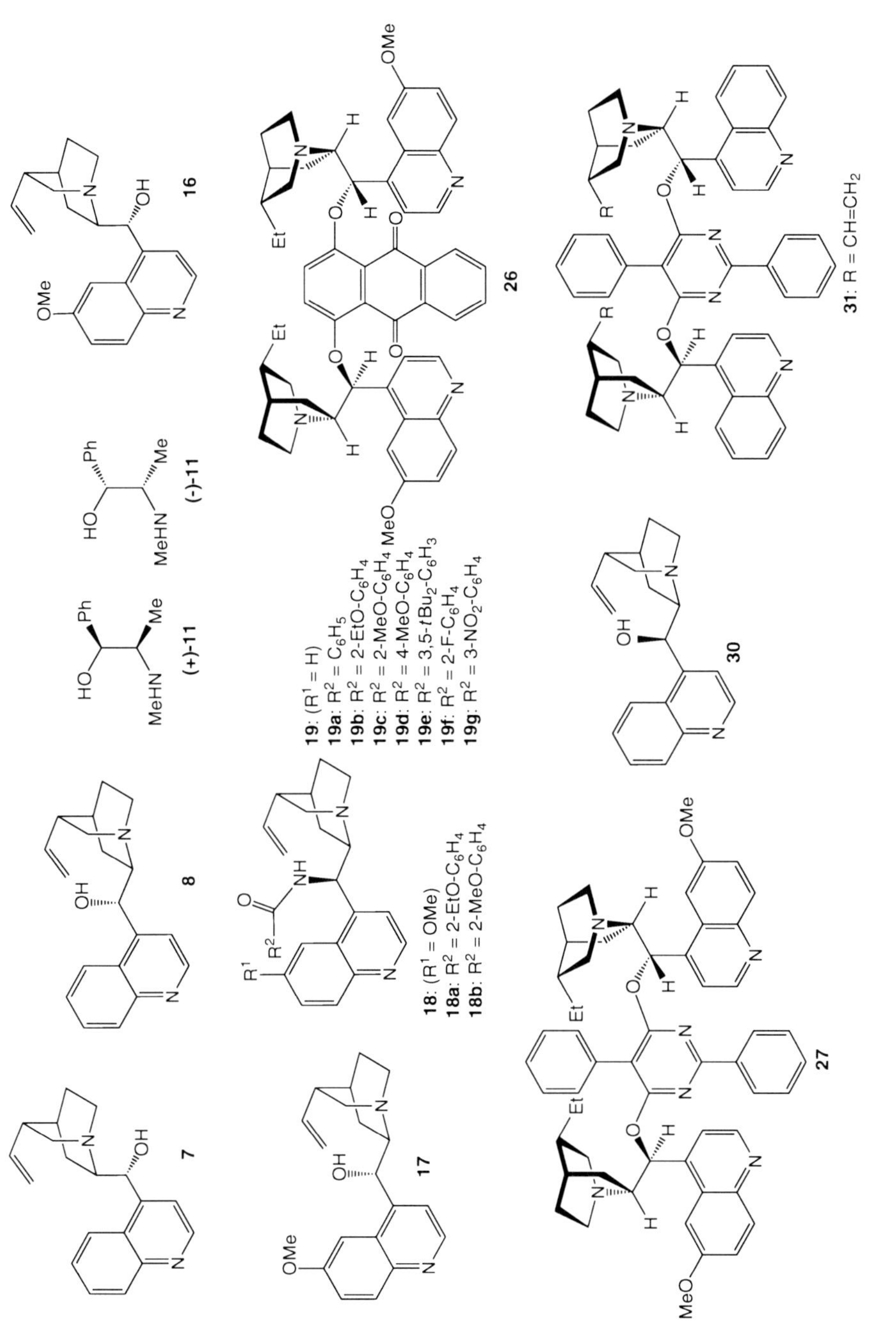

7

8

16

(+)-11

(-)-11

17

18: (R^1 = OMe)
18a: R^2 = 2-EtO-C_6H_4
18b: R^2 = 2-MeO-C_6H_4

19: (R^1 = H)
19a: R^2 = C_6H_5
19b: R^2 = 2-EtO-C_6H_4
19c: R^2 = 2-MeO-C_6H_4
19d: R^2 = 4-MeO-C_6H_4
19e: R^2 = 3,5-tBu$_2$-C_6H_3
19f: R^2 = 2-F-C_6H_4
19g: R^2 = 3-NO_2-C_6H_4

26

27

30

31: R = CH=CH_2

Fig. 3.1 List of catalysts

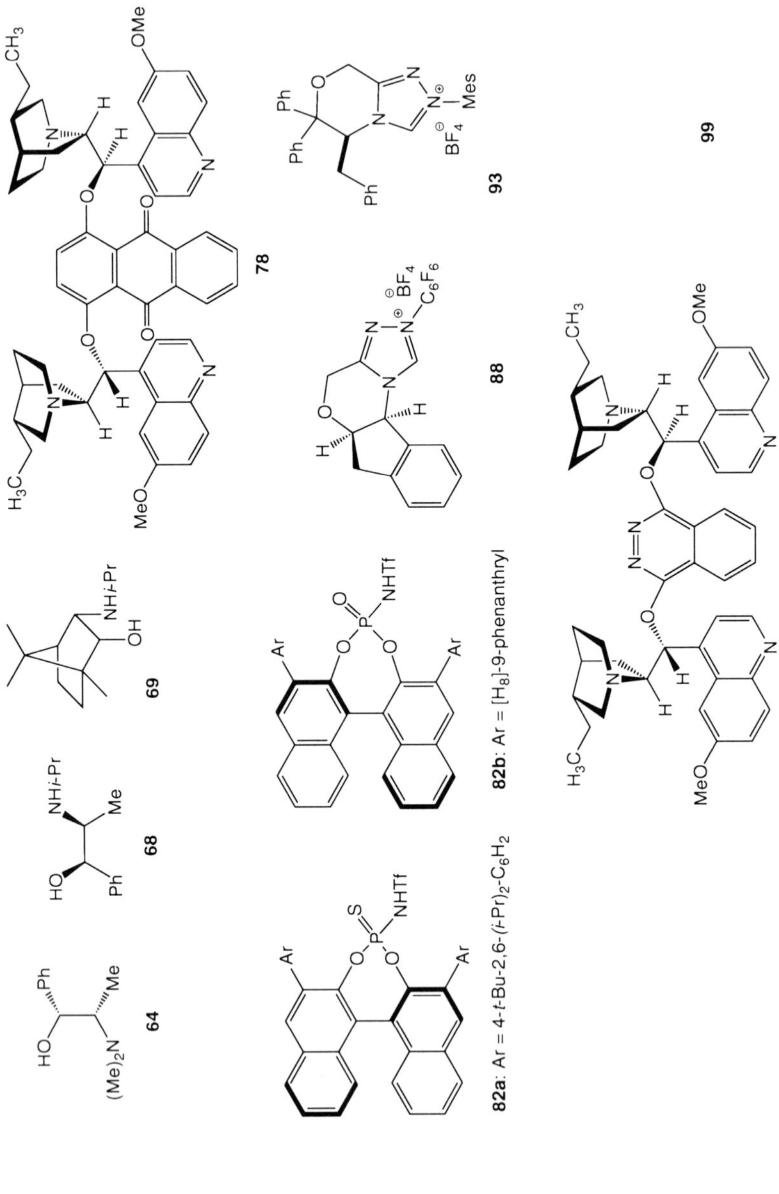

64 **68** **69** **78**

82a: Ar = 4-*t*-Bu-2,6-(*i*-Pr)$_2$-C_6H_2

82b: Ar = [H_8]-9-phenanthryl

88 **93** **99**

Fig. 3.1 (continued)

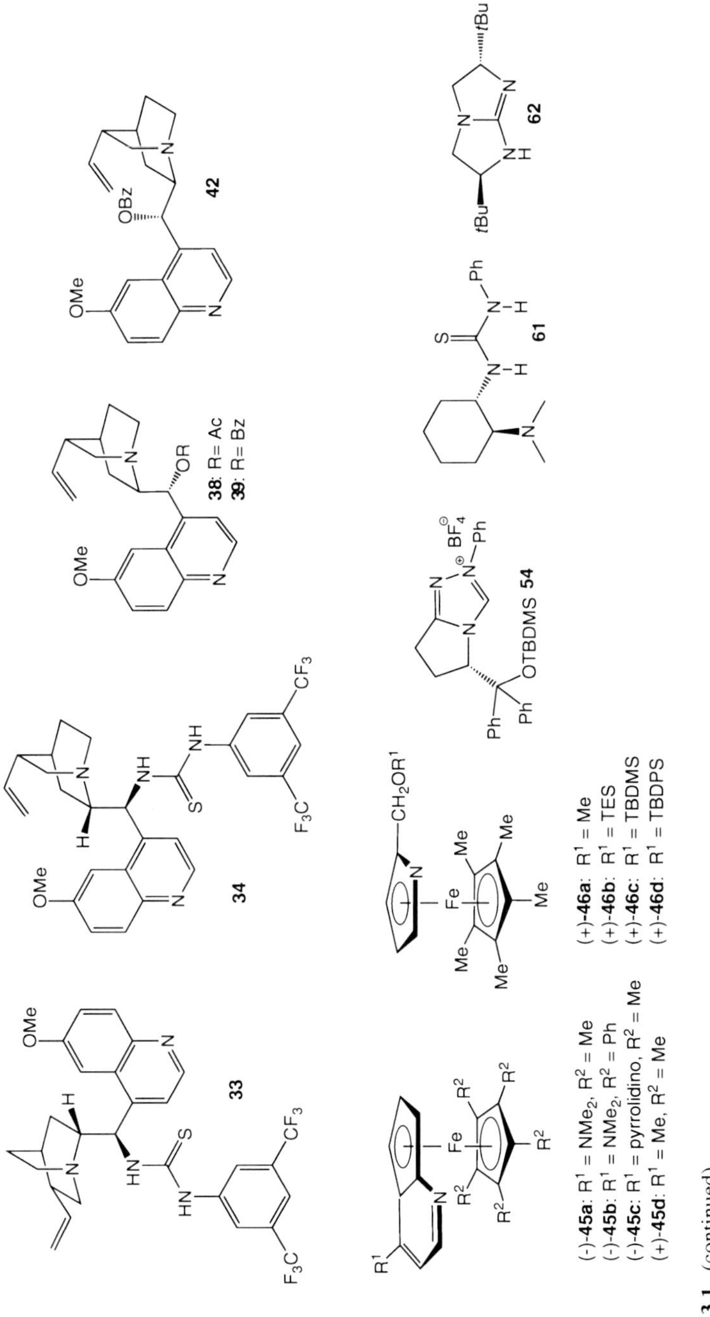

33 34 38: R= Ac 39: R= Bz 42

(-)-**45a**: R^1 = NMe_2, R^2 = Me
(-)-**45b**: R^1 = NMe_2, R^2 = Ph
(-)-**45c**: R^1 = pyrrolidino, R^2 = Me
(+)-**45d**: R^1 = Me, R^2 = Me

(+)-**46a**: R^1 = Me
(+)-**46b**: R^1 = TES
(+)-**46c**: R^1 = TBDMS
(+)-**46d**: R^1 = TBDPS

54 61 62

Fig. 3.1 (continued)

3.2 Enantioselective Decarboxylative Protonation of Malonates

As early as 1904, Marckwald et al. [4] reported the thermal decarboxylation of ethyl(methyl)malonic acid **1** at 170°C in the presence of brucine affording (*S*)-2-methylbutyric acid **2** in 10% ee (Scheme 3.2). While this report might probably be considered as the first asymmetric transformation, this result could not be successfully reproduced several decades later [6]. In 1975, Verbit et al. [7] reported the decarboxylation of ethylphenylmalonic acid **3** in cholesteryl benzoate at 160°C giving rise to (*R*)-phenylbutanoic acid **4** in 18% ee (Scheme 3.2). However, Kagan et al. [8] failed to obtain any enantioenriched phenylbutanoic acid **4** under the same reaction conditions.

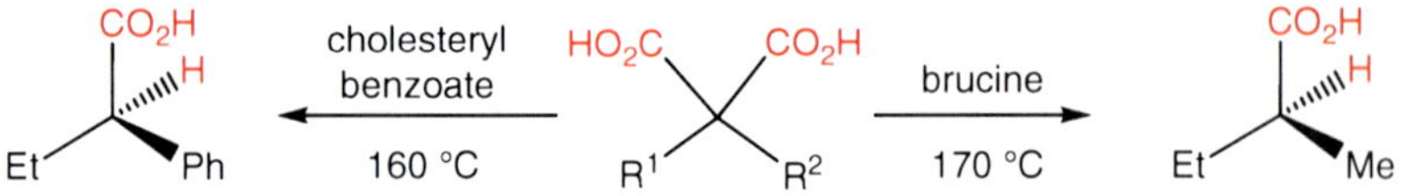

CO_2H H Et Ph — cholesteryl benzoate, 160 °C ← HO_2C CO_2H R^1 R^2 → brucine, 170 °C — CO_2H H Et Me

R-**4** 18% ee | **3**: R^1 = Et, R^2 = Ph | **1**: R^1= Me, R^2= Et | *S*-**2** 10% ee

Scheme 3.2 First enantioselective decarboxylative protonation under thermal conditions

In 1987, Maumy et al. [9] reported the first enantioselective decarboxylative protonation of malonic acid **5** carried out under mild conditions. *R*-configured carboxylic acid **6** was obtained in 27% ee by using CuCl (70 mol%) and cinchonidine **7** (1.4 eq) in acetonitrile at 60°C (Scheme 3.3). The role of copper (I) was then minimized by Brunner et al. [10] by revealing that decarboxylation could be carried out with a significant reduced amount of CuCl (3 mol%) and an excess of cinchonine **8** (3.3 eq) affording *S*-configured carboxylic acid **6** in 36% ee (Scheme 3.3). The same group showed later that the reaction is not copper- but base-catalyzed, paving the way for further development in organocatalytic decarboxylation [11]. Thus, malonic acid **5** was decarboxylated in the presence of a catalytic amount of cinchonine **8** (10 mol%) in THF at room temperature, providing carboxylic acid **6** with comparable enantioselectivity to that obtained at 85°C in acetonitrile in the presence of CuCl (Scheme 3.3).

Shortly afterwards, Muzart et al. [12] reported the decarboxylation of racemic β-ketoacid **9** in acetonitrile at room temperature in the presence of cinchonine **8** or (+)-ephedrine **(+)-11** (20 mol%). Both chiral bases exhibited similar level of enantioselection (30–35% ee), providing 2-methyltetralone **10** with opposite sense of stereoinduction. While cinchonine **8** gave preferentially (*R*)-2-methyltetralone **10**, (+)-ephedrine **(+)-11** furnished (*S*)-2-methyltetralone **10** (Scheme 3.4).

In an attempt to develop a straightforward stereoselective access to β-hydroxyisobutyric acid derivatives, both academically and industrially important chiral synthons, Kim et al. [13] investigated the enantioselective decarboxylative

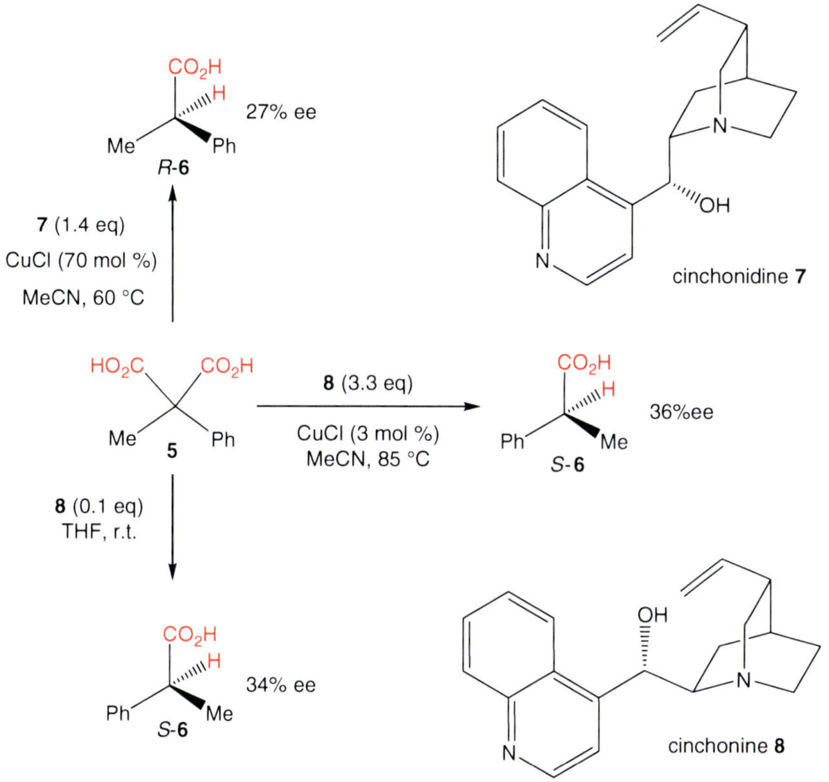

Scheme 3.3 Pioneering works in enantioselective decarboxylative protonation

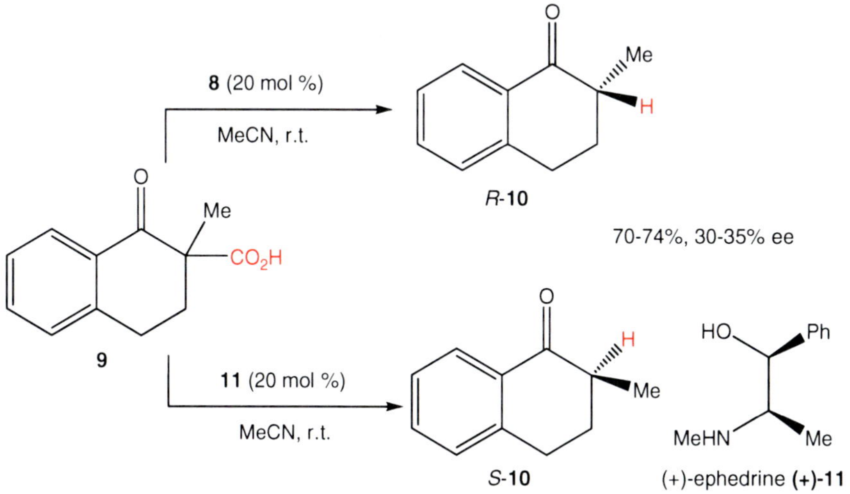

Scheme 3.4 Enantioselective, decarboxylative protonation of β-ketoacid in tetralone series

protonation of 2-hydroxymethyl-2-methyl hemimalonate **12**. The desired β-hydroisobutyric acid ester **13** was obtained in modest yield and 18% ee by conducting the reaction in refluxing acetonitrile in the presence of stoichiometric amount of cinchonidine **7** (Scheme 3.5).

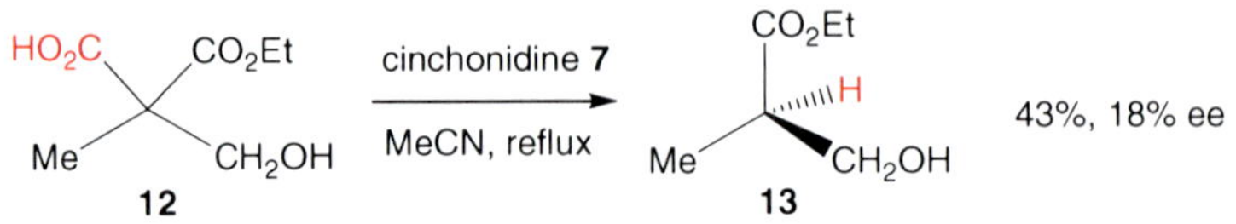

HO₂C CO₂Et / Me CH₂OH **12** —cinchonidine **7**, MeCN, reflux→ CO₂Et, H, Me CH₂OH **13** 43%, 18% ee

Scheme 3.5 Enantioselective decarboxylative protonation of 2-hydroxymethyl-2-methyl hemimalonate

In 2000, Brunner et al. [14] reported the first synthetic application with a satisfactory enantioselectivity (Scheme 3.6). The decarboxylation of 2-cyanopropionic acid **14**, an immediate precursor of Naproxen®, was investigated by screening various chiral bases as catalyst. Although common cinchona alkaloids **7–8**, **16–17** gave modest enantiomeric excesses ranging from 6% to 34%, a breakthrough was achieved when various amides were installed at position C9 of cinchona alkaloids. More than 40 cinchona alkaloid analogues were prepared in the *epi*-quinidine **18** and *epi*-cinchonine **19** series [14]. A selection of the results obtained from this catalyst library is shown in Scheme 3.6. Most of these cinchona-benzamides displayed higher selectivity than parent cinchona alkaloids. A general trend seems to be emerging which suggests that benzamides bearing an ortho alkoxy substituent exhibit better performance. According to molecular modelling studies [15] conformational stabilisation of the catalyst through an intramolecular hydrogen bonding between the amide group and the alkoxy substituent has been proposed to account for the better performances observed. The best catalyst was found to be 9-*epi*-cinchonine-benzamide **19b**, affording the desired (*S*)-Naproxen analogue **15** in 70% ee (10 mol% of **19b**, THF, 15°C, 24 h). Interestingly, no erosion of the selectivity was observed by carrying out the reaction with 5 mol% of the catalyst, pointing out the good catalytic activity of **19b** (Scheme 3.6).

The decarboxylation of 2-cyano-2-phenylpropionic acid **20**, a model compound of **14** was also investigated by the same authors [14]. While cinchonine **8** provided (*S*)-2-phenyl propionitrile **21** in only 13% ee, 9-*epi*-cinchona-benzamides **19b** and **18b** confirmed their superiority over the parent cinchona alkaloids affording **21** with up to 61% ee (Scheme 3.7).

The decarboxylation of aminomalonic acid derivatives constitutes a classical route to proteinogenic and non-proteinogenic amino acids. In 2003, Brunner et al. [16] explored the potential of the 9-*epi*-cinchonine-benzamides previously designed [14], in the decarboxylation of *N*-acetylaminomalonic acid derivatives **22a–c**. Selected examples presented in Scheme 3.8 indicate that a number of 9-*epi*-cinchonine-benzamides reached acceptable levels of enantioselectivity, up to 71%. In contrast to 2-cyano-2-aryl propionic acids **14** and **20**, 9-*epi*-cinchonine-benzamides **19b**, **c** and **f** gave rather disappointing results with *N*-acetylaminomalonic acids

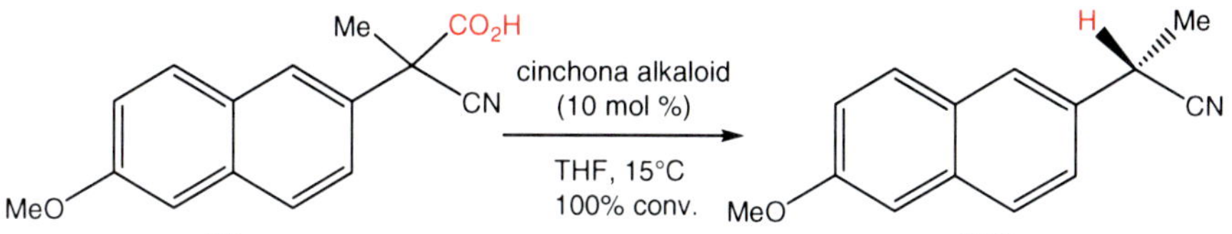

cinchona alkaloid (10 mol %)
THF, 15°C
100% conv.

14 → S-15

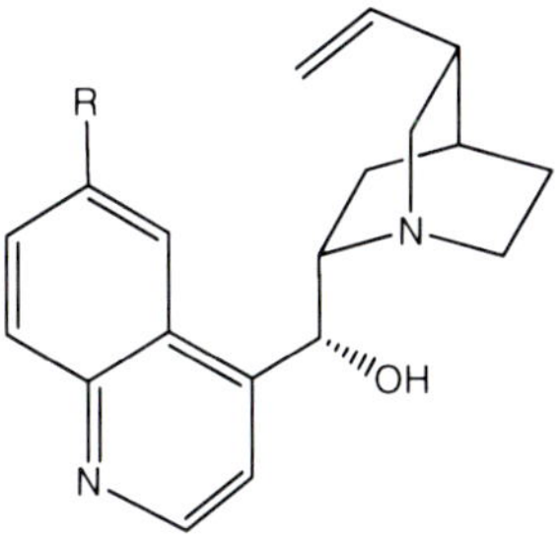

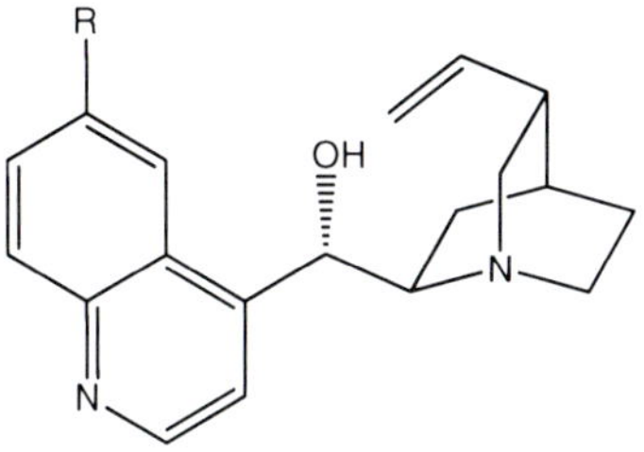

7: cinchonidine (R = H), 6% ee (R)
16: quinine (R = OMe), 13% ee (R)

8: cinchonine (R = H), 17% ee
17: quinidine (R = OMe), 34% ee

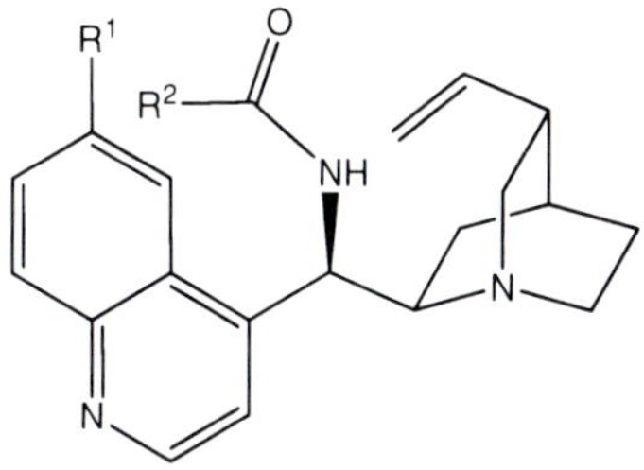

18: (R^1 = OMe)
18a: R^2 = 2-EtO-C_6H_4, 61% ee
18b: R^2 = 2-MeO-C_6H_4, 65% ee

19: (R^1= H)
19a: R^2 = C_6H_5, 51% ee
19b: R^2 = 2-EtO-C_6H_4, 70% ee
19c: R^2 = 2-MeO-C_6H_4, 65% ee
19d: R^2 = 4-MeO-C_6H_4, 43% ee
19e: R^2 = 3,5-*t*-Bu-C_6H_3, 46% ee
19f: R^2 = 2-F-C_6H_4, 43% ee
19g: R^2 = 3-NO_2-C_6H_4, 43% ee

Scheme 3.6 Stereoselective access to Naproxen by enantioselective decarboxylative protonation of 2-cyanopropionic acid

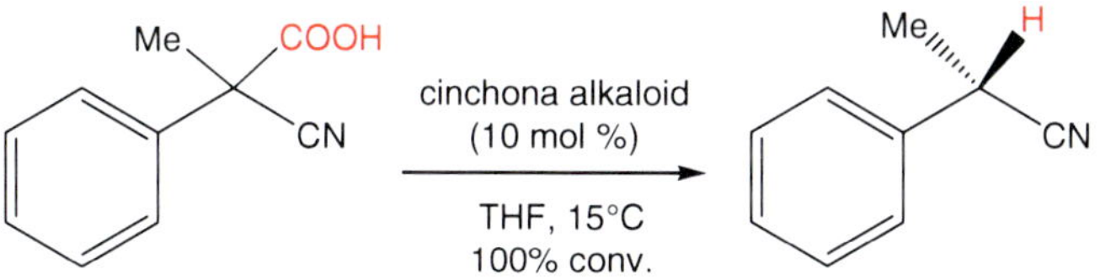

cinchona alkaloid (10 mol %)
THF, 15°C
100% conv.

20 → S-21

cinchona alkaloid
8: 13% ee
19b: 60% ee
18b: 61% ee

Scheme 3.7 Enantioselective decarboxylative protonation of 2-cyano-2-phenylpropionic acid

22a–c. The highest enantiomeric excesses were obtained with 9-*epi*-cinchonine-benzamides **19a, d** and **e** having unsubstituted ortho positions. As already observed with substrates **14** and **20**, parent cinchona alkaloids **7–8** and **16–17** exhibited moderate enantiomeric excesses ranging from 4% to 10% (Scheme 3.8).

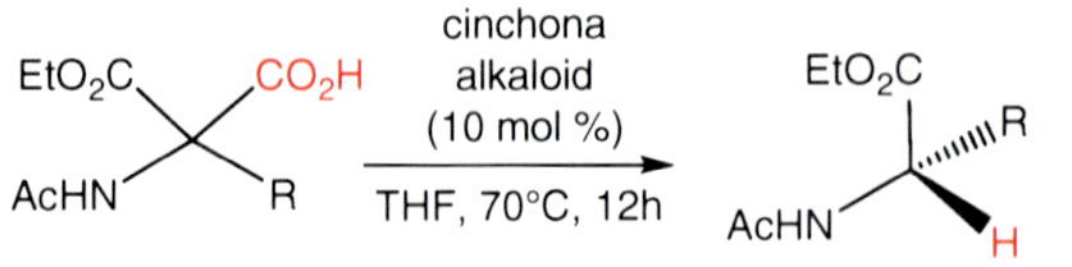

cinchona alkaloids **7-8** and **16-17**

23a: **7**, 5% ee

23a, **8**, 10% ee (*R*)

23a: **17**, 4% ee

23a: **16**, 6% ee (*R*)

9-amino cinchona alkaloids **19**

23a: **19a**, 52% ee

23c: **19d**, 71% ee

23a: **19e**, 60% ee (*R*)

23b: **19e**, 48% ee (*R*)

23a: **19b**, 19% ee

23b: **19c**, 0% ee

23a: **19f**, 27% ee

Scheme 3.8 Enantioselective decarboxylative protonation of *N*-acetylaminomalonic acid derivatives

A mechanistic study reported by Brunner et al. [16] provided important insight regarding the stereochemical course of the decarboxylation, while ruling out a concerted mechanism proposed elsewhere [15]. A kinetic study of the decarboxylation of substrates **14** and **22a** has revealed in both cases that the enantioselectivity is roughly constant during the whole course of the reaction, making the occurrence of a kinetic resolution process impossible during the reaction. In addition, decarboxylation of the two isolated enantiomers (*S*)-**14** and (*R*)-**14** was conducted independently in the presence of catalyst **19b**. In both cases, the same major enantiomer was obtained with the same level of enantioselectivity and the same sense of stereoselection as those observed when conducting the same experiment from racemic **14**. All these data are incompatible with a concerted mechanism and argue in favour of a two-step decarboxylation/protonation mechanism. A general catalytic pathway depicted in Scheme 3.9 may reasonably be proposed to account for the catalytic activity of the chiral base. After deprotonation of the carboxylic acid, decarboxylation would occur to generate an enolate intermediate that is then subjected to selective protonation while regenerating the chiral base (Scheme 3.9).

The enantioselective decarboxylative protonation of aminomalonic acid derivatives has been also extensively investigated by Rouden et al. [17]. Their first contribution in this area aimed at preparing enantioenriched pipecolic esters by decarboxylation of *N*-acetyl piperidinohemimalonate **24** [17]. Among the different cinchona alkaloids investigated, the best result was obtained by means of 9-*epi*-cinchonine-benzamide **19b,** previously developed by Brunner [14, 16]. Pipecolic ester **25** was obtained in up to 52% ee when conducting the reaction in THF at room temperature for 24 h. Various bis-cinchona alkaloids such as $(DHQD)_2$AQN **26** or $(DHQD)_2$Pyr **27** were also evaluated providing modest enantiomeric excesses not exceeding 24% ee. Most of the results outlined in this study were obtained in the presence of a stoichiometric amount of the chiral base. The few attempts to carry out experiments under catalytic conditions seem to

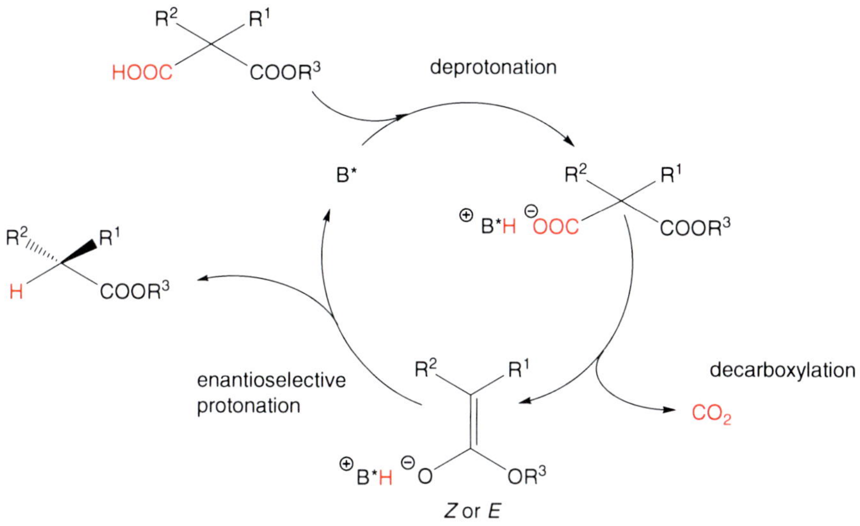

Scheme 3.9 General catalytic pathway for the enantioselective decarboxylative protonation of malonates catalyzed by a chiral Brønsted base (B*)

indicate that the selectivity of the reaction is highly dependent on the nature of the catalyst. Thus, although the use of catalytic amounts of $(DHQD)_2$Pyr **27** did not affect the level of enantioselectivity, it did significantly with $(DHQD)_2$AQN **26** (Scheme 3.10).

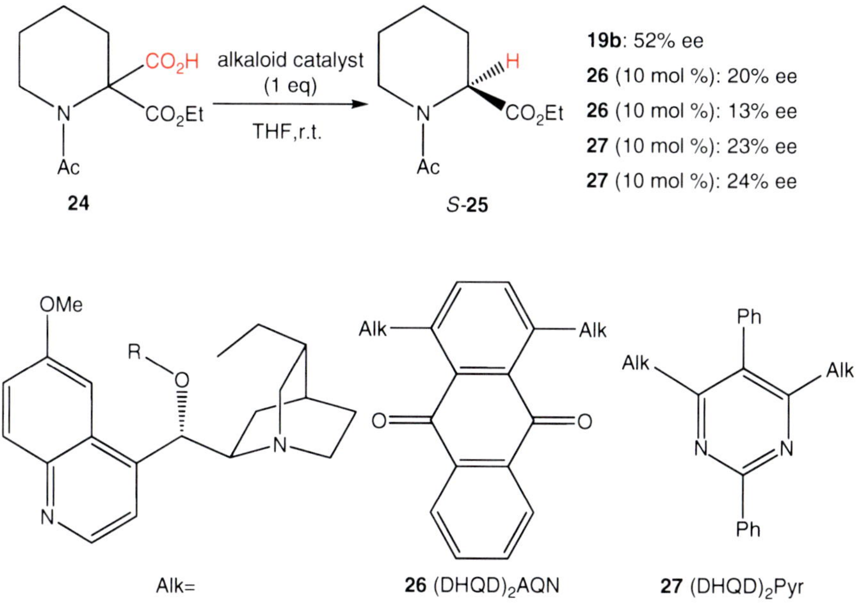

Scheme 3.10 Enantioselective decarboxylative protonation of *N*-acetyl piperidinohemimalonate

The same authors demonstrated that changing *N*-acetyl piperidinohemimalonate **24** to *N*-benzoyl substituted piperidinohemimalonates **28** had a beneficial effect on both the enantioselectivity and catalytic activity of the process [17]. This highlights the importance of the aromatic nature of the *N*-substituent for promoting cooperative π-stacking interactions with the catalyst. Among the various alkaloids tested, cinchonine **8** and the *epi*-cinchonine **30** gave the highest enantioselectivity, up to 72%. Interestingly, inversion of the configuration at C9 in cinchonine **8** does not change the sense of the stereoinduction of the catalyst. One can notice the good performances of the two bis-alkaloids $(DHQD)_2$Pyr **27** and $(CN)_2$Pyr **31** during the decarboxylation of *N*-benzoyl piperidinohemimalonate **28a**. The choice of the solvent revealed to be crucial to get an optimal enantioselectivity, carbon tetrachloride and toluene being superior to more polar solvents such as THF or acetonitrile. The temperature has a significant effect on the reaction rate but not on the enantioselectivity. Lastly, decarboxylation of **28a** could be conducted on a multi-gram scale under catalytic conditions without any detrimental effect on the enantioselectivity (Scheme 3.11).

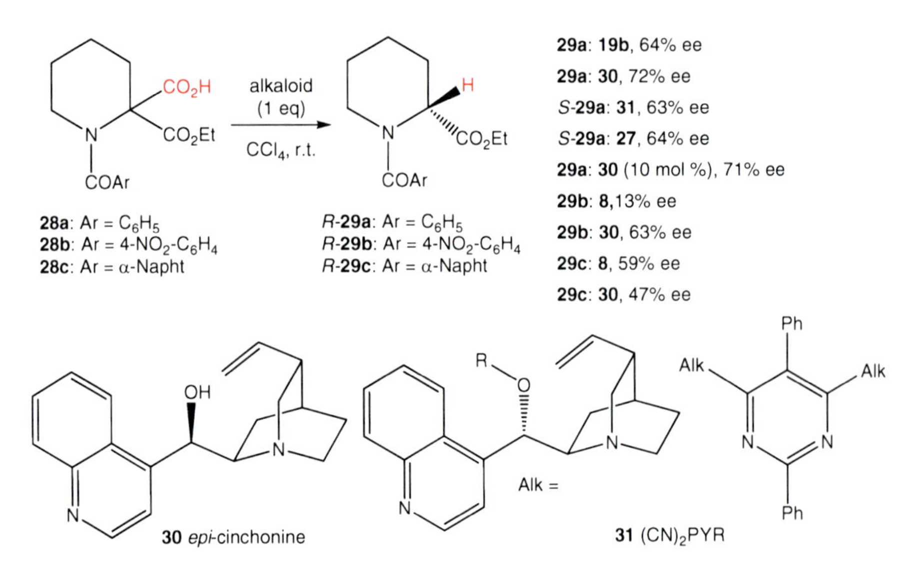

Scheme 3.11 Enantioselective decarboxylative protonation of *N*-benzoyl substituted piperidinohemimalonates

A further important advance was also achieved by Rouden et al. [17] who developed the use of thiourea cinchona alkaloids **33**–**34** in the enantioselective decarboxylative protonation of α-aminomalonates (Scheme 3.12). The basic idea in using these bifunctional catalysts was to take advantage of the good hydrogen-bond donor properties of the thiourea moiety to promote further interactions between the chiral proton source and the prochiral enolate intermediate. Bifunctional catalyst **33** in quinidine series turned out to be especially efficient with a large range of substrates affording in most cases the corresponding aminoesters in high yields and excellent enantioselectivities up to 93%. A synthetically useful feature of this approach is that the pseudoenantiomer **34** in quinine series lead to the opposite absolute configuration with the same level of enantioselectivity. Despite the high level of enantioselectivity attained, this process still suffers from the stoichiometric use of the chiral base. An attempt to use catalytic amounts of the chiral base **33** (20 mol%) resulted in a substantial drop in the enantiomeric excess. Selected examples given in Scheme 3.12 show that high enantiomeric excesses could be reached not only with acyclic substrates such as **23b** but also with cyclic substrates **25** and **32** in piperidine and isoquinoline series respectively (Scheme 3.12).

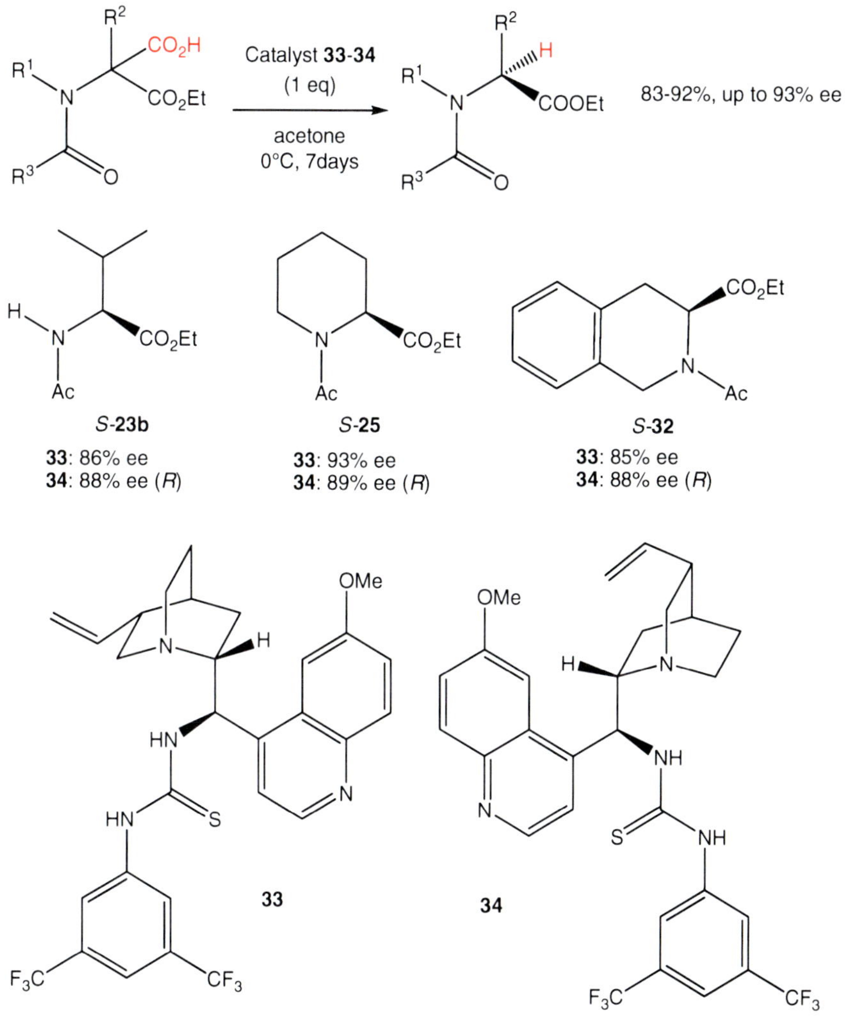

Scheme 3.12 Enantioselective decarboxylative protonation of α-aminomalonates

3.3 Enantioselective Protonation of Ketenes

Enantioselective protonation of ketenes, first pioneered by Pracejus et al. [18] in the 1960s is considered to be one of the earliest contributions in organocatalysis. During the alcoholysis of simple disubstituted ketenes in the presence of various cinchona alkaloids, a remarkable level of enantioselectivity was obtained with *O*-acetyl quinine **38** and *O*-benzoyl quinine **39** (Scheme 3.13). The reaction of phenylmethylketene **35a** with MeOH proceeded smoothly at −110°C in toluene using as low as 1 mol% of *O*-acetylquinine **38** to yield the corresponding ester **37** in up to 74% ee. However, the enantioselectivity proved to be highly dependent on the temperature. When the reaction was conducted at −40°C, the organocatalytic process was plagued by a significant competitive background reaction leading to an almost racemic ester **37**. The proposed mechanism suggests a first addition of MeOH to the ketene facilitated by the assistance of the quinuclidine nitrogen of quinine which would form a 1:1 aggregate with MeOH. The resulting ammonium enolate **36** would then undergo an enantioselective proton transfer from the chiral ammonium to the enolate to yield the desired ester (Scheme 3.13).

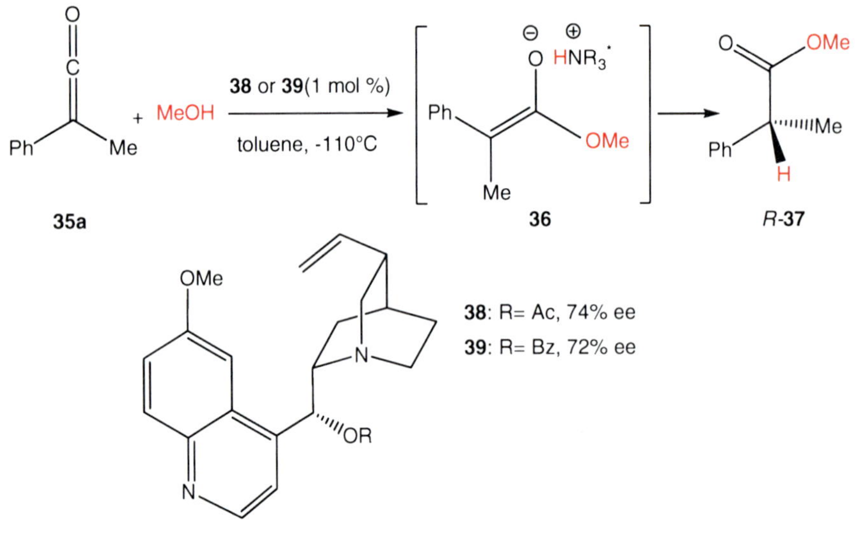

Scheme 3.13 Pracejus' pioneering works in enantioselective protonation of ketenes

Several decades later, Simpkins et al. [19] successfully applied Pracejus' pioneering work to the stereoselective preparation of α-silylthioesters **41**. For example, addition of thiophenol to readily available silylketenes **40a–d** in the presence of *O*-benzoyl quinine **39** (2–10 mol%) afforded the corresponding α-silylthioesters **41a–d** in good yields and enantioselectivities up to 93% (Scheme 3.14). The use of the pseudoenantiomer *O*-benzoyl quinidine **42** led to the formation of the opposite enantiomer with a comparable level of asymmetric induction during addition of thiophenol to silylketenes **40a–d**.

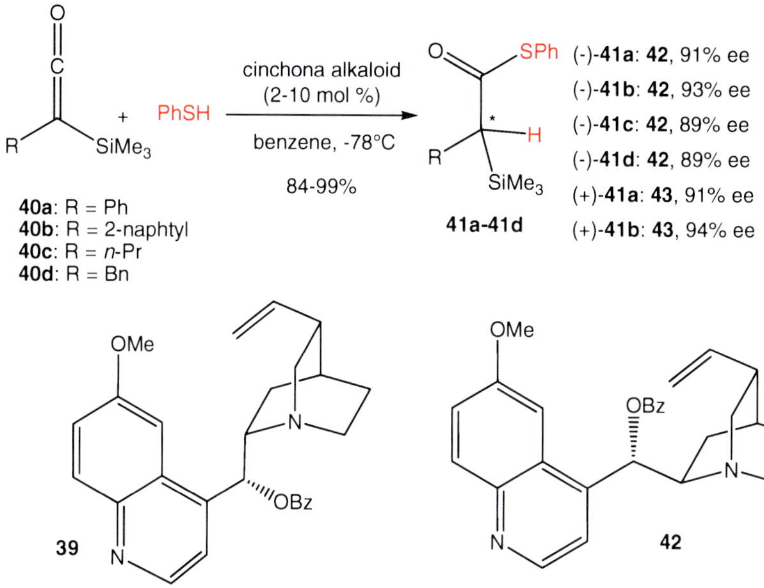

Scheme 3.14 Enantioselective addition of thiophenol to silylketenes

While a Brønsted base activation of methanol mediated by cinchona alkaloids proposed by Pracejus in Scheme 3.13 remains conceivable with silylketenes, Simpkins et al. did not rule out that the reaction might proceed through a nucleophilic pathway as presented in Scheme 3.15. In a first step, the high nucleophilicity of the quinuclidine nitrogen would promote addition of the catalyst to the silylketenes **40a–d** generating a chiral zwitterionic enolate intermediate **43**. Diastereoselective protonation of **43** by thiophenol furnishes the *N*-acyl ammonium **44** and subsequent attack of the latter compound by the thiolate would result in the formation of the desired α-silylthioesters **41a–d** together with the release of the alkaloid catalyst.

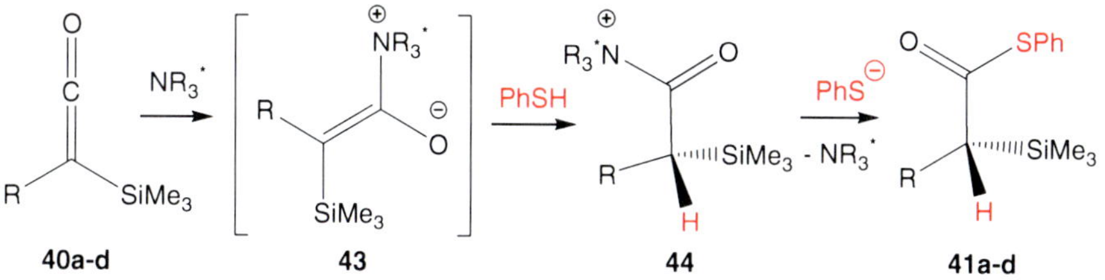

Scheme 3.15 Plausible mechanism for addition of thiophenol to ketenes

Initially reported for the resolution of racemic alcohols, Fu's planar-chiral derivatives of 4-dimethylaminopyridine (DMAP) **45–46** have been also successfully exploited as catalyst in diverse asymmetric transformations [20], including enantioselective protonation of ketenes [21]. These planar-chiral ferrocenes **45–46** are considered in this account as organocatalysts, the ferrocene moiety taking part only in the design of the planar chirality element and not formally in the catalytic process. In 1999, Fu et al. [21] examined the addition of methanol to methylphenylketene **35a** in the presence of various planar-chiral heterocycles **45–46** (Scheme 3.16). Whereas both chiral DMAP derivatives **45a** and **45b** were unsuccessful, azaferrocenes **46a–d** exhibited enantioselectivities ranging from 27% to 77% ee. An assortment of alkylarylketenes, not represented herein, was subjected to methanol addition in the presence of azaferrocene catalyst **46c** furnishing the corresponding esters **37** in satisfactory enantiomeric excesses ranging from 68% to 80% ee.

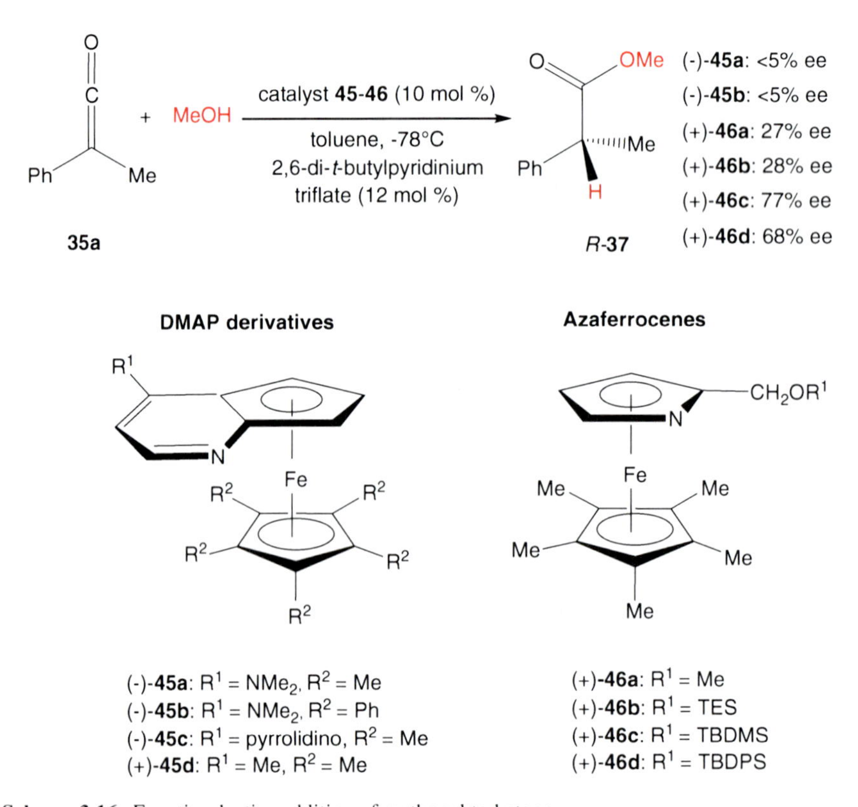

Scheme 3.16 Enantioselective addition of methanol to ketene

As can be noticed in Scheme 3.16, optimized reaction conditions made use of an additional proton source, *i.e.* 2,6-di-*t*-butylpyridinium triflate which notably enhances the enantioselectivity. Although the role of this extra proton source is not clearly established, an explanation consistent with the nucleophile-catalyzed pathway outlined in Scheme 3.17 can be reasonably proposed. This hindered proton source might protonate the zwitterionic enolate **47** with higher diastereoselectivity than does methanol, while generating a non-nucleophilic species, *i.e.* 2,6-di-*t*-butylpyridine unable to compete thereafter with methanol in the final step (Scheme 3.17).

Scheme 3.17 Proposed mechanism for the enantioselective addition of alcohols to ketene

From the early 2000's, a succession of papers by Fu et al. [21] reported the reaction of ketenes with diverse protic nucleophiles, namely amines, phenols, enolizable aldehydes and hydrazoic acid, offering a straightforward route to optically active α-substituted amides, esters and amines. In 2002, Fu et al. [21] firstly investigated the enantioselective addition of nitrogen nucleophiles to ketenes (Scheme 3.18). One of the challenges in developing a catalytic process lies in the difficulty to suppress the background addition of amines to ketenes in the absence of the catalyst. This competitive background addition could be thwarted by using pyrroles, a class of nitrogen nucleophiles sufficiently weak to not react at room temperature with ketenes, while manifesting an excellent reactivity in the presence of a planar-chiral DMAP derivative. A survey of various pyrroles highlighted a strong dependence of the enantioselectivity on the structure of the pyrrole, 2-cyanopyrrole affording the best selectivity when reacting with ethylphenylketene **35b** in the presence of the planar-chiral DMAP (−)-**45c**. A selection of the results obtained with a variety of ketenes **35** is presented in Scheme 3.18 showing that high enantioselectivities up to 98% were achieved with good yields at room temperature. This approach provides a simple stereoselective access to *N*-acylpyrroles **48** bearing a α−stereocenter which can be further transformed into other functional groups (alcohol, aldehyde, amide, ester) under mild conditions and importantly without significant loss of the enantioselectivity.

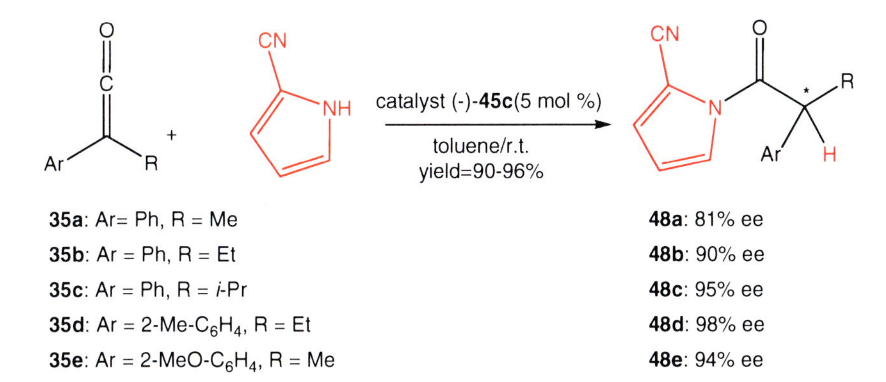

Scheme 3.18 Enantioselective addition of 2-cyanopyrrole to ketenes

While the mode of action of planar-chiral heterocycles **45** is frequently related to a nucleophilic catalysis (see Scheme 3.15), a mechanistic investigation argues strongly in support of a Brønsted acid/base catalysis as illustrated in Scheme 3.19. In a first step, the protonation of the catalyst by 2-cyanopyrrole gives rise to the formation of an ion pair composed of a nucleophilic pyrrole anion and a chiral Brønsted acid. The nucleophilic pyrrole anion then reacts with the ketene **35** to form an enolate intermediate **49** which undergoes enantioselective protonation by its counter anion, *i.e.* the protonated catalyst, generating the desired *N*-acylpyrrole **48** while liberating the Brønsted base catalyst (Scheme 3.19). This scenario is supported by the following experimental data. The protonation of the catalyst **45c** takes place by addition of 2-cyanopyrrole, while no interaction could be detected by ^{1}H NMR between ketene **35b** and the catalyst **45c**. A kinetic study showed that the reaction is first-order in ketene, first-order in catalyst **45c** and zero-order in 2-cyanopyrrole. Lastly, a primary kinetic isotope effect of kH/kD = 5 was observed suggesting that the enantio-determining protonation step is also the turnover limiting step.

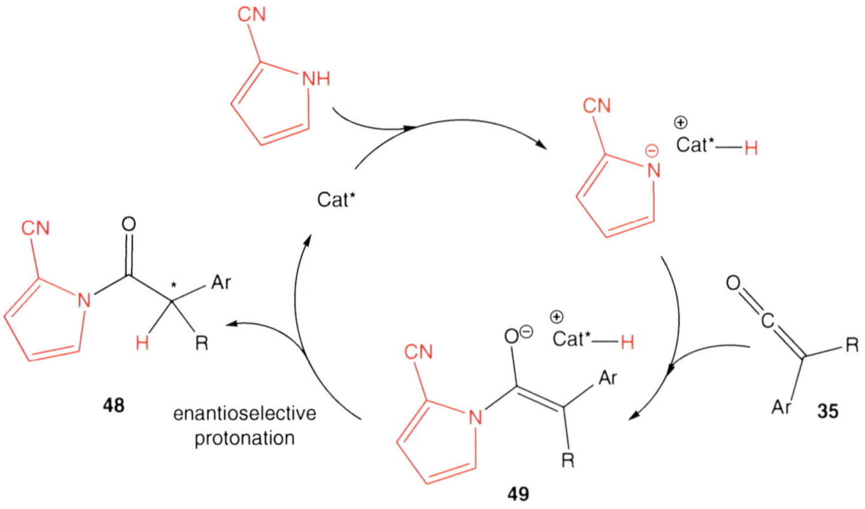

Scheme 3.19 Proposed mechanism for the enantioselective addition of 2-cyanopyrrole to ketenes (Brønsted acid/base catalysis)

Following this work, Fu et al. [21] reconsidered the catalytic addition of alcohols to ketenes by using planar-chiral DMAP derivatives in a Brønsted acid/base catalysis process. In order to favor the protonation of the catalyst, a prerequisite for a Brønsted acid/base pathway, the more acidic phenols were employed. After having checked that phenol protonates DMAP **45c**, different other phenols were reacted with ethylphenylketene **35b** in the presence of chiral DMAP **45c** (3 mol%) at room temperature in toluene. The best level of stereoselection was obtained with the more sterically hindered 2-*tert*-butylphenol affording ester **50b** in 91% ee.

Selected results obtained from the reaction of 2-*tert*-butylphenol with various ketenes **36** are presented in Scheme 3.20. In most cases, esters **50** were obtained in much higher enantioselectivities than those observed under nucleophilic catalysis during the addition of methanol to ketene **35** (see Scheme 3.16).

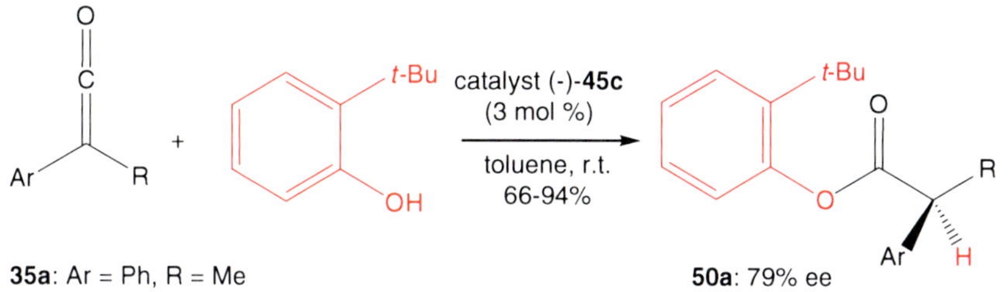

catalyst (-)-**45c** (3 mol %)
toluene, r.t.
66-94%

35a: Ar = Ph, R = Me — **50a**: 79% ee
35b: Ar = Ph, R = Et — **50b**: 91% ee
35c: Ar = Ph, R = *i*-Pr — **50c**: 91% ee
35d: Ar = 2-Me-C_6H_4, R = Et — **50d**: 92% ee
35e: Ar = 2-MeO-C_6H_4, R = Me — **50e**: 94% ee
35f: Ar = 3-thienyl, R = *i*-Pr — **50f**: 79% ee

Scheme 3.20 Enantioselective addition of 2-*tert*-butylphenol to ketenes

In 2005, a new variant of this enantioselective protonation of ketenes was reported by Fu et al. [21] wherein the protic nucleophile is provided by an enolizable aldehyde. After screening a variety of enolizable aldehydes and ketones, diphenylacetaldehyde was found to be the best candidate when reacting with various arylalkylketenes **35** in the presence of 10 mol% of the planar-chiral DMAP (–)-**45c**. A selection of representative examples is shown in Scheme 3.21. Enol esters **51** were obtained in good yields (74–99%) and high enantioselectivities (78–98%) when conducting the reaction in chloroform at 0°C. A synthetic advantage of these enol esters **51** over phenolic esters **50** is their higher reactivity allowing a number of transformations (reduction, hydrolysis) under milder conditions while minimizing the risk of racemisation.

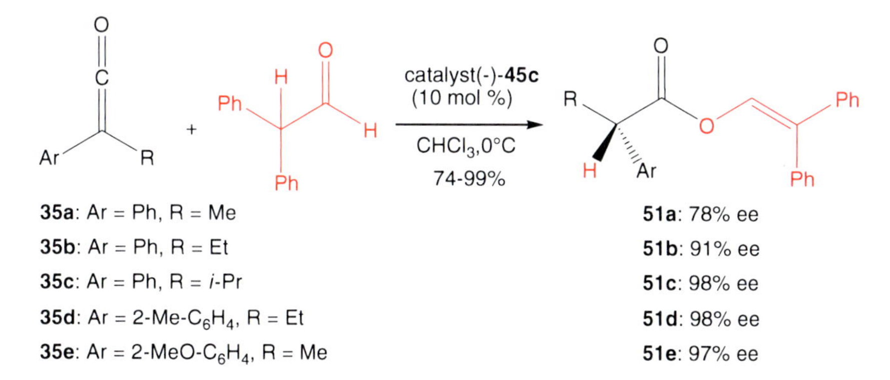

Scheme 3.21 Enantioselective addition of diphenylacetaldehyde to ketenes

Both nucleophilic and Brønsted acid/base catalyzed pathways previously discussed were suggested as possible relevant mechanisms for the formation of enol esters **51**. From a stereochemical point of view, the two mechanisms differ mainly in that the nucleophilic pathway would involve a diastereoselective protonation of the chiral zwitterionic enolate **47** (Scheme 3.22), while in the Brønsted acid/base mechanism, the catalyst may be considered as a chiral proton shuttle taking part in both deprotonation of diphenylacetaldehyde and enantioselective protonation of the prochiral enolate **52** (Scheme 3.23). Although a number of experiences have been conducted to gain insights into the mechanism, no clear-cut evidence was obtained to decide between these two relevant mechanisms.

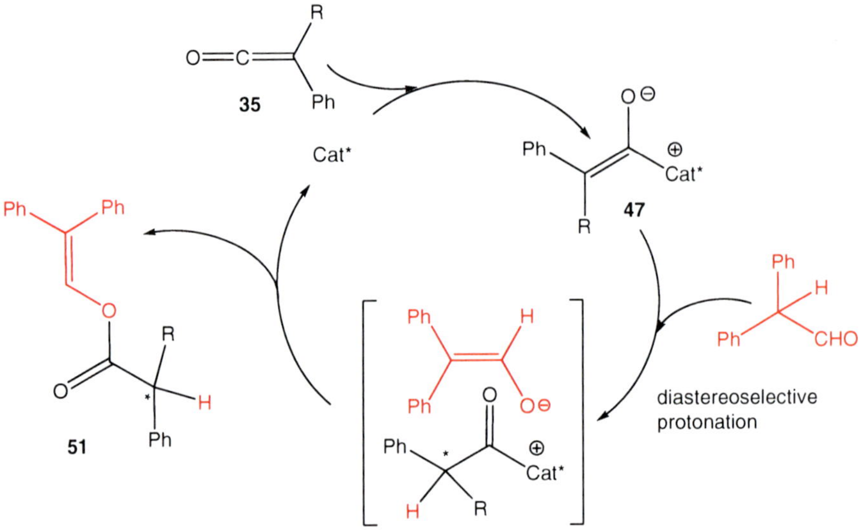

Scheme 3.22 Relevant mechanistic pathways for the addition of diphenylacetaldehyde to ketenes via nucleophilic catalysis

Chiral *N*-heterocyclic carbenes (NHC), well known as potent nucleophilic organocatalysts in numerous transformations such as benzoin condensation and Stetter reaction [22], have also recently been reported to catalyze asymmetric esterification of ketenes [23]. By surveying different alcohols and optimizing the reaction conditions, it was found that the addition of benzhydrol to a range of ketenes was efficiently catalyzed by a catalytic amount of NHC **54'**, the later being formed *in situ* from the precatalyst **54** (12 mol%) and cesium carbonate (10 mol%). Some selected results presented in Scheme 3.24 show that enantioenriched esters **53** could be obtained in up to 95% ee, although both the enantioselectivity and the yield appeared to be highly substrate dependent.

The mechanism proposed in Scheme 3.25 draws close parallels to the nucleophilic catalysis pathway already encountered with planar-chiral DMAP derivatives **45**. The ketene **35** firstly reacts with the *in situ* generated nucleophilic NHC **54'** giving rise to a chiral triazolium enolate **55**. Subsequent diastereoselective protonation

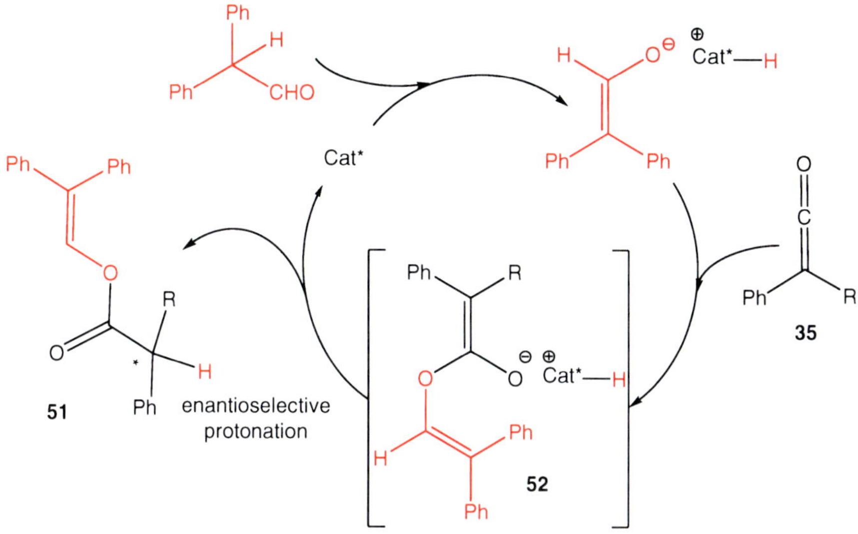

Scheme 3.23 Relevant mechanistic pathways for the addition of diphenylacetaldehyde to ketenes via Brønsted acid/base catalysis

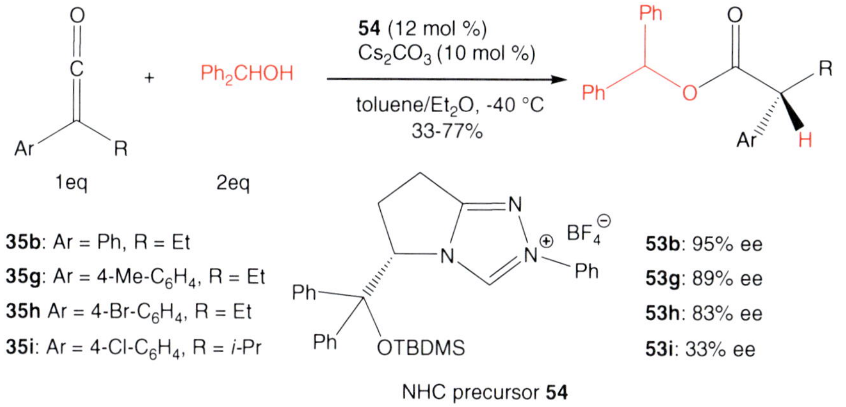

Scheme 3.24 *N*-heterocyclic carbenes (NHC) catalyzed enantioselective addition of alcohols to ketenes

of **55** with diphenylmethanol furnishes an acyl triazolium intermediate which is then substituted by its counter ion, *i.e.* diphenylmethoxide to provide the desired ester while regenerating NHC catalyst **54'**.

And last, but not least, Fu et al. [21] reported that the planar-chiral heterocycle **45d** catalyzed the enantioselective addition of hydrazoic acid to ketenes. This result in the formation of an acyl azide intermediate, which then can be subjected to Curtius rearrangement to supply a straightforward stereoselective access to chiral amines. As can be seen from Scheme 3.26, amine derivatives **56** were obtained in

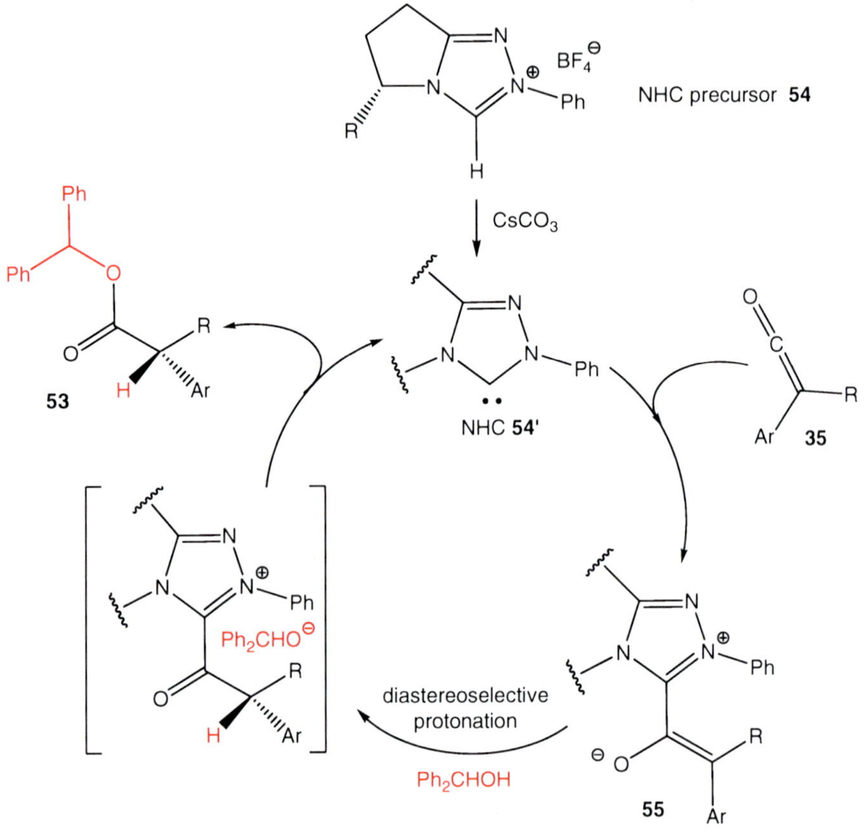

Scheme 3.25 Mechanistic proposal for the enantioselective addition of alcohols to ketenes catalyzed by *N*-heterocyclic carbenes

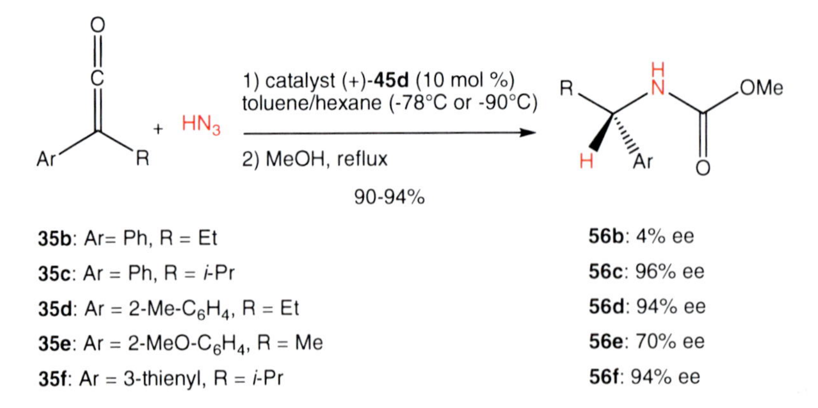

Scheme 3.26 Enantioselective addition of hydrazoic acid to ketenes

excellent enantioselectivities (up to 96% ee) in numerous cases, provided that hindered ketenes are employed to prevent the uncatalyzed addition of HN_3. Control experiments revealed that, in contrast to the hindered phenylisopropylketene **35c**, the background addition of HN_3 with the more reactive unhindered phenylethylketene **35b** occurred at an appreciable rate and prevailed over the catalyzed pathway to yield **56b** nearly racemic.

To account for the catalytic activity of catalyst **45d**, several experimental data led the authors to privilege a Brønsted acid/base catalysis over a nucleophilic catalysis. In particular, the full protonation of catalyst **45d** observed by simply adding HN_3 lends strong support to this point of view. The protonated catalyst **45d** would promote the nucleophilic addition of azide counter anion to ketene followed by an enantioselective protonation of the resulting enolate **57** (Scheme 3.27).

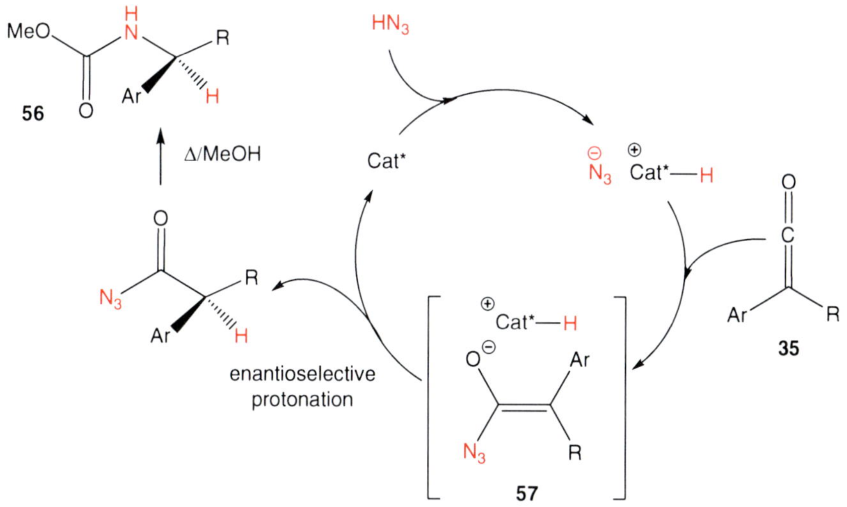

Scheme 3.27 Proposed catalytic pathway for the addition of hydrazoic acid to ketenes (Brønsted acid/base catalysis)

3.4 Conjugated Addition of Nucleophiles

Organocatalyzed enantioselective protonation can also be accomplished by conjugated addition of protic nucleophiles (NuH) to various α,β-unsaturated carbonyl compounds **58** in the presence of nitrogen chiral bases. The general mode of action of the catalyst is related to a Brønsted acid/base catalysis. The key steps of the catalytic process include a protonation of the catalyst resulting in the formation of a chiral nucleophilic ion pair which then adds to the Michael acceptor **58**. Subsequent enantioselective protonation of the resulting achiral enolate **59** furnishes the Michael adduct **60** along with the regeneration of the chiral Brønsted base catalyst (Scheme 3.28).

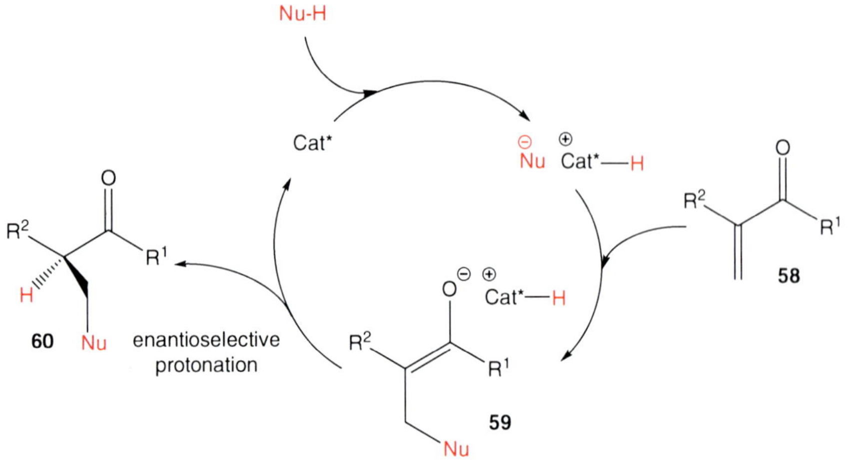

Scheme 3.28 General mechanism for conjugated addition of protic nucleophiles (NuH) to unsaturated carbonyl compounds

This tandem conjugate addition-enantioselective protonation approach was first investigated by Pracejus et al. [24] in 1977. A large panel of chiral amines derived from natural products, including various cinchona alkaloids, were screened during the thia-Michael addition of benzylthiol to methyl 2-phthalimidoacrylate **58a**. The highest level of enantioselectivity was achieved with quinidine **17** (5 mol%) affording cysteine ester derivative **60a** in 54% ee (Scheme 3.29). Parent cinchona alkaloids **7–8** and **16–17** were also evaluated as catalysts in the thia-Michael addition of thiophenol to various α-arylacrylates such as **58b** [25]. Whereas the pseudoenantiomers quinine **16** and quinidine **17** afforded modest enantioselectivity in a range of 45–50% ee and displayed quasienantiomeric behavior, the two other pseudoenantiomers **7** and **8** devoid of a methoxy group at C6 of the quinoline ring supplied at best 32% ee. In the selected example shown in Scheme 3.30, this strategy was applied to the stereoselective preparation of (*S*)-Naproxen®. In the presence of quinine **16** (20 mol%), thiophenol was reacted with α-arylacrylate **58b** to give Michael adduct **60b** in 85% yield and 46% ee. Subsequent desulfurization and ester hydrolysis led to (S)-Naproxen® in 72% overall yield and 85% ee (after a single recrystallization).

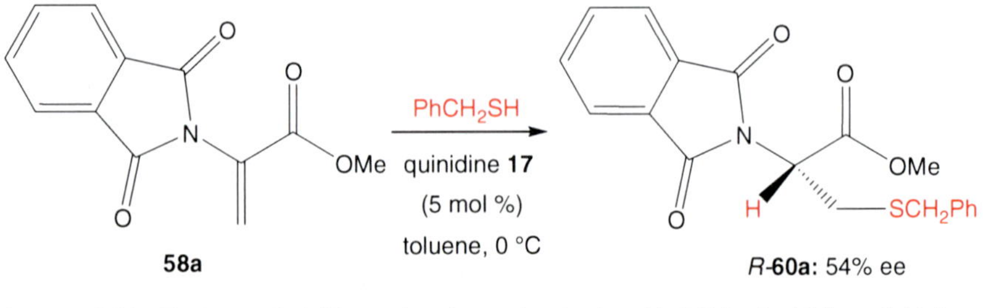

Scheme 3.29 Cinchona alkaloids catalyzed enantioselective thia-Michael addition of thiols to methyl 2-phthalimidoacrylate

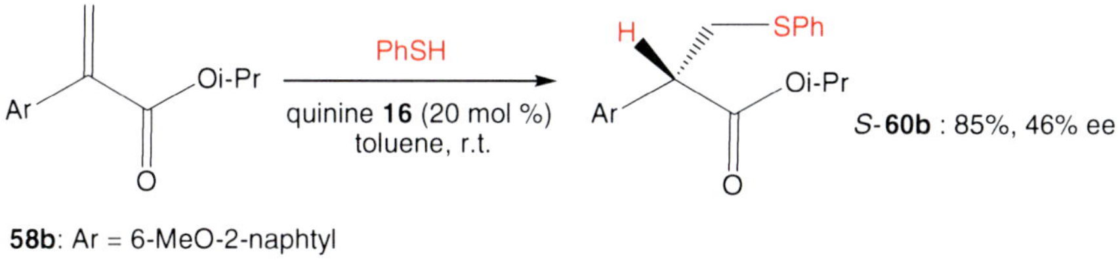

Scheme 3.30 Cinchona alkaloids catalyzed enantioselective thia-Michael addition of thiols to α-arylacrylate

Chen et al. [26] reported the use of a bifunctional thiourea catalyst **61** during the organocatalyzed thia-Michael addition of thiophenol to unsaturated imide **58c** (Scheme 3.29). Michael adduct **60c** was obtained in 60% ee and 97% yield by conducting the reaction in dichloromethane at −78°C. The authors speculated that while the tertiary amine of the bifunctional catalyst **61** would act as a proton shuttle according to a Brønsted acid/base catalysis, the presence of the thiourea moiety might possibly cooperate in the stabilization of the more stable *Z*-enolate intermediate via hydrogen bond formation as illustrated in Scheme 3.31.

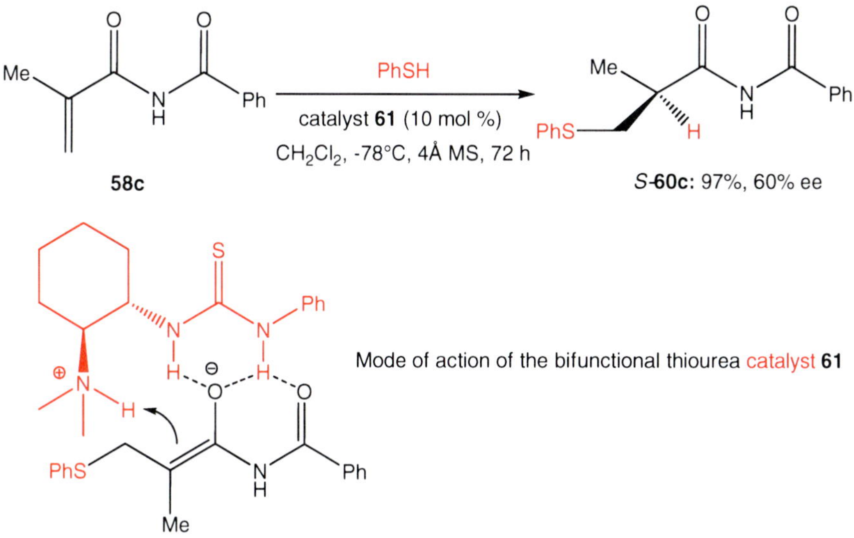

Scheme 3.31 Bifunctional thiourea catalyst catalyzed enantioselective thia-Michael addition of thiophenol to unsaturated imide

Chiral guanidines have emerged as promising organocatalytic tools in a wide range of asymmetric transformations [27]. Recently, Tan et al. [28] reported the use of a chiral bicyclic guanidine **62** in the enantioselective protonation of transient enolates generated *in situ*, by conjugated addition of thiols to *tert*-butyl 2-phthalimidoacrylates (Scheme 3.32) and secondary phosphine oxides to *N*-aryl itaconimides (Scheme 3.33). As can be seen in these schemes, high enantioselectivities ranging from 84% to

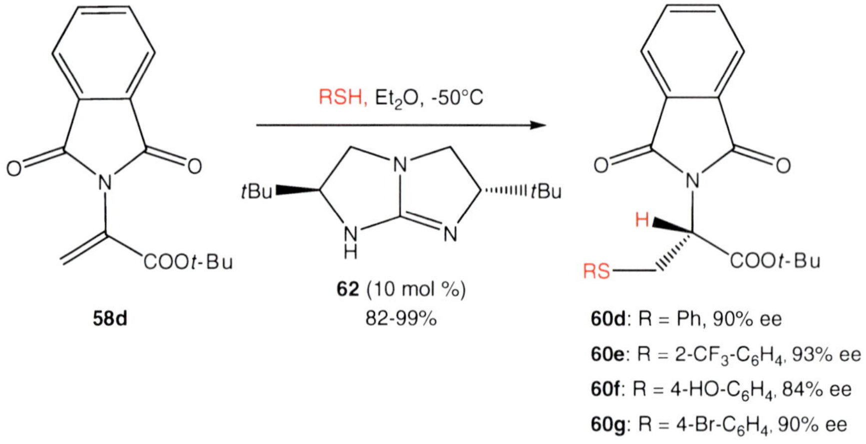

RSH, Et_2O, -50°C

62 (10 mol %)
82-99%

58d

60d: R = Ph, 90% ee
60e: R = 2-CF_3-C_6H_4, 93% ee
60f: R = 4-HO-C_6H_4, 84% ee
60g: R = 4-Br-C_6H_4, 90% ee

Scheme 3.32 Conjugated addition of thiols to *tert*-butyl 2-phthalimidoacrylates catalyzed by chiral bicyclic guanidine

Ar_2(PO)H, PhMe, 0°C

62 (10 mol %)
79-94%

58e

60h: Ar = 3-Cl-C_6H_4, 92% ee
60i: Ar = 4-F-C_6H_4, 91% ee
60j: Ar = 3-Cl-C_6H_4, 90% ee
60k: Ar = 1-naphthyl, 98% ee

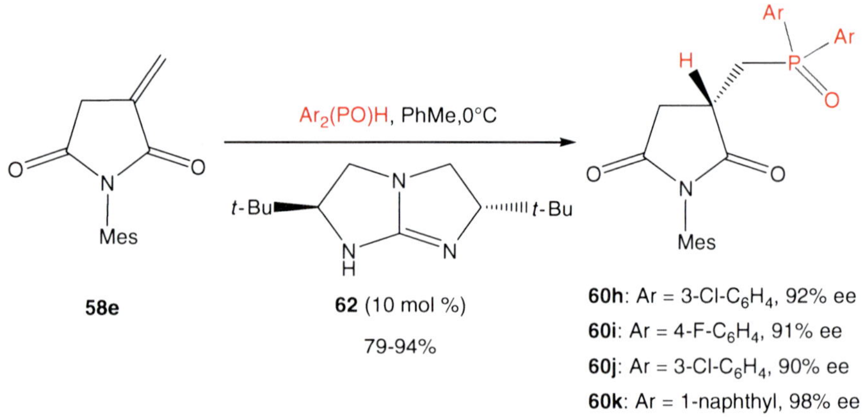

Scheme 3.33 Conjugated addition of secondary phosphine oxides to *N*-Mes itaconimides catalyzed by chiral bicyclic guanidine

98% ee could be observed with both thiols and secondary phosphine oxides. One should notice that methyl 2-phthalimidoacrylate **58a**, already used by Pracejus (Scheme 3.29), afforded modest ee values with guanidine **62**. The higher performances observed with **58d** would arise from the presence of a more sterically hindered ester group offering better control of the *E/Z* selectivity in the transient enolate. In contrast, the nature of the thiol had only little effect on the selectivity; all Michael adducts **60d–g** exhibiting comparable and high enantioselectivities (Scheme 3.32). The same trend was found with Michael adducts **60h–k**; the selectivity being moderately affected by changes of the substitution pattern at the aryl moiety of the diaryl phosphine oxide (Scheme 3.33). Thia-Michael adducts were further exploited in the preparation of optically pure analogues of cysteine, while Michael adducts from **58e** provided original chiral cyclic structures incorporating an α,γ-aminophosphine oxide motif.

Duhamel et al. [29] reported the conjugated addition of thiobenzoic acid to 2-benzyl acroleine **58f** resulting in the formation of a metastable enol **63** with high *Z/E* selectivity (*Z/E* >95%). Subsequent enantioselective tautomerization of **63** was achieved in the presence of a chiral protonating agent. By conducting the tautomerization in the presence of 1 eq of (−)-*N*-methyl ephedrine **64**, the desired aldehyde **60l** was obtained in 58% ee, while the enantioselectivity could be raised to 71% ee by using cinchonidine **7** as chiral inductor (Scheme 3.34).

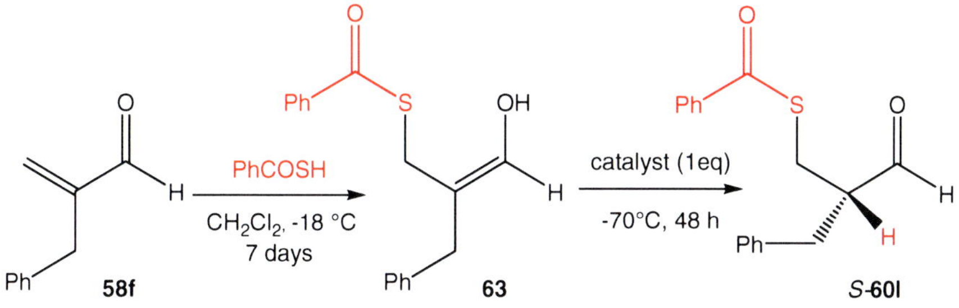

PhCOSH

CH_2Cl_2, -18 °C
7 days

catalyst (1eq)

-70°C, 48 h

58f **63** *S*-**60l**

(-)-*N*-methyl ephedrine **64**: 58% ee

cinchonidine **7**: 71% ee

Scheme 3.34 Enantioselective tautomerisation of enol obtained by conjugated addition of thiobenzoic acid to 2-benzyl acroleine

3.5 Tautomerisation of Enols

The enantioselective tautomerisation of enols, already briefly addressed above, is usually based on photochemical processes to generate the required transient enolic species. From the middle 1980s, a series of papers by Pete et al. [30] reported the enantioselective photodeconjugation of α,β-unsaturated esters and lactones in the presence of a catalytic amount of chiral protonating agents. The irradiation of α,β-unsaturated esters at 254 nm caused a Norrish type II rearrangement resulting in the formation of a photodienol **66** which could then undergo enantioselective protonation to give the corresponding optically active photodeconjugated ester (Scheme 3.35). The screening of a panel of chiral proton sources designated β-amino alcohols as a privileged class of organocatalysts for this enantioselective tautomerisation of enols. Some representative examples are depicted in Scheme 3.35. A hypothesis was proposed to account for the better induction noted with β-amino alcohols wherein abstraction of the enolic proton by the amine would be concerted with protonation of the double bound by the alcohol function as illustrated in Scheme 3.35 [30]. The photodeconjugation of the α,β-unsaturated ester **65** in the presence of (+)-ephedrine **(+)-11** or cinchonidine **7** furnished ester **67** with rather modest selectivities of 37% and 41% ee respectively, while the use of the β-amino alcohol **68** resulted in a somewhat higher enantiomeric excess of 70% [30]. A significant improvement was observed by using the more rigid β-amino alcohol **68**

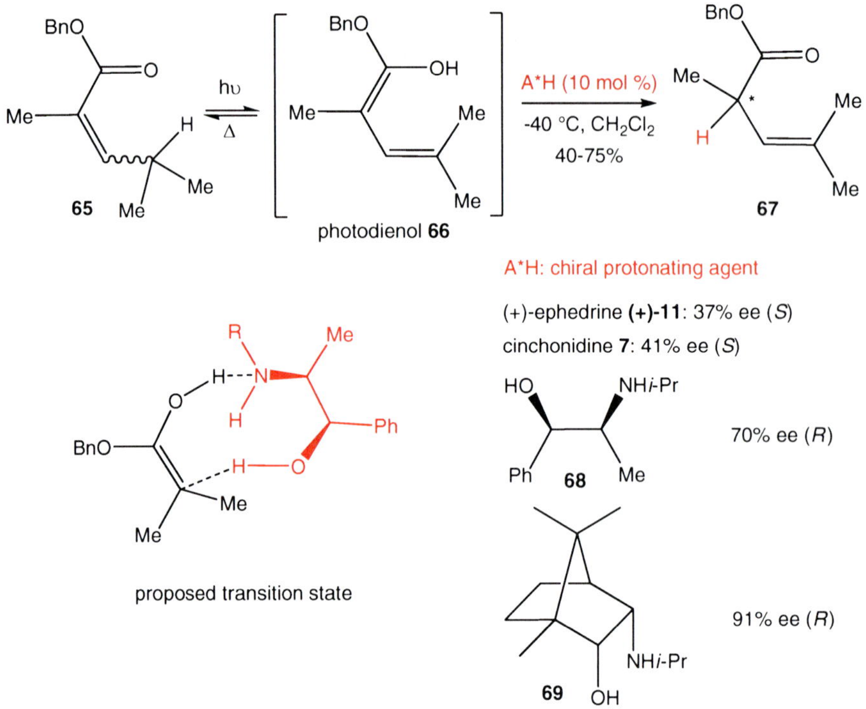

Scheme 3.35 Enantioselective protonation of photodienols organocatalyzed by β-amino alcohols

derived from camphor affording ester **67** in 91% ee [30]. The nature of the *N*-alkyl substituent in both β-amino alcohols **68** and **69** seemed to be crucial, smaller N-alkyl substituents seriously affecting the performance of the organocatalyst. Optimum results were obtained under well-defined reaction conditions. In particular, low temperature (−40°C to −78°C), absence of any moisture and non-polar solvents are prerequisites for this reaction.

Alternatively, Henin et al. [31] reported the photodeconjugation of the chiral ammonium carboxylate **70** formed between the corresponding carboxylic acid and equimolecular amount of the β-amino alcohol **69** (Scheme 3.36). Simultaneously the resulting (*E*)-dienolate intermediate **71** undergoes enantioselective protonation to furnish the deconjugated carboxylate **72** which was then esterified with benzyl alcohol affording ester **73** in 65% yield and in up to 85% ee. The best results were obtained at −46°C in CH_2Cl_2; at lower temperatures the ammonium carboxylate **70** precipitated, whereas photolysis of the solvent took place at room temperature leading to the formation of HCl. While strong ionic interactions between the chiral ammonium and the dienolate species within **71** might have had a beneficial effect on the selectivity, the level of induction did not exceed those obtained by photodeconjugation of ester **65** with the same β-amino alcohol **69** under catalytic conditions (see Scheme 3.35).

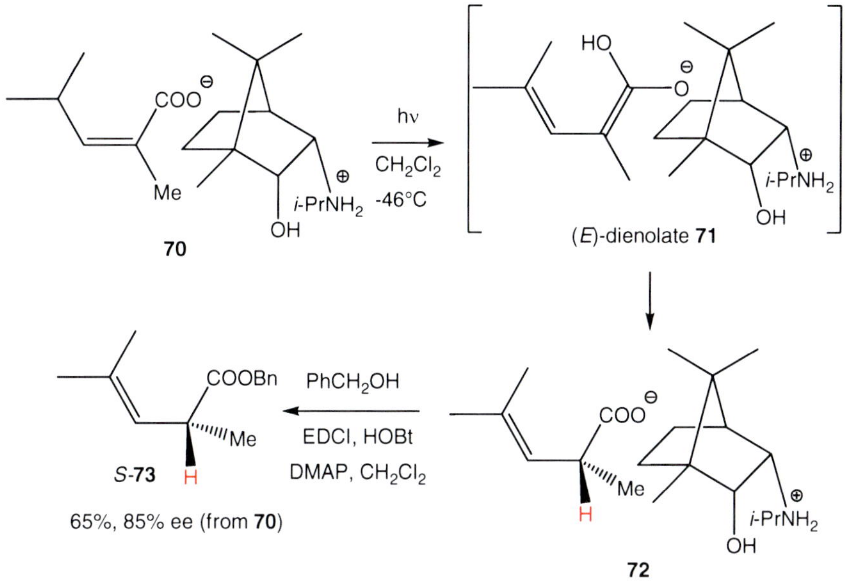

Scheme 3.36 Photodeconjugation of the chiral ammonium carboxylate

In 1991, Muzart et al. [32] extended this photochemical approach to the enantioselective protonation of simple enols photogenerated via Norrish type II photoelimination of ketones (Scheme 3.34). Photolysis of 2-methyl-2-isobutylindan-2-one **74a** at 366 nm in the presence of (–)-ephedrine **(–)-11** (10 mol%) afforded 2-methyl indanone **75a** with modest ee values of up to 47%. It is worth noting that the reaction could be carried out at lower catalyst loading (1 mol%) without compromising the enantioselectivity [32]. The use of amino alcohol **69** (10 mol%) resulted in significantly higher ee values during the photolysis of 2-methyl-2-isobutyl tetralone **74b**, affording 2-methyl tetralone **75b** in 89% ee. In both cases, the isolated yield showed to be rather modest, essentially due to the formation of side products resulting from undesired photochemical fragmentation reactions (Scheme 3.37).

3.6 Protonation of Silyl Enolates

Quite surprisingly, whereas the chemistry of silyl enolates is considered as one of the cornerstones in organic synthesis, only a few papers have been devoted to organocatalytic enantioselective protonation of this class of substrates. The first organocatalyzed process was reported by Levacher et al. [33] by making use of cinchona alkaloids as catalysts and a latent source of hydrogen fluoride. The mechanism postulated by the authors is illustrated in Scheme 3.38. Basically, the reaction between benzoyl fluoride and ethanol in the presence of a catalytic amount of

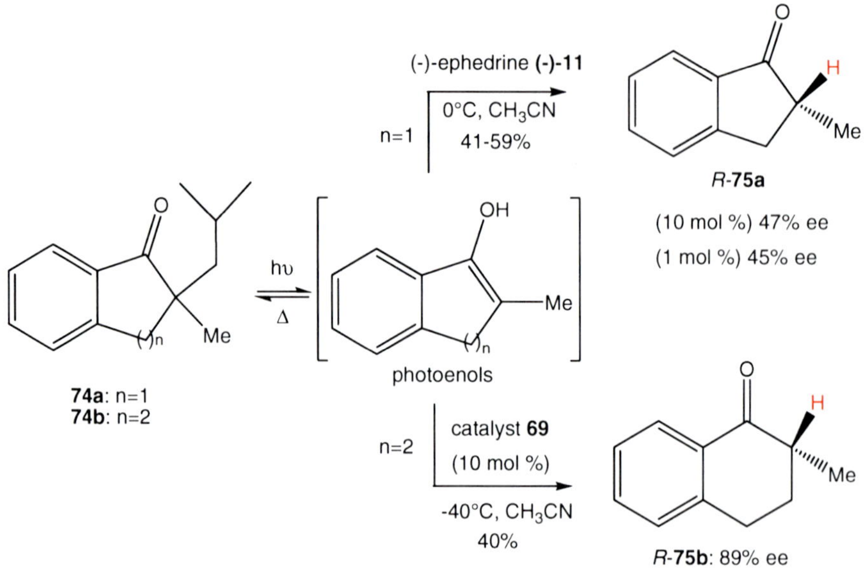

(-)-ephedrine **(-)-11**
0°C, CH_3CN
41-59%
n=1
R-**75a**
(10 mol %) 47% ee
(1 mol %) 45% ee
hυ
Δ
OH
Me
photoenols
74a: n=1
74b: n=2
n=2
catalyst **69**
(10 mol %)
-40°C, CH_3CN
40%
R-**75b**: 89% ee

Scheme 3.37 Enantioselective protonation of simple photoenols organocatalyzed by β-amino alcohols

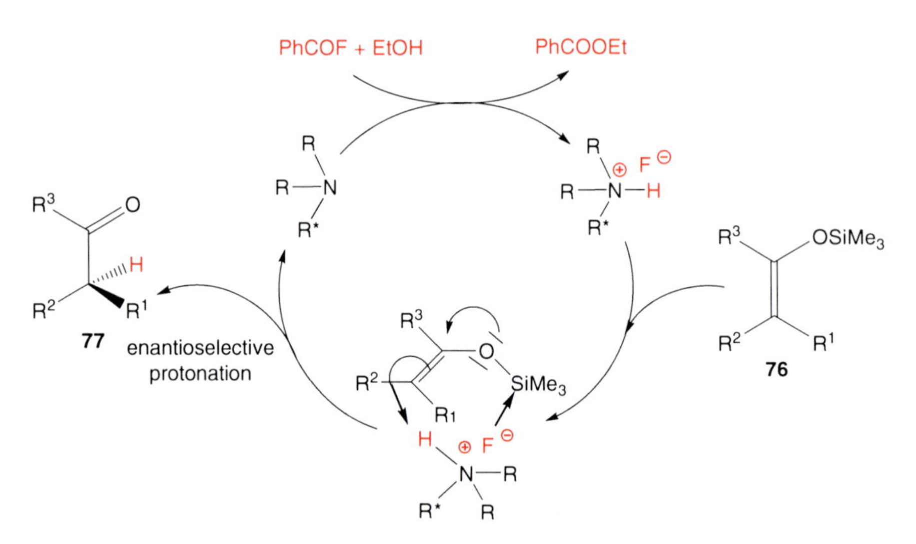

Scheme 3.38 Mechanism for the enantioselective protonation of silyl enolates by means of cinchona alkaloids and a latent source of hydrogen fluoride

cinchona alkaloid delivers the corresponding cinchonium fluoride "at will". The fluoride anion can then exert a specific silyl enolate activation due to the high silicon-fluoride affinity to promote the protonation of the silyl enolate **76**, providing the desired ketone **77** together with the regeneration of cinchona alkaloid.

After screening different cinchona alkaloids under various reaction conditions, it was found that the best results were obtained with bis-cinchona alkaloid $(DHQ)_2AQN$ **78** (10 mol%) at room temperature in DMF. As can be seen from Scheme 3.39, this organocatalytic process proved to be more performant in the tetralone series, affording **77a–c** with satisfactory to good enantioselectivities of up to 92% ee, while indanones **77d, e** were obtained in somewhat lower ee values ranging from 64% to 74% ee. One can notice that cyclohexanone **79** was isolated in 30% ee, although this rather low enantioselectivity could be improved to 58% ee by conducting the reaction at −10°C.

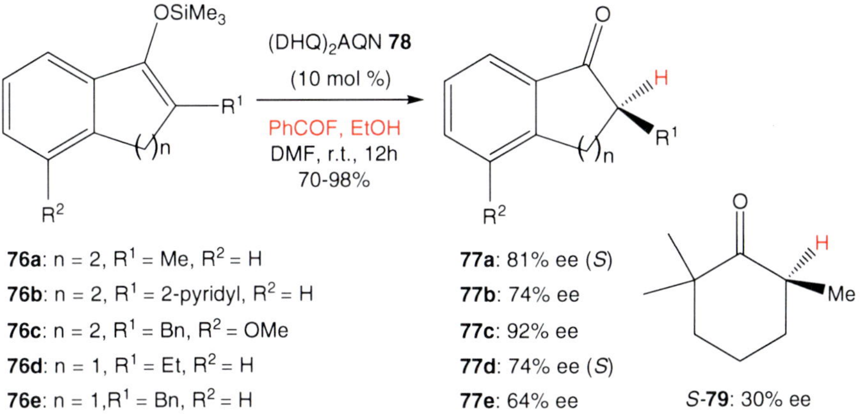

$OSiMe_3$ R[1] R[2] n

$(DHQ)_2AQN$ **78** (10 mol %)
PhCOF, EtOH
DMF, r.t., 12h
70-98%

O H R[1] R[2] n

76a: n = 2, R^1 = Me, R^2 = H
76b: n = 2, R^1 = 2-pyridyl, R^2 = H
76c: n = 2, R^1 = Bn, R^2 = OMe
76d: n = 1, R^1 = Et, R^2 = H
76e: n = 1, R^1 = Bn, R^2 = H

77a: 81% ee (*S*)
77b: 74% ee
77c: 92% ee
77d: 74% ee (*S*)
77e: 64% ee

O H Me
S-**79**: 30% ee

Scheme 3.39 Enantioselective protonation of silyl enolates by means of cinchona alkaloids and a latent source of hydrogen fluoride

Shortly after, a simplified procedure reported by Levacher et al. [34] made use of a simple carboxylic acid as proton source. The initial protonation of $(DHQ)_2AQN$ **78** by carboxylic acid gives rise to the corresponding hydrogen carboxylate salt $(DHQ)_2AQN\text{-}RCO_2H$. The resulting chiral ion pair is presumed to be a better protonating agent than the more acidic carboxylic acid itself, owing to the cooperative activation of the silyl enolate by the carboxylate counterion. A plausible mechanistic pathway accounting for this scenario is depicted in Scheme 3.40.

In a survey of different carboxylic acids, citric acid has emerged as the best candidate to reach optimum enantioselectivity. The protonation of various silyl enolates **76** in the presence of $(DHQ)_2AQN$ **78** (10 mol%) afforded the corresponding ketones **77** with fair levels of enantioinduction up to 75% (for example see **76a,c,e** → **77a,c,e** Scheme 3.41). Although this simplified process exhibits somewhat lower performances than that using a latent source of hydrogen fluoride [33], it shows clear advantages in view of implementation simplicity and cost (Scheme 3.41).

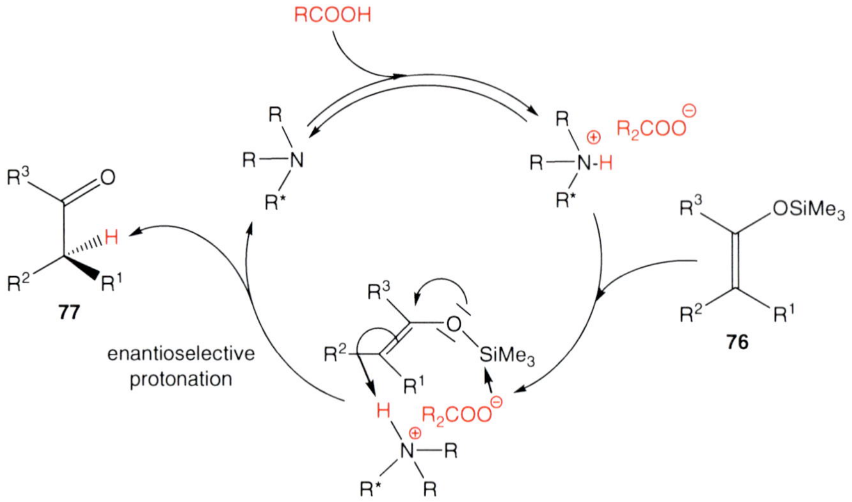

Scheme 3.40 Mechanism for the enantioselective protonation of silyl enolates using cinchona alkaloids and carboxylic acids

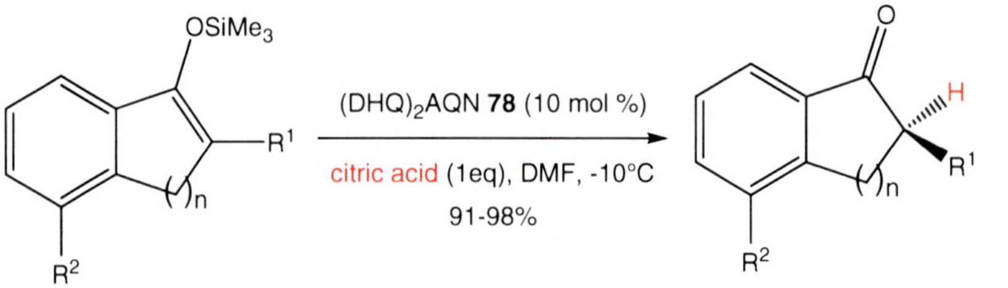

76a: n = 2, R^1 = Me, R^2 = H — **77a**: 70% ee (*S*)

76c: n = 2, R^1 = Bn, R^2 = OMe — **77c**: 75% ee

76e: n = 1, R^1 = Bn, R^2 = H — **77e**: 45% ee

Scheme 3.41 Enantioselective protonation of silyl enolates using cinchona alkaloids and carboxylic acids

The highly acidic chiral *N*-triflyl thiophosphoramide catalyst **82a** was also successfully employed as a Brønsted acid during the enantioselective protonation of silyl enolates derived from cyclohexanone or cycloheptanone [35]. This proto-desilylation was achieved in the presence of phenol as the stoichiometric proton source. The highest levels of enantioselectivity were obtained with ketones bearing a α-aryl substituent, with slightly better results being observed in cycloheptanone series (for examples, see **80→81**, Scheme 3.42). Remarkably, catalyst loadings as low as 0.005 mol% could be used without damaging neither the enantioselectivity nor the yield.

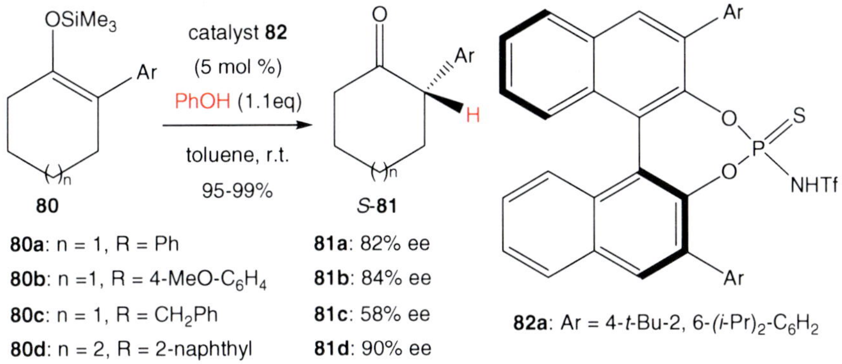

Scheme 3.42 Chiral *N*-triflyl thiophosphoramides derived from BINOL organocatalyzed enantioselective protonation of silyl enolates

Further control experiments gave some insights into the mechanism of this organocatalytic process. Informatively, the reaction did not occur in the absence of the phenol, suggesting that the active catalytic species would result from a pre-association between the *N*-triflyl thiophosphoramide **82a** and phenol. The authors speculated that the formation of a chiral oxonium ion pair would promote the proto-desilylation of the silyl enolate **80** through a two-step mechanism. Initial enantioselective protonation of the silyl enolate **80** would then be followed by desilylation of the resulting intermediate to furnish the desired ketone **81** along with silylated phenol (Scheme 3.43).

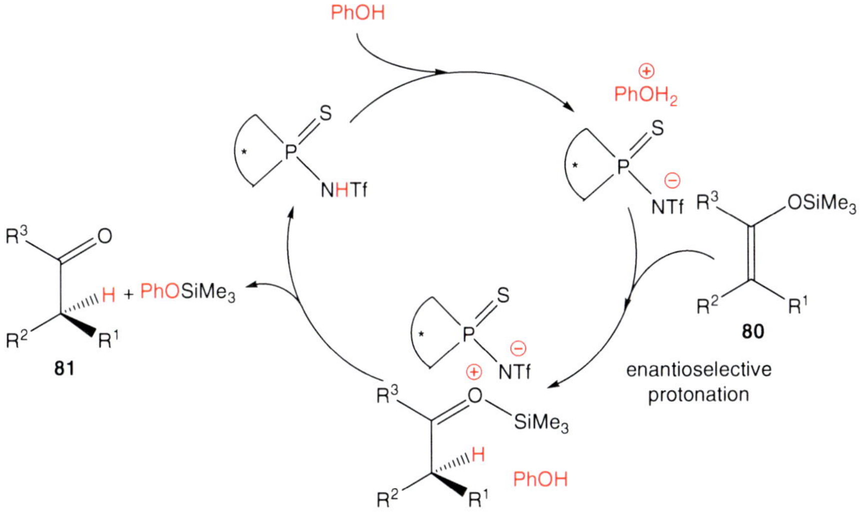

Scheme 3.43 Proposed mechanism for the enantioselective protonation of silyl enolates assisted by chiral *N*-triflyl thiophosphoramide catalysts

3.7 Miscellaneous

Recently, Rueping et al. [36] disclosed the first enantioselective Nazarov cyclization reaction organocatalyzed by a chiral Brønsted acid. The proposed reaction pathway involves a conrotatory 4π electrocyclization of the divinyl ketone **83** leading to the formation of an enol intermediate which is then subjected to enantioselective protonation by the chiral Brønsted acid **82b**. This electrocyclization-protonation reaction was conducted with various divinyl ketones **83** under optimized conditions, *i.e.* in the presence of the chiral *N*-triflyl phosphoramide **82b** (5 mol%) in chloroform at −10°C, affording the corresponding cyclopentenones **84** with 67–78% ee (for examples, see **83a–c** → **84a–c**, Scheme 3.44).

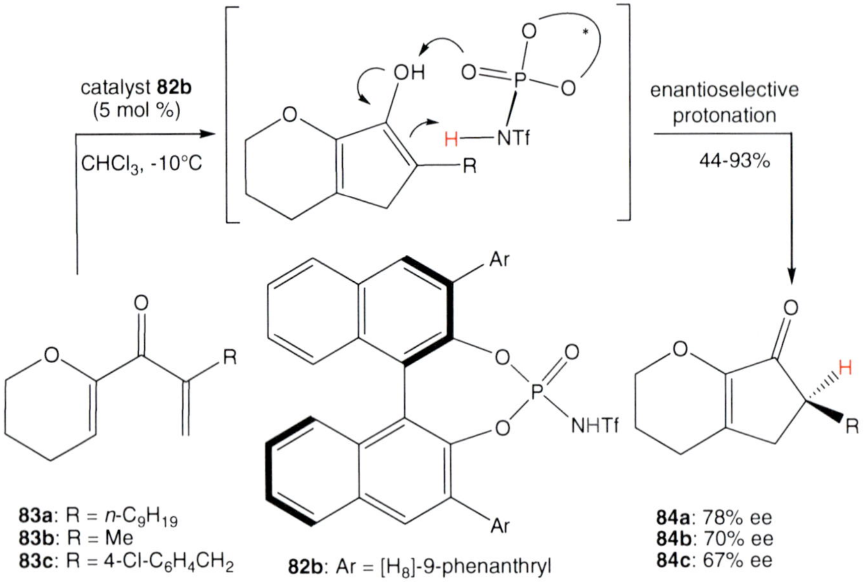

catalyst **82b** (5 mol %)
$CHCl_3$, -10°C
enantioselective protonation
44-93%
Ar
NHTf
Ar
83a: R = n-C_9H_{19}
83b: R = Me
83c: R = 4-Cl-$C_6H_4CH_2$
82b: Ar = $[H_8]$-9-phenanthryl
84a: 78% ee
84b: 70% ee
84c: 67% ee

Scheme 3.44 Chiral Brønsted acid-catalyzed enantioselective Nazarov reaction

Rovis et al. [37] developed an efficient stereoselective access to α-chloro esters **87** by reaction of 2,2-dichloroaldehydes **85** with phenol catalyzed by the chiral triazolinylidene carbene **88'**, the latter being generated *in situ* upon deprotonation of the carbene precursor **88**. The key steps of the reaction sequence are the addition of the carbene **88'** to the 2,2-dichloroaldehyde **87** followed by the elimination of HCl and diastereoselective protonation of the resulting chiral triazolium chloroenolate **86** by phenol. The reactive acyl azolinium intermediate is then trapped either with its phenolate counter anion or potassium phenolate to give rise to the desired α-chloro esters **87** while regenerating carbene **88'**. Under the optimized reaction conditions depicted in Scheme 3.45, a number of α-chloro esters **87** could be prepared in high yields and excellent enantioselectivities ranging from 84% to 93% ee (for examples, see **85a–c → 87a–c**, Scheme 3.45). The conversion of the resulting α-chloro phenyl esters **87** into a large panel of useful derivatives was accomplished by hydrolysis, *trans*-esterification or reduction of the ester function as well as by substitution reaction of the chloride by sodium azide. All these transformations were achieved without alteration of the enantioselectivity.

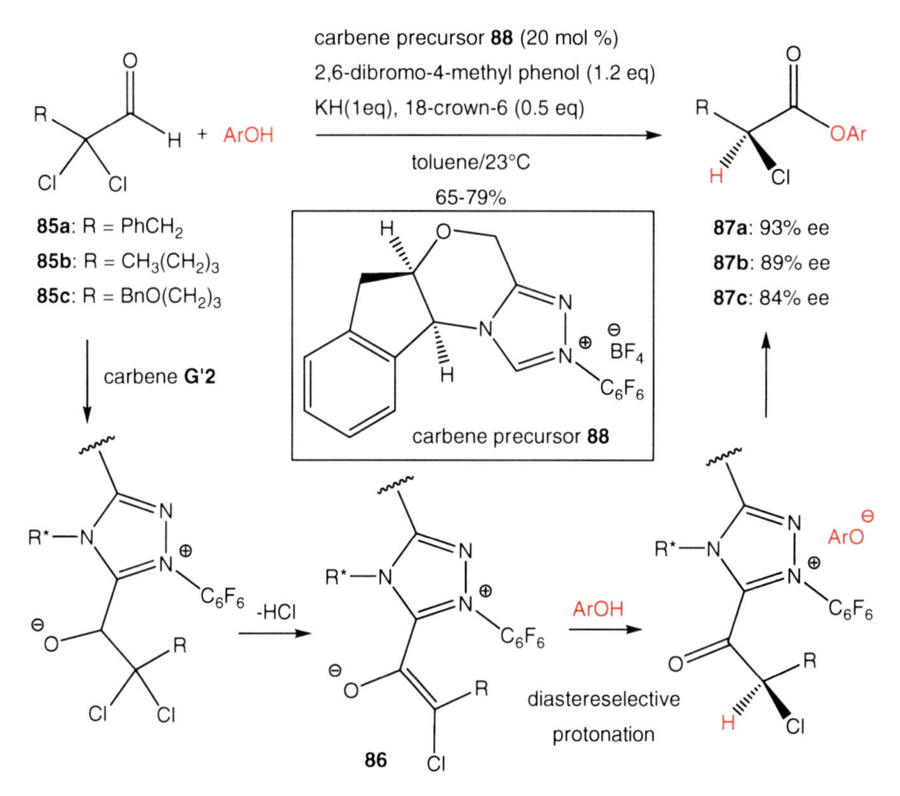

Scheme 3.45 Stereoselective protonation of catalytically generated chiral enolates by means of chiral *N*-heterocyclic carbenes

N-heterocyclic carbenes (NHC) are also efficient organocatalytic tools for generating homoenolate equivalents from α,β-unsaturated aldehydes. These reactive intermediates display a versatile reactivity in a number of catalytic transformations attesting to an important synthetic potential [38]. Recently, Scheidt et al. [39a] accomplished the first enantioselective protonation of a homoenolate species generated by a chiral NHC precursor **93** in the presence of DIEA and an excess of ethanol as the achiral proton source (Scheme 3.46). The suggested mechanism involves an initial addition of NHC **93'** to the enal **89** followed by a formal 1,2-proton shift resulting in the formation of the chiral homoenolate equivalent **91**. A diastereoselective β-protonation/tautomerization sequence leads to the acyl triazolinium intermediate wherein the activated carbonyl would be trapped by ethanol to furnish ester **92** while liberating the catalyst **93'**. Although the protonation product **92** exhibited a moderate enantioselectivity of up to 55% ee, this result remains remarkable in that the prostereogenic center and the chiral appendage are arranged relatively far one from each other. The major limitation of this approach is that the efficiency of the protonation route proved to be seriously hampered by the competitive formation of an oxidation by-product **90**. A detailed mechanistic study [39b] demonstrated that the side-formation of **90** could be minimized by carrying out the reaction in a non-polar solvent such as toluene, affording the desired protonation product **92** in 53% ee and 81% yield (Scheme 3.43).

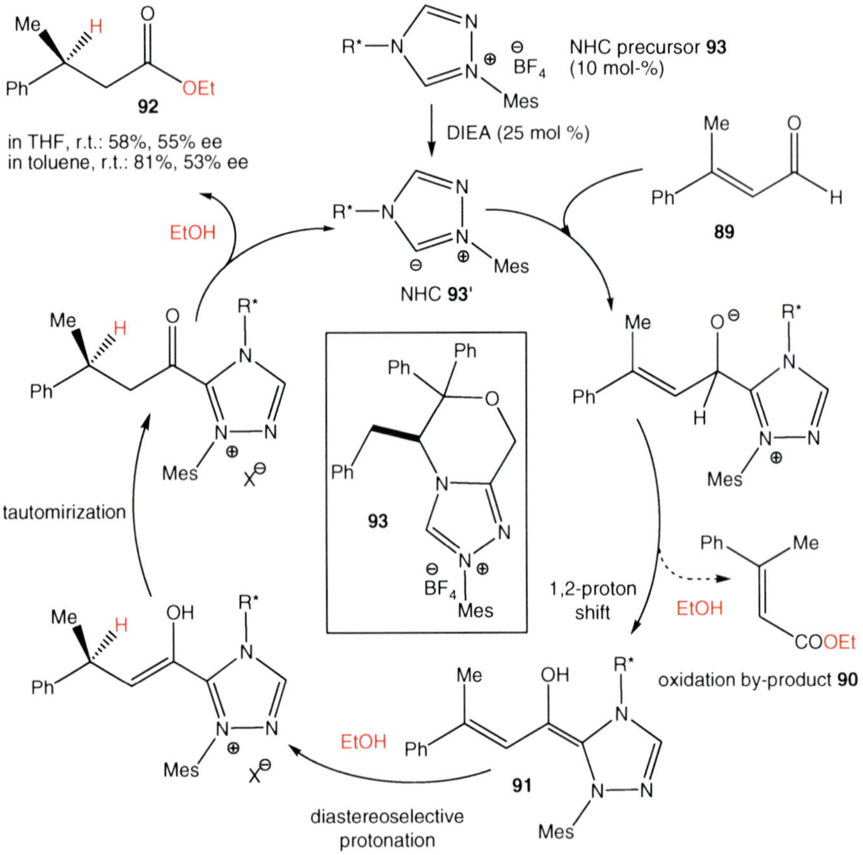

Scheme 3.46 Chiral *N*-heterocyclic carbene-catalyzed enantioselective β-protonation of homoenolate equivalents

Catalytic asymmetric transamination of β-keto carboxylic esters **94** catalyzed by a chiral base was firstly reported by Soloshonosk et al. [40a] providing an original biomimetic approach to β-fluoroalkyl-β-amino acids. This transformation involves the formation of *N*-benzyl enamines **96** in tautomeric equilibrium with *N*-benzyl imines **97** which undergo enantioselective [1, 3]-proton transfer. Hydrolysis of resulting thermodynamically favored *N*-benzylidenes **98** furnished the desired free β-amino acids. The presence of an electron-withdrawing perfluoroalkyl chain in β-keto carboxylic esters **94** proved to be crucial to driving the reaction in favor of the transamination process [40b]. When conducting the reaction at 100°C under free-solvent condition in the presence of cinchonidine **7** (10 mol%), Soloshonosk et al. [40a] obtained aldimine **98a** in 36% ee and 71% yield. Plaquevent et al. [40c] then improved these preliminary results by making use of the bis-cinchona $(DHQ)_2$PHAL **99** (10 mol%) and a more acidic benzylic proton. The aldimine **98b** could be obtained in 71% ee and 90% yield by carrying out the reaction in toluene at 80°C (Scheme 3.47). Performing the transamination process from deuterated benzylamine along with further mechanistic investigations led the authors to suggest the deprotonation step of *N*-benzyl imines **97** at the benzylic position as both rate- and enantio-determining step.

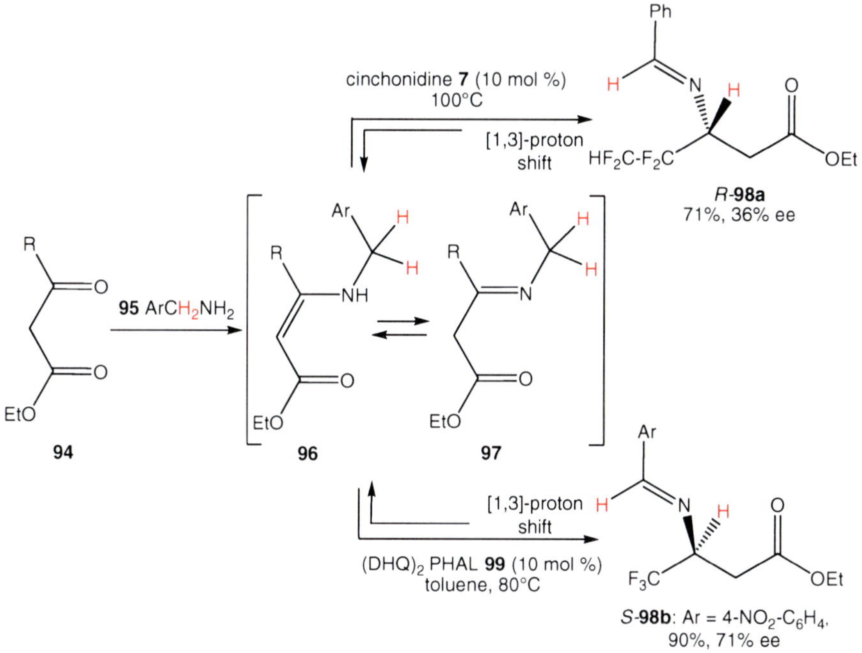

Scheme 3.47 Cinchona alkaloids-catalyzed asymmetric transamination of β-keto carboxylic esters with benzylamine derivatives

3.8 Conclusions

The stereochemical outcome of numerous fundamental chemical transformations such as decarboxylation of malonates, tandem hetero-Michael addition and nucleophilic addition to ketenes, photodeconjugation of unsaturated esters, protodesilylation of silyl enolates, besides some other miscellaneous transformations, may be viewed as arising from a simple proton transfer. As presented in this chapter, much effort has been devoted to developing organocatalytic processes as attractive alternatives to the protonation of metal-enolates and enzymatic approaches. Parent cinchona alkaloids and their synthetic analogues are by far the most frequently employed class of organocatalysts. This is particularly true for the enantioselective decarboxylative protonation of malonates, the cinchona alkaloid playing the role of a proton shuttle in the reaction. While modest to good enantioselectivities were obtained, a number of organocatalyzed decarboxylation processes still suffer from low catalytic performances leading to use stoichiometric amounts of the cinchona alkaloid. Cinchona alkaloids have also been reported to catalyze nucleophilic addition on ketenes, nonetheless the most remarkable advances in this field were obtained with Fu's planar-chiral heterocycles providing high degrees of enantiocontrol with a variety of protic nucleophiles. Depending on both the nature of the protic nucleophile and the planar-chiral heterocycle, the catalytic activity observed may stem from either a Brønsted acid/base or nucleophilic catalysis. Although rarer, protic nucleophiles have also been used in tandem conjugated addition-enantioselective protonation catalyzed by chiral nitrogen bases. Chiral guanidines were found to be manifestly superior over cinchona alkaloids for promoting high level of enantioselectivity. The tautomerisation of enol species photochemically generated in the presence of chiral β-aminoalcohols may also lead to efficient enantioselective proton transfers. However, this original photochemical approach is restricted to carbonyl compounds able to undergo Norrish type II reaction and appears to be highly substrate-dependent in terms of yield and stereoselectivity. Organocatalyzed enantioselective protonation of silyl enolates has been disclosed only recently. Although high level of enantioselectivity could be reported in numerous cases, the difficulty in controlling the silyl enolate geometry limits badly the substrate scope of this approach to cyclic silyl enolates, namely principally in tetralone, indanone and cyclohexanone series. Finally, the stereochemical outcome of a number of organocatalytic transformations which do not appear *prima facie* to be related in a straightforward manner to an enantioselective protonation is nevertheless the result of a proton transfer. These include biomimetic Nazarov cyclization and transamination of β-keto carboxylic esters. While still in its infancy, organocatalyzed enantioselective protonation is emerging as a valuable tool among the panel of organocatalytic processes already available in the literature.

References

1. (a) Matsumoto K, Tsutsumi S, Ihori T, Ohta H (1990) J Am Chem Soc 112:9614; (b) Hirata T, Shimoda K, Kawano T (2000) Tetrahedron: Asymmetry 11:1063; (c) Sakai T, Matsuda A, Tanaka Y, Korenaga T, Ema T (2004) Tetrahedron : Asymmetry 15:1929
2. (a) Ohta H (1999) Adv Biochem Eng Biotechnol 63:1; (b) Ijima Y, Matoishi K, Terao Y, Doi N, Yanagawa H, Ohta H (2005) Chem Commun 877; (c) Terao Y, Ijima Y, Miyamoto K,

Ohta H (2007) J Mol Catal B:Enzym 45:15; (d) Matoishi K, Ueda M, Miyamoto K, Ohta H (2004) J Mol Catal B:Enzym 27:161; (e) Miyamoto K, Hirokawa S, Ohta H (2007) J Mol Catal B:Enzym 46:14

3. (a) Fehr C (1996) Angew Chem Int Ed 35:2566; (b) Yanagisawa A, Ishihara K, Yamamoto H (1997) Synlett 411; (c) Eames J, Weerasooriya N (2001) Tetrahedron: Asymmetry 12:1; (c) Duhamel L , Duhamel P, Plaquevent JC (2004) Tetrahedron: Asymmetry 15:3653; (d) Berkessel A, Gröger H (2005) Protonation of enolates and tautomerization of enols. In: Asymmetric organocatalysis – from biomimetic concepts to applications in asymmetric synthesis, Wiley-VCH, Weinheim; (e) Blanchet J, Baudoux J, Amere M, Lasne MC, Rouden J (2008) Eur J Org Chem 5493; (f) Mohr JT, Hong AY, Stoltz BM (2009) Nature Chemistry 1:359
4. (a) Marckwald W (1904) Ber Dtsch Chem Ges 37:349; (b) Marckwald W (1904) Ber Dtsch Chem Ges 37:1368
5. Pracejus H (1960) Justus Liebigs Ann Chem 634:9
6. (a) Kenyon J, Ross WA (1951) J Chem Soc 3407; (b) Kenyon J, Ross WA (1951) J Chem Soc 2307
7. Verbit L, Halbert TR, Patterson RB (1975) J Org Chem 40:1649
8. Eskenazi C, Nicoud JF, Kagan HB (1979) J Org Chem 44:995
9. Toussaint O, Capdevielle P, Maumy M (1987) Tetrahedron Lett 28:539
10. Brunner H, Kurzwart M (1992) Monatsh Chem 123:121
11. Brunner H, Mueller J, Spitzer J (1996) Monatsh Chem 127:845
12. Henin F, Muzart J, Nedjma M, Rau H (1997) Monatsh Chem 128:1181
13. Ryu SU, Kim YG (1998) J Ind Eng Chem (Seoul) 4:50
14. Brunner H, Schmidt P (2000) Eur J Org Chem 11:2119
15. (a) Brunner H, Schmidt P, Prommesberger M (2000) Tetrahedron: Asymmetry 11:1501; (b) Drees M, Kleiber L, Weimer M, Strassner T (2002) Eur J Org Chem 14:2405
16. Brunner H, Baur MA (2003) Eur J Org Chem 2854
17. (a) Rogers LMA, Rouden J, Lecomte L, Lasne MC (2003) Tetrahedron Lett 44:3047; (b) Seitz T, Baudoux J, Bekolo H, Cahard D, Plaquevent JC, Lasne MC, Rouden J (2006) Tetrahedron 62:6155; (c) Amere M, Lasne MC, Rouden J (2007) Org Lett 9:2621
18. (a) Pracejus H (1960) Justus Liebigs Ann Chem 634:9; (b) Pracejus H, Maetje H (1964) J Prakt Chem 24:195; (c) Pracejus H, Kohl G (1969) Justus Liebigs Ann Chem 722:1
19. Blake AJ, Friend CL, Outram RJ, Simpkins NS, Whitehead AJ (2001) Tetrahedron Lett 42:2877
20. (a) Fu GC (2004) Acc Chem Res 37:542; (b) Wurz RP (2007) Chem Rev 107:5570
21. (a) Hodous BL, Ruble JC, Fu GC (1999) J Am Chem Soc 121:2637; (b) Hodous BL, Fu GC (2002) J Am Chem Soc 124:10006; (c) Sheryl LW, Fu GC (2005) J Am Chem Soc 127:6176; (d) Schaefer C, Fu GC (2005) Angew Chem Int Ed 44:4606; (e) Dai X, Nakai T, Romero JAC, Fu GC (2007) Angew Chem Int Ed 46:4367
22. Enders D, Niemeier O, Henseler A (2007) Chem Rev 107:5606
23. Wang XN, Lv H, Huang XL, Ye S (2009) Org Biomol Chem 7:346
24. Pracejus H, Wilcke FW, Hanemann K (1977) J Prakt Chem 319:219
25. Kumar A, Salunkhe, RV, Rane RA, Dike, SY (1991) J Chem Soc Chem Commun 485
26. Li BJ, Jiang L, Liu M, Chen YC, Ding LS, Wu Y (2005) Synlett 603
27. Leow D, Tan CH (2009) Chem Asian J 4:488
28. Leow D, Lin S, Chittimalla SK, Fu X, Tan CH (2008) Angew Chem Int Ed 47:5641
29. Henze R, Duhamel L, Lasne MC (1997) Tetrahedron: Asymmetry 8:3363
30. (a) Henin F, Mortezaei R, Muzart J, Pete JP (1985) Tetrahedron Lett 26:4945; (b) Mortezaei R, Henin F, Muzart J, Pete JP (1985) Tetrahedron Lett 26:6079; (c) Pete JP, Henin F, Mortezaei R, Muzart J, Piva O (1986) Pure Appl Chem 58:1257; (d) Mortezaei R, Piva O, Henin F, Muzart J, Pete JP (1986) Tetrahedron Lett 27:2997; (e) Piva O, Henin F, Muzart J, Pete JP (1986) Tetrahedron Lett 27:3001; (f) Piva O, Henin F, Muzart J, Pete JP (1987) Tetrahedron Lett 28:4825; (g) Piva O, Mortezaei R, Henin F, Muzart J, Pete JP (1990) J Am Chem Soc 112:9263; (h) Piva O, Pete JP (1990) Tetrahedron Lett 31:5157
31. Henin F, Létinois S, Muzart J (2000) Tetrahedron: Asymmetry 11:2037

32. (a) Henin F, Muzart J, Pete JP, M'Boungou-M'Passi A, Rau H (1991) Angew Chem Ed Int. 30:1416; (b) Henin F, M'Boungou-M'Passi A, Muzart J, Pete JP (1994) Tetrahedron 50:2849
33. Poisson T, Dalla V, Marsais F, Dupas G, Oudeyer S, Levacher V (2007) Angew Chem Int Ed 46:7090
34. Poisson T, Oudeyer S, Dalla V, Marsais F, Levacher V (2008) Synlett 2447
35. Cheon CH, Yamamoto H (2008) J Am Chem Soc 130:9246
36. (a) Rueping M, Ieawsuwan W, Antonchick, AP, Nachtsheim , BJ (2007) Angew Chem Int Ed 46:2097; (b) Rueping M, Ieawsuwan W (2009) Adv Synth Catal 351:78
37. Reynolds NT, Rovis T (2005) J Am Chem Soc 127:16406
38. Nair V, Vellalath S, Babu BP (2008) Chem Soc Rev 37:2691
39. (a) Maki, BE, Chan A, Scheidt KA (2008) Synthesis 1306; (b) Maki BE, Patterson EV, Cramer CJ, Scheidt KA (2009) Org Lett 11:3942
40. (a) Soloshonok VA, Kirilenko AG, Galushko SV, Kukhar VP (1994) Tetrahedron Lett 35:5063; (b) Soloshonok VA, Kukhar VP (1996) Tetrahedron 52:6953; (c) Michaut V, Metz F, Paris JM, Plaquevent JC (2007) J Fluorine Chem 128:500

Chapter 4
Enantioselective α-Heterofunctionalization of Carbonyl Compounds

Diego J. Ramón and Gabriela Guillena

Abstract Enantioselective organocatalytic processes have reached maturity over recent years with an impressive number of new applications each day. The application of these advantageous methodologies to the construction of chiral α-hetereofunctionalized carbonyl compounds gave important chiral building blocks such as α-amino acids, α-amino alcohols, aziridines, epoxides, 1,2-diols, and α-sulfenylated, selenenylated and halogenated carbonyl derivatives. Proline, imidazolidinone derivatives, cinchona alkaloids and their ammonium salts, as well as Brønsted acid derivatives have been used as chiral catalysts for these purposes. A survey of contributions in this field will be discussed throughout this chapter.

4.1 Introduction

Inspired by Nature, where enantioselective reactions are efficiently performed by enzymes [1], chemists have developed new strategies to enantioselective synthesize of chiral compounds. While the last century has been dominated by the use of transition metal catalysts to achieve this goal, the use of organocatalyst is reaching its *golden age* in this twenty-first century [2].

Only in the last decades, the use of organic molecules as catalysts has emerged as an important area of research with spectacular numbers of contributions in every journal issue. The aim of this chapter is to cover some important achievements in the enantioselective α-heterofunctionalization of carbonyl compounds using the organocatalyst strategy, in which a new carbon-heteroatom bond is

D.J. Ramón (✉) • G. Guillena
Instituto de Síntesis Orgánica (ISO) and Departamento de Química Orgánica,
Facultad de Ciencias, Universidad de Alicante, Apdo. 99, E-03080 Alicante, Spain
e-mail: djramon@ua.es; gabriela.guillena@ua.es

R. Mahrwald (ed.), *Enantioselective Organocatalyzed Reactions I: Enantioselective Oxidation, Reduction, Functionalization and Desymmetrization*,
DOI 10.1007/978-90-481-3865-4_4,

formed [3]. The employment of this methodology leads to a wide range of very important chiral building blocks, including α-amino acids, α-amino alcohols, epoxides, 1,2-diols, α-sulfenylated, α-selenenylated and α-halogenated carbonyl derivatives. Many of them are very difficult to obtain by other strategies since, for some purposes, the presence of hazardous metallic traces are inadmissible in the final product, as it is for agrochemical and pharmaceutical companies. The organocatalysis represents a very simple, easy and straightforward method for the synthesis of these compounds.

4.2 Enantioselective α-Amination of Carbonyl Compounds

4.2.1 Aziridination of α,β-Unsaturated Carbonyl Compounds

The first reported aziridination reaction was performed using cinchona salt derivatives under phase transfer catalysis conditions [4]. The reaction of different hydroxamic acids **1** as nitrogen source with acrylate derivatives **2** in the presence of cinchoninium salts **3** gave the expected aziridines **4** (Scheme 4.1). The ideal biphasic mixture was obtained after different trials, finding that more polar solvents than toluene gave worse results due to the possible disruption of the coulombic interaction between the ammonium cation and the enolate. Other bases gave either lower chemical yield or enantioselectivity. Somewhat more interesting was the study of the influence of the structural modifications of reagents. As electrophilic partner of the reaction was chosen the corresponding *tert*-butyl acrylate since, although the enantioselectivity was slightly lower, the chemical yield was sensible higher than using the related methyl acrylate. The best results were obtained using pivaloyl derivative as nucleophilic nitrogen source, since the use of the related benzoyl compound or the bromo substituted reagent dropped the enantioselectivity. Among nine different cinchoninium quaternary salts tested, catalyst **3b** gave the best results. Catalysts with less electron withdrawing group such as **3a** gave worse results. Even, the counterion atom has a great impact on the result, bromide salts giving better result than the related chlorides. Surprisingly, the use of the so called pseudoenantiomeric catalyst cinchonidinium salt **5a** did not change the absolute configuration of the final aziridine **4** (50% yield and 28% ee).

Better enantioselectivities were obtained with modified cinchoninium salt **6a** and its pseudoenantiomer cinchonidinium salt **6b**, in which the hydroxy group is derivatized as tosyl ester. In this case, both catalysts acted, as was expected, leading to both different aziridine enantiomers with similar enantioselectivity (up to 95% ee), with the enantiomeric excess being highly influenced by the presence of substituents on the aromatic ring of the hydroxamic acid **1** [5].

This methodology has been extended to the aziridination of different cyclic α,β-unsaturated ketones, in which the phase transfer catalyst **3b** seemed to be the best catalyst. It should be pointed out that the best conditions for this process were the biphasic solid/liquid phase and the source of nucleophilic nitrogen nosyloxycarbamate [6].

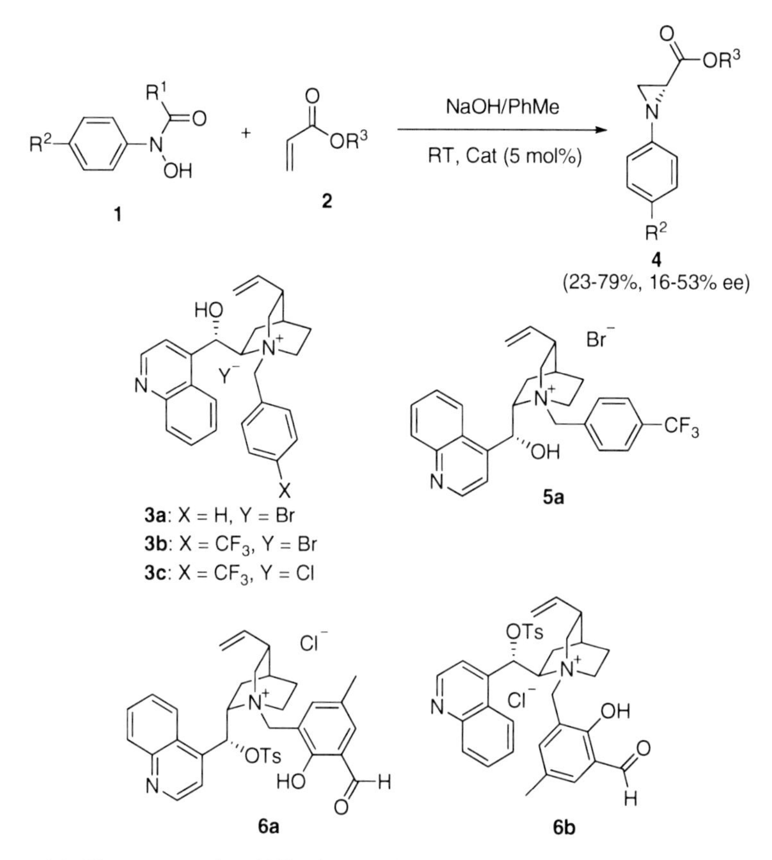

Scheme 4.1 First asymmetric aziridination reactions

Chalcones **7** could be aziridinated using a one-pot process, which involves the in situ generation of a hydrazinium salt derived from *O*-mesitylenesulfonylhydroxylamine **8**, deprotonation of a hydrazinium salt to form an aminimide, and subsequent aziridination. Thus, using the chiral Tröger's base as catalyst (**9**), $CsOH{\cdot}H_2O$ (3 equiv) at 0 °C, aziridine **10a** was obtained in moderate enantioselectivity (Scheme 4.2). A slightly better enantioselectivity was obtained by using 4-chlorochalcone (**7b**) as starting material affording the corresponding aziridine **9b**. This ee value can be improved up to 98% upon recrystallization [7]. Similar results (64% yield and 56% ee for **10a**) were obtained in the aziridination of chalcone **7a**, using 105 mol% of quinine (**11**) as catalyst, *O*-(diphenylphosphinyl)hydroxylamine (1.05 equiv) as a nitrogen source and NaH (2 equiv) as base in a mixture of isopropanol-dichloromethane as solvent at room temperature [8].

Different carbamates have been also used as nitrogen source to perform this type of transformation under different reaction conditions giving increased level of enantioselectivity. So, dimethylpyrazole acrylates could be aziridinated using *N*-chloro-*N*-sodium benzylcarbamate under solid-liquid phase transfer conditions using 10 mol% of cinchoninium salt of type **3** in dichloromethane, affording the corresponding aziridine in 84% yield and 70% ee [9]. As it was expected, using the pseudo-enantiomeric cinchonidinium salt, the aziridine with the opposite absolute configuration was obtained.

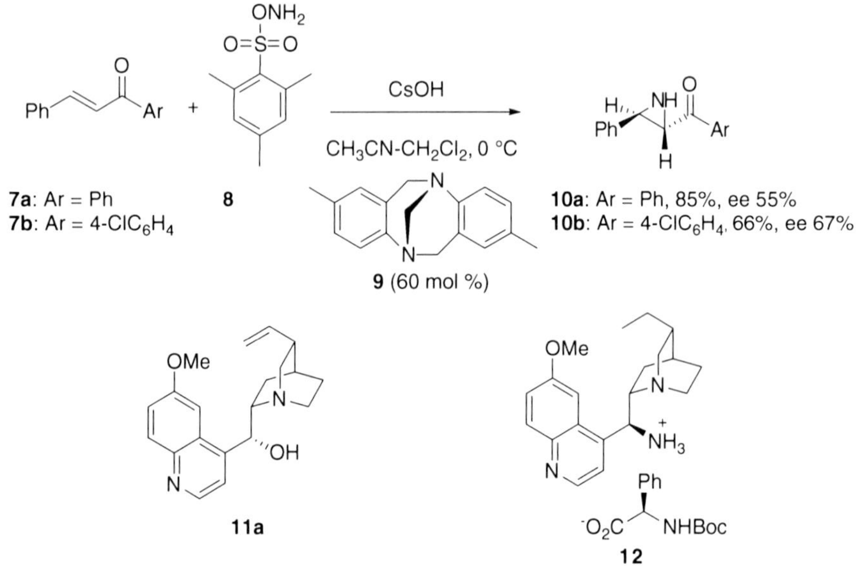

Scheme 4.2 Aziridination of aromatic α,β-unsaturated ketones

Salt **12** (20 mol%) was able to promote the aziridination of aliphatic and aromatic α,β-unsaturated ketones, using tosylated carbamate derivatives and Na_2CO_3 as base at room temperature, yielding the corresponding aziridines in high yields, diastereo- and enantioselectivities [10]. The reaction seems to be tolerant towards steric and electronic demands on the starting olefin systems. These reaction conditions were also applied to cyclohexenone and β-methyl cyclohexenone, leading to the corresponding aziridine with 98% and 73% ee, respectively.

Prolinol derived organocatalyst have been successfully used in the asymmetric aziridination of α,β-unsaturated aldehydes. Thus, compound **15a** catalyzed the formation of aliphatic 2-formylaziridines **16** in moderated to high yields and in high levels of diastereo- and enantioselectivities (Scheme 4.3) [11]. Products **16** decomposed during prolonged reaction times, therefore temperatures of 40°C were used to perform faster the reaction. The achieved absolute configuration for 2-formylaziridines **16** were attributed to the efficient shielding of the *Si* face of the chiral iminium intermediate by the aryl groups of the catalyst, leading to the stereoselective *Re*-facial nucleophilic conjugate attack of the β carbon atom of the electrophile by the nitrogen source **14**. The formed chiral enamine intermediate gave a 3-*exo*-tet nucleophilic attack on the now electrophilic nitrogen atom, releasing acetic acid. Catalyst **15b** was able to promote the aziridination of aliphatic and aromatic α,β-unsaturated ketones with similar results to catalyst **15a** [12]. However in this case, the presence of a base was mandatory to obtain the corresponding 2-formylaziridines **16**. This catalyst was also used for the aziridination of aromatic α,β-unsaturated aldehydes. *N*-4-methoxybenzenesulfonyloxy-carbamate, instead of *N*-4-methylbenzenesulfonyloxycarbamate, could be used as an alternative nitrogen source to give the corresponding aziridine **16** in moderated yields, which was sensitive to silica gel purification. Other aromatic aldehydes bearing electron-withdrawing groups were also suitable substrates to yield products **16** in high yield and enantioselectivity. Further oxidation transformations of products **16** allowed the synthesis of chiral β-amino esters derivatives.

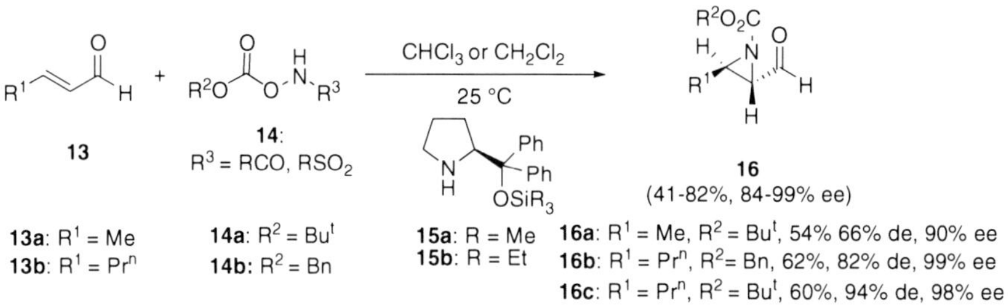

$CHCl_3$ or CH_2Cl_2, 25 °C

13; **14**: R^3 = RCO, RSO_2; **16** (41-82%, 84-99% ee)

13a: R^1 = Me
13b: R^1 = Pr^n

14a: R^2 = Bu^t
14b: R^2 = Bn

15a: R = Me
15b: R = Et

16a: R^1 = Me, R^2 = Bu^t, 54% 66% de, 90% ee
16b: R^1 = Pr^n, R^2 = Bn, 62%, 82% de, 99% ee
16c: R^1 = Pr^n, R^2 = Bu^t, 60%, 94% de, 98% ee

Scheme 4.3 Aziridination of aliphatic α,β-unsaturated aldehydes

4.2.2 *Direct α-Amination of Carbonyl Compounds*

Almost simultaneously, two groups reported for the first time the direct organocatalyzed α-amination of aldehydes using (*S*)-proline (**20**) and different azodicarboxylate derivatives as the electrophilic nitrogen source (Scheme 4.4). In the work of Jørgensen's group [13], the reaction was carried out in methylene chloride at room temperature using diethyl azodicarboxylate (DEAD, **18a**, R^2=Et) as electrophile,

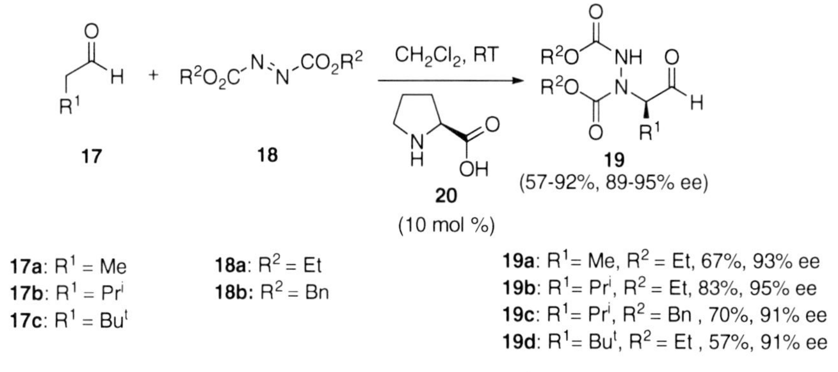

CH_2Cl_2, RT

17; **18**; **20** (10 mol %); **19** (57-92%, 89-95% ee)

17a: R^1 = Me
17b: R^1 = Pr^i
17c: R^1 = Bu^t

18a: R^2 = Et
18b: R^2 = Bn

19a: R^1= Me, R^2 = Et, 67%, 93% ee
19b: R^1= Pr^i, R^2 = Et, 83%, 95% ee
19c: R^1= Pr^i, R^2 = Bn , 70%, 91% ee
19d: R^1= Bu^t, R^2 = Et , 57%, 91% ee

Scheme 4.4 Direct organocatalyzed α-amination of aldehydes

achieving up to a 93% chemical yield and 95% ee. In this case, the obtained product could be easily isolated just by the addition of water and simple extraction with diethyl ether. Furthermore, the yield and enantioselectivity did not change performing the reaction on gram scale. All these facts make this procedure very convenient for an industrial large scale synthesis. Due to the relative high acidity of the hydrogen atom placed in the stereogenic center of product **19** and to avoid its racemization, an in situ reduction of aldehyde **19** can be done using $NaBH_4$, with the final product after basic treatment being the corresponding oxazolidinones. In a similar way, the aldehydes **17** could be transformed into the corresponding chiral α-amino acids by a multiple step procedure involving the oxidation of aldehyde and the final N-N bond cleavage using nickel-Raney.

The same reaction was reported later [14]. In this case, the reaction was performed in acetonitrile as solvent at 0°C and using dibenzyl and di-*tert*-butyl azodicarboxylate (**18**, R^2=Bn or Bu^t) as electrophiles. Under these conditions, the end point of the reaction could be easily detected by the disappearance of the typical yellow color of the azodicarboxylate and after in situ reduction; the corresponding amino alcohols could be isolated in practically quantitative yields with very high enantioselectivities (92–97% ee). Due to the crystalline character of these amino alcohols, their enantiomeric excess could be further improved by simple recrystallization.

The possible mechanism of the direct amination of aldehydes was studied by means of kinetic measurements indicating that the whole processes seems to go through two parallel catalytic cycles [15]. These investigations revealed that the reaction exhibited an autoinductive effect. The reaction product interacts with the original catalyst to form a more active and efficient catalyst. Thus, when an amino aldehyde of type **19** was mixed with (*S*)-proline (**20**) in a 1:1 molar ratio before the introduction of the reactants, the reaction was notably faster than with the original protocol (only using catalyst **20**). Surprisingly, this rate enhancement was independent of the absolute configuration of the amino aldehyde (either **19** or its enantiomer), with the outcome of the reaction being only determined by proline. On the basis of computational calculations, the hydrogen-bonded product-proline complex resembles an open proline conformation. The approximation of the diazocarboxylic ester **18** was governed by the formation of a new hydrogen bond, which is responsible of the outcome of the reaction. The observed nonlinear effect is consistent with a kinetic resolution of proline in this autoinductive process.

Not only proline but also other catalysts have been used in the direct α-amination of aldehydes, many of them being designed with the aim of increasing the solubility and, therefore, the turnover number of the catalyst. Thus, 2-(arylsulfonyl-aminomethyl) pyrrolidines **21** fulfill all the requirements to act as organocatalysts since they possess a secondary amine, which allows the formation of an enamine by reaction with the corresponding aldehyde, and the presence of a sulfonyl group increases the acidity of hydrogen of the amide, facilitating the hypothetically required hydrogen bonding between the enamine and the diazocarboxylate derivative in the transition state. Independently of the aryl substituent, all catalyst of type **21** gave similar results (enantiomeric excess never higher than 87%), with these enantioselectivities being clearly inferior to those obtained using proline [16].

Catalyst **22** was introduced to improve the poor enantioselectivity obtained in reactions between aldehydes and electrophiles with low hydrogen-bond acceptor character. This catalyst would provide a high enantioselectivity by controlling the enamine geometry and through a very efficient face biasing [17]. When this catalyst was used in the α-amination of aldehydes **17** with dialkyl azodicarboxylates **18**, the corresponding oxazolidinones obtained after a reduction step showed a very high enantiomeric excess (90–97% ee). This catalyst was more efficient and its reaction faster than those using proline. Remarkably, the absolute configuration of final products was the opposite to the obtained using proline (**20**).

The generation of molecules containing quaternary stereocenters is even a more difficult and challenging task in organic chemistry. The enantioselective amination of α,α-disubstituted aldehydes is an useful indirect entry to the synthesis of this type of compounds [18]. Thus, the reaction of α-alkyl-α-aryl disubstituted aldehydes with either DEAD (**18a**, R^2=Et) or DBAD (**18b**, R^2=Bn) catalyzed by either proline (**20**) or (*S*)-azetidine (**23**) carboxylic acid yielded the expected aldehydes with good enantioselectivity and yield, for the case of using catalyst **20** and moderated for azetidine catalyst (**23**). The same reaction but using α,α-dialkyl aldehydes gave, however, moderated enantioselectivities (up to 28% ee). As in the aforementioned cases, aldehydes could be easily reduced to the corresponding alcohols, cyclized to render the related oxazolidinones, as well as oxidized and deprotected to give the expected α,α-disubstituted amino acids. This strategy has been successfully applied to the synthesis of important medicinal amino acids such as (*S*)-AIDA and (*S*)-APICA, which are metabotropic glutamate receptor ligands and therefore are associated to the treatment of several neurodegeneratives diseases. For this purpose, indanone carbaldehyde derivatives were submitted to the direct α-amination process using DBAD (**18b**) and (*S*)-proline (**20**) [19], achieving the expected products as a single enantiomer and excellent chemical yields. The high catalyst loading was further reduced by using microwave conditions [20]. By using constant microwave power (200 W, 60 min, 60°C) in acetonitrile as solvent, an increase of the yields and enantioselectivities were observed (54–97%, 52–90% ee) compared to those obtained at room temperature using α-alkyl aryl aldehydes as substrates. In addition, a considerable reduction on the reaction time (from days to 60 min) was achieved under these new reaction conditions. However, the catalyst loading (50 mol%) could not be further decreased without affecting the yield and enantioselectivity (Fig. 4.1).

Tetrazole proline derivative **24** has shown to be superior to proline as catalyst in the reaction of substituted aldehydes with DBAD (**18b**) to yield the expected α-amino aldehyde with good enantioselectivity (65–88% ee). In the case of using proline, the enantioselectivity was only 44% and the time increased from 3 h to 5 days. The higher reactivity and selectivity of catalyst **23** compared to proline was attributed to the lower pK_a (8 in DMSO, 12 for proline) and its higher steric hindrance [21].

The usefulness of the α-amination of aldehydes organocatalyzed by (*S*)-proline (**20**) has been demonstrated by its application as the key step for the synthesis of several interesting chiral products. For instance, the enantioselective synthesis of the

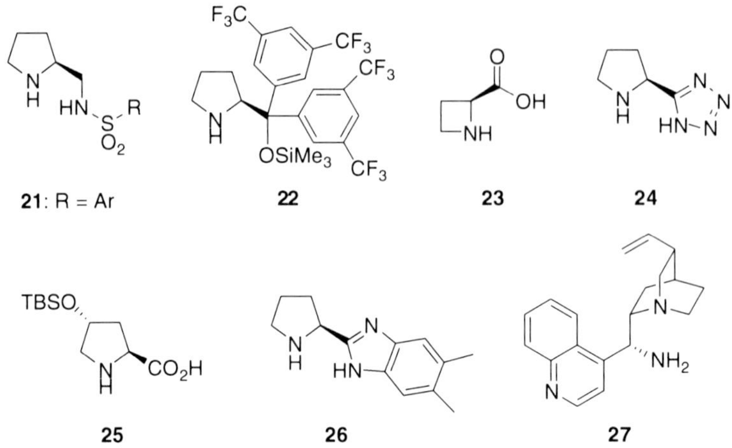

Fig. 4.1 Catalysts for direct organocatalyzed α-amination of aldehydes

tuberculostatic antibiotic (*S*,*S*)-ethambutol [22], an effective drug for the treatment of Parkinson's and Alzheimer's diseases; (*R*)-selegiline [23]; proteogenic and non-proteogenic amino acids such as (*R*)-pipecolic acid and (*R*)-proline [24]; piperazic acid derivative [25], and the antibiotic (−)-anisomycin [26] was also accomplished.

The sequential use of the organocatalytic α-amination of aldehydes by azodicarboxylates with another process has provided a useful tool for the synthesis of complex molecules. Thus, the combination of the amination of linear aldehydes catalyzed by (*S*)-proline (**20**) and the Passerini reaction allowed the rapid access to norstatine based peptidomimetics albeit with low diastereoselectivities (36–98% yield, up to 4:1 dr) [27]. Moreover, the enantioselective synthesis of γ-amino-α,β-unsaturated esters can be performed via the sequential α-amination-Horner-Wadsworth-Emmons olefination of aldehydes catalyzed by (*S*)-proline (**20**,10 mol%) affording the expected products in high yields and enantioselectivities (85–90% yield, ee 92–99%) [28].

Following a similar α-amination-intramolecular Wittig olefination sequential process, chiral 3,6-dihydropyridazines were obtained in high yields and enantioselectivities using pyrrolidinyl tetrazole (**24**, 10 mol%) as catalyst [29]. While bulkier substituents at the α-position of aldehydes provided higher yields and enantioselectivities, unbranched aldehydes rendered the expected products with lower results, probably due to the facile homo-aldol coupling displayed by these substrates. With respect to the nitrogen source, di-*tert*-butyl azodicarboxylate gave improved selectivities compared to others.

The use of a sequential α-amination-allylation one-pot sequence of aldehydes followed a ring-closing metathesis strategy has permitted the synthesis of chiral cyclic hydrazines [30]. Thus, the α-amination of several unbranched aldehydes using (*S*)-proline as catalyst (10 mol%) and dibenzylic azodicarboxylate (**18b**) in acetonitrile at −20°C for 12 h, followed by the addition of indium powder and allyl bromide at room temperature gave the corresponding 1,2-allylic amino alcohols in high yields (67–88%) and diastereoselectivities (62–98% de) and excellent enantioselectivities (98–99% ee). These compounds were further transformed to the corresponding RCM substrates, by reaction with Grubb's catalyst giving the corresponding chiral 8-membered cyclic hydrazines in moderated to high yield (59–92%).

This strategy was further used with ketones [31] and (*S*)-proline (**20**) as catalyst. The best results were obtained using DEAD (**18a**) as the electrophilic source of nitrogen and, although for some ketones the enantioselectivity was higher in other solvents different from acetonitrile, it should be pointed out that, in general, this was the solvent of choice (84–99% ee). All reactions were very regioselective, the major isomer was that in which the amination occurred on the most substituted α-carbon atom to the carbonyl of ketone, with the larger substitutent giving the higher enantioselectivity. Azetidine carboxylic acid (**23**) has also been used for this reaction [32]. However, the enantioselectivity found was in general somewhat lower than using proline (**20**). In constrast to the previous case, the increase of the length of substituents had a detrimental effect, not only in the enantioselectivity but also in the reaction rate.

4-Silyloxyproline **25** (10 mol%) was able to catalyzed the reaction of several cyclic and acyclic ketones with dibenzylic azodicarboxylate (**18b**). In order to achieve high yields and enantioselectivities, nine equivalents of water were added to the reaction mixture, although this addition retards the reaction [33]. When these conditions were applied to the reaction with cyclopentanone, the major product was the α,α´-diaminated cyclopentanone in 46% yield and 99% ee. In this process, contrary to that observed for the case of aldehydes, a linear correlation between the enantiomeric excess of the product versus the catalyst was found. Quantum mechanical calculations suggested that the most stable transition state was that in which the *anti*-configured enamine reacts with the azodicarboxylate in the *trans* configuration.

Benzimidazole-pyrrolidine **26** in combination with trifluoroacetic acetic (10–20 mol%) was also applied as catalysts for this transformation in dichloromethane a room temperature affording the aminated ketones in low yields and enantioselectivities (65–92%, 66–71% ee) [34].

Aryl ketones can be aminated also using as catalyst the amine **27** derived from cinchona alkaloids (20 mol%) [35]. For this reaction, the addition of *p*-toluenesulfonic acid (40 mol%), and 4 Å molecular sieves was compulsory in order to achieve good enantioselectivities. Under similar conditions, aryl ketones, such as propiophenones, butyrophenones, heteroaryl ketones or even cyclic 1-tetralone, and diethyl azodicarboxylate (**18a**) in isopropanol as solvent, yielded at 40°C the aminated product in good results (39–77%, 88–97% ee).

The direct organocatalyzed α-amination reaction has been also applied to 1,3-dicarbonyl compounds and related compounds. So, the reaction of aryl cyanoacetates (**28**, R^1 = aryl and heteroaryl) with diazo compounds **18** catalyzed by tertiary amine **30** (Scheme 4.5) gave the expected products **29**. The enantioselectivity of the reaction seems to be dependent on the bulkiness of the diazo compound used. The more bulkier compound **18**, the higher enantioselectivity was obtained. This sentence was also true for the ester moiety in **28**. The presence of substituents at the aryl group at either *meta*- or *ortho*-position did not have any accountable influence on results. Only a slight decrease in the enantioselectivity was accomplished with

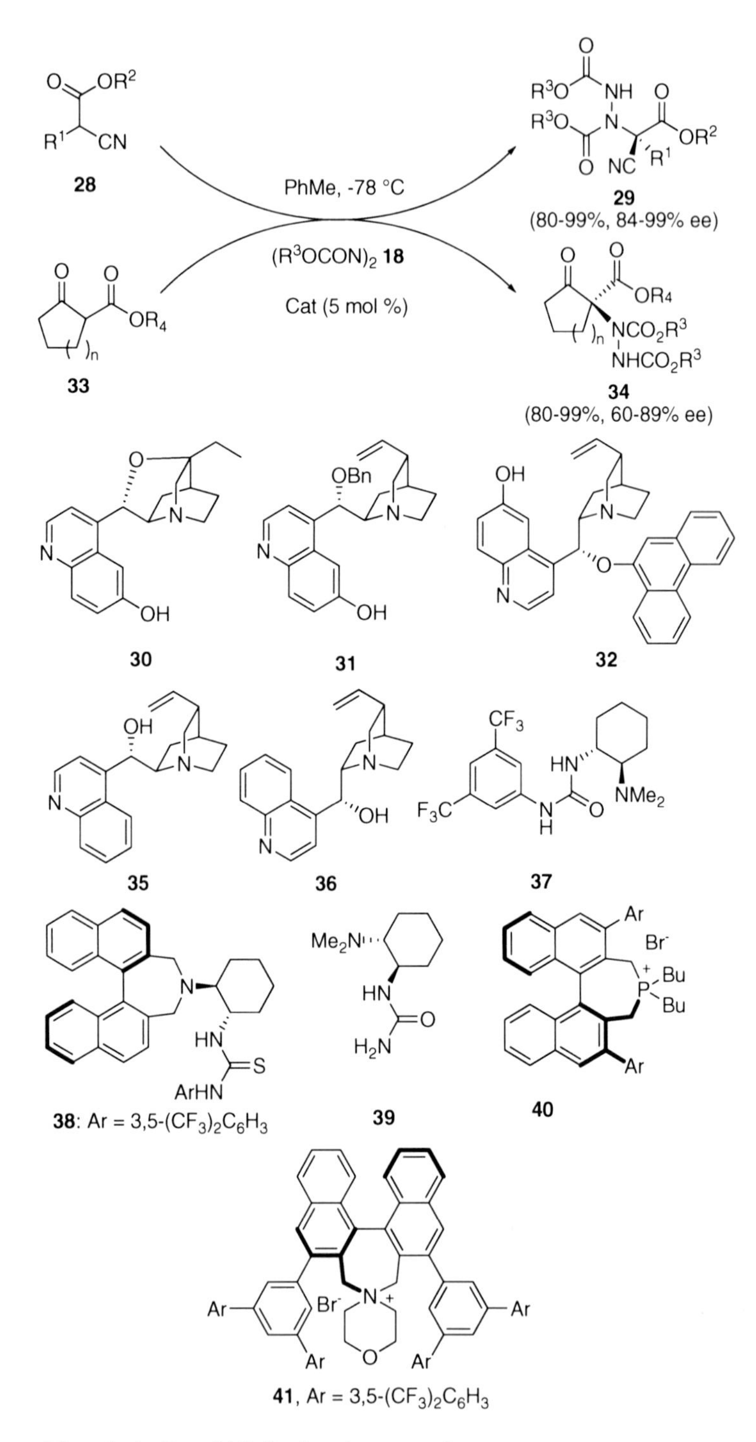

Scheme 4.5 α-Amination of 1,3-dicarbonyl compounds

substrates bearing electron withdrawing or donating groups at *para*-position. The cleavage of N-N bond of the final products **29** was easily performed using trifluoroacetic acid and SmI_2, being an indirect entry to the synthesis of α-amino acids bearing quaternary stereocenters [36]. Catalysts **31** and **32** have been introduced as alternative organocatalysts for the aforementioned amination, achieving similar results [37]. These catalysts showed also a broad substrate scope, being successfully applied to *para*-, *meta*- and *ortho*-substituted aromatic α-cyanoacetate derivatives **28**. However, all attempts to extend the reaction to alkyl α-cyanoacetate (**28**, R^1 = alkyl) failed, giving very low enantioselectivities.

The substrate scope of catalyst **30** has been also tested with different β-ketoesters (**33**, Scheme 4.5), obtaining good enantioselectivities not only for cyclic compounds but also for the related acyclic ones [36]. Cinchonine (**35**) and cinchonidine (**36**) has been also used for the same reaction rendering the expected functionalized compounds **34** with somewhat lower enantioselectivities but using a higher catalyst loading and temperature. The behavior of these two catalysts is like pseudoenantiomers, since the absolute configuration of the final product in one is the opposite of the other. It should be pointed out that the enantioselectivity dropped considerably when acylic substrates where submitted to the amination process [38].

The urea derivative of type **37** has been used, being clearly less efficient than previous *cinchona* derived catalysts **31** to **36** [39]. Better results (85–95%, 93–99% ee) were obtained by using bifunctional thiourea **38** (10 mol%) for the amination of cyclic *tert*-butyl β-ketoesters, in toluene at −30°C [40]. DFT studies have been performed for the related bifunctional urea **39** in the reaction between 2-acetylcyclopentanone and dimethyl azodicarboxylate [41]. Based on the obtained results, a mechanism involving initial nucleophile activation via protonation of the amino group and electrophile activation through substrate binding to the urea moiety and then C-N bond formation between these two activated components was proposed. The catalytic cycle finished with the proton transfer from the protonated amino group to the adduct, followed by its dissociation to give the final product. Four reaction channels corresponding to the different coordinate modes of two isomers of the enolate of the β-ketoester and the diazocarboxylate to the catalyst were characterized in detail, founding that the enantioselectivity of the reaction was controlled by the C-N bond-forming step, which was the rate-determining step.

Binaphthyl phosphonium chiral salt **40** (3 mol%) has been used also used under phase-transfer conditions at −20°C [42]. The reaction was relatively insensitive to the electronic effects of the substituents on the aromatic ring (77–91% ee), but rendered worse yields when six membered ring cyclic β-diketone was used as substrate (87% ee). Better results in terms of yields and enantioselectivities for compounds **34** (99%, 76–97% ee) were obtained by using spiro-type chiral ammonium salt **41** (1 mol%) at −40ºC under similar phase-transfer reaction conditions [43].

Another strategy to perform an electrophilic amination of a carbonyl compound is through the reaction of an enamine or enolate with a nitroso derivative. As the nitroso compound contains two nucleophilic sites two possible processes could take place. The first one, generally promoted by the presence of carboxylic groups in the catalysts, is the C-O bond forming reaction (aminoxylation or *O*-nitroso aldol reaction) which will be discussed in Sect. 4.3.3. On the other hand, the reaction through the nitrogen atom of the nitroso compound is the so-called hydroxyamination, oxyamination or *N*-nitroso aldol reaction and it has been less developed by organocatalytic methods. The first example was obtained when α-methyl aldehdyes **42** were reacted with amide **45** (Scheme 4.6). The reason for this unusual behavior was attributed to the low acidity of the proton of prolinamide, which is not able to protonate the nitrogen atom but does the oxygen atom of the nitroso derivative **43**, therefore making more electrophilic the nitrogen atom yielding the corresponding alcohols **44**, after final reduction [44].

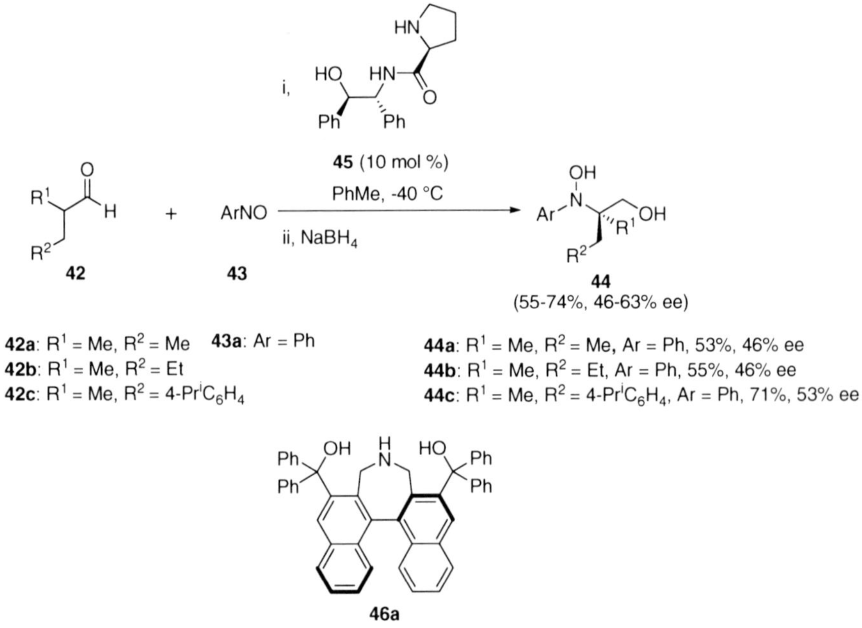

i, **45** (10 mol %), PhMe, -40 °C; ii, $NaBH_4$

42 + ArNO **43** → **44** (55-74%, 46-63% ee)

42a: R^1 = Me, R^2 = Me
42b: R^1 = Me, R^2 = Et
42c: R^1 = Me, R^2 = 4-$Pr^iC_6H_4$

43a: Ar = Ph

44a: R^1 = Me, R^2 = Me, Ar = Ph, 53%, 46% ee
44b: R^1 = Me, R^2 = Et, Ar = Ph, 55%, 46% ee
44c: R^1 = Me, R^2 = 4-$Pr^iC_6H_4$, Ar = Ph, 71%, 53% ee

46a

Scheme 4.6 *N*-Nitroso aldol reaction

Also, *N*-hydroxy-β-amino alcohols **44** (R^1 = H, Scheme 4.6) were achieved in high yields and enantioselectivities using axially chiral secondary amine catalyst **46a** (10 mol%) to perform the hydroxyamination of linear aldehydes **42** with nitrosobenzene (**43a**, Ar = Ph), with excellent results (77–90%, 97–99% ee) [45]. Due to the instability of the resulting hydroxyamination product, in situ reduction to the corresponding alcohol was carried out using $NaBH_4$.

Catalyst **15a** (20 mol%) provided products **44** (R^1 = H) in moderated yields and high enantiomeric excesses (40–75%, 91–99% ee) in the reaction of nitrosobenzene in dichloromethane at 0°C. Room temperatures and longer reaction times were needed for long-chain aldehydes to afford the expected products. DFT calculations showed that the free activation energy $\Delta G^{\neq}$ for the non-catalyzed hydroxyamination was higher than the $\Delta G^{\neq}$ values obtained in the presence of water or

N,*N*-dimethylhydroxyamine, which mimics product **44** or its precursor. In addition, the encountered $\Delta G^{\neq}$ values for the catalyzed aminoxylation process are higher than those of the hydroxyamination. In this case, an enamine intermediate with a steric control approach from the *Si* face was proposed to rationalized the observed (*S*)-preference for the product [46]. However, further theoretical investigations on the mechanism suggested that the reaction proceeds via an enol intermediate that has the lowest energy barrier among the initial reactions between the aldehyde and the catalyst [47]. The proposed transition state corresponded to two simultaneous hydrogen-exchanges between the aldehyde and the catalyst to form a *trans* enol. Immediately, the *trans* enol formed a complex with the catalyst via hydrogen bonding and subsequently, the nitrosobenzene attacks from the *Si* face of the enol via a new transition state. This transition state consisted in a simultaneous C-N bond formation between the enol and the nitrosobenzene; a hydrogen-shift for the oxygen of the enol to the nitrogen of the catalyst and a hydrogen-shift from the nitrogen of the catalyst to the oxygen of the nitrosobenzene, with the last hydrogen shift dictating the stereochemical outcome of the reaction.

Instead of carbonyl compounds **42**, there is another alternative to perform this transformation. This implies the use of achiral enamines, nitrosoarenes and the use as chiral organocatalyst of different alcohols, such as TADDOL derivatives [48].

4.3 Enantioselective α-Chalcogenation of Carbonyl Compounds

4.3.1 Darzens Reaction

The Darzens condensation is an old methodology for the construction of α,β-epoxy carbonyl compounds **49** with control of two formed stereogenic centers. This reaction includes an aldol reaction (C-C bond formation), which normally requires stoichiometric amounts of base to achieve good yields. Only phase transfer catalysts, usually chiral ammonium halides have shown to be efficient to perform this transformation in a catalytic and enantioselective form.

In this way, cinchona ammonium salts have been used as efficient catalysts for the enantioselective Darzens condensation. Although, the first reported catalytic example [49] using quininium salt **5b** and NaOH as base gave a poor enantioselectivity (lower than 8% ee), the use of cinchoninium quaternary salt **3b** improved the enantioselection up to 79% ee when sterically demanding aldehydes [50] were used in combination with chloroacetophenone (**47**: R^1=Ph, X=Cl, Scheme 4.7). However, when the reaction was performed using either less congested or aromatic aldehyde, the reaction proceeded with moderate enantioselectivity. The modification on the benzyl moiety of ammonium salt did not produce improvement on the results, with the protection of hydroxy group as ether giving very low enantiomeric excess [50]. Catalyst **3b** promoted the enantioselective Darzens condensation of α-chloro ketones derived from α-tetralone under similar reaction conditions, yielding the corresponding spiro epoxides with enantiomeric excess up to 86% and as a single diastereoisomer. In this context, a mechanistic study was performed in order to establish whether the observed stereoselectivity was controlled by the aldol condensation between enolate and aldehyde [50]. Findings explained that the stereoselection was determined by a further kinetic resolution of aldol type intermediate.

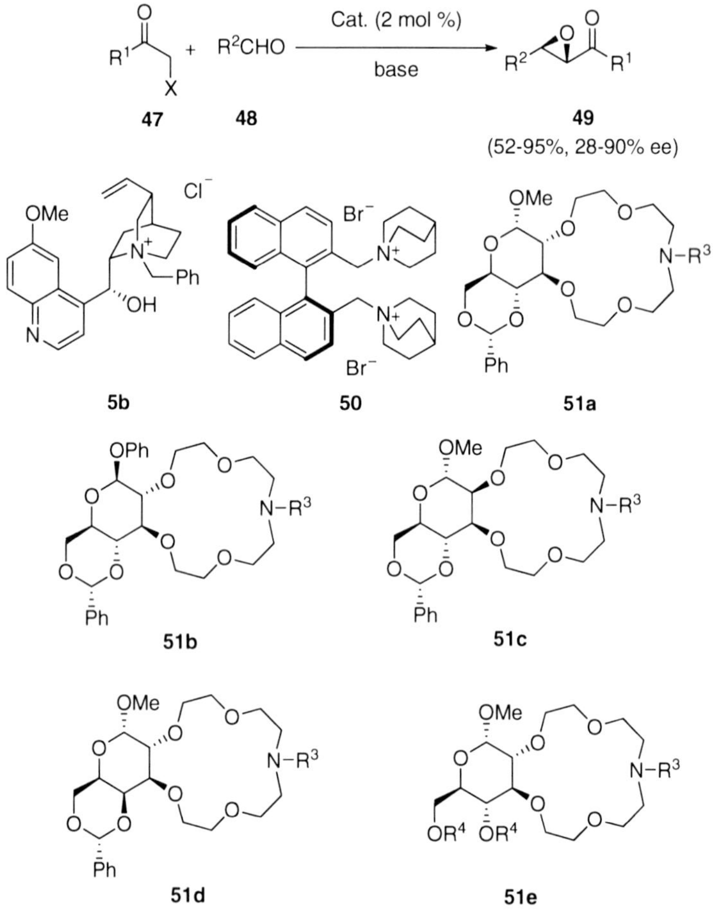

Scheme 4.7 Asymmetric Darzens reaction

Phenylacetamide (**47**: $R^1 = Ph_2N$, X = Br) can be also used as starting material for the aforementioned reaction (Scheme 4.7). However, results were slightly lower compared to the use of chloroacetophenone with diastereoselectivity being lower than 80%. In this case, the chiral ammonium catalyst used was the binaphthyl derivative **50**. Best result was obtained with RbOH as base, giving homogeneous enantioselectivities for compounds **49** when substituted benzaldehydes were used as electrophilic partner of the reaction [51].

Besides the use of chiral ammonium salts as phase transfer catalyst, different azacrown ethers **51a–e** derived from different sugars has been proposed as an alternative [52]. For all systems the influence of the length of the R^3 chain on nitrogen atom was evaluated. For compounds **51a**, **b** (glucose derivatives) the best results were obtained when R^3 was 2-hydroxyethyl. Whereas for compounds **51c–e**, the best results were obtained for the corresponding 3-hydroxypropyl derivative. In neither case the enantiomeric excess was higher than 74%.

4.3.2 Epoxidation of α,β-Unsaturated Carbonyl Compounds

Although methods for the enantioselective epoxidation of alkenes have been admirably performed in the last 30 years, achievements in the epoxidation of electron poor alkenes, such as enones, with high results have been less developed. In this case, a nucleophilic oxygen donor molecule is necessary to carry out this transformation, with the Michael-type addition being the first step of the process. Recently a number of useful combinations of different types of organocatalysts and oxidative reagents have been elaborated. Since a pioneer work [53], the epoxidation of chalcone derivative **7a** using *N*-benzylquinine derivative **5b** as the phase chiral transfer catalyst, a 30% aqueous solution of H_2O_2 as oxidant and NaOH as base in toluene at room temperature (92%, 34% ee), chiral phase transfer agents occupy a unique place for this transformation. An inverse relationship between the dielectric constant of solvent and the enantioselectivity was detected in these early works, with this correlation being attributed to the formation of chiral ionic pair. Under similar conditions but using LiOH as base and the catalyst **3d** (X = I and Y = Cl), results were highly improved (97%, 84% ee). Enantioselectivities for other enones showed to be strongly dependent on the substituents, giving the best results for 1,3-diarylpropenone derivatives [54]. NaClO as oxidant improved highly the enantiomeric excess of the product in the reaction using other related ammonium derivatives [55].

Dimeric cinchona derivatives catalyzed the epoxidation of (*E*)-diarylenones with high levels of enantioselectivities [56]. The length of the space between the two cinchona units and the use of different surfactants influences yields and selectivities of the reaction, with the best achieved results being obtained using surfactant Span 20.

Another type of catalyst successfully used in the enantioselective epoxidation of (*E*)-enones **52** is guanidine (Scheme 4.8). Thus, derivative **54a** was able to perform the epoxidation of chalcone by using NaClO in toluene at 0°C [57]. Pentacyclic guanidine **54b** (10 mol%), which have a characteristic closed-type cavity, gave products in moderated yields (22–99%) and enantioselectivities (35–60% ee) using *tert*-butyl hydroperoxide as oxidant in a liquid-liquid dichloromethane and aqueous potassium hydroxide two-phase system at 0°C. In contrast to that, tricyclic guanidine **54c** (10 mol%) afforded the corresponding epoxides almost quantitatively with the same levels of enantioselectivity in similar reaction conditions [58].

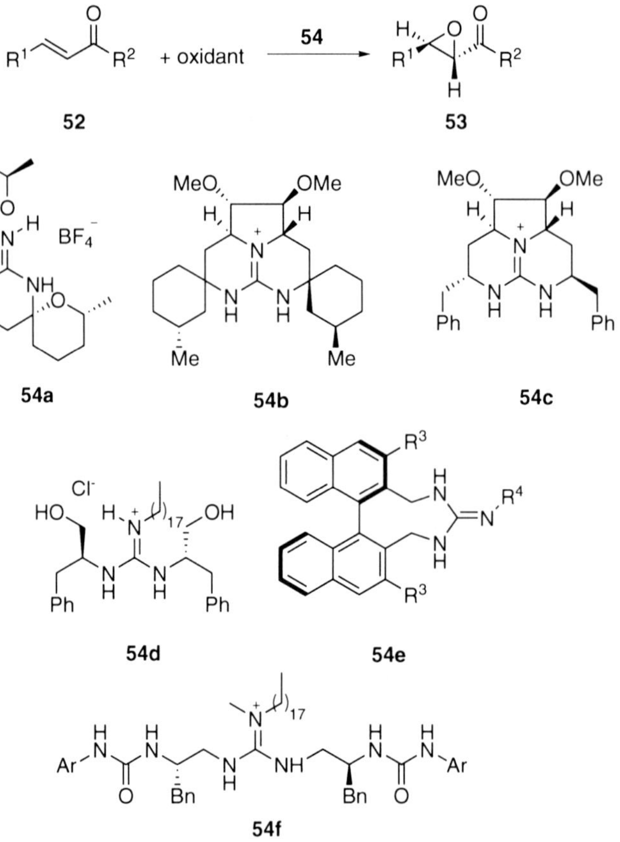

Scheme 4.8 Asymmetric epoxidation of (*E*)-diarylenones

Higher enantioselectivities (60–73% ee) were obtained by using acyclic hydroxy-guanidine **54d** (10 mol%), in which a plausible transition state involving the coordination of the carbonyl group with the guanidine through hydrogen bonding and an interaction of the hydroxy group with the *tert*-butylhydroperoxide anion was proposed to explain the stereososelectivity observed [59]. Hydrogen peroxide was also used as oxidant in reactions involving guanidines as catalysts. Thus, several axial stereogenic guanidines **54e** (10 mol%) were deployed in the enantioselective epoxidation of chalcone in the presence of 30% aqueous hydrogen peroxide in toluene at room temperature [60]. The highest catalytic activity was obtained when R^3 was a phenyl moiety bearing a *tert*-butyl substituent at the *para* position. The achieved enantioselectivities being very modest. In order to increase them, a chiral substituent was incorporated to the nitrogen atom of the guanidine moiety, with the best results being obtained for the (*S*)-phenethyl and (*S*)-naphthylethyl groups (50 and 52% ee, respectively). The highest enantioselectivities were achieved when guanidine-urea bifunctional catalyst **54f** (5 mol%) with aryl being 3,5-ditrifluoromethylphenyl groups was used for the epoxidation of (*E*)-enones in the presence of 30% aqueous hydrogen peroxide and sodium hydroxide as a base in toluene at −10°C [61]. Under these reaction conditions a variety of (*E*)-diarylenones, having electron-withdrawing or electron donating groups could be used affording the corresponding epoxides **53** in 91–99% yield and 85–96% of enantiomeric excesses.

The Juliá-Colonna epoxidation uses different polyamino acid derivatives as catalysts [62]. However, owing to their very high molecular weight (close to the enzymes) this reaction is not a topic of this article.

Besides the success obtained in the epoxidation of enones by either phase transfer catalysts or polyamino acid derivatives, there was not any example of the related reaction with aldehydes **13**. Recently, chiral amine **22** was deployed as a soluble catalyst for the enantioselective epoxidation of α,β-unsaturated aldehydes **13** to give the expected epoxide **55** (Scheme 4.9).

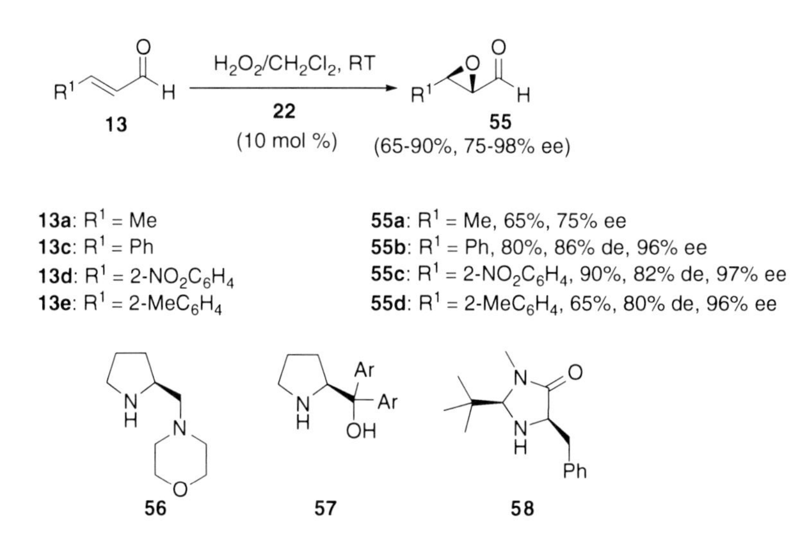

Scheme 4.9 Enantioselective epoxidation of α,β-unsaturated aldehydes

As far as oxidants is concerned, it could be used with similar results not only H_2O_2 but also urea-hydrogen peroxide and even organic peroxides. As far as solvents is concerned, it should be pointed out that the reaction could be performed in other solvents different from CH_2Cl_2 with only a marginal erasing of the enantioselectivity, even in the case of using the environmentally safe mixture of water-ethanol. The proposed mechanism consists in the formation of the corresponding iminium ion by condensation of amine **22** with aldehyde **13**, which suffers the nucleophilic attack of peroxide derivative at the α-position, leading to an enamine. The intramolecular nucleophilic attack of this enamine to the peroxide moiety gave the corresponding iminium epoxide, which after hydrolysis liberates the final epoxide and the starting catalyst [63]. The use of other organocatalyst such as **15a**, **56** and **57**, under similar conditions, did not improve the aforementioned results [64]. However, the change of the oxidant to either solid sodium percarbonate or *tert*-butyl hydroperoxide as well as the solvent to $CHCl_3$ permitted to reach the enantioselectivity higher than 95%.

The aforementioned epoxidation has been also performed using imidazolidinone **58** in combination with perchloric acid and (nosylimino)iodo benzene as hypervalent iodine oxidant. In this manner, several α,β-unsaturated aliphatic aldehydes were epoxidized in high yields and enantioselectivities (72–93%, 87–97% ee) [65], with only enals possessing an electron-withdrawing group being unreactive under these reaction conditions.

The organocatalysts of type **57** have been also applied in the enantioselective epoxidation of different chalcones **7** using *tert*-butyl hydroperoxide as oxidant and non polar solvents such as hexane, affording the expected chiral epoxides **53** with good results [66].

Not only oxidants with nucleophilic character but also intermediates of electrophilic nature can be used in the enantioselective epoxidation of α,β-unsaturated carbonyl compounds. Among the possible candidates, dioxirane reagents have been successfully used for this purpose. Contrary to the usual nucleophilic oxidants, dioxiranes add to double bonds in a concerted manner. These dioxiranes could be easily prepared in situ by reaction of oxone ($2KHSO_5 \cdot KHSO_4 \cdot K_2SO_4$) with chiral ketones. Chiral ketones derived from quinic acid such as compounds **61** [67], which have been successfully used in the enantioselective epoxidation of electron rich olefins, have been also applied to the epoxidation of electron poor olefins such as chalcone **7** or α,β-unsaturated esters **59** (Scheme 4.10) to give the corresponding

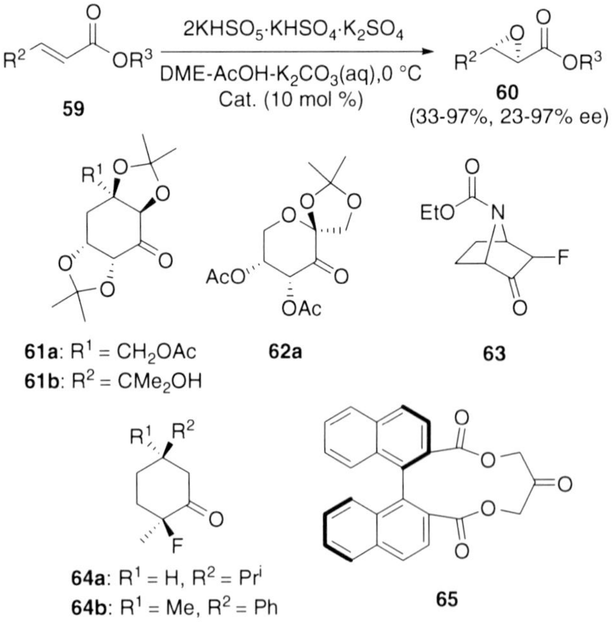

Scheme 4.10 Enantioselective epoxidation of α,β-unsaturated esters

epoxides with good enantioselectivity, albeit with modest chemical yields. This strategy has been successfully applied in the synthesis of an intermediate for the preparation of dipeptide isosteres with HIV inhibition and blood pressure modulation properties and also in the asymmetric epoxidation of *cis*-1-propenylphosphoric acid affording the antibiotic fosfomycin [67].

Owing to the low reactivity of the in situ formed dioxiranes, they suffer decomposition processes to give the corresponding Bayer-Villiger oxidation products. Therefore, new ketones with enhanced stability have been introduced to perform the aforementioned epoxidation. One of these ketones was the fructose derivative **62a** [68], which efficiently catalyzed the oxidation of different cinnamate esters with higher yields (40–96%) and improved enantioselectivities (up to 97% ee). Surprisingly, the epoxidation of the related ethyl (*Z*)-cinnamate gave a lower result (84%, 44% ee).

The use of α-fluoroketone derivatives, as catalysts for the aforementioned epoxidation, has been less effective. So, the epoxidation of methyl cinnamate using the tropinone derivative **63** [69] gave the expected epoxide with modest results (33%, 64% ee). The use of organocatalysts **64** [70] did not improve the enantioselectivity but did improve chemical yields (up to 97%).

The chiral ketone **65** showed a very high reactivity, and therefore the loaded amount could be reduced from the typical 10 mol% to 0.3 mol%, with the enantioselectivity being up to 95% [71]. Whereas the stereoselectivity was explained as a consequence of a π-π interaction between the aromatic rings of the ligand **65** and of the substrate **59**, the high chemical yield was attributed to the higher stability of the chiral ketone owing to the presence of two electron withdrawing ester moieties. The study of substrate scope showed that the bulkiness of ester moiety (R^3 in **59**) did not play an important role on the enantiomeric excess of product **60**. However, the steric hindrance of substituent of double bond (R^2 in **59**), as well as the geometry of double bond, had a great impact on the enantioselectivity, with the best results being obtained for *E*-olefins with aryl phenyl substituents. The protocol was successfully applied also to primary cinnamides. This chiral in situ formed dioxirane has been used as catalyst in a large-scale preparation of key intermediate in the synthesis of anti-hypertensive agent diltiazem, showing the great potential of this methodology for industrial purposes.

4.3.3 Aminoxylation Processes

Among the existing methods for the asymmetric synthesis of chiral α-hydroxy carbonyl compounds, the direct enantioselective α-aminoxylation of carbonyl compounds is one of the most important strategies to achieve this purpose. Although its nitrogen versus oxygen electrophilic reactivity in nitroso compounds should be carefully controlled through the election of appropriate catalysts and reaction conditions.

Almost simultaneously and independently, three research groups reported the first direct α-oxidation of aldehydes **17** using natural proline (**20**) and nitrosobenzene (**43a**) with high yields and excellent enantiomeric excess (Scheme 4.11), differing only in the reaction conditions. Whereas the first group performed the reaction in DMSO as solvent and at room temperature using 20 mol% of organocatalyst, the second and the third groups carried out the same reaction at a lower temperature (4°C and −20°C, respectively) in order to minimize the possible homo aldol reaction, and using less polar solvents such as $CHCl_3$ and CH_3CN [72]. The optimum catalyst loading was 5 mol% for the last two protocols. However, this amount could be reduced to 0.5 mol%, maintaining the high enantioselectivity but increasing the reaction time. The substrate scope was very broad, being applied to alkyl and aryl substituted aldehydes. In order to facilitate the isolation of products **66**, they were reduced to the corresponding primary alcohols just by an one-pot addition of $NaBH_4$. Related chiral diols could be obtained by a further either hydrogenation on palladium-charcoal or with treatment with substoichiometric amounts of copper sulfate.

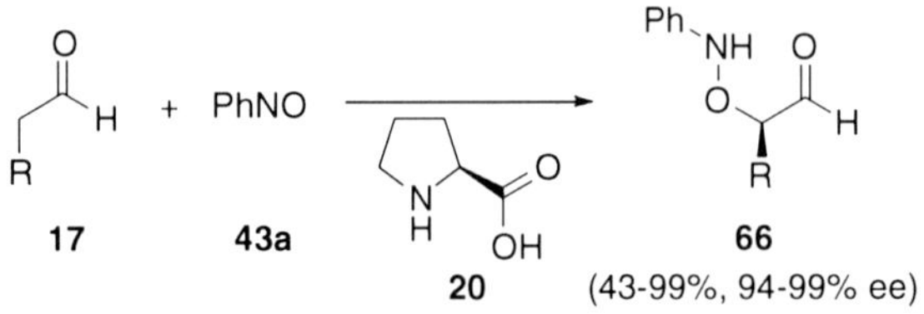

17 + PhNO (43a) → [20] → 66 (43-99%, 94-99% ee)

17a: R = Me
17b: R = Pr^i
17d: R = Ph

66a: R = Me, 60%, 97% ee (ref. 72a)
66b: R = Pr^i, 85%, 99% ee (ref. 72b)
66c: R = Ph, 62%, 99% ee (ref. 72c)

Scheme 4.11 Enantioselective α-aminoxylation of aldehydes

The easy detection of the end of the reaction by color changing from green to orange permitted a further in situ transformation of aldehydes **66** by a second reaction process. So, after the α-aminoxylation reaction process, the in situ formed aldehyde **66** was submitted to an allylation process promoted by indium, giving the expected diols as mixture of *ca* 4:1 diastereomeric ratio, and enantiomeric excess up to 99% for both diastereoisomers [73]. Aldehydes **66** could be also trapped by a Wadsworth-Emmons-Horner reaction with diethyl (2-oxopropyl) phosphonate under basic conditions (LiOH) to give the corresponding γ-hydroxy-α,β-unsaturated ketone [74]. An efficient and general route to prepare optically active α-hydroxyketones has been achieved by a one-pot tandem aminoxylation of aldehydes using nitrosobenzene and chemoselective diazomethane homologation [75].

The aforementioned protocol has been used in the synthesis of many interesting products. So, the formal synthesis of HRV 3C-protease inhibitor (1*R*,2*S*)-thysanone has been accomplished with a 98% enantiomeric excess using (*S*)-proline catalyzing the asymmetric α-aminoxylation of an aldehyde and oxa-Pictet-Spengler cyclization as key steps [76]. Following the same strategy, an efficient route for the synthesis of atorvastatin side chain, a building block present in the statin family which acted as cholesterol regulator, was performed with high enantioselectivity [77]. Not only

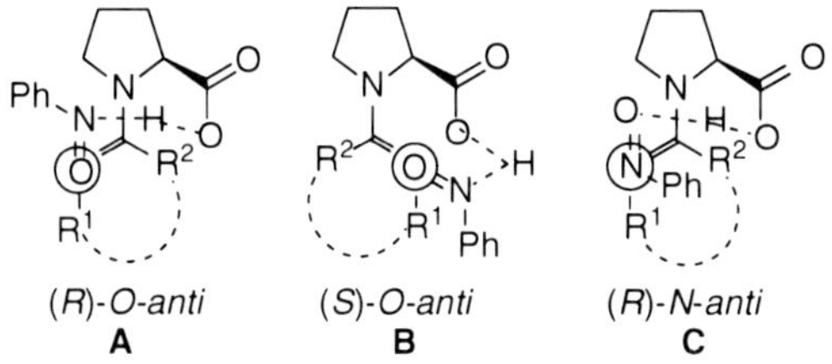

(*R*)-*O*-*anti*
A
(*S*)-*O*-*anti*
B
(*R*)-*N*-*anti*
C

Fig. 4.2 Proposed transition state models

formal synthesis, but also total natural product or drug syntheses have been made using of this useful reaction. For instance, (*S*,*S*)-ethambutol [22], (*R*)-selegiline [23] and (−)-anisomycin [26] can be prepared with similar results by using this transformation. In a similar way, an anti-inflammatory such as halipeptin A [78], antitumor agents such as tarchonanthuslactone [79], (+)-cytotrienin A [80] and (+)-harzialactone A and (*R*)-(+)-4-hexanolide [81], antibacterials such as linezolid and eperezolid [82], an antiepileptic drugs such as levetiracetam [83], pheromones such as (−)-*endo*-brevicomin [84], (+)-disparlure [85] and (2*S*,3*S*)-2-hydrohexylcyclopentanone [86], an enzyme inhibitors such as (+)-panephenanthrin [87], and β-adrenergic blockers such as (S)-propranolol and (*S*)-naftopidil [88] have been synthesized using the α-aminoxylation of aldehydes as a key step.

Even different chiral dihydro-1,2-oxazine derivatives were obtained when the enantioselective α-aminoxylation reaction was catalyzed by tetrazole proline derivative **24**, in which the in situ formed aldehyde **66** reacted subsequently with vinylphosphonium salt derivatives or Grignard reagents [89].

A quantum mechanical computational study was carried out in order to establish the origin of the regioselectivity displayed by nitrosobenzene and the stereoselectivity of obtained aldehydes **66** [90]. From the different proposed pseudo-six-membered transition state structures (Fig. 4.2), the most stable seems to be *O*-*anti*-transition state. This transition state involves the attack of (*E*)-*anti* enamine to the oxygen of the nitrosobenzene (**43a**), adopting its phenyl group a *pseudo* axial position and *anti* respect to the carbonyl group of proline. These calculations predicted a 99% ee for the products, which are in agreement with the obtained experimental value (97% ee). The heteroatom selectivity could be explained by the higher basicity of the nitrogen atom versus oxygen one, which led to its preferential protonation, with the oxygen atom becoming more electrophilic. Therefore, the energies of the transition state for the nitrogen attack are generally higher than those for the oxygen one.

Kinetic studies of the aforementioned transformation (Scheme 4.11) showed a high auto acceleration in the reaction rate, suggesting an improvement of the catalyst over the time [91]. Once the substrate inhibition was discarded, a proline-product adduct (enamine-like intermediate) was suggested as the improved catalyst. This new catalyst reacts with the initial aldehyde to form an intermediate bearing two enamine moieties, one of them formed by the proline and the final hydroxy aldehyde **66** and the second one formed by the previous aminohydroxy derivative and the initial aldehyde **17**. Presumably, the reaction of nitrosobenzene (**43a**) takes place on this double enamine intermediate.

A new protocol, involving the addition of a bifunctional urea to the reaction mixture, increased significantly the reaction rate [92]. The origin of the increased reaction rate was due to a hydrogen bond interaction between the bifunctional thiourea and the possible oxazolidine intermediate formed [93] which enhanced the rate of the enamine formation. Moreover, the addition of urea removed the aforementioned autoinduction behavior, and therefore, an alternative catalytic cycle could take place in the presence of the urea.

The scope of this protocol was further extended to ketones, obtaining the expected hydroxy ketones with excellent enantioselectivities, even in the presence of small amounts of water (<25%) using DMSO as solvent [94]. In order to improve the chemical yield and to prevent the formation of the related dihydroxy ketone derivative, the addition of nitrosobenzene should be done very slowly. In the case of using acyclic ketones, a small amount of racemic α-amino ketone derivative was always accomplished. Remarkably, when the reaction was performed using methyl ketones, the reaction took place exclusively at the methylene moiety. The formed hydroxy ketones could be easily transformed into the corresponding diols by diastereoselective reduction of carbonyl group (diastereoselectivity around 40%) and hydrogenolysis using PtO_2 as catalyst. The absence of non linear effect was attributed to the presence of only one chiral proline in the mechanism pathway of the reaction. Moreover, DFT calculations for the cyclohexanone case showed that the most probable transition state was the *O-anti*-transition state **A** (Fig. 4.2) as for the case of the one calculated for aldehydes.

In a new protocol the solvent used was DMF at 0°C, decreasing the loaded amount of proline to 10 mol%. The regioselectivity was almost completed for cyclic ketones and for acyclic ketones a mixture *ca* 1:1 of α-hydroxy and α-amino ketones were obtained. This methodology was also applied successfully to the enantioselective desymmetrization of 4-monosubstituted and 3,5-disubstituted cyclohexanones, with ee up to 99% [95]. This methodology has been used as the key step in the total synthesis of fumagillol, RK-805, FR658145-demethylovalicin and ovalicin [96].

Another organocatalyst proposed as alternative to proline was the 4-silyloxy proline derivative **25**, which was able to catalyze the enantioselective α-aminoxylation of ketones very efficiently, yielding the final products after only 1 min and with good chemical yields and excellent enantioselectivities [97]. Remarkably, in this case neither α-amino ketones nor dihydroxy ketone derivatives could be detected in the crude reaction mixture. Moreover, this catalyst could perform the reaction with substrates, which failed with the initial proline protocol, such as cycloheptanone (50%, >99% ee). Also, the triflamide **21b** ($R=CF_3$) has been used as organocatalyst in the reaction of ketones, as well as aldehydes, with nitrosobenzene to give the expected hydroxylated carbonyl compounds with high level of enantioselectivity [98].

Not only derivatives from proline are able to catalyze this transformation but also binaphthyl-based chiral secondary amine **67** (5 mol%) has been applied in the α-aminoxylation of both aldehyde and ketone substrates [99].

Hydroxy acids such as 1-naphthyl glycolic acid **68** can be used as organocatalyst in the reaction of nitrosobenzene with, in this case, enamines to yield, after work up,

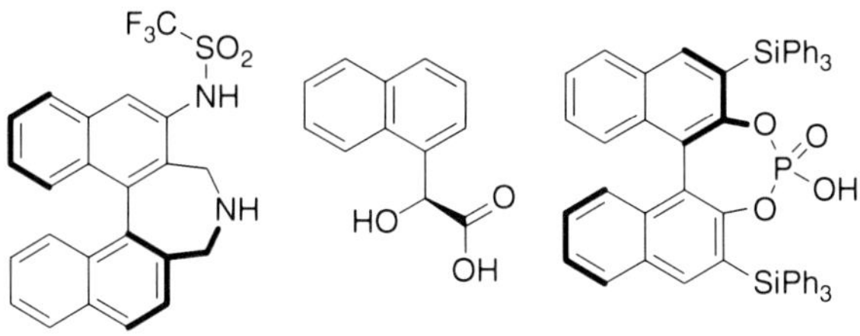

67 68 69

Fig. 4.3 Organocatalysts for α-aminoxylation

the corresponding hydroxy ketone with good chemical yields and enantiomeric excess ranging from 70% to 92% (Fig. 4.3) [48].

The binaphthyl phosphoric acid **69** (1 mol%) was used in the α-hydroxylation of cyclic 1,3-dicarbonyl compounds of type **33** with 4-chloronitrosobenzene (**43b**), affording the corresponding α-hydroxy β-ketoesters in high *O*/*N* selectivity (94:6), good yields and enantioselectivities (46–88%, 68–98% ee). Several ester derivatives such as methyl, ethyl, isopropyl or *tert*-butyl were well tolerated [100], with the present protocol being extended to the related β-diketones.

4.3.4 Miscellaneous Oxidations

The direct asymmetric synthesis of α-hydroxy carbonyl derivatives is possible by the use of other methodologies. For instance, the oxidation of silyl enol ethers with oxone ($2KHSO_5{\cdot}KHSO_4{\cdot}K_2SO_4$) in the presence of excess of fructose derivative **62a** lead directly to the corresponding chiral α-hydroxy ketones with good enantioselectivities [101].

Not only oxone but also other oxidants can be used in the preparation of hydroxy carbonyl derivatives. So, *tert*-butyl hydroperoxide has been used in the enantioselective oxidation of 2-alkoxycarbonyl indanone derivatives of type **33** catalyzed by cinchonine **35** to give the expected compound with moderated enantiomeric excess (50%). This chiral compound serves as the starting material for the synthesis of pyrazoline-type insecticide indoxacarb [102]. The use of chiral dihydroquinine (double bond hydrogenated compound **11**) in combination with cumyl hydroperoxide at room temperature seems to be a more promising protocol, since the expected hydroxy carbonyl compounds could be obtained with excellent chemical yields and good enantiomeric excess [103].

The recently reported use of benzoylperoxide (**70**) as oxidant for aldehydes has been more efficient (Scheme 4.12). When 2-tritylpyrrolidine catalyst **72** was used in this reaction, hydroquinone (10 mol%) was added in order to increase the yield of products **71**, which were obtained with high enantioselectivities. The obtained α-benzoyloxyladehydes could be converted into the corresponding chiral diols or

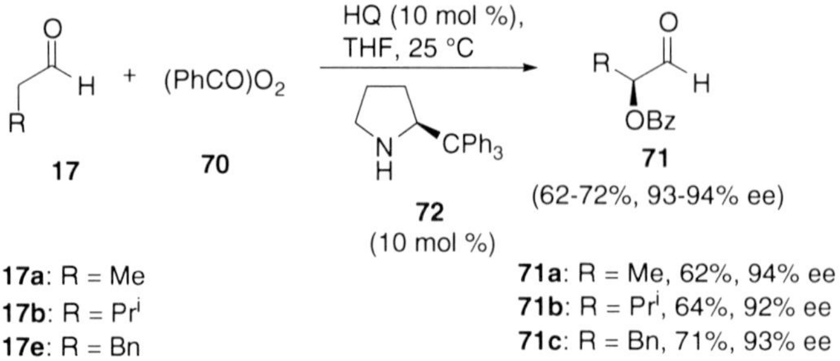

17 + $(PhCO)_2O_2$ 70 → HQ (10 mol %), THF, 25 °C; 72 (10 mol %) → 71 (62-72%, 93-94% ee)

17a: R = Me
17b: R = Pr^i
17e: R = Bn

71a: R = Me, 62%, 94% ee
71b: R = Pr^i, 64%, 92% ee
71c: R = Bn, 71%, 93% ee

Scheme 4.12 Enantioselective oxidation of aldehydes

monoprotected diols, by reduction using $NaBH_4$ and further DBU treatment, respectively. Although, radical intermediates cannot be ruled out for the mechanism process, two possible ionic pathways were suggested, in which oxidant **70** would approach to the s-*trans* enamine, in which one face is shielded by the trityl group, providing the (*S*)-isomer **71** predominantly [104].

Catalyst **15c** (R=Bu^n, 20 mol%) showed to be effective [105] for this reaction, with the use of additives (acid or bases) not increasing the obtained yield (54–78%). However, in order to achieve good enantioselectivities (90–94% ee), the catalyst has to be added in two portions, one at the beginning of the reaction and the second portion after 4 h of reaction time. Under these conditions, aliphatic aldehydes and aldehydes containing aromatic, silyl groups or double bonds could be oxidized with good results.

The use of molecular oxygen considered as a green oxidant to carry out the preparation of α-hydroxy carbonyl compounds has been already reported. Aldehydes can be hydroxylated with molecular oxygen in the presence of α-methyl proline to give, after final reduction, the corresponding diols with moderate enantioselectivities. The excitation of molecular oxygen by ultraviolet light (UV) in the presence of tetraphenylporphine permitted the generation of single molecular oxygen in appreciable amounts, which is the real reactive agent. Better results were obtained when catalyst **15a** (R=Me, 20 mol%) was employed in this reaction at 0°C in $CHCl_3$. After reduction of the aldehyde moiety, the corresponding 1,2-diols were obtained in good yields (50–71%) with 74–98% ee [106]. This methodology has been further extended to cyclic ketones with similar results. In this case, the reaction was carried out in DMSO as solvent and with the natural amino acid L-alanine giving the best enantioselectivities [107]. These results have an important impact on our knowledge of the possible prebiotic chemistry, since it seems to indicate that terrestrial amino acids could catalyze the asymmetric introduction of molecular oxygen in organic compounds, serving as a plausible first step in the homochirality transfer to other bioorganic molecules.

4.3.5 *Sulfenylation Processes*

Despite the synthetic potential of α-sulfenylated carbonyl compounds, their enantioselective synthesis has been scarcely developed, although this tendency is going to change. So, the reaction of aldehydes **17** with electrophilic sulfur source triazole **73** in the presence of substoichiometric amounts of catalyst **22** gave after final reduction and hydrolysis chiral 2-benzylsulfanyl alcohols **74** with excellent results (Scheme 4.13) [17, 108]. Upon prolongated reaction times, the obtained aldehyde racemizes and/or led to the α,α-disulfenylation products, therefore the reaction time was a parameter to be strictly controlled. This protocol has been also expanded to the synthesis of alcohols bearing quaternary stereocenters, just by using the corresponding dialkyl substituted aldehyde, with the enantioselectivity dropping up to 61%.

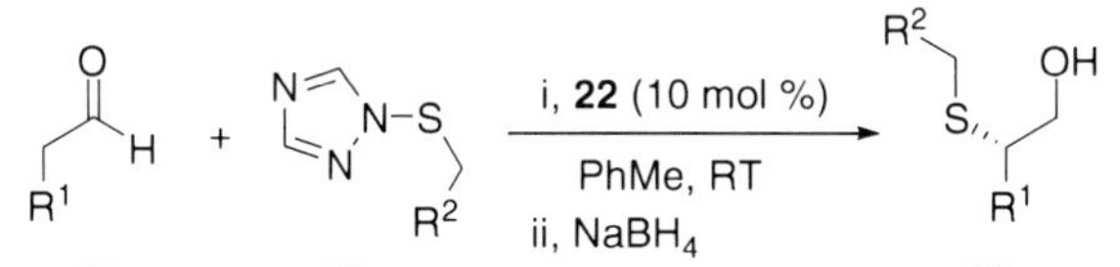

i, **22** (10 mol %)
PhMe, RT
ii, $NaBH_4$

17 **73** **74**
(52-94%, 88-98% ee)

17a: R^1 = Me
17b: R^1 = Pr^i
17e: R^1 = Bn

73a: R^2 = Bn

74a: R^1 = Me, R^2 = Bn, 60%, 95% ee
74b: R^1 = Pr^i, R^2 = Bn, 81%, 98% ee
74c: R^1 = Bn, R^2 = Bn, 94%, 97% ee

Scheme 4.13 Enantioselective sulfenylation of aldehydes

The combination of the organocatalytic α-sulfenylation of an aldehyde with a stereospecific [2, 3]-sigmatropic rearrangement has provided a useful protocol for the enantioselective synthesis of vinyl glycines [109]. For this purpose, the first step was the one-pot synthesis of chiral allylic sulfides by organocatalyzed α-sulfenylation (Scheme 4.13), followed by the in situ Horner-Wadsworth-Emmons olefination, giving the expected α-alkylsulfanyl-α,α-unsaturated ester. Subsequent reaction of this compound with an *N*-Boc-oxaziridine resulted in a [2, 3]-sigmatropic rearrangement with complete diastereoselectivity, to give the expected glycine derivatives. Using a similar protocol, different chiral allenamides were prepared by a [2, 3]-sigmatropic rearrangement of chiral propargylic sulfimides, which could be easily obtained [109].

The sulfenylation of 1,3-dicarbonyl compounds, such as pyrrolidone **75**, as well as cyclic β-keto esters derivatives of type **33**, could be carried out using substoichiometric amounts of dihydroquinidine derived catalyst **77** and different triazole derivatives **73**, affording the expected chiral products with moderate to high enantioselectivity (Scheme 4.14). It should be pointed out that the nature of R substituent on the reagent **73** did not have any important effect on the enantioselectivity of the reaction. However, the bulkiness of the ester moiety of the 1,3-dicarbonyl compound had an accountable effect, the more crowded the ester moiety the higher enantioselectivity. Other alkaloids tested as catalyst gave lower results [110].

In the case of cyclic β-keto esters **33**, a similar reaction could be performed using *N*-(phenylthio)phtalimide and catalyst **22** (20 mol%), which gave the best result among all tested α,α-diaryl prolinol derivatives [111]. Thus, several sulfur phtalimide derivatives with electron-withdrawing or electron-donating groups on

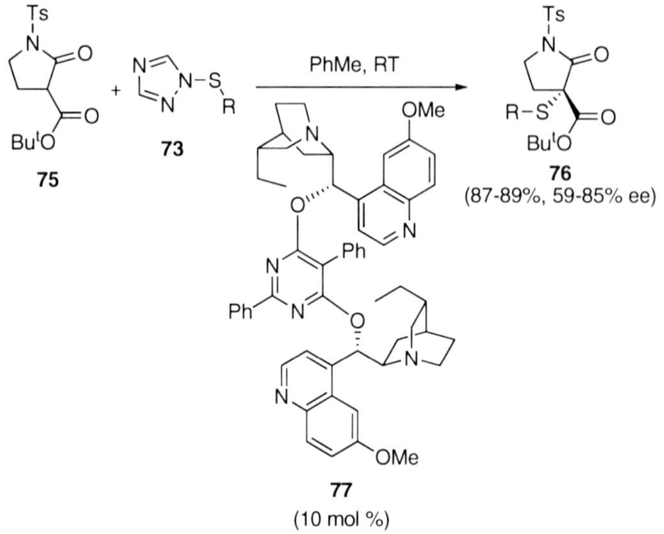

Scheme 4.14 Sulfenylation of 1,3-dicarbonyl compounds

the aromatic ring were well tolerated in the reaction, with *N*-(4-chlorophenyl)phtalimide displaying the higher reactivity. Moreover, aliphatic five-membered ring β-keto esters were also suitable substrates for this reaction affording the functionalized products in good yields (88–89%) and enantioselectivities (79–86% ee).

4.3.6 Selenylation Reaction

The α-selenylation of carbonyl compounds is in its first preliminary stage. In fact, there are only few examples, in which different aldehydes reacted with *N*-(phenylseleno)phthalimide **78** to yield the expected α-functionalized aldehyde in the presence of 4-imidazolidinone **58**, with very moderated results. This result could be slightly improved by the use of 2-(tosylaminomethyl)pyrrolidine **21b** (Ar = 4-MeC_6H_4) up to 60% ee [112]. A similar reaction using cyclohexanone gave the expected α-seleno cyclohexanone with a good chemical yield but very low enantioselectivity (88%, 18% ee).

Alternatively, α,α-diaryl prolinol catalysts **22** (5 mol%) in combination with *p*-nitrobenzoic acid or catalyst **15a** (20 mol%) could be used to promote the α-selenenylation of aldehydes using *N*-(phenylseleno)phtalimides as electrophile [113]. For this reaction, toluene was the solvent of choice and the in situ reduction of the obtained product to the corresponding alcohol was needed in order to preserve the achieved levels of enantioselectivities (Scheme 4.15). A wide range of aldehydes, including alkyl, alkenyl and heterosubstituted aldehydes were suitable substrates, affording the expected products **79** with excellent enantioselectivities.

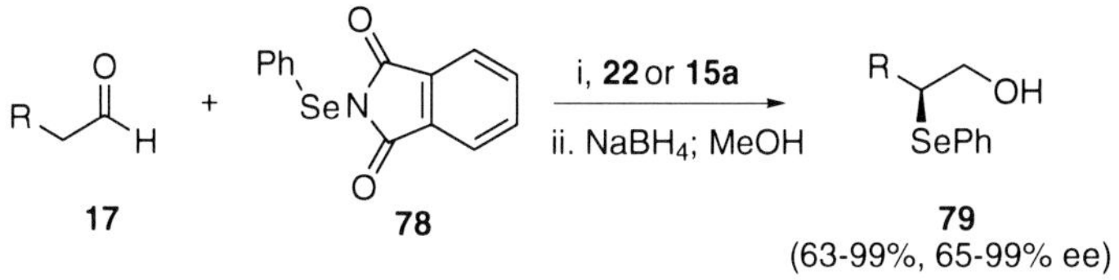

17 + 78 → 79

i, **22** or **15a**
ii. $NaBH_4$; MeOH

79
(63-99%, 65-99% ee)

Scheme 4.15 Asymmetric selenylation of aldehydes

The α-selenylated aldehydes could easily react through an aldol-type reaction with electron withdrawing-stabilized anions (such as sulfones or esters) rendering the corresponding γ-selenyl-β-hydroxy sulfones or carboxylic esters. They were submitted to a one-pot selenide oxidation, in situ epoxide formation and final ring opening to afford the corresponding γ-hydroxy-α,β-unsaturated sulfones or esters in 50% overall yields and 95% enantiomeric excesses [114]. This methodology was applied to the formal synthesis of alkaloid (+)-α-conhydrine.

4.4 Enantioselective α-Halogenation of Carbonyl Compounds

The enantioselective organocatalyzed α-halogenation of carbonyl compounds is a very interesting approach to the synthesis of this type of compounds. This could only be achieved when milder sources of electrophilic halogen source were developed, since the diatomic dihalides, commonly used as halogenating reagents, are far too reactive for asymmetric synthesis. An important approach to the synthesis of these compounds implied the multicomponent reaction between a α,α-unsaturated carbonyl compound with a nucleophile (Michael-type addition) and a mild halogenating agent [115], which is out of this chapter.

4.4.1 Fluorination Reactions

The reaction of silyl enol ethers **80** with selectfluor (**81**) as initial source of electrophilic fluorine atom catalyzed by stoichiometric amounts of dihydroquinine ester **83** gave the expected fluorinated cyclic ketones **82** with moderate to good enantioselectivities (Scheme 4.16), being the first example of this type. Better results were found for indanone derivatives (n = 1) compared to their related tetralones (n = 2). The substitution also had an important effect on the enantioselectivity, with compounds bearing benzyl substituents giving the highest enantiomeric excess. The obtained enantiomeric excesses were surprisingly higher when dihydro quinidine acetate **84** was used as catalyst. This permitted the extension of the above methodology to other ketones such as cyclic α-keto ester of type **33** and their benzocondensated derivatives, as well as to aryl cyanoacetates **28**, achieving the α-fluorinated derivatives with enantioselectivities up to 87% [116]. As the mechanism is concerned, it seems that the cinchona alkaloid reacted firstly with selectfluor (**81**) to generate the corresponding *N*-fluoro alkaloid derivative, which is the real electrophilic fluorinating agent. ^{19}F-NMR experiments of the mixture of both reagents and crystallographic studies support the existence of this intermediate. Other *N*-fluoro cinchona alkaloid derivatives have been proposed as an alternative [117]. However, the enantioselectivity found was significant lower [118].

Scheme 4.16 Fluorination of silyl enol ethers

Better results for the preparation of chiral α-fluoroketones were obtained in the fluorodesilylation of silyl enol ethers **80** mediated by bis-cinchona alkaloid **77** (10 mol%) [119]. For this purpose, *N*-fluorobenzenesulfonimide (**85**) was used as a fluorinating agent, carrying out the reaction in the presence of potassium carbonate as a base in CH_2Cl_2 as solvent at −40°C, to yield the corresponding α-fluoroketones **82** with good results (R = $ArCH_2$, 74–90%, 76–84% ee). Lower enantioselectivities were found using ethyl substituted silyl enol ether (95%, 67% ee). This procedure could be extended to the enantioselective fluorination of oxindoles using cesium hydroxide as a base, affording the expected chiral fluoroderivatives in good yields (89–99%) and high enantioselectivities (79–87% ee).

The fluorination of β-keto ester **86** was accomplished by the use of *N*-fluorobenzenesulfonimide (**85**) and substoichiometric amounts of quaternary ammonium salt **6c**, which acts as a phase transfer catalyst (Scheme 4.17). Although the enantiomeric excess of the final product **87** was modest, this work showed the possibilities of the substoichiometric version of this approach [120].

Not only phase transfer catalyst but also proline derivatives can be used as organocatalyst for the fluorination of carbonyl compounds. Thus, the reaction of cyclohexanone with selectfluor (**80**) in the presence of substoichiometric amounts

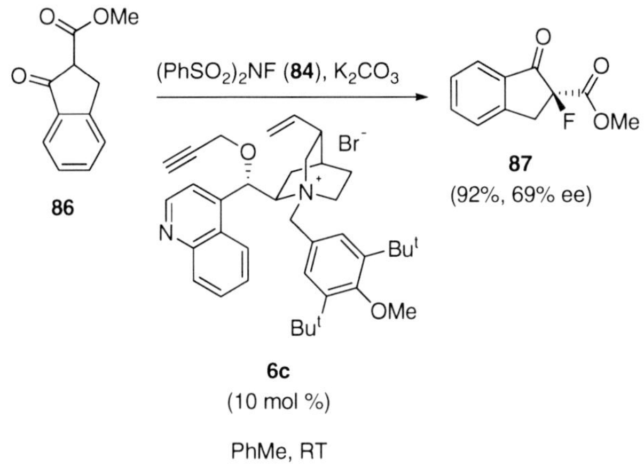

Scheme 4.17 Fluorination of keto esters

of trifluoroacetic acid and 4-silyloxyproline **25** gave the expected α-fluoroketone derivative with rather low enantioselectivity (32% ee). This result could not be improved by the use of other systems such as L-proline (**20**) or prolinol **57a** (Ar=Ph) [121].

However, the fluorination of aldehydes **17** is much more successful. This goal was achieved almost simultaneously by the three different research groups. In all cases, the selected fluorinating agent was *N*-fluorobenzenesulfonimide (**85**) [122]. Thus, when used with catalyst **22** the expected chiral 2-fluoro aldehydes were achieved with moderate to good chemical yields and excellent enantioselectivities. It should be pointed out that the fluorinating agent **85** reacted with catalyst **22** to give the corresponding desilylated prolinol of type **57** in typical solvents such as methylene chloride or acetonitrile, this prolinol derivative was inactive for the reaction and therefore products were obtained in low yields. This side reaction did not take place in methyl *tert*-butyl ether at room temperature, allowing to reduce the amount of catalyst to 1 mol%. The use of another fluorinating agent such as selectfluor (**81**) gave worse results since the desilylation reaction could not be avoided. In order to prevent the possible racemization of final aldehydes upon isolation by silica gel, they were in situ reduced with $NaBH_4$ to the corresponding alcohols, with the whole process giving excellent enantioselectivities independently of the nature of the aldehyde. Initial mechanistic studies showed the presence of a small non linear effect [122]. Imidazolidinone **58** in dimethyl formamide at 4°C has shown to be also efficient for this type of transformation. In this case, the amount of catalyst had to be increased up to 30 mol%, but the chemical yields were excellent [122]. The dichloroacetate ammonium salt derived from imidazolidinone **58** was also used. In this case, the reaction medium was a mixture of THF and isopropanol at −10°C, achieving good chemical yields and excellent enantioselectivities for the corresponding fluorinated aldehydes. Although the standard catalyst loading used in this study was 20 mol%, this amount could be reduced to only 2.5 mol% without erosion on the enantioselectivity but prolonging the reaction time. The protocol permits the fluorination of different aldehydes with a broad type of functionalities maintaining a very high enantioselectivity [122].

The use of the organocatalyzed α-fluorination reaction of aldehydes in combination with other synthetic processes has provided a valuable tool for the asymmetric synthesis of chiral fluorine compounds. Thus, the one-pot α-fluorination of aldehydes mediated by catalyst **22** (1 mol%), followed by in situ trapping and homologation of the aldehyde using the Ohira-Bestmann reagent (**88**) in combination of methanol and potassium carbonate, afforded chiral 3-fluoro terminal alkynes **89** in moderated yields and excellent enantioselectivities (Scheme 4.18) [123]. Similar results were obtained by in situ generation of Ohira-Bestmann reagent. The isolated propargylic fluorides were further used as starting materials in the synthesis of optically active fluoro-containing azides through a click-chemistry process and for the generation of chiral fluorinated peptides and peptide-mimics through a multicomponent reaction. Not only, propargylic fluorides could be obtained by one-pot procedures but also allylic fluorides could be successfully achieved in 43–47% yield and 93–96% enantiomeric excesses by α-fluorination followed by elongation through a Wittig reaction.

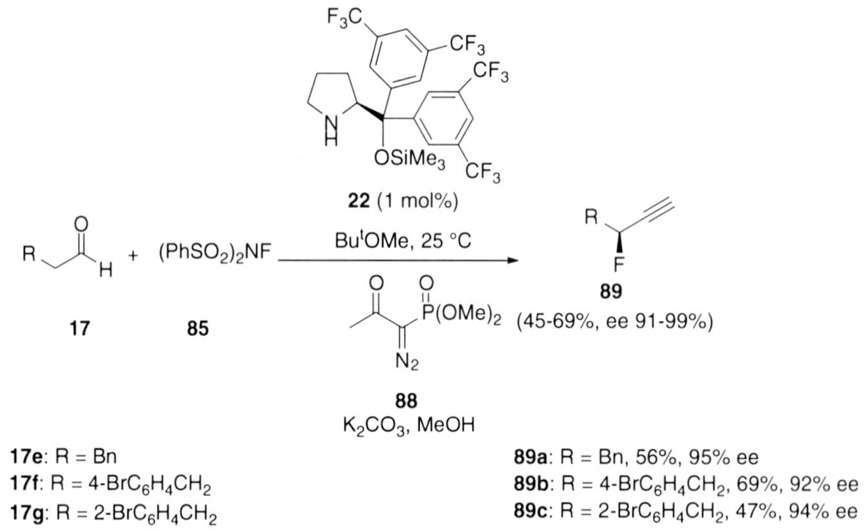

22 (1 mol%)
Bu^tOMe, 25 °C
17 + $(PhSO_2)_2NF$ **85** → **89** (45-69%, ee 91-99%)
88
K_2CO_3, MeOH

17e: R = Bn
17f: R = 4-$BrC_6H_4CH_2$
17g: R = 2-$BrC_6H_4CH_2$

89a: R = Bn, 56%, 95% ee
89b: R = 4-$BrC_6H_4CH_2$, 69%, 92% ee
89c: R = 2-$BrC_6H_4CH_2$, 47%, 94% ee

Scheme 4.18 α-Fluorination of aldehydes

By application of catalytic amounts of dichloroacetate ammonium salt derived from imidazolidinone **58** (20 mol%), the obtained α-fluoroaldehydes were subjected to a reductive amination protocol using primary and secondary amines in a two-pot process providing the corresponding β-fluoroamines in 65–82% yields and enantioselectivities up to 99%. Using this methodology, more challenging tertiary β-fluoroamines could be prepared in good yields (74–93%), albeit with low enantioselectivities

(12–40% ee). The synthesis of chiral β-fluoroamines was also possible using a one-pot procedure affording the expected products with lower yields compared to the two-step procedure but with similar levels of enantioselectivity [124].

α-Branched aldehydes **42** are suitable substrates for the α-fluorination reaction using the fluorinating agent **85** and the chiral catalyst **90** in a mixture of hexane/isopropanol as solvent in order to solubilize the substrates and thus to achieve the best results (Scheme 4.19) [125]. Good enantioselectivities were obtained when R^1 was an aromatic substituent without substitution or with electron-withdrawing substituents (78–90% ee), whereas the enantioselectivities dropped significantly with two aliphatic substituents (7–31% ee). Unfortunately, when the same procedure was applied to linear aldehydes or to ketones such as acetone or cyclohexanone, the reaction did not take place.

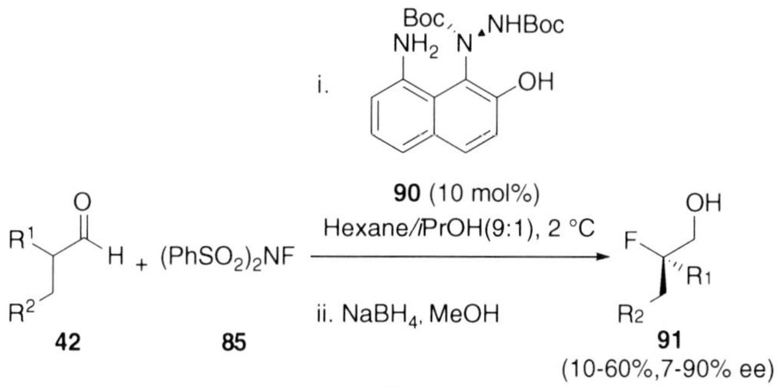

i. **90** (10 mol%), Hexane/iPrOH(9:1), 2 °C
ii. $NaBH_4$, MeOH

42 + $(PhSO_2)_2NF$ **85** → **91** (10-60%,7-90% ee)

42c: R^1 = Me, R^2 = 4-$Pr^iC_6H_4$
42d: R^1 = Ph, R^2 = H
42e: R^1 = Me, R^2 = Pr

91a: R^1 = Me, R^2 = 4-$Pr^i C_6H_4$, 29%, 30% ee
91b: R^1 = Ph, R^2 = H, 36%, 90% ee
91c: R^1 = Me, R^2 =Pr, 27%,7% ee

Scheme 4.19 α-Fluorination of aldehydes

4.4.2 Chlorination Reactions

The first chlorination of carbonyl compounds was accomplished by a double process of enantioselective chlorination and esterification [126]. The ketene generated by deprotonation of acyl chloride **92** with a strong base reacts with quinine benzoate **11b** to give a chiral zwitterionic enolate, which reacts with the electrophilic chlorine source **93** to render an chlorinate acyl ammonium salt. This intermediate undergoes, in turn, a transacetylation with in situ formed perchlorophenol to give the final product of type **94**, regenerating the starting quinine benzoate **11b**. Results are relatively sensitive to stoichiometric base used. Thus, when a proton sponge was used, only moderate yields of final halogenated ester were obtained, albeit with high enantiomeric excess. This low yield was partially due to the easy ring-chlorination of aromatic base and the consequent liberation of pentachlorophenol, which competes with the base for the starting acyl chloride **92**. As an alternative, the ketene was quantitatively generated when a THF solution of acyl chloride **92** was passed through a funnel at −78°C containing the basic 2-*tert*-butylimino-2-diethylamino-1,3-dimethylperhydro-1,3,2-diazaphosphorine (BEMP)-type resin. This ketene solution was added to a solution containing all reagents, affording the expected product **94** with good results (Scheme 4.20).

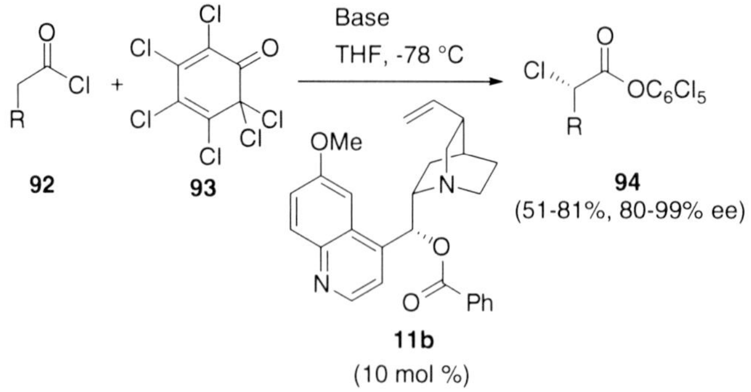

Scheme 4.20 α-Chlorination of carbonyl compounds

Other tested bases, such as NaH and $NaHCO_3$, gave similar results; although the latter reduces the economical reaction cost and facilitates the purification. It should be pointed out that the deprotonation of the starting acyl chloride did not take place when $NaHCO_3$ was used alone. This implies that this base only intervenes as a final scavenger of hydrogen chloride, with the chiral amine being the initial base. The reaction rates of both the NaH and $NaHCO_3$ protocols are proportional to the concentration of chlorinating reagent **93**. These results were explained on the basis of a reversible enolate formation which did not vary its concentration over the time.

Aldehydes **17** could also serve as substrates for the α-chlorination process (Scheme 4.21). Thus, ammonium trifluoroacetate catalyst **97** (5 mol%) was able to catalyze the chlorination of aldehydes using the aforementioned chlorinated reagent **93** in acetone at −30°C, affording the corresponding products **96** with high yields and enantioselectivities, independently of bulkiness of aldehyde substitution [127]. The mild conditions permitted the presence of acid sensitive functionalities.

The aforementioned transformation also can be performed with simple and inexpensive *N*-chloro succinimide (**95a**). In this case either C_2-symmetric pyrrolidine **98** or prolinamide **99** in 10 mol% could be successfully used with the reaction conditions being milder than in the previous case (methylene chloride at room temperature, Scheme 4.21). The obtained chiral α-chloro aldehydes **96** are configurationally stable to pH neutral silica purification, although they can be easily transformed into a wide variety of different compounds, such as epoxides (reduction with $NaBH_4$ and

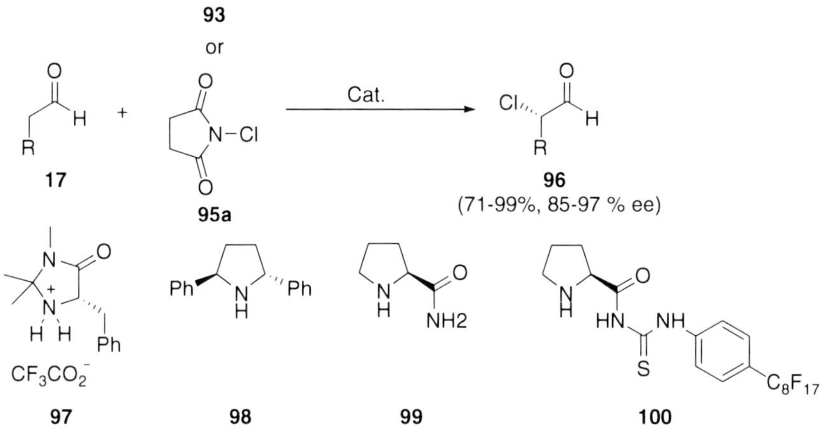

Scheme 4.21 α-Chlorination of aldehydes

KOH treatment), or acids after (oxidation with $KMnO_4$). All these transformations took place without decreasing the initial enantiomeric excess, showing the versatility of these compounds [128]. Recently, new fluorous (*S*)-pyrrolidine-thiourea recoverable catalyst **100** (10 mol%) has been allowed the synthesis of products **96** in high yields (91–99%) and excellent enantioselectivities (85–92%), with the recovery of the catalysts being possible by fluorous solid-phase extraction [128].

Less reactive starting carbonyl compounds such as ketones have been successfully chlorinated with inexpensive reagent **95a** by using optically pure (4*R*,5*R*)-4,5-diphenylimidazolidine as the catalyst (20 mol%). The initial trials gave moderated chemical yields due to the polychlorination of the starting ketone, as well as the organocatalyst. However, these processes could be suppressed by the use of acetonitrile as solvent. The addition of different carboxylic acids had a beneficial effect not only on the reaction rate, as in the above example, but also on the enantioselectivities and chemical yield. With *ortho*-nitrobenzoic acid, the best results were obtained. Under these conditions different ketones were monochlorinated with moderate to good chemical yields and excellent enantioselectivities, the latter being practically independent of ketone as well as the presence of functional groups [129].

4.4.3 Bromination Reactions

The enantioselective construction of chiral α-bromo carbonyl compounds has been achieved by using two main processes. The first one involves a double α-bromination esterification process of acyl chlorides **92** catalyzed by chiral alkaloid **11b**, similar to reaction outlined in Scheme 4.20. When used with as brominating agent quinone **101a**, esters were obtained with excellent enantioselectivities [130].

The second one obtained higher yields and enantioselectivities and it was characterized by deployment of catalyst **98** for the α-bromination of aldehydes (Scheme 4.22). In this case the quinone **101b** was used as the best brominating agent, since the cheaper agent *N*-bromo succinimide (**95b**) led to worse yields and enantiomeric excesses. The obtained results seemed to be strongly dependent on the chosen solvent, with a mixture 1:1 of methylene chloride and pentane giving optimal results. As in previous cases, the addition of benzoic acid and water was necessary in order to accelerate the reaction. The lowering of the temperature, which prevents undesirable reactions such as polybromination process and bromination of catalyst [131]. Under these conditions, not only linear but also branched and cyclic aldehydes were brominated with good enantioselectivities. The application of catalyst **22** to the same transformation gave similar results [17]. The advantage of the use of only methylene chloride as solvent and the presence of additives being avoided.

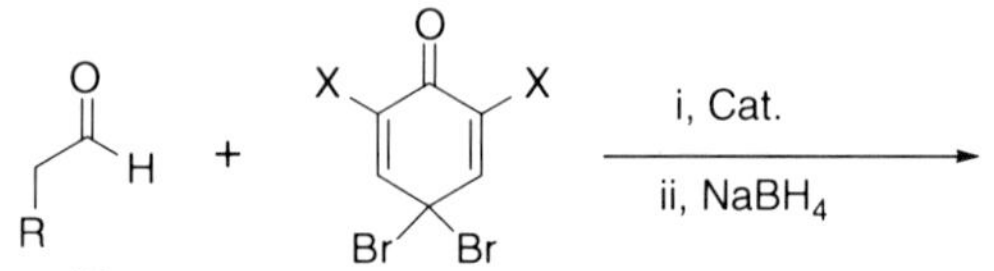

17 + **101a**: X = Br; **101b**: X = Bu^t → i, Cat.; ii, $NaBH_4$ → **102** (72-95%, 68-96% ee)

Scheme 4.22 α-Bromination of aldehydes

As in the chlorination case, (4*R*,5*R*)-4,5-diphenylimidazolidine has been applied for the α-bromination of ketones. The best brominating agent was again the compound **101b**, with the presence of benzoic acid in the reaction mixture being compulsory. The brominated ketones were isolated with up to 91% ee, after reduction to the corresponding alcohols [131].

4.4.4 Iodination Reactions

The first example of iodination reaction was a simple expansion of the above protocol [131], using *N*-iodo succinimide (**95c**) and catalyst **98** to give the corresponding product **103** (R = *i*Pr) with a good result (78%, 89% ee).

However, there is another approach for the rare direct asymmetric α-iodination of aldehydes (Scheme 4.23). In this reaction, catalyst **46b** and benzoic acid as co-catalyst were used to promote the α-iodination of several aldehydes **17** using *N*-iodo succinimide (**95c**) as a iodinating agent, affording the corresponding chiral α-iodoaldehydes **102** in moderated isolated yields (30–80%) and with excellent enantioselectivities (90–99% ee) [132].

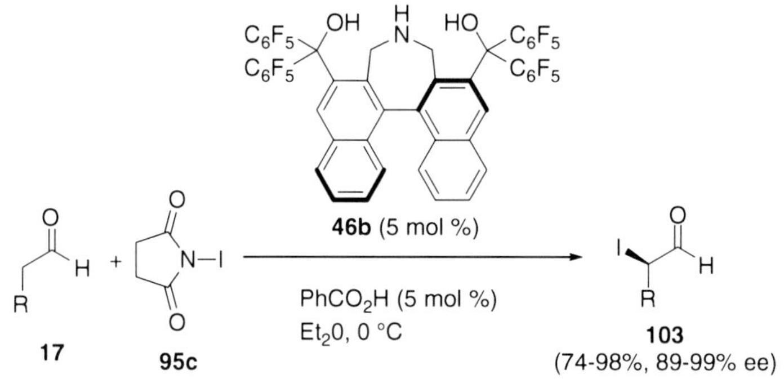

46b (5 mol %)

PhCO$_2$H (5 mol %)
Et$_2$O, 0 °C

17 95c 103 (74-98%, 89-99% ee)

17b: R = Pr^i
17e: R = Bn
17h: R = $CH_2=CHCH_2$

103a: R = Pr^i, 93%, 99% ee
103b: R = Bn, 81%, 92% ee
103c: R = $CH_2=CHCH_2$, 98%, 95% ee

Scheme 4.23 α-Iodination of aldehydes

4.5 Conclusions and Perspectives

The application of organocatalysts to α-heterofunctionalization permitted the preparation of very valuable chiral compounds with exclusion of any trace of hazardous transition metals and with several advantages from an economical, healthy and environmental point of view. Among these advantages, the superior atom efficiency, simple procedures, easy equipment and manipulation make these protocols very attractive for the industry. Despite the great number of contributions and results obtained in a relatively short period of time, many challenges remain. Generally a high catalyst loading should be used for these transformations. Thus, a fine tuning properties of these organocatalysts by structural modifications would allow a decrease in the amount of catalyst, favoring their incorporation to industry. Also, the recoverability of catalysts should be taken in account, all these facts making necessary a better understanding of the mechanism involved in these transformations.

References

1. (a) Wong C-H, Withesides G (1994) Enzymes in organic chemistry. Pergamon Press, Oxford; (b) (2002) In: Drauz K, Waldmann H (eds.). Enzyme catalysis in organic synthesis: a comprehensive handbook. Wiley-VCH, Weinheim, vol. 1–3; (c) (2005) In: Keinan E (ed.) Catalytic antibodies. Wiley-VCH, Weinheim
2. (a) (2005) In: Berkessel A, Gröger H (eds.) Asymmetric organocatalysis: from biomimetic concepts to applications in asymmetric synthesis. Wiley-VCH, Weinheim; (b) (2007) In: Dalko PI (ed.) Enantioselective organocatalysis: reactions and experimental procedures. Wiley-VCH, Weinheim

3. (a) Marigo M, Jørgensen KA (2006) Chem Comm 2001; (b) Guillena G, Ramón DJ (2006) Tetrahedron: Asymm 17:1465
4. (a) Aires-de-Sousa J, Lobo AM, Prabhakar S (1996) Tetrahedron Lett 37:3183; (b) Aires-de-Sousa J, Prabhakar S, Lobo AM, Rosa AM, Gomes MJS, Corvo MC, Williams DJ, White AJP (2001) Tetrahedron: Asymm 12:3349
5. Murugan E, Siva A (2005) Synthesis 2022
6. Fioravanti S, Mascia MG, Pellacani L, Tardella PA (2004) Tetrahedron 60:8073
7. Shen Y-M, Zhao M-X, Xu J, Shi Y (2006) Angew Chem Int Ed 45:8005
8. Armstrong A, Baxter CA, Lamont SG, Pape AR, Wincewicz R (2007) Org Lett 9:351
9. Minakata S, Murakami Y, Tsuruoka R, Kitanaka S, Komatsu M (2008) Chem Comm 6363
10. Pesciaioli F, De Vicentiis F, Galzerano P, Bencivenni G, Bartoli G, Mazzanti A, Melchiorre P (2008) Angew Chem Int Ed 47:8703
11. Vesely J, Ibrahem I, Zhao G-L, Rios R, Córdova A (2007) Angew Chem Int Ed 46:778
12. Arai H, Sugaya N, Sasaki N, Makino K, Lectard S, Hamada Y (2009) Tetrahedron Lett 50:3329
13. Bøgevig A, Juhl K, Kumaragurubaran N, Zhuang W, Jørgensen KA (2002) Angew Chem Int Ed 41:1790
14. List B (2002) J Am Chem Soc 124:5656
15. (a) Iwamura H, Mathew SP, Blackmond DG (2004) J Am Chem Soc 126:11770; (b) Iwamura H, Well DH, Mathew SP, Klussmann M, Armstrong, A, Blackmond DG (2004) J Am Chem Soc 126:16312; (c) Mathew SP, Klussmann M, Iwamura H, Wells DH, Armstrong A, Blackmond DG (2006) Chem Comm 4291
16. Dahlin N, Bøgevig A, Adolfsson H (2004) Adv Synth Catal 346:1101
17. Franzén J, Marigo M, Fielenbach D, Wabnitz TC, Kjærsgaad A, Jørgensen KA (2005) J Am Chem Soc 127:18296
18. (a) Vogt H, Vanderheiden S, Bräse S (2003) Chem Comm 2448; (b) Baumann T, Vogt H, Bräse S (2007) Eur J Org Chem 266
19. Suri JT, Steiner DD, Barbas CF III (2005) Org Lett 7:3885
20. Baumann T, Bächle M, Hartmann C, Bräse S (2008) Eur J Org Chem 2207
21. Chowdari NS, Barbas CF III (2005) Org Lett 7:867
22. Kotkar SP, Sudalai A (2006) Tetrahedron: Asymm 17:1738
23. Talluri SK, Sudalai A (2006) Tetrahedron 63:9758
24. Kalch D, De Rycke N, Moreau X, Greck C (2009) Tetrahedron Lett 50:492
25. Chandrasekhar S, Parimala G, Tiwari B, Narsihmulu C, Sarma GD (2007) Synthesis 1677
26. Chouthaiwale PV, Kotkar SP, Sudalai A (2009) ARKIVOC ii:88
27. Umbreen S, Brockhaus M, Ehrenberg H, Schmidt B (2006) Eur J Org Chem 4585
28. Kotkar SP, Chavan VB, Sudalai A (2007) Org Lett 9:1001
29. (a) Oelke AJ, Kumarn S, Longbottom DA, Ley SV (2006) Synlett 2548; (b) Kumarn S, Oelke AJ, Shaw DM, Longbottom DA, Ley SV (2007) Org Biomol Chem 5:2678
30. Lim A, Choi JH, Tae J (2008) Tetrahedron Lett 49:4882
31. Kumaragurubaran N, Juhl K, Zhuang W, Bøgevig A, Jørgensen KA (2002) J Am Chem Soc 124:6254
32. Thomassigny C, Prim D, Greck C (2006) Tetrahedron Lett 47:1117
33. Hayashi Y, Aratake S, Imai Y, Hibino K, Chen Q-Y, Yamaguchi J, Uchimaru T (2008) Chem Asian J 3:225
34. Lacoste E, Vaique E, Berlande M, Pianet I, Vicent J-M, Landais Y (2007) Eur J Org Chem 167
35. Liu T-Y, Cui H-L, Zhang Y, Jiang K, Du W, He Z-Q, Chen Y-C (2007) Org Lett 9:3671
36. Saaby S, Bella M, Jørgensen KA (2004) J Am Chem Soc 126:8120
37. Liu X, Li H, Deng L (2005) Org Lett 7:167
38. Pihko PM, Pohjakallio A (2004) Synlett 2115
39. Xu X, Yabuta T, Yuan P, Takemoto Y (2006) Synlett 137
40. Jung SH, Kim DY (2008) Tetrahedron Lett 49:5527
41. Zhu R, Zhang D, Wu J, Liu C (2007) Tetrahedron: Asymm 18:1655

42. (a) He R, Wang X, Hashimoto T, Maruoka K (2008) Angew Chem Int Ed 47:9466; (b) He R, Maruoka K (2009) Synthesis 2289
43. Lan Q, Wang X, He R, Ding C, Maruoka K (2009) Tetrahedron Lett 50:3280
44. Guo H-M, Cheng L, Cun L-F, Gong L-Z, Mi A-Q, Jiang Y-Z (2006) Chem Comm 429
45. Kano T, Ueda M, Maruoka K (2006) J Am Chem Soc 128:6046
46. Palomo C, Vera S, Velilla I, Mielgo A, Gómez-Bengoa E (2007) Angew Chem Int Ed 46:8054
47. Wong CT (2009) Tetrahedron Lett 50:811
48. Momiyama N, Yamamoto H (2005) J Am Chem Soc 127:1080
49. Hummelen JC, Wynberg H (1978) Tetrahedron Lett 19:1089
50. (a) Arai S, Shioiri T (1998) Tetrahedron Lett 39:2145; (b) Arai S, Shirai Y, Ishida T, Shioiri T (1999) Chem Comm 49; (c) Arai S, Shirai Y, Ishida T, Shioiri T (1999) Tetrahedron 55:6375
51. Arai S, Tokumaru K, Aoyama T (2004) Tetrahedron Lett 45:1845
52. (a) Bakó P, Szöllõsy Á, Bombicz P, Tõke L (1997) Synlett 291; (b) Bakó P, Szöllõsy Á, Vizvárdi K, Tõke L (1998) Chem Comm 1193; (c) Bakó P, Vizvárdi K, Toppet S, van der Eycken E, Hoornaert G, Tõke L (1998) Tetrahedron 54:14975; (d) Bakó P, Czinege E, Bakó T, Czugler M, Tõke L (1999) Tetrahedron: Asymm 10: 539; (e) Bakó P, Makó A, Keglevich G, Kubinyi M, Pál K (2005) Tetrahedron: Asymm 16:1861; (f) Novák T, Bakó P, Keglevich G, Greiner I (2007) Phosphorous, Sulfur Silicon Relat Elem 182:2449; (g) Makó A, Szöll sy A, Keglevich G, Menyhárd D K, Bakó P, T ke L (2008) Monatsh Chem 139:525
53. (a) Helder R, Hummelen JC, Laane RWPM, Wiering JS, Wynberg H (1976) Tetrahedron Lett 17:1831; (b) Wynberg H, Greijdanus B (1978) J Chem Soc, Chem Comm 427; (c) Marsman B, Wynberg H (1979) J Org Chem 44:2312
54. (a) Arai S, Tsuge H, Shioiri T (1998) Tetrahedron Lett 39:7563; (b) Arai S, Tsuge H, Oku M, Miura M, Shioiri T (2002) Tetrahedron 58:1623
55. (a) Lygo B, Wainwright PG (1998) Tetrahedron Lett 39:1599; (b) Lygo B, Wainwright PG (1999) Tetrahedron 55:6289; (c) Lygo B, Gardiner SD, To DCM (2006) Synlett 2063; (d) Lygo B, Gardiner SD, McLeod MC, To DCM (2007) Org Biomol Chem 5:2283
56. Jew S-s, Lee J-H, Jeong B-S, Yoo M-S, Kim M-J, Lee Y-J, Choi S-h, Lee K, Lah MS, Park H-g. (2005) Angew Chem Int Ed 44:1383
57. Allingham MT, Howard-Jones A, Murphy PJ, Thomas DA, Caulkett PWR (2003) Tetrahedron Lett 44:8677
58. Kita T, Shin B, Hasimoto Y, Nagasawa K (2007) Heterocycles 73:241
59. Shin B, Tanaka S, Kita T, Hasimoto Y, Nagasawa K (2008) Heterocycles 76:801
60. Terada M, Nakano M (2008) Heterocycles 76:1049
61. Tanaka S, Nagasawa K (2009) Synlett 667
62. (a) Pu L (1998) Tetrahedron: Asymm 9:1457; (b) Porter M J, Roberts SM, Skidmore J (1999) Bioorg Med Chem 7:2145; (c) Lauret C, Roberts SM (2002) Aldrichimica Acta 35:47
63. (a) Marigo M, Franzén J, Poulsen T B, Zhuang W, Jørgensen K A (2005). J Am Chem Soc 127:6964; (b) Zhuang W, Marigo M, Jørgensen K A (2005) Org Biomol Chem 3:3883; (c) Zhao G-L, Dziedzic P, Ibrahem I, Córdova A (2006) Synlett 3521; (d) Zhao G-L, Ibrahem I, Sundén H, Córdova A (2007) Adv Synth Catal 349:1210
64. Sundén H, Ibrahem I, Córdova A (2006) Tetrahedron Lett 47:99
65. Lee S, MacMillan DWC (2006) Tetrahedron 62:11413
66. (a) Lattanzi A (2005) Org Lett 7: 2579; (b) Lattanzi A (2006) Adv Synth Catal 348:339; (c) Liu X, Li Y, Wang G, Chai Z, Wu Y, Zhao G (2006) Tetrahedron: Asymm 17:750; (d) Lattanzi A, Russo A (2006) Tetrahedron 62:12264; (e) Russo A, Lattanzi A (2008) Eur J Org Chem 2767; (f) Russo A, Lattanzi A (2009) Synthesis 1551; (g) Cui H, Li Y, Zheng C, Zhao G, Zhu S (2008) J Fluor Chem 129:45
67. (a) Wang Z-X, Shi Y (1997) J Org Chem 62:8622; (b) Wang Z-X, Miller SM, Anderson OP, Shi Y (1999) J Org Chem 64:6443; (c) Ager DJ, Anderson K, Oblinger E, Shi Y, VanderRoest J (2007) Org Proc Res Devolp 11:44; (d) For the use of the related α,β-unsaturated alkenylphophoric acid derivatives as starting materials, see: Zhang Z, Tang J, Wang X, Shi H (2008) J Mol Catal A: Chem 285:68

68. Wu X-Y, She X, Shi Y (2007) J Am Chem Soc 124:8792
69. Armstrong A, Hayter BR (1998) Chem Comm 621
70. (a) Solladié-Cavallo A, Bouérat L (2000) Org Lett 2:3531; (b) Solladié-Cavallo A, Bouérat L, Jierry L (2001) Eur J Org Chem 4557
71. (a) Imashiro R, Seki M (2004) J Org Chem 69: 4216; (b) Seki M (2008) Synlett 164
72. (a) Zhong G (2003) Angew Chem Int Ed 42:4247; (b) Brown SP, Brochu MP, Sinz CJ, MacMillan DWC (2003) J Am Chem Soc 125:10808; (c) Hayashi Y, Yamaguchi J, Hibino K, Shoji M (2003) Tetrahedron Lett 44:8293
73. Zhong G (2004) Chem Comm 606
74. Zhong G, Yu Y (2004) Org Lett 6:1637
75. Yang L, Liu R-H, Wang B, Weng L-L, Zheng H (2009) Tetrahedron Lett 50:2628
76. Sawant RT, Waghmode SB (2009) Tetrahedron 65:1599
77. George S, Sudalai A (2007) Tetrahedron Lett 48:8544
78. Hara S, Makino K, Hamada Y (2006) Tetrahedron Lett 47:1081
79. George S, Sudalai A (2007) Tetrahedron: Asymm 18:975
80. Hayashi Y, Shoji M, Ishikawa H, Yamaguchi J, Tamura T, Imai H, Nishigaya Y, Takabe K, Kakeya H, Onose R, Osada H (2008) Angew Chem Int Ed 47:6657
81. Kotkar SP, Suryanvanshi GS, Sudalai A (2007) Tetrahedron: Asymm 18:1795
82. Narina SV, Sudalai A (2006) Tetrahedron Lett 47:6799
83. Kotkar SP, Sudalai A (2006) Tetrahedron Lett 47:6813
84. Kim SG, Park T-H, Kim BJ (2006) Tetrahedron Lett 47:6369
85. Kim S-G (2009) Synthesis 2418
86. Kondekar NB, Kumar P (2009) Org Lett 11:2611
87. Matsuzawa M, Kakeya H, Yamaguchi J, Shoji M, Onose R, Osada H, Hayashi Y (2006) Chem Asian J 1:845
88. Panchgalle SP, Gore RG, Chavan SP, Kalkote UR (2009) Tetrahedron: Asymm 20:1767
89. (a) Kuman S, Shaw DM, Longbottom DA, Ley SV (2005) Org Lett 7:4189; (b) Jiao P, Kawasaki M, Yamamoto H (2009) Angew Chem Int Ed 48:3333
90. Cheong PH-Y, Houk KN (2004) J Am Chem Soc 126:13912
91. Mathew SP, Iwamura H, Blackmond DG (2004) Angew Chem Int Ed 43:3317
92. Poe SL, Bogdan AR, Mason BP, Steinbacher JL, Opalka SM, MacQuade DT (2009) J Org Chem 74:1574
93. Seebach D, Beck AK, Badine DM, Limbach M, Eschenmoser A, Tresurywala AM, Hobi R, Prikoszovich W, Linder B (2007) Helv Chim Acta 90:425
94. (a) Bøgevig A, Sundén H, Córdova A (2004) Angew Chem Int Ed 43:1109; (b) Córdova A, Sundén H, Bøgevig A, Johansson M, Himo F (2004) Chem Eur J 10:3673
95. (a) Hayashi Y, Yamaguchi J, Sumiya T, Shoji M (2004) Angew. Chem. Int. Ed. 43:1112; (b) Hayashi Y, Yamaguchi J, Sumiya T, Hibino K, Shoji M (2004) J Org Chem 69:5966; (c) Bickley JF, Evans P, Meek A, Morgan BS, Roberst SM (2006) Tetrahedron: Asymm 17:355
96. Yamaguchi J, Toyoshima M, Shoji M, Kakeya H, Osada H, Hayashi Y (2006) Angew Chem Int Ed 45:789
97. Hayashi Y, Yamaguchi J, Hibino K, Sumiya T, Urushima T, Shoji M, Hashizume D, Koshino H (2004) Adv Synth Catal 346:1435
98. Wang W, Wang J, Li H, Liao L (2004) Tetrahedron Lett 45:7235
99. (a) Kano T, Yamamoto A, Mii H, Takai J, Shirakawa S, Maruoka K (2008) Chem Lett 37:250; (b) Kano T, Yamamoto A, Maruoka K (2008) Tetrahedron Lett 49:5369; (c) Kano T, Yamamoto A, Shirozu F, Maruoka K (2009) Synthesis 1557
100. Lu M, Zhu D, Lu Y, Zeng X, Tan B, Xu Z, Zhong G (2009) J Am Chem Soc 131:4562
101. Adam W, Fell RT, Saha-Möller CR, Zhao C-G (1998) Tetrahedron: Asymm 9:397; (b) Zhu Y, Tu Y, Shi Y (1998) Tetrahedron Lett. 39:7819; (c) Zhu Y, Shu L, Tu Y, Shi Y (2001) J Org Chem 66:1818
102. McCann SF, Annis GD, Shapiro R, Piotrowski DW, Lahm GP, Long JK, Lee KC, Hughes MM, Myers BJ, Griswold SM, Reeves BM, March RW, Sharpe PL, Lower P, Barnette WE, Wing KD (2001) Pest Manag Sci 57:153

103. Acocella MR, García Mancheño O, Bella M, Jørgensen KA (2004) J Org Chem 69:8165
104. Kano T, Mii H, Maruoka K (2009) J Am Chem Soc 131:3450
105. Gotoh H, Hayashi Y (2009) Chem Comm 3083
106. Córdova A, Sundén H, Engqvist M, Ibrahem I, Casas J (2004) J Am Chem Soc 126:8914; (b) Ibrahem I, Zhao G-L, Sundén H, Córdova A (2006) Tetrahedron Lett 47:4659
107. Sundén H, Engqvist M, Casas J, Ibrahem I, Córdova A (2004) Angew Chem Int Ed 43:6532
108. Marigo M, Wabnitz TC, Fielenbach D, Jørgensen KA (2005) Angew Chem Int Ed 44:794
109. (a) Armstrong A, Challinor L, Cooke RS, Moir JH, Treweeke NJ (2006) J Org Chem 71:4028; (b) Armstrong A, Challinor L, Moir JH (2007) Angew Chem Int Ed 46:5369; (c) Armstrong A, Emmerson D PG (2009) Org Lett 11:1547
110. Sobhani S, Fielenbach D, Marigo M, Wabnitz TC, Jørgensen KA (2005) Chem Eur J 11:5689
111. Fang L, Lin A, Hu H, Zhu C (2009) Chem Eur J 15:7039
112. Wang J, Li H, Mei Y, Lou B, Xu D, Xie D, Guo H, Wang W (2005) J Org Chem 70:5678
113. (a) Tiecco M, Carlone A, Sternativo S, Marini F, Bartoli G, Melchiorre P (2007) Angew Chem Int Ed 46:6882; (b) Sundén H, Rios R, Córdova A (2007) Tetrahedron Lett 48:7865
114. Petersen KS, Posner GH (2008) Org Lett 10:4685
115. Guillena G, Ramón DJ, Yus M (2007) Tetrahedron: Asymmetry 18:693
116. (a) Shibata N, Suzuki E, Takeuchi Y (2000) J Am Chem Soc 122:10728; (b) Shibata N, Suzuki E, Asahi T, Takeuchi Y (2001) J Am Chem Soc 123:7001; (c) Fukuzumi T, Shibata N, Sugiura M, Nakamura S, Toru T (2006) J Fluor Chem 127:548
117. (a) Cahard D, Audouard C, Plaquevent J-C, Toupet L, Roques N (2001) Tetrahedron Lett 42:1867; (b) Mohar B, Baudoux J, Plaquevent C, Cahard D (2001) Angew Chem Int Ed 40:4214
118. Cahard D, Audouard C, Plaquevent J-C, Roques N (2000) Org Lett 2:3699
119. Ishimaru T, Shibata N, Horikawa T, Yaasuda N, Nakamura S, Toru T, Shiro M (2008) Angew Chem Int Ed 47:4157
120. Kim DY, Park EJ (2002) Org Lett 4:545
121. Enders D, Hüttl MRM (2005) Synlett 991
122. (a) Marigo M, Fielenbach D, Braunton A, Kjærsgaad A, Jørgensen KA (2005) Angew Chem Int Ed 44:3703; (b) Steiner DD, Mase N, Barbas CF III (2005) Angew Chem. Int Ed 44: 3706; (c) Beeson TD, MacMillan DWC (2005) J Am Chem Soc 127:8826
123. Jiang H, Falcicchio A, Jensen KL, Paixão MW, Bertelsen S, Jørgensen KA (2009) J Am Chem Soc 131:7153
124. Fadeyi OO, Lindsley CW (2009) Org Lett 11:943
125. Brandes S, Niess B, Bella M, Prieto A, Overgaard J, Jørgensen KA (2006) Chem Eur J 12:6039
126. (a) Wack H, Taggi AE, Hafez AM, Drury WJ III, Lectka T (2001) J Am Chem Soc 123:531; (b) Taggi AE, Wack H, Hafez AM, France S, Lectka T (2002) Org Lett 4:27; (c) France S, Wack H, Taggi AE, Hafez AM, Wagerle TR, Shah MH, Dusich CL, Lectka T (2004) J Am Chem Soc 126: 245
127. Brochu MP, Brown SP, MacMillan DWC (2004) J Am Chem Soc 126:108
128. (a) Halland N, Braunton A, Bachmann S, Marigo M, Jørgensen KA (2004) J Am Chem Soc 126:790; (b) Halland N, Lie MA, Kjærsgaad A, Marigo M, Schiøtt B, Jørgensen KA (2005) Chem Eur J 11:7083; (c) Kang B, Britton R (2007) Org Lett 9:5083
129. Marigo M, Bachmann S, Halland N, Braunton A, Jørgensen KA (2004) Angew Chem Int Ed 43:5507
130. (a) Hafez AM, Taggi AE, Wack H, Esterbrook J, Lectka T (2001) Org Lett 3:2049; (b) Dogo-Isonagie C, Bekele T, France S, Wolfer J, Weatherwax A, Taggi AE, Lectka T (2006) J Org Chem 71:8946; (c) Dogo-Isonagie C, Bekele T, France S, Wolfer J, Weatherwax A, Taggi AE, Paull DH, Dudding T, Lectka T (2007) Eur J Org Chem 1091
131. Bertelsen S, Halland N, Bachmann S, Marigo M, Braunton A, Jørgensen KA (2005) Chem Comm 4821
132. Kano T, Ueda M, Maruoka K (2008) J Am Chem Soc 130:3728

Chapter 5
Chiral Primary Amine Catalysis

Liujuan Chen and Sanzhong Luo

Abstract Aminocatalysis is a versatile catalytic motif for both nature and organocatalysis. While Nature's aminoenzymes, e.g. Type I aldolase and decarboxylase, normally employ the primary amine group of lysine residue for catalysis, disproportionate emphasis has been placed on chiral secondary amines in organocatalysis due to the observed low activity in the initial attempt. Recently, there have been numerous exciting discoveries about the simple primary-amine organocatalysts. Beside their relevance in biomimetic catalysis and biogenesis, chiral primary aminocatalysis facilitates a range of transformation beyond the reach of secondary aminocatalysis. This review focuses on chiral primary amine catalysts, their history, catalytic features and applications as enamine, imminium and non-covalent catalysts.

5.1 Introduction

In the blossom of organocatalysis, the development with primary amine catalysts has been rather slow and largely neglected prior to 2005 despite of the remarkable successes with secondary aminocatalyts [1]. This fact is mainly ascribed to the relatively stabilized imine/iminium intermediate (I) and the assumed unfavorable iminium (I)-enamine (II) transformation in primary amine catalysis (Scheme 5.1). In fact,

L. Chen • S. Luo (✉)
Beijing National Laboratory for Molecular Sciences (BNLMS),
CAS Key Laboratory of Molecular Recognition and Function, Institute of Chemistry,
the Chinese Academy of Sciences, Beijing 100190, P.R. China
e-mail: luosz@iccas.ac.cn; chenliujuan@iccas.ac.cn

R. Mahrwald (ed.), *Enantioselective Organocatalyzed Reactions I: Enantioselective Oxidation, Reduction, Functionalization and Desymmetrization*,
DOI 10.1007/978-90-481-3865-4_5, © Springer Science+Business Media B.V. 2011

primary amine catalysis is a very ancient catalytic motif with its origin deeply rooted in Nature [2]. Over the past century, chemists have been exploring primary aminocatalysis and most of the earlier work focused on using simple primary amines as chemical mimics for aminocatalytic enzymes such as aldolases and decarboxylases, wherein ε-amino group of lysine is the ubiquitous catalytic functional group [3].

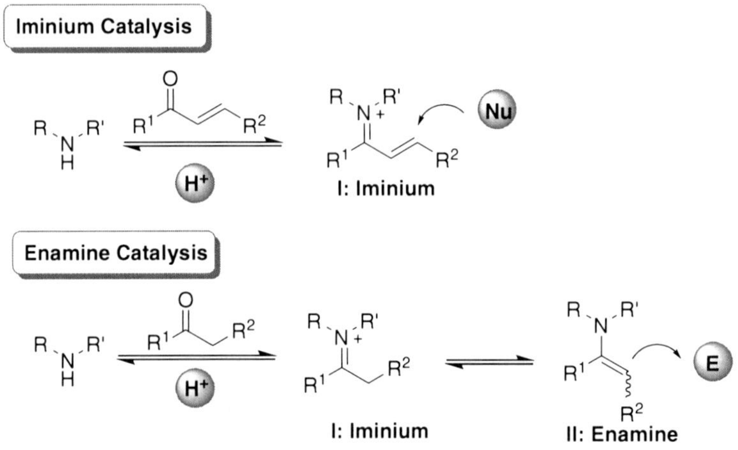

Scheme 5.1 Iminium and enamine catalysis

In 1930s, Westheimer formulated the mechanism of Type I aldolase via simple primary amine model studies [4]. Similar imine-enamine mechanism has also been proposed in primary amine mediated decarboxylation of acetoacetic acid by Peterson in early 1930s [5]. In 1970s, Hine and coworkers performed extensive studies on primary amine-catalyzed α-deuteration of carbonyl compounds, demonstrating the catalytic power of primary amines in rate acceleration and substrate activation [6]. In the classical Hajos-Parrish-Eder-Sauer-Wiechert reaction, primary amino acid phenylalanine was found to be an alternative catalyst besides proline and this process was later successfully utilized in the synthesis of steroids [7]. Collectively, these much earlier studies clearly indicated that primary amine should be viable aminocatalysts. Indeed, previous studies by Reymond found that primary benzyl amine catalysts promoted the typical direct aldol reaction of acetone and aldehyde in water ten times faster than the corresponding *N*-methylated secondary amine catalysts [8]. Computation studies from Houk suggested that hydrogen-bonding arisen from the secondary enamine *N*-H contributes favorably in the catalysis of primary amine [9]. However, it is not until 2005 when chiral primary amines were reinstated in the adolescence of organocatalysis [10]. Since then, chiral primary amines have been shown to be powerful aminocatalysts for a range of asymmetric transformations beyond the reach of typical chiral secondary amine catalysts such as chiral pyrrolidines. In this chapter, we will summarize the major advances along this line since the year of 2005 (Scheme 5.2).

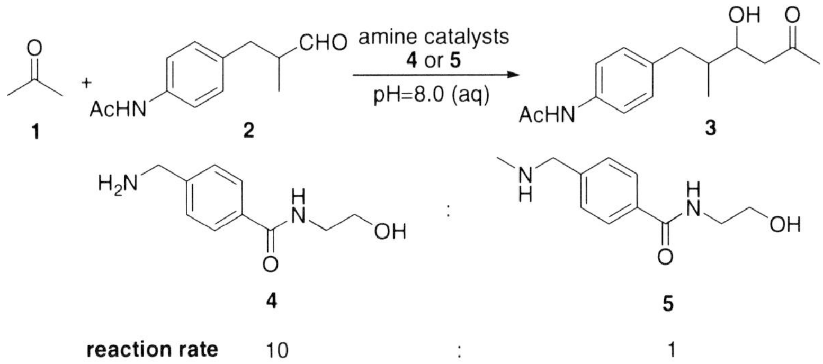

amine catalysts **4** or **5**

pH=8.0 (aq)

1 + AcHN **2** CHO → AcHN **3**

4 : **5**

reaction rate 10 : 1

Scheme 5.2 Primary and secondary amine catalysis

5.2 Enamine-Based Chiral Primary Amine Catalysis

5.2.1 Aldol Reaction

Aldol reaction is one of the fundamental C-C bond forming reactions. Though extensively developed in chiral metal-, bio- and organo- catalysis [11], there remains challenging problems with respects to scopes, efficiency and selectivity in asymmetric direct aldol reactions. Chiral primary amines represent the most recently developed catalysts to address those problems. Primary amino acids are certainly the first and also the most extensively explored primary amine catalysts for asymmetric direct aldol reactions. In 2004, Weber reported that simple primary amino acids such as alanine and isovaline catalyzed the aldol reaction of glycolaldehyde in water to generate tetroses with low yields but significant enantioselectivity, disclosing an intriguing relevance in the origin of homochirality [12]. Subsequently, Amedjkouh [10d], Cordova [10a–c] and Lu [13a–c] independently reported primary amino acids such as L-valine, L-alanine and L-tryptophan catalyzed asymmetric direct aldol reactions of ketones in the presence of water. The reactions proceeded particularly well with cyclohexanone donor with good yields and stereoselectivity, results comparable with those obtained with L-proline catalysis (Scheme 5.3). Later on, small peptides with primary amino acid terminals have also been examined in asymmetric direct aldol reactions [14].

The real catalytic potential of primary amino acids was disclosed by Barbas in the asymmetric direct aldol reactions of α-hydroxyketones such as hydroxyacetone and dihydroxyacetone (DHA) catalyzed by *O*-*t*-Bu-L-threonine **14** [15]. The reactions gave unprecedentedly *syn*-selectivity, to note that most chiral secondary amine catalysts like L-proline are *anti*-selective in these reactions (Scheme 5.4). A *Z*-enamine was proposed to account for the *syn*-diastereoselectivity and in this model the secondary enamine *N*-H would participate in intramolecular hydrogen-bonding with OH group, contributing the *Z*-enamine (Scheme 5.4). Very recently,

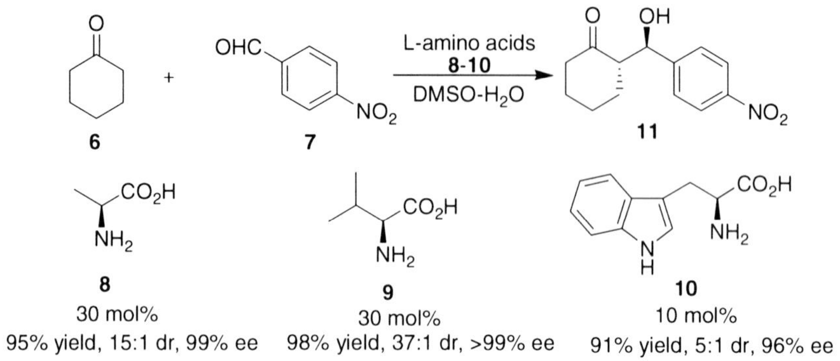

Scheme 5.3 Primary amino acid catalysis

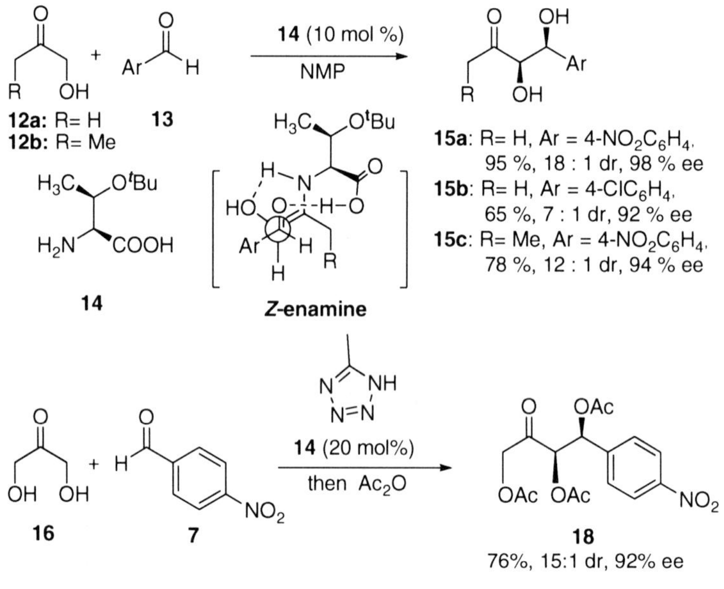

Scheme 5.4 *Syn*-selective aldol additions

Mahrwald reported that simple L-histidine could promote cross aldol reactions of aldehydes in water with good activity and moderate to good stereoselectivity (Scheme 5.5) [16].

Inspired by the primary aminocatalytic motif in nature, our group is interested in developing simple and efficient chiral primary amine catalysts. With Hine's pioneering contribution as a starting point, we developed simple primary-tertiary diamine catalysts derived from chiral *trans*-cyclohexanediamine such as **24** for asymmetric aldol reaction with excellent efficiency and enantioselectivity.

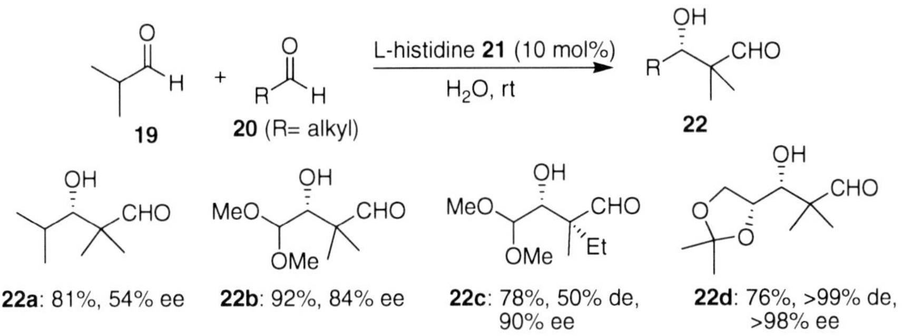

L-histidine **21** (10 mol%)
H_2O, rt

19 + **20** (R= alkyl) → **22**

22a: 81%, 54% ee
22b: 92%, 84% ee
22c: 78%, 50% de, 90% ee
22d: 76%, >99% de, >98% ee

Scheme 5.5 Histidine-catalyzed aldol additions

The catalysis of **24**-TfOH conjugate not only reached much improved efficiency and stereoselectivity over L-proline in the typical direct aldol reactions of acetone, but enabled the first *syn*-diastereoselective aldol reactions of linear aliphatic ketones (Scheme 5.6) [17a].

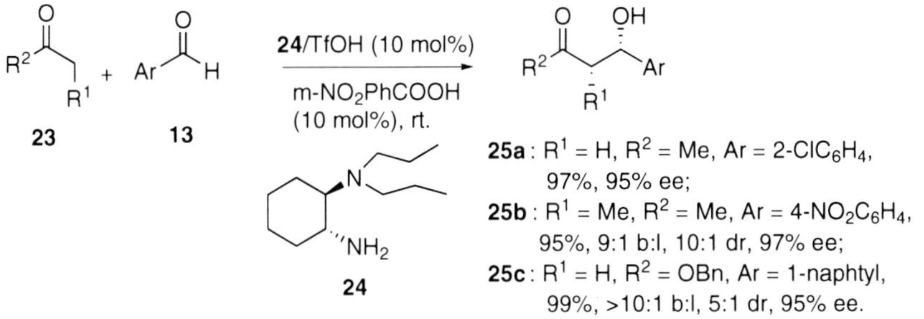

24/TfOH (10 mol%)
m-NO_2PhCOOH (10 mol%), rt.

23 + **13** → **25**

25a : R^1 = H, R^2 = Me, Ar = 2-ClC_6H_4, 97%, 95% ee;
25b : R^1 = Me, R^2 = Me, Ar = 4-$NO_2C_6H_4$, 95%, 9:1 b:l, 10:1 dr, 97% ee;
25c : R^1 = H, R^2 = OBn, Ar = 1-naphtyl, 99%, >10:1 b:l, 5:1 dr, 95% ee.

Scheme 5.6 Primary, tertiary 1.2-diamine catalysis

The simple primary-tertiary diamine salts can be successfully applied in the aldol reactions of α-hydroxyketones with good activity and excellent stereoselectivity. Notably, the catalyst enabled the reaction of dihydroxyacetone (DHA), a versatile C3-building block in the chemical and enzymatic synthesis of carbonhydrates. By employing either free or protected DHA, *syn*- or *anti*-diols could be selectively formed with excellent enantioselectivity (Scheme 5.7). Since enantiomers of diamine **26** and **29** are readily available, this class of chiral primary amine catalysts thus functionally mimics four types of DHA aldolases in nature [17b]. Later, simple chiral primary-tertiary diamine **27** derived from amino acid was also found to be a viable catalyst for the *syn*-selective aldol reactions of hydroxyacetone and free DHA (Scheme 5.7) [18].

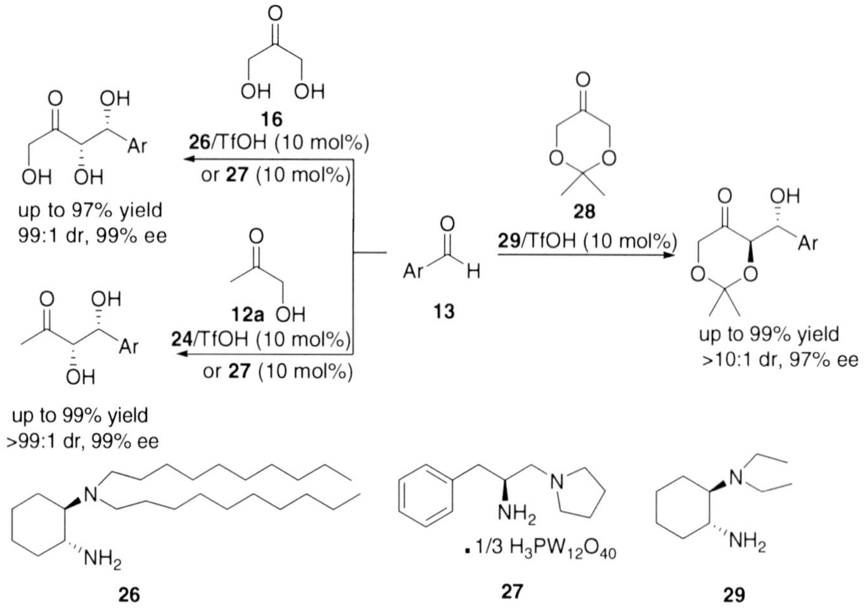

Scheme 5.7 Primary, tertiary 1.2-diamine catalysis

The obtained *syn* diastereoselectivity with acyclic ketone donors could be explained by *Z*-enamine transition state (Scheme 5.8), whereas the reactions of cyclic ketones such as protected DHA, which are capable only of forming *E*-enamine due to ring size constraints, give preferentially *anti*-aldol adducts (Scheme 5.8). These models can be applied to other primary-tertiary diamine catalyzed aldol reactions.

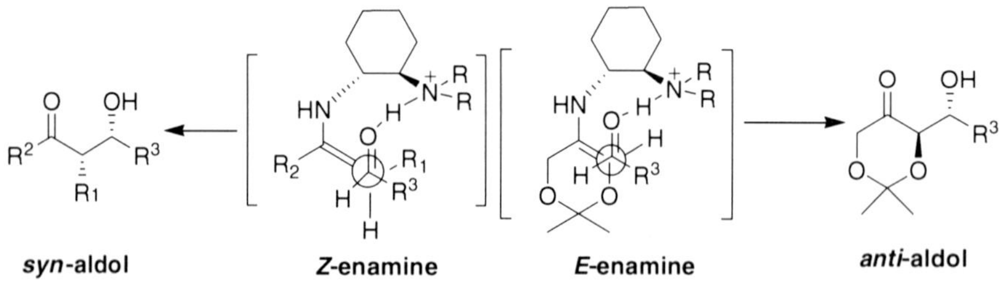

Scheme 5.8 Proposed transition state models

Subsequently, the catalytic potentials of chiral primary-tertiary diamines have been further explored in the direct aldol reactions of pyruvic donors. Primary–tertiary diamine-TfOH conjugate can effectively catalyze the coupling of pyruvic aldehyde acetals with aldehyde with excellent *syn*-diastereoselectivities and enantioselectivities (Scheme 5.9). This process functionally mimics the pyruvate-dependent type I aldolase [19].

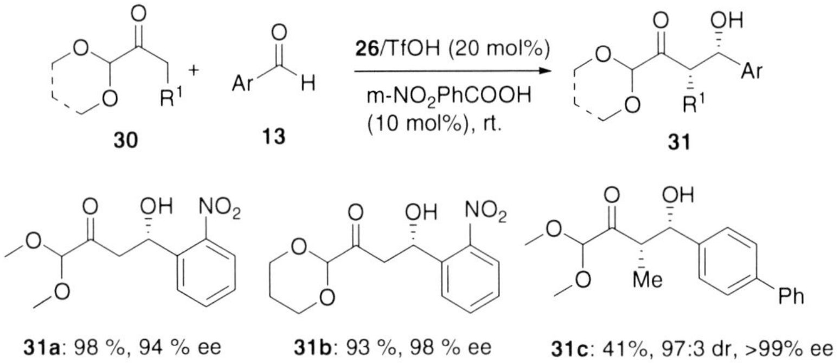
30
13
26/TfOH (20 mol%)
m-NO2PhCOOH
(10 mol%), rt.
31
31a: 98 %, 94 % ee
31b: 93 %, 98 % ee
31c: 41%, 97:3 dr, >99% ee

Scheme 5.9 Primary-tertiary diamine catalysis

Very recently, a similar primary-tertiary diamine **33**-DNBS conjugate was identified to catalyze the first direct aldol reactions of acetoacetal. The reactions afforded vinylogous-type aldol adducts with good yields and excellent enantioselectivity (Scheme 5.10). In cases of substituted acetoacetals, good *syn*-diastereoselectivity was obtained again [20].

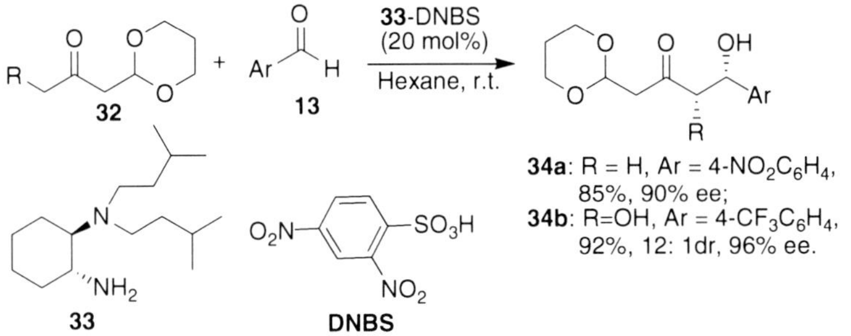
32
13
33-DNBS
(20 mol%)
Hexane, r.t.
34a: R = H, Ar = 4-NO2C6H4,
85%, 90% ee;
34b: R=OH, Ar = 4-CF3C6H4,
92%, 12: 1dr, 96% ee.
33
DNBS

Scheme 5.10 Primary-tertiary diamine catalysis

Since 2007, other chiral primary amines such as **35–39** (Fig. 5.1) have been synthesized and explored in the catalysis of asymmetric direct aldol reactions [21–23]. Maruoka developed primary amine catalysts **36** and **37** using a common chiral skeleton. Interestingly, these two catalysts demonstrated opposite chiral

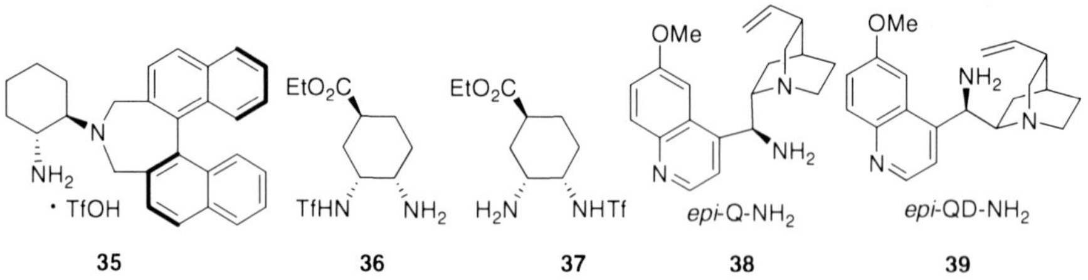

Fig. 5.1 1,2-Diamine catalysts

induction in aldol reactions of cyclohexanone and aromatic aldehydes [22]. Using chiral primary-tertiary diamine type catalysts **38** and **39** derived from cinchona alkaloids, List and coworkers realized highly enantioselective desymmerizations of 2, 6-heptadione via intramolecular aldol reactions (Scheme 5.11) [23].

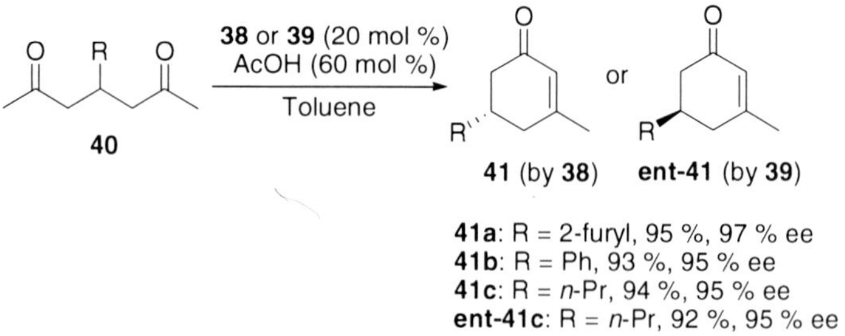

41a: R = 2-furyl, 95 %, 97 % ee
41b: R = Ph, 93 %, 95 % ee
41c: R = *n*-Pr, 94 %, 95 % ee
ent-41c: R = *n*-Pr, 92 %, 95 % ee

Scheme 5.11 Enantioselectice desymmetrization

Gong group designed chiral primary amine-amide type catalyst for the aldolization of hydroxyacetone [24] with excellent *syn*-aldol selectivity (up to >20:1 dr.) and stereoselectivity (up to 98% ee) (Scheme 5.12). Recently, Zhao and Da have further explored this type of primary amine catalysts [25]. Feng developed bispidine-derived chiral primary amine catalyst **45** for the aldolization of activated ketone acceptors with excellent enantioselectivity (Scheme 5.12) [26].

Aldol reaction is well known to be an intrinsically reversible process. Microscopically, asymmetric retro aldol reaction would address those challenging substrates which are normally sluggish under the typical asymmetric aldol conditions. Unfortunately, though principally conceivable, such asymmetric retro aldol processes remain basically underdeveloped in asymmetric synthesis. In their further explorations on chiral primary amine catalysis, Luo group found that simple chiral primary–tertiary diamine such as **24, 26, 27, 29, 33** catalyze unprecedentedly stereoselective retro aldol reactions (Scheme 5.13).

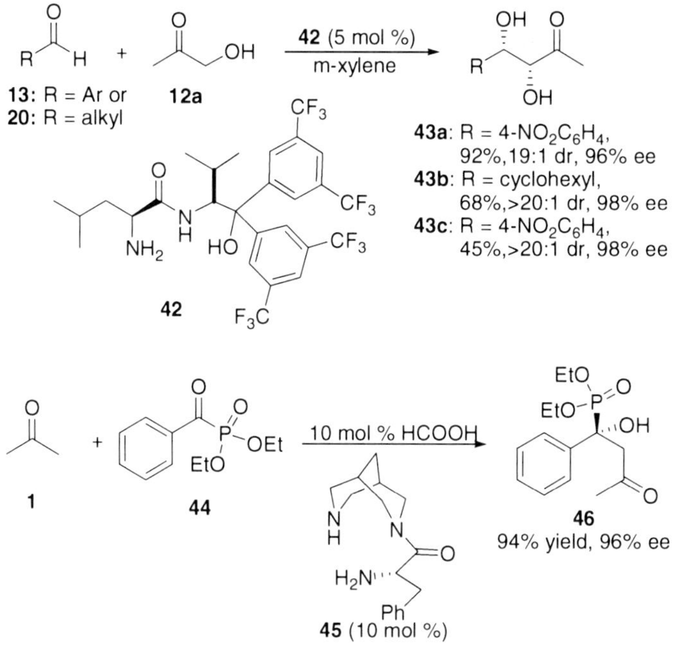

13: R = Ar or **20:** R = alkyl + **12a** → **42** (5 mol %), m-xylene →

43a: R = 4-$NO_2C_6H_4$, 92%, 19:1 dr, 96% ee
43b: R = cyclohexyl, 68%, >20:1 dr, 98% ee
43c: R = 4-$NO_2C_6H_4$, 45%, >20:1 dr, 98% ee

42

1 + **44** → 10 mol % HCOOH, **45** (10 mol %) → **46**, 94% yield, 96% ee

Scheme 5.12 Amid catalyzed aldol additions

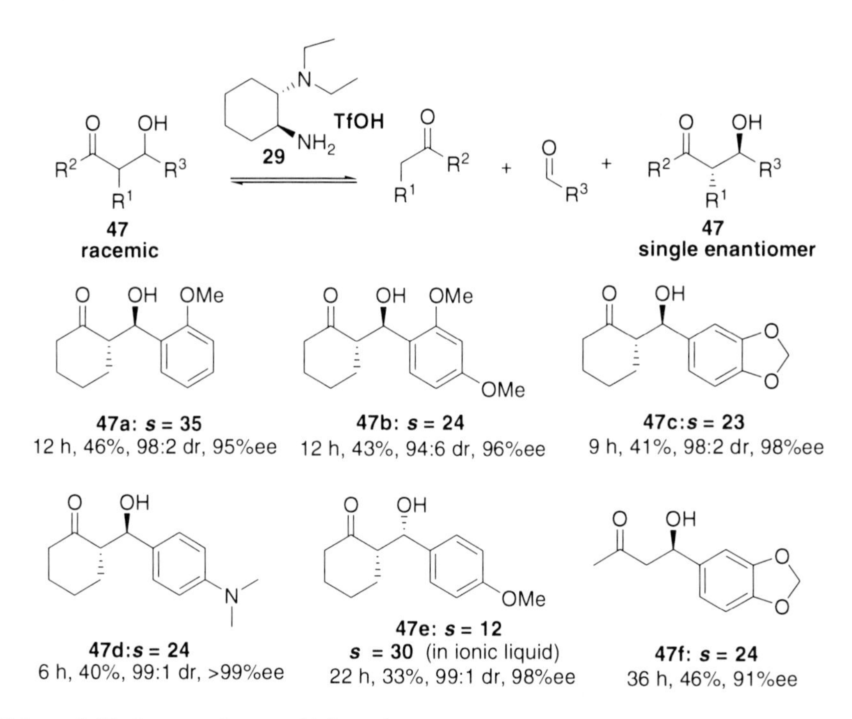

Scheme 5.13 Asymmetric retro aldol reaction

This retro aldol protocol enabled an ideal catalytic kinetic resolution of racemic aldol adducts that are usually difficult to be obtained through forward processes as illustrated by the resolutions of cyclohexanone aldol adducts. An intriguing feature of this process is that one chiral primary amine (e.g. **29**) could catalyze stereoselectively the resolution of both *anti-* and *syn*-configured aldol adducts, whereas the forward reactions with the same catalyst yield selectively *anti*-configured aldol products. In addition, the catalytic power of **29**-TfOH on both aldol and retro aldol reactions has also made possible an unprecedented asymmetric transfer aldol reaction that can generate two enantioenriched aldol adducts with opposite chiral induction from a single chiral catalyst (e.g. aldol products **48** and **49** in Scheme 5.14) [27].

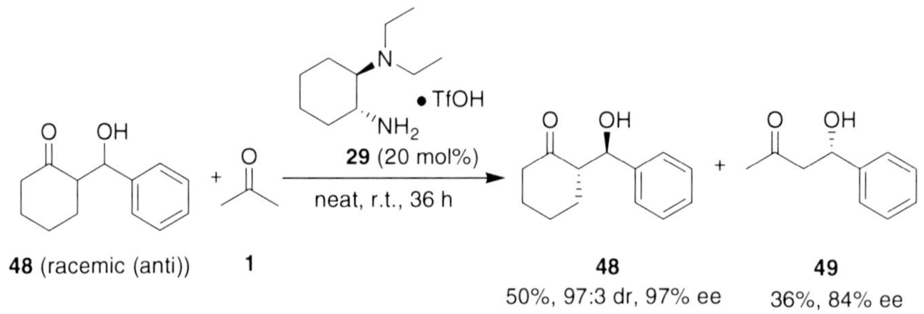

48 (racemic (anti)) + **1** → [**29** (20 mol%), •TfOH; neat, r.t., 36 h] → **48** 50%, 97:3 dr, 97% ee + **49** 36%, 84% ee

Scheme 5.14 Asymmetric transfer aldol reaction

5.2.2 *Mannich Reaction*

Following immediately the initial efforts on primary amino acids catalyzed aldol reactions, the application of primary amine acid in Mannich reaction has also been attempted. Cordova reported that simple primary amino acids and their derivatives could catalyze the asymmetric Mannich reactions of ketones with comparable results to those obtained in the catalysis of proline[28]. Later, Barbas [29] and Lu [30] independently reported that L-Trp or *O*-protected L-Thr could catalyze *anti*-selective asymmetric Mannich reactions of α-hydroxyacetones with either pre-formed or *in-situ* generated imines. The preference of *anti*-diastereoselectivity was ascribed to the formation of a *Z*-enamine, with the assistance of an intramolecular H-bond (Scheme 5.15).

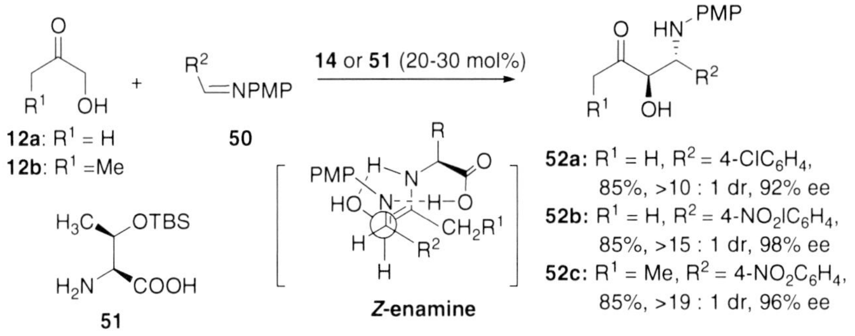
14 or 51 (20-30 mol%)
12a: R1 = H
12b: R1 =Me
50
51
Z-enamine
52a: R1 = H, R2 = 4-ClC6H4, 85%, >10 : 1 dr, 92% ee
52b: R1 = H, R2 = 4-NO2lC6H4, 85%, >15 : 1 dr, 98% ee
52c: R1 = Me, R2 = 4-NO2C6H4, 85%, >19 : 1 dr, 96% ee

Scheme 5.15 Asymmetric direct Mannich reactions (PMP: p-methoxyphenyl)

5.2.3 *Michael Addition Reactions*

Asymmetric Michael addition reaction represents one of the most extensively explored organoctalytic transformations. Chiral secondary amines such as pyrrolidines have been proved to very effective catalysts for the reactions of Michael donors such as cyclohexanone and aldehydes [31]. Ala-Ala dipeptide **57** [32] was firstly found to be viable catalyst for the reaction of cyclohexanone and nitrostyrene. Later, some primary amine-amide type catalysts such as **55** [33a], **56** [33b–c] and **58** [33d] have also been indentified for the same reactions (Scheme 5.16) with slightly lower activity compared with typical chiral pyrrolidine catalysis.

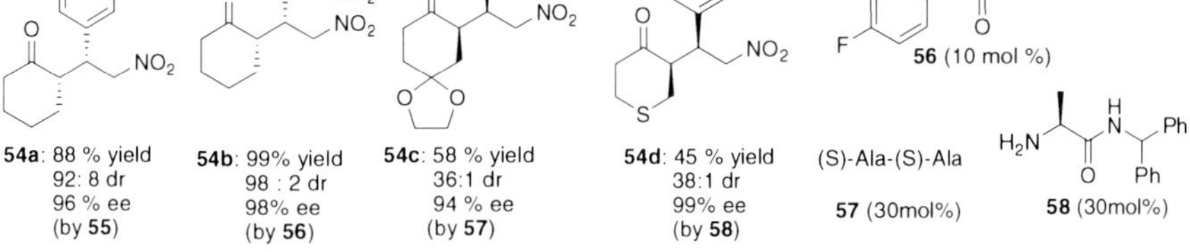
23
53: R3 = Ar
54
55(15 mol %)
56 (10 mol %)
54a: 88 % yield 92: 8 dr 96 % ee (by 55)
54b: 99% yield 98 : 2 dr 98% ee (by 56)
54c: 58 % yield 36:1 dr 94 % ee (by 57)
54d: 45 % yield 38:1 dr 99% ee (by 58)
(S)-Ala-(S)-Ala
57 (30mol%)
58 (30mol%)

Scheme 5.16 Asymmetric Michael addition

Interestingly, though primary amino acids were unable to catalyze the Michael addition to nitrostyrene, their alkaline salts such as lithium phenylalanine were found to be viable catalysts for the reaction (Scheme 5.17) [33a,34].

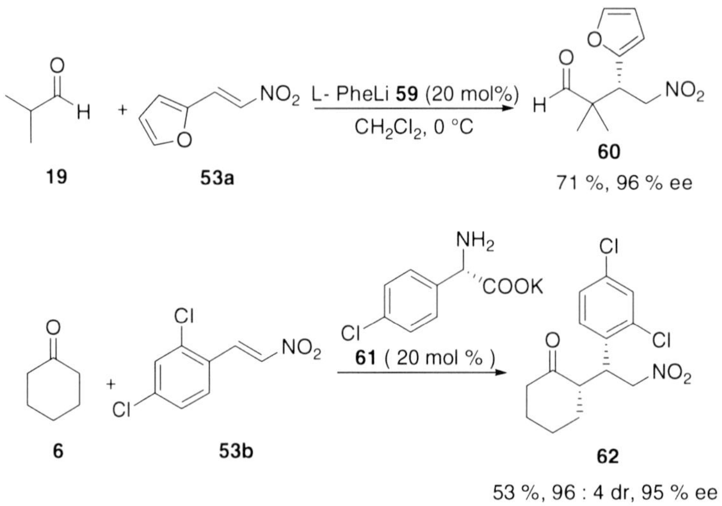

Scheme 5.17 Asymmetric Michael addition

In the realm of secondary amino catalysis, acyclic aliphatic ketone donors have remained as challenging substrates in asymmetric Michael additions to nitrostyrene. Breakthroughs came with chiral primary amine-thioureas catalysts. Tsogoeva group [35] and Jacobsen group [36] have independently developed chiral primary amine-thiourea catalysts for the Michael reaction of acyclic ketones with nitroolefins (Schemes 5.18 and 5.19). In both cases, good activity and enantioselectivity have been achieved in the reactions of acetones. Notably, *anti*-stereoselectivities were obtained in

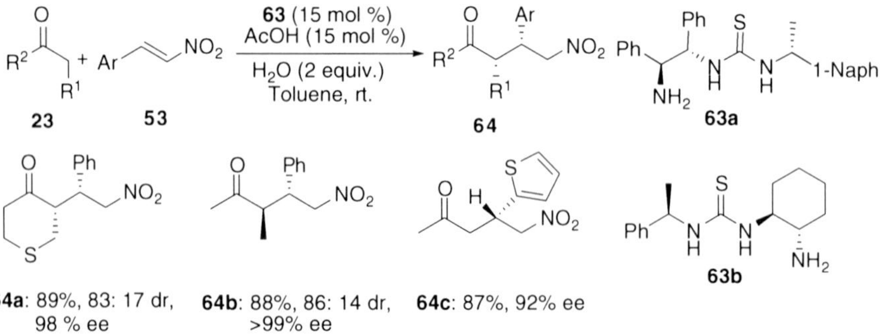

Scheme 5.18 Tsogoeva's primary amine thiourea catalyst

cases of larger acyclic ketones such as 2-butanone and hydoxyacetone (Schemes 5.18 and 5.19). Jacobson's thiourea catalysts such as **66** can also effectively catalyze the reactions of α-substituted aldehydes to nitroolefins with good *anti*-diastereoselectivity and excellent enantioselectivity (Scheme 5.19). *Anti*-selective Michael addition of protected glycolaldehyde to nitrostyrene has also been realized using a similar primary-thiourea catalyst based on cyclohexanediamine[37].

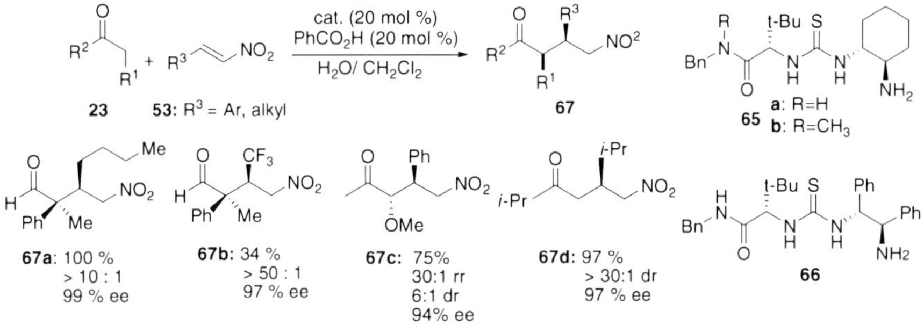

Scheme 5.19 Jacobsen's primary amine thiourea catalysts

Recently, carbohydrates and steroids derived primary-thioureas catalysts **69** and **70** have been reported for the Michael reactions of aryl ketones and aryl nitrolefins with high yield and enantioselectivity (Scheme 5.20)[38]. Primary amine-sulfonamide catalysts can also be applied to similar reactions with good to high enantioselectivity [39].

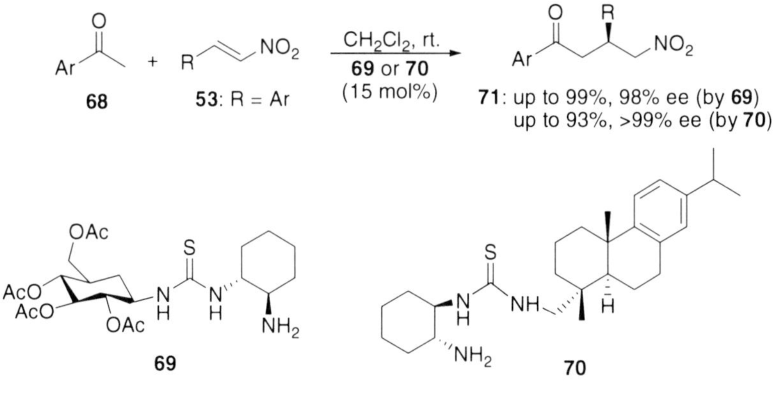

Scheme 5.20 Asymmetric Michael addition

Chiral primary amines based on cinchona alkaloids are versatile primary aminocatalysts and have been attempted in the asymmetric Michael additions of aliphatic ketones to nitro-olefins with moderate to high yield and good stereoselectivity [40].

This type of catalysts has also enabled the Michael reaction to vinyl sulfone [41] (up to 93% yield, 6:1 dr and 97% ee, Scheme 5.21) and asymmetric Michael reaction of aromatic ketones to alkylidenemalononitriles with moderate enantioselectivity [42] (Scheme 5.21).

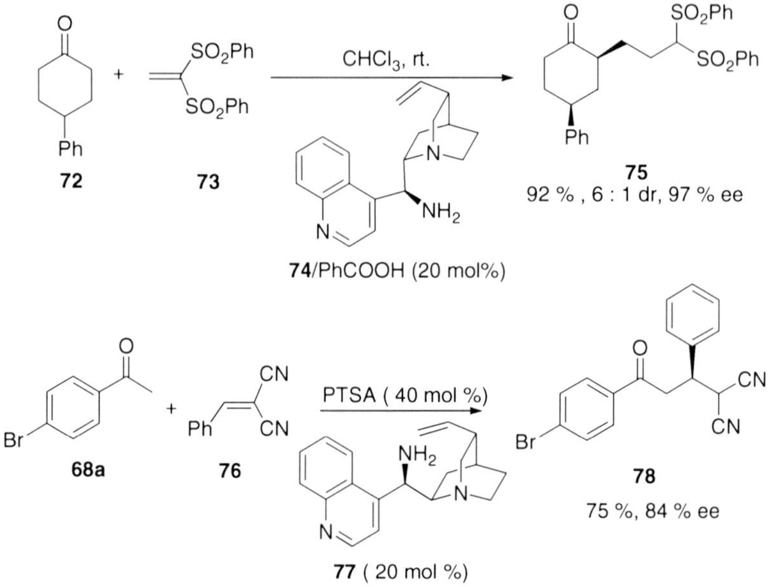

Scheme 5.21 Asymmetric Michael addition

5.2.4 α-Amination

Chen group developed first asymmetric amination of aryl ketones by employing primary amine catalyst derived from cinchona alkaloid [43]. In the catalysis of **77**, an array of aryl ketones react with diethyl azodicarboxylate to afford α-amino products with up to 99% ee (Scheme 5.22).

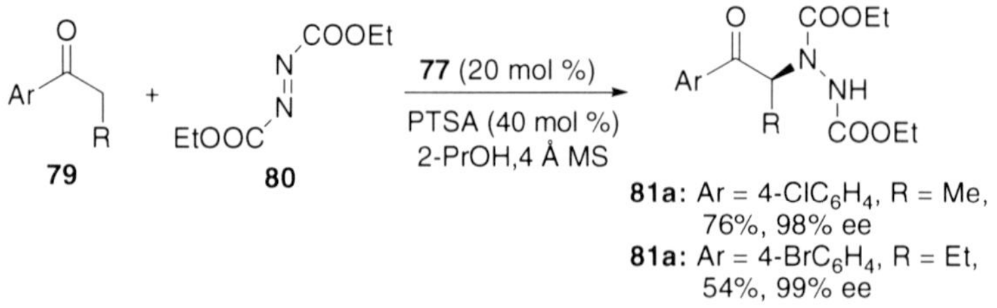

Scheme 5.22 Asymmetric amination catalyzed by cinchona alkaloid derivative

5.2.5 *α-Fluorination*

The product of asymmetric amination, diamine **84**, was reported to be able to catalyze asymmetric fluorination of α-branched aldehydes, yielding 1, 2-fluorine alcohols with up to 90% ee (Scheme 5.23) [44]. Worthy to note that the chiral center of the catalyst is orientated on the nitrogen atom and it is the first example of asymmetric organocatalyst with enantiomeric 'nitrogen' as the only source of chirality.

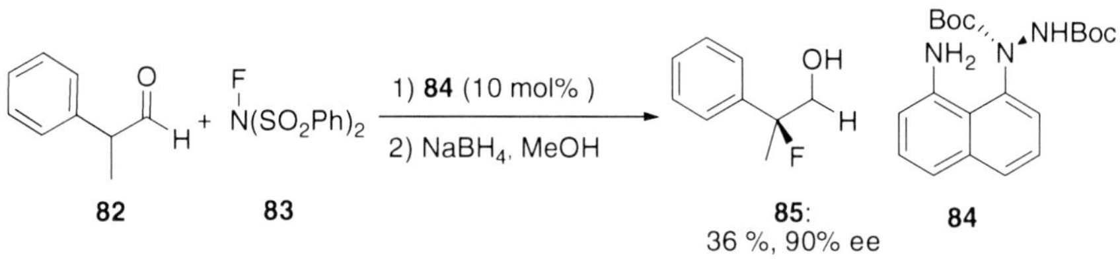

82 + 83 ($FN(SO_2Ph)_2$) → 1) **84** (10 mol%); 2) $NaBH_4$, MeOH → **85**: 36 %, 90% ee

84: Boc, NHBoc, NH_2, N

Scheme 5.23 Asymmetric fluorination catalyzed by primary amine catalyst

5.3 Iminium-Based Chiral Primary Amine Catalysis

The potential of chiral secondary amines in iminium catalysis [45], exemplified by MacMillan's imidazoline catalyst, [46, 47] have been well demonstrated in the reactions involving α,β-unsaturated aldehydes. However, the success with deployment of α,β-unsaturated ketones or α-substituted enals have been quite limited, presumably due to the steric hindrance in forming iminium intermediates with secondary amines. In this context, chiral primary amines have recently emerged as viable and powerful catalysts for these challenging substrates.

5.3.1 *Asymmetric Conjugate Additions with Carbon Nucleophiles*

Asymmetric Friedel-Crafts alkylations of indoles with α,β-unsaturated carbonyl compounds have been and continue to be of significant interests in synthesizing chiral indole alkaloids. Following the very successful iminium-catalysis with enals by MacMillan's catalyst [47], Chen and Melchiorre have independently reported asymmetric Friedel-Crafts alkylation of indoles with α,β-unsaturated aryl ketones [48] using similar cinchona-alkaloid derived catalysts **77** and **91**, respectively (Scheme 5.24) [49]. In both cases, the proper choice of an acidic additive has been shown to be essential for catalytic activity and stereoselectivity.

1, 3-Dicarbonyl compounds are versatile stabilized enolate-type nucleophiles in asymmetric Michael addition reactions. In particular, the reaction of 4-hydroxy-2H-chromen-2-one and α,β-enones provided a direct and efficient synthesis of warfarin, a widely prescribed anticoagulant. Recently, Chin reported a simply vicinal

MeO
86
Ph
Et
87
77 (30 mol%)
TfOH (60 mol %)
Ph
Et
MeO
NH
88: 91%, 85% ee

89
R^1
R^2
90
91 (20 mol %)
Boc-D-Phg-OH
(40 mol %)
$PhCH_3$, 0.2 M
OMe
N
NH_2
91
R^1
R^2
92a: R^1 = COOEt, R^2 = Me, 99 %, 95 % ee
92a: R^1 = Me, R^2 = Et, 76 %, 95 % ee
92a: R^1 = *n*-Pen, R^2 = Me, 84 %, 94 % ee

Scheme 5.24 Asymmetric Friedel-crafts alkylation

diamine such as **95** could effectively catalyze the synthesis of warfarin with up to 92% ee [50a–b]. A diimine intermediate from the vicinal diamine and ketone was proposed to be involved in the reaction coordinates (Scheme 5.25). Later, Chen and coworkers explored the use of quinine-based primary [50a, 57] amine catalyst such as **38** in the similar Michael addition reaction with quite broad substrate scope and excellent enantioselectivity (Scheme 5.25) [51].

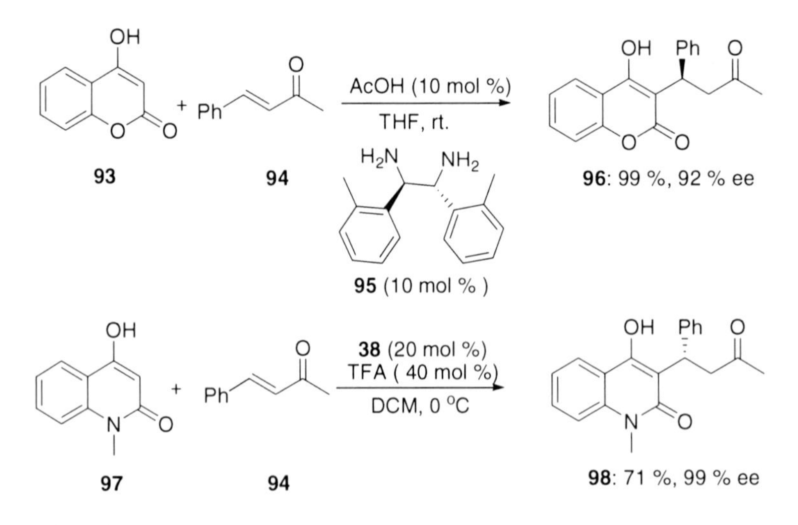

Scheme 5.25 Asymmetric Michael addition

α,α-Dicyanoalkenes have been recently established as a useful vinylogous synthons via the works by Jorgenson [52a–b] and Chen [52c]. Chen and coworkers reported first asymmetric direct vinylogous Michael addition of α,α-Dicyanoalkenes to α,β-unsaturated ketones using a quinine-based primary amine catalyst **38** (Scheme 5.26) [53]. The reactions accommodate a range of enones and dicyanoalkenes with excellent stereoselectivity. Cinchona-based primary amine catalyst **39** has also reported by Deng and coworkers to catalyze the asymmetric addition of malononitrile to α,β-unsaturated ketones (Scheme 5.26) [42].

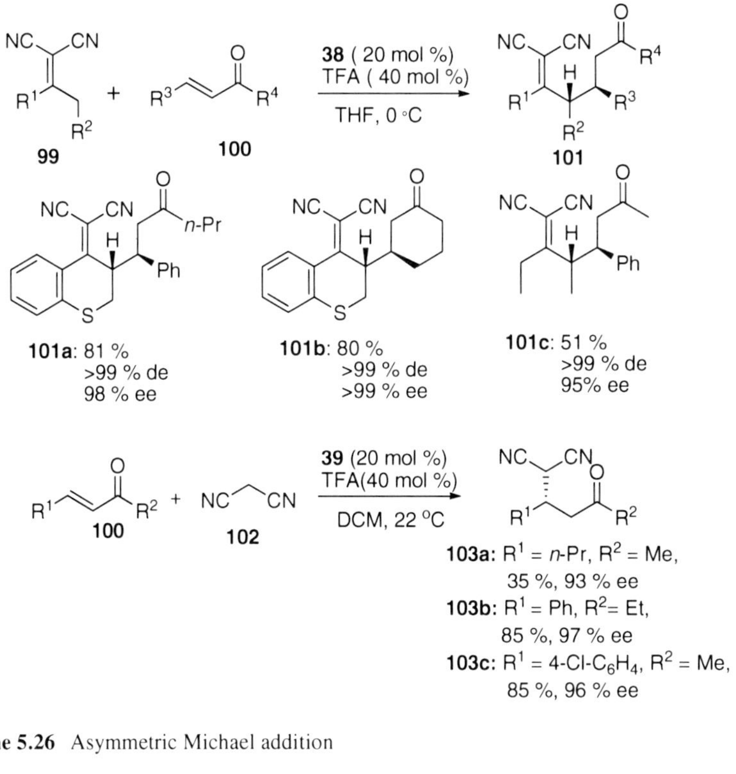

99 + **100** → **101**: **38** (20 mol %), TFA (40 mol %), THF, 0 °C

101a: 81 %, >99 % de, 98 % ee

101b: 80 %, >99 % de, >99 % ee

101c: 51 %, >99 % de, 95% ee

100 + **102** → **103**: **39** (20 mol %), TFA(40 mol %), DCM, 22 °C

103a: R^1 = *n*-Pr, R^2 = Me, 35 %, 93 % ee

103b: R^1 = Ph, R^2= Et, 85 %, 97 % ee

103c: R^1 = 4-Cl-C_6H_4, R^2 = Me, 85 %, 96 % ee

Scheme 5.26 Asymmetric Michael addition

Zhao and co workers [54] developed simply primary-secondary diamine catalysts derived from primary amino acids. This type of catalysts such as **105** was found to catalyze the asymmetric Michael addition of malonates to acyclic α,β-unsaturated ketones with good activity and excellent enantioselectivity (Scheme 5.27). Liang and coworkers designed a new primary amine catalyst combining two privileged skeletons, cinchona and cylohexanediamine [55]. The obtained optimal catalysts **107** and **110** were applicable to the Michael additions reactions of malonate or nitroalkanes to α,β-unsaturated ketones. The reactions worked well with both cyclic and acyclic enones (Scheme 5.28).

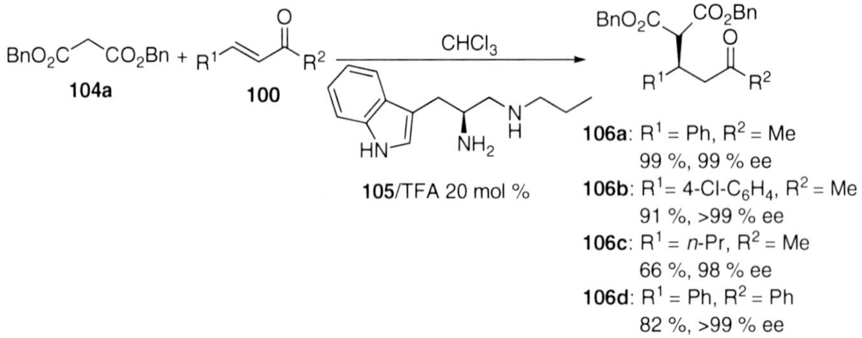

BnO_2C–CH_2–CO_2Bn (104a) + R^1–CH=CH–CO–R^2 (100) → ($CHCl_3$, 105/TFA 20 mol %) → 106

106a: R^1 = Ph, R^2 = Me
99 %, 99 % ee
106b: R^1= 4-Cl-C_6H_4, R^2 = Me
91 %, >99 % ee
106c: R^1 = *n*-Pr, R^2 = Me
66 %, 98 % ee
106d: R^1 = Ph, R^2 = Ph
82 %, >99 % ee

Scheme 5.27 Asymmetric Michael addition

100 + EtO_2C–CH_2–CO_2Et (104b), EtOAc, rt., 107 (10 mol %) → 108

100 + $R^3R^4CHNO_2$ (109), EtOAc, rt., 110 (10 mol %) → 111

108a: 80 %, 97 % ee
108b: 90 %, 97 % ee

107

111a: 70%, 96% ee
111b: 92%, 3:2 dr, 91/88% ee

110

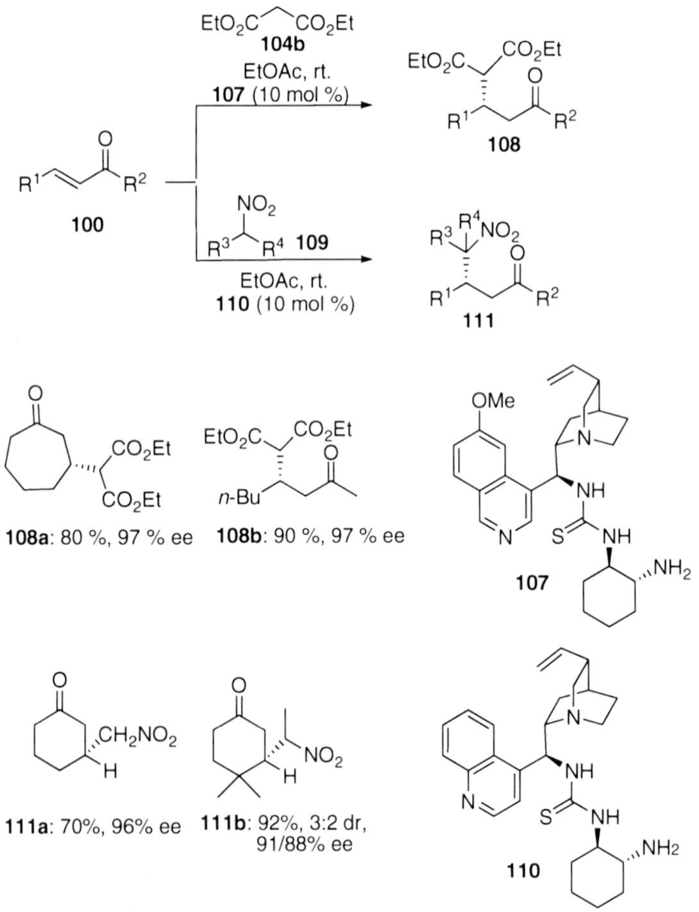

Scheme 5.28 Asymmetric Michael addition of malonates

Asymmetric catalytic method to generate C-C bond with adjacent quaternary-tertiary stereocenters remains a challenging synthetic task. In this regard, the use of 3-substituted oxindoles as carbon-nucleophiles has attracted intensive interests due to their relevance in synthesizing bioactive indole alkaloids. Melchiorre recently reported asymmetric Michael addition of 3-substituted oxindoles to α,β-unsaturated aldehydes catalyzed by a primary amine thiourea catalyst **114** (Scheme 5.29). Good diastereoselectivity and enantioselectivity have been achieved in most cases. However, the reactions were generally sluggish [56].

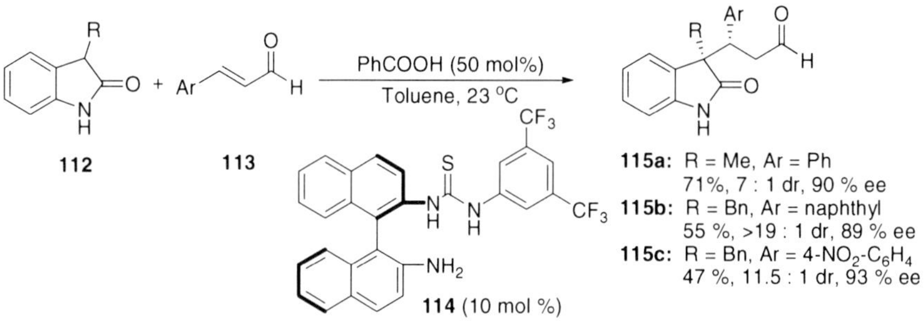

112 + **113** → PhCOOH (50 mol%), Toluene, 23 °C, **114** (10 mol %)

115a: R = Me, Ar = Ph, 71%, 7 : 1 dr, 90 % ee
115b: R = Bn, Ar = naphthyl, 55 %, >19 : 1 dr, 89 % ee
115c: R = Bn, Ar = 4-NO_2-C_6H_4, 47 %, 11.5 : 1 dr, 93 % ee

Scheme 5.29 Asymmetric Michael addition

5.3.2 *Asymmetric Conjugate Additions with Heteroatom Nucleophiles*

In a manner largely complementary to secondary amine-catalyzed asymmetric conjugate addition to enals with heteroatom nucleophiles, chiral primary amines were recently found to be the catalysts of choice for similar Michael addition to α,β-unsaturated ketones. With their previously developed cinchona-type catalyst **91**, Melchiorre and coworkers achieved the asymmetric sulfa-Michael addition to α,β-unsaturated ketones with either benzyl or *tert*-butyl mercaptane (Scheme 5.30) [58a]. The same catalyst could be further extended to oxa-Michael addition to enones by optimizing the ratio of acidic additive and solvents (Scheme 5.30) [58b].

Very recently, Deng and coworkers reported a highly enantioselective aza-Michael addition to enones with *N*-Boc-benzyloxyamine by employing a cinchona alkaloid-derived primary amine-TFA salt (Scheme 5.31) [59].

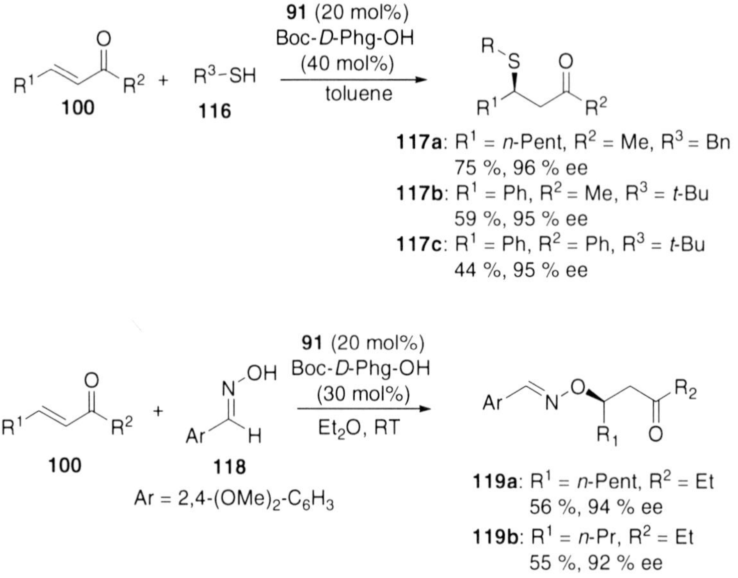

91 (20 mol%)
Boc-*D*-Phg-OH (40 mol%)
toluene

100 + R^3-SH **116** → **117**

117a: R^1 = *n*-Pent, R^2 = Me, R^3 = Bn
75 %, 96 % ee
117b: R^1 = Ph, R^2 = Me, R^3 = *t*-Bu
59 %, 95 % ee
117c: R^1 = Ph, R^2 = Ph, R^3 = *t*-Bu
44 %, 95 % ee

91 (20 mol%)
Boc-*D*-Phg-OH (30 mol%)
Et_2O, RT

100 + **118** → **119**

Ar = 2,4-$(OMe)_2$-C_6H_3

119a: R^1 = *n*-Pent, R^2 = Et
56 %, 94 % ee
119b: R^1 = *n*-Pr, R^2 = Et
55 %, 92 % ee

Scheme 5.30 Asymmetric thio-Michael addition

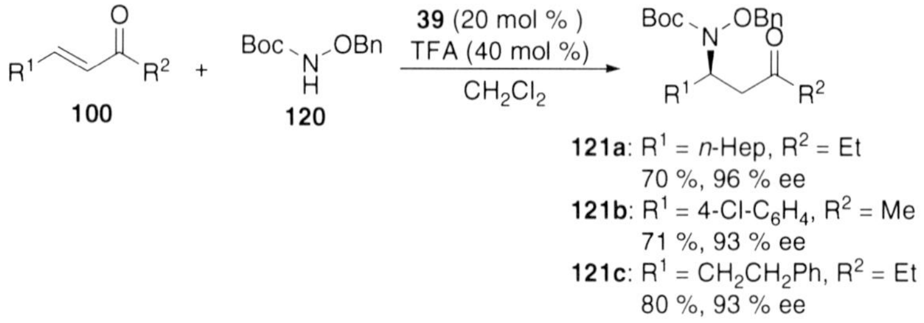

39 (20 mol %)
TFA (40 mol %)
CH_2Cl_2

100 + **120** → **121**

121a: R^1 = *n*-Hep, R^2 = Et
70 %, 96 % ee
121b: R^1 = 4-Cl-C_6H_4, R^2 = Me
71 %, 93 % ee
121c: R^1 = CH_2CH_2Ph, R^2 = Et
80 %, 93 % ee

Scheme 5.31 Asymmetric aza-Michael addition

5.3.3 Diels-Alder Reaction

Asymmetric Diels-Alder reaction represents one of the earliest successful organocatalytic reactions [46, 60]. Though extraordinary progresses have been achieved in this field, there are quite a few challenges remained to be addressed. For example, the reactions have been limited with enals and successful examples with enones are rare; the exo/endo selectivity is generally poor in many cases and a great number of α-substituted enals remain difficult substrates. Chiral primary amine such as **124** has been attempted in the Diels-Alder reactions of enals and cyclopentadiene. Though the reaction scope is quite limited, notable improvements on endo/exo ratio was obtained (Scheme 5.32) [61].

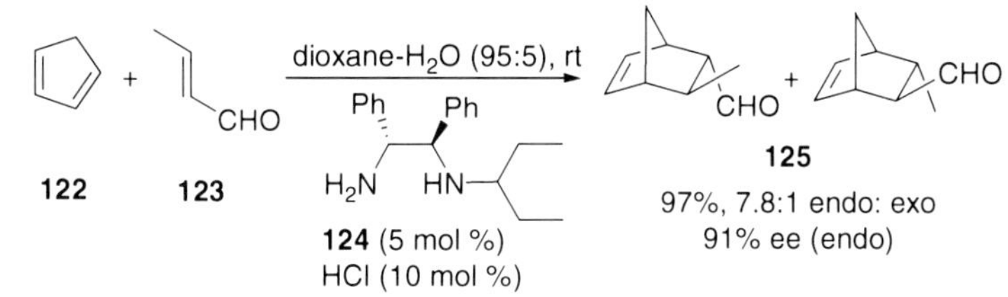

Scheme 5.32 Asymmetric Diels-Alder reaction

The first iminium-based asymmetric Diels-Alder reaction of α-acyloxyacrolein was realized by using a primary amine catalyst **128** derived from amino acids [62a]. The same catalyst could also be applied to the Diels-Alder reactions of α-phthalimidoacroleins with dienes [62b] with up to 99:1 *endo*:*exo*-selectivity and 96% ee (Scheme 5.33). Two factors are proposed to account for the high stereocontrol: π-π interaction between the phenyl group of the catalyst and the R group in the acroleins and the steric hindrance resulted from the bulky ion pair (Scheme 5.33).

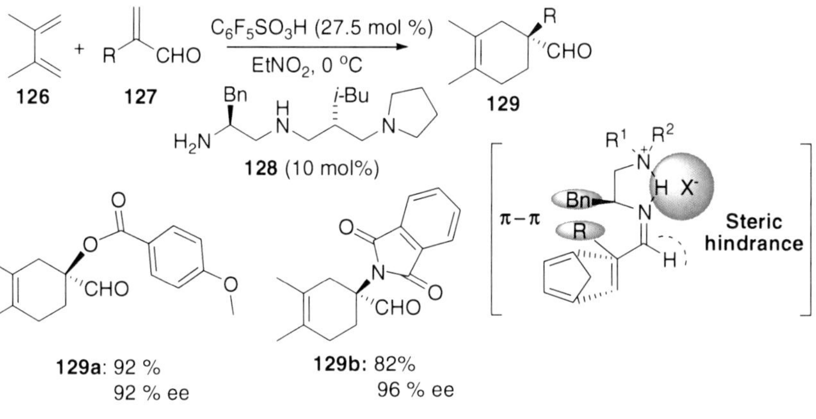

Scheme 5.33 Asymmetric Diels-Alder reaction

Axially chiral diamine **132** in combination with Tf_2NH has also been reported to promote the asymmetric Diels-Alder reaction of α-acyloxyacroleins [63]. Improvements on stereoselectivity were observed in the reactions with cyclopentadiene over those in catalysis with **128** (Scheme 5.34). With a more advanced analogue of **132**, i.e. diamine **134**, Maruoka reported highly enantioselective Diels-Alder reactions with α-alkyl substituted acroleins (Scheme 5.35) [64].

Deng and co-workers developed the first asymmetric Diels-Alder reactions of α,β-unsaturated ketones and 2-pyrones (Scheme 5.36) [65] with cinchona-based primary amine catalysts **38** or **39**. The substrate scopes are substantial and in most cases excellent diastereoselectivity and enantioselectivity were obtained.

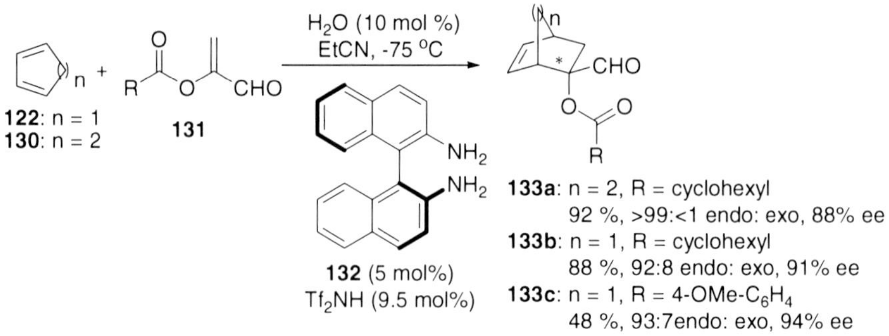

H_2O (10 mol %)
EtCN, -75 °C

122: n = 1
130: n = 2
131

132 (5 mol%)
Tf_2NH (9.5 mol%)

133a: n = 2, R = cyclohexyl
92 %, >99:<1 endo: exo, 88% ee
133b: n = 1, R = cyclohexyl
88 %, 92:8 endo: exo, 91% ee
133c: n = 1, R = 4-OMe-C_6H_4
48 %, 93:7endo: exo, 94% ee

Scheme 5.34 Asymmetric Diels-Alder reaction of α-acyloxyacroleins

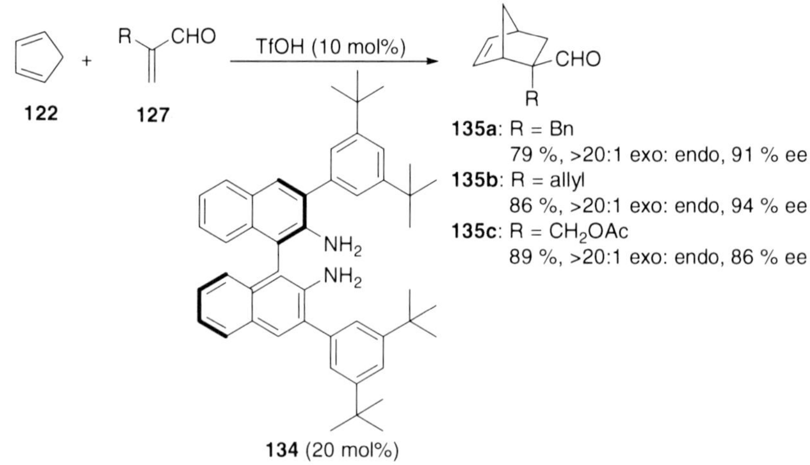

TfOH (10 mol%)

122 **127**

134 (20 mol%)

135a: R = Bn
79 %, >20:1 exo: endo, 91 % ee
135b: R = allyl
86 %, >20:1 exo: endo, 94 % ee
135c: R = CH_2OAc
89 %, >20:1 exo: endo, 86 % ee

Scheme 5.35 Asymmetric Diels-Alder reaction of acroleins

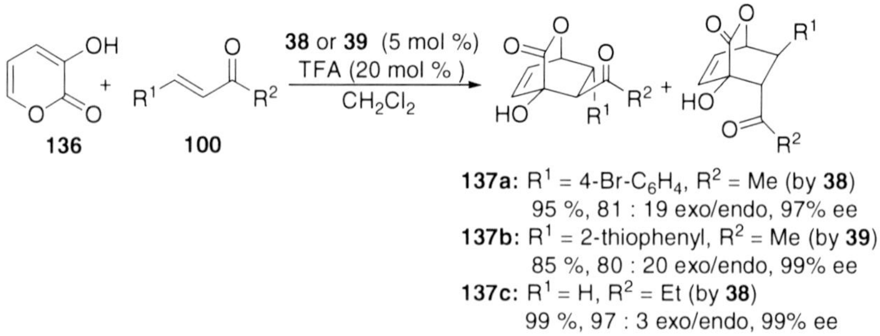

38 or **39** (5 mol %)
TFA (20 mol %)
CH_2Cl_2

136 **100**

137a: R^1 = 4-Br-C_6H_4, R^2 = Me (by **38**)
95 %, 81 : 19 exo/endo, 97% ee
137b: R^1 = 2-thiophenyl, R^2 = Me (by **39**)
85 %, 80 : 20 exo/endo, 99% ee
137c: R^1 = H, R^2 = Et (by **38**)
99 %, 97 : 3 exo/endo, 99% ee

Scheme 5.36 Asymmetric Diels-Alder reactions of α,β-unsaturated ketones

5.3.4 [2+2] Cycloaddition

Using their previously developed triamine catalyst **128**, Ishihara and coworkers reported an unprecedented [2+2] cycloaddition between α-acyloxyacroleins and tris-substituted alkenes (Scheme 5.37) [66]. Presumably, the reaction occurs via stepwise Michael addition and S_N1 type cyclization. A similar transition state as in the Diels-Alder reaction (Scheme 5.33) was proposed to account for the observed stereoselectivity.

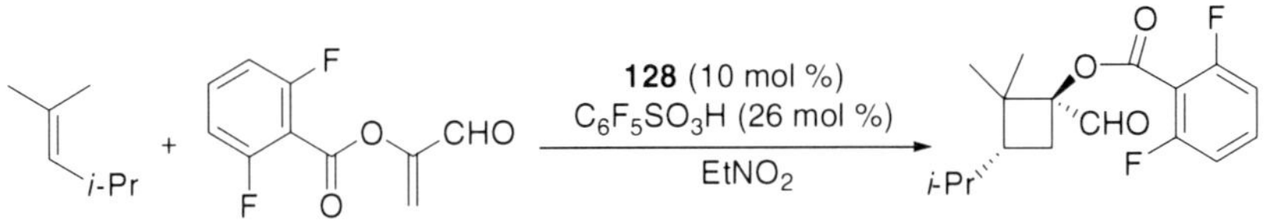

Scheme 5.37 Asymmetric [2+2] cycloaddition

5.3.5 1, 3-Dipolar Cycloaddition

Asymmetric 1, 3-dipolar cycloaddition is a powerful tool for synthesizing chiral heterocyclopentane compounds. Chen group demonstrated that primary amine **143** derived from cinchona-alkaloid was an effective catalyst for the 1, 3-dipolar cycloaddition of α,β-enones and azomethine imines (Scheme 5.38) [67]. It was proposed that multi-hydrogen-bonding, particularly those between the free OH and the substrate are critical for stereocontrol (Scheme 5.38). The reaction went smoothly with a wide range of substrates with up to 99% yield, 99:1 dr and 95% ee.

5.3.6 Aziridinations

Melchiorre and co-workers reported the first asymmetric aziridination of α,β-unsaturated ketones using the primary amine-amino acid salt **91** (Scheme 5.39) [68]. The reaction occurred via first nucleophilic addition of *N*-centered nucleophile (iminium catalysis) followed by intramolecular cyclization (enamine catalysis). Essential factors for the success of the current reaction included the proper selection of nitrogen nucleophile and the addition of solid $NaHCO_3$. Under the optimized conditions, a range of cyclic or acyclic enones can be incorporated in the protocol to afford either *N*-CBz or *N*-Boc aziridines with excellent diastereo- and enantiostereocontrol (Scheme 5.39).

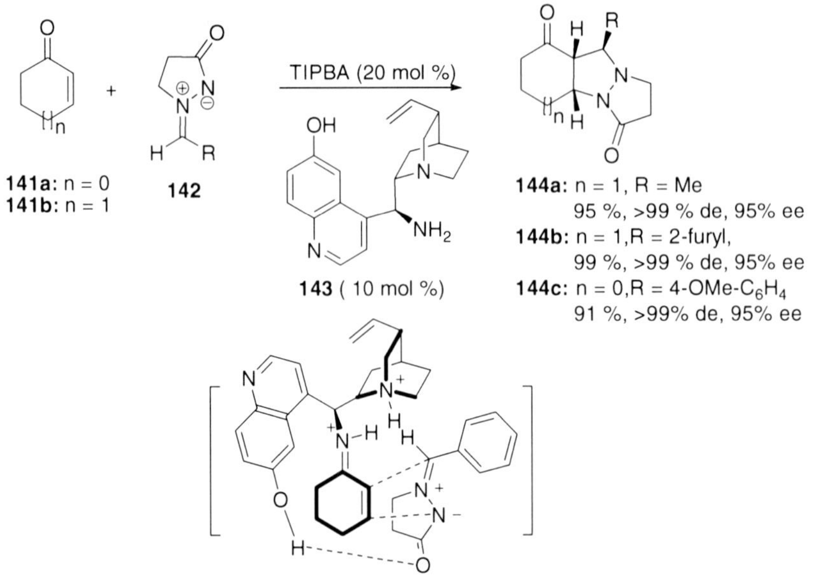

TIPBA (20 mol %)

141a: n = 0
141b: n = 1

142

143 (10 mol %)

144a: n = 1, R = Me
95 %, >99 % de, 95% ee
144b: n = 1,R = 2-furyl,
99 %, >99 % de, 95% ee
144c: n = 0,R = 4-OMe-C_6H_4
91 %, >99% de, 95% ee

Scheme 5.38 (TIPBA = 2,4,6-triisopropylbenzenesulfonic acid)

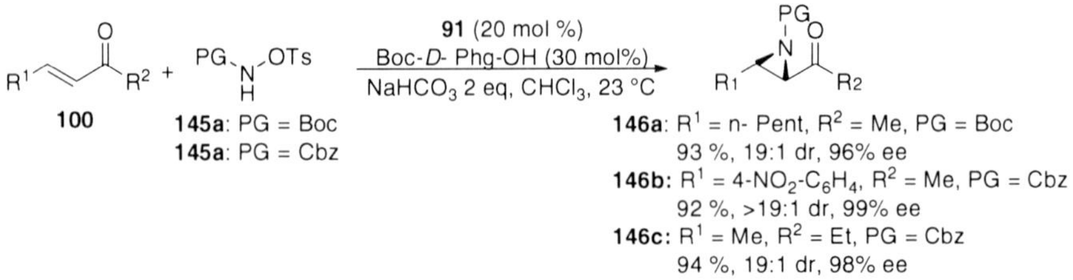

91 (20 mol %)
Boc-*D*- Phg-OH (30 mol%)
$NaHCO_3$ 2 eq, $CHCl_3$, 23 °C

100

145a: PG = Boc
145a: PG = Cbz

146a: R^1 = n- Pent, R^2 = Me, PG = Boc
93 %, 19:1 dr, 96% ee
146b: R^1 = 4-NO_2-C_6H_4, R^2 = Me, PG = Cbz
92 %, >19:1 dr, 99% ee
146c: R^1 = Me, R^2 = Et, PG = Cbz
94 %, 19:1 dr, 98% ee

Scheme 5.39 Asymmetric aziridination

5.3.7 *Epoxidation, Peroxidation and Hydroperoxidation*

Iminium catalysis has been quite successful for asymmetric epoxidation of α,β-unsaturated carbonyl compounds, particularly, enals. Enones have remained difficult substrates. Recently, List and coworkers reported an enantioselective epoxidation of cyclic enones with either cinchona-based primary amine **38** or a counter-anion catalytic system **149** combining a chiral vicinal diamine and a chiral phosphoric acid [69]. High enantioselectivities could be achieved in a number of cyclic enones (Scheme 5.40).

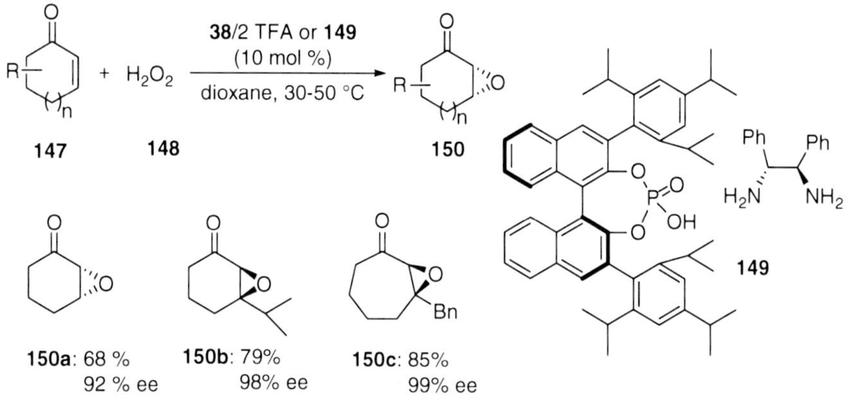

Scheme 5.40 Asymmetric epoxidation

The asymmetric epoxidation of acyclic enones have also been attempted using chiral primary aminocatalysis. Deng [70] and List independently reported an interesting peroxidation or hydroperoxidation of simple α,β-enones using the same quinine-derived primary amine **38** (Schemes 5.41 and 5.42). In Deng's studies, alkyl peroxides such as *t*-butyl peroxide and t-amylperoxide were used as the oxidizing reagents and the reactions afforded the peroxidation products instead of epoxide products (Scheme 5.41). List and coworkers utilized hydroperoxide in their studies and hydroperoxidation products were predominantly isolated (Scheme 5.42) [71]. In both cases, the peroxidation products could be further transformed into the corresponding epoxides by either increasing the reaction temperature (Deng) or treating the hydroperoxide products with base (List).

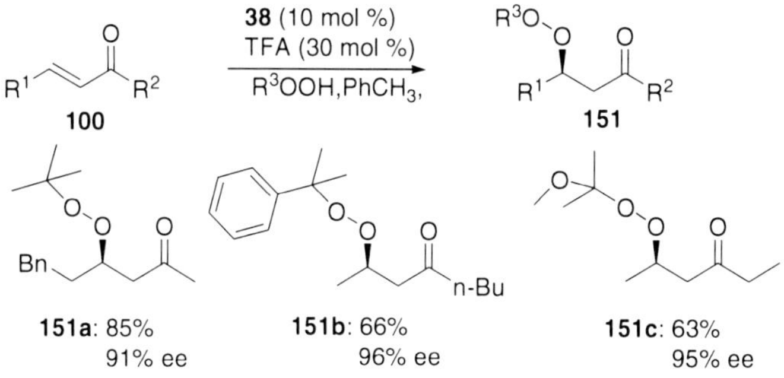

Scheme 5.41 Asymmetric peroxidation

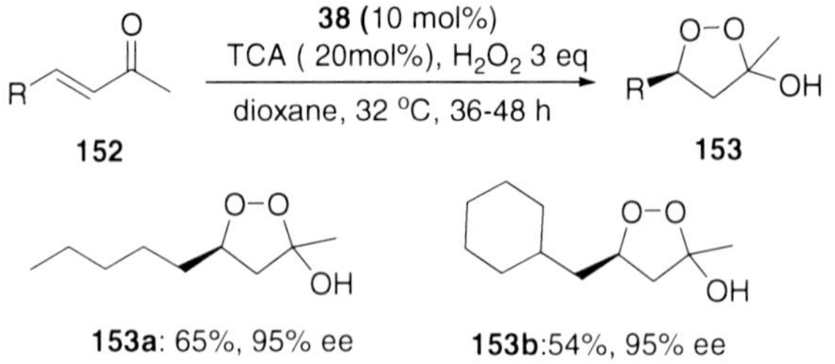

Scheme 5.42 (TCA = trichloroacetic acid)

5.3.8 Transfer Hydrogenation

Asymmetric hydride reduction using Hantzsch ester has recently been extensively explored in organocatalysis using iminium-based catalysts or Brønsted acid catalysts [72a–c]. As an advance to their asymmetric conterion-directed catalysis (ACDC), List and coworkers found that the combination of simple primary amino acids such as L-valine with a chiral phosphoric acid led to an effective primary aminocatalyst for asymmetric transfer hydrogenation of α,β-unsaturated ketones (Scheme 5.43) [72d]. The catalysis could be applied to a range of substrates with good yields and excellent enantioselectivity.

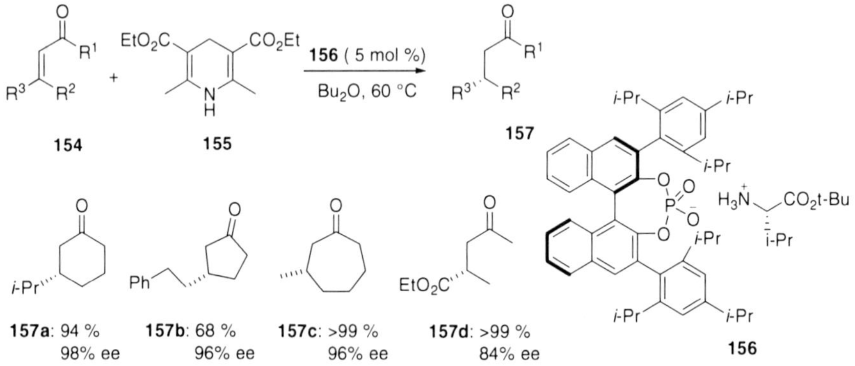

Scheme 5.43 Asymmetric transfer hydrogenation

5.3.9 Semipinacol Rearrangement

Recently, Tu group reported an unprecedented vinylogous α-ketol rearrange via semipinacol-type 1,2-carbon migration with cinchona-type catalyst **143**, giving products in good yields and up to 97% ee values [73]. In the reaction, *cis* isomer was observed to be the favored diasteroisomer. Worthy to note that this represent a very convenient procedure for constructing spiral quaternary stereogenic centers (Scheme 5.44).

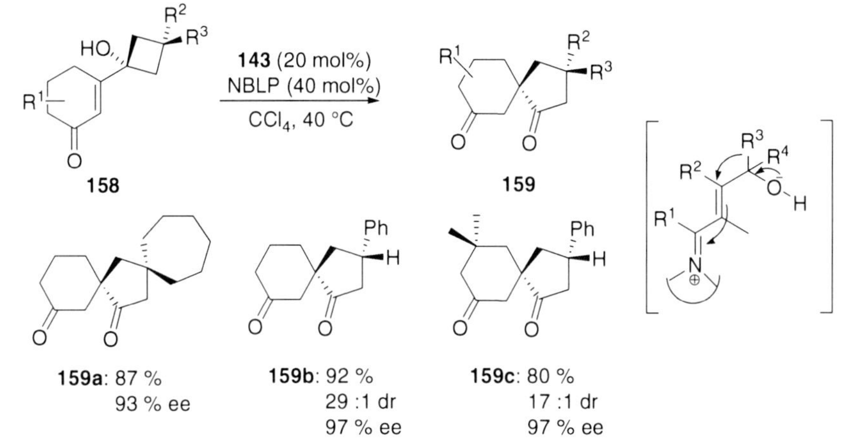

158

143 (20 mol%)
NBLP (40 mol%)
CCl_4, 40 °C

159

159a: 87 %
93 % ee

159b: 92 %
29 :1 dr
97 % ee

159c: 80 %
17 :1 dr
97 % ee

Scheme 5.44 (NBLP = *N*-Boc-L-phenylglycine)

5.4 Enamine-Iminium Bis-Catalysis with Vicinal Diamines

Iminium and enamine, like the Ying and Yang in Chinese philosophy, are two divergent, but interdependent and complementary catalytic modes in aminocatalysis. The use of two or multi enamine and iminium activation modes in tandem sequence is an attractive concept that is actively pursued in organocatalysis. In particular, cascade catalysis with one catalyst has become a powerful approach for generating molecular complexity in a fast manner. Recently, there are interesting examples that require simultaneous substrate activation via both enamine and iminium within one catalyst skeleton to achieve effective catalysis. In their pursuing of iminium catalysis in Nazarov reaction, Tius and coworkers found that chiral vicinal diamine **161** reacted with α-ketoenones to afford Nazarov product with high enantioselectivity [74a]. Though stoichiometric amount of **161** was required, the reaction demonstrated a distinctive feature of primary aminocatalysis where an enamine-iminium ion was probably involved (Scheme 5.45).

Li and coworkers found that 20 mol% of catalyst **164** served as effective catalyst for the asymmetric Michael addition of cylcopentanone to chalcones [74b]. Moderate diastereoselectivities but excellent enantioselectivities were achieved with a range of chalcones. The reaction was proposed to occur via an enamine-iminium transition state (Scheme 5.46).

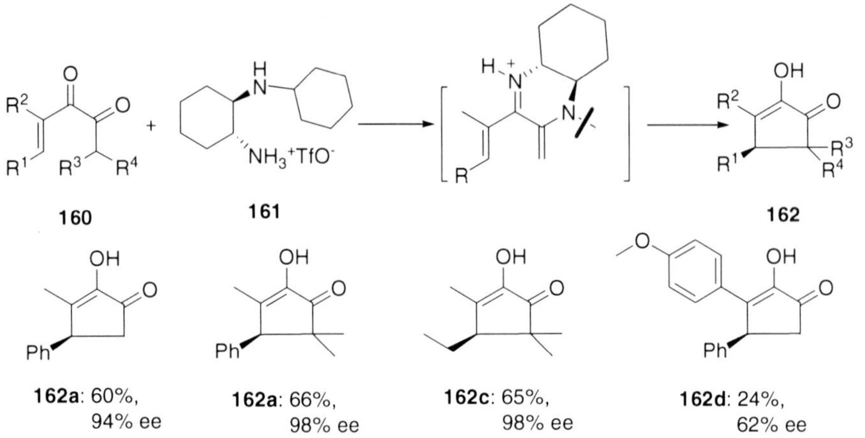

Scheme 5.45 Asymmetric Nazarov reaction

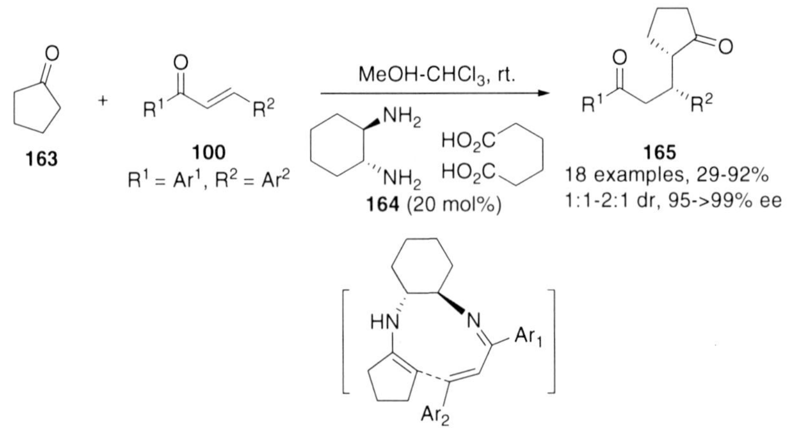

Scheme 5.46 Asymmetric Michael addition

Very recently, Kotsuki and coworkers reported an enantioselective Robinson annulation reaction for the synthesis of cyclohexenone derivatives bearing a quaternary center. Chiral vicinal diamine-chiral Bronsted acid conjugate **168** was found to be the optimal catalyst. The reactions afforded chiral cyclohexenone with moderate yields and good enantioselectivity [75]. It was proposed that simultaneous enamine activation of donor and iminium activation of acceptor were involved in the catalytic cycle (Scheme 5.47).

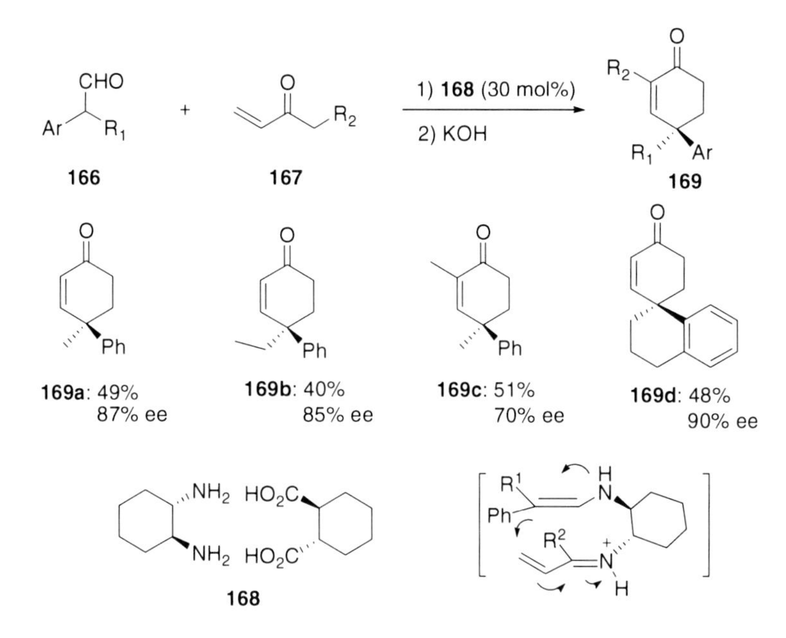

Scheme 5.47 Enantioselective Robinson annulation

5.5 Non-Covalent Catalysis with Chiral Primary Amines

Distinct from the classical aminocatalysis, Tsogoeva and co-workers documented the first example of enol activation in the typical Mannich-type reaction catalyzed by primary amine-thiourea **63** (Scheme 5.48) [76]. The O^{18}/O^{16} isotope experiments exclude the involvement of a covalent enamine/iminium intermediate and theoretical calculation proves an enol type transition state in this reaction.

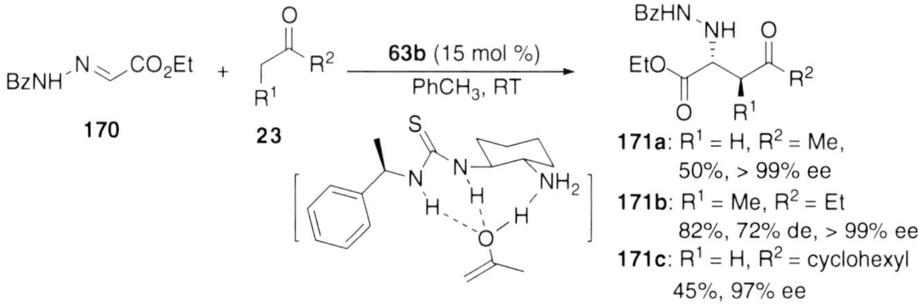

BzNH–N=CH–CO2Et (**170**) + R²C(O)CH2R¹ (**23**) → **63b** (15 mol %), PhCH3, RT → EtO2C–CH(NHNHBz)–CH(R¹)–C(O)R²

171a: R¹ = H, R² = Me, 50%, > 99% ee
171b: R¹ = Me, R² = Et 82%, 72% de, > 99% ee
171c: R¹ = H, R² = cyclohexyl 45%, 97% ee

Scheme 5.48 Asymmetric Mannich reaction

Zhong group reported that cinchona-derived primary amine **39** could promote tandem Michael-Henry reaction [77a] to give chiral cyclohexanes and cyclopentanes [77b] bearing four neighboring chiral centers (Scheme 5.49). In these reactions, two new C-C bonds and two quaternary stereocenters were formed in a single step with great efficiency and excellent selectivities. The same catalyst was also capable of mediating domino Michael-Michael reactions in a similar manner [77c]. In all these cases, multi H-bonds involving the cinchona-alkaloid derived diamine **39** were proposed to account for the effective catalysis (Scheme 5.49).

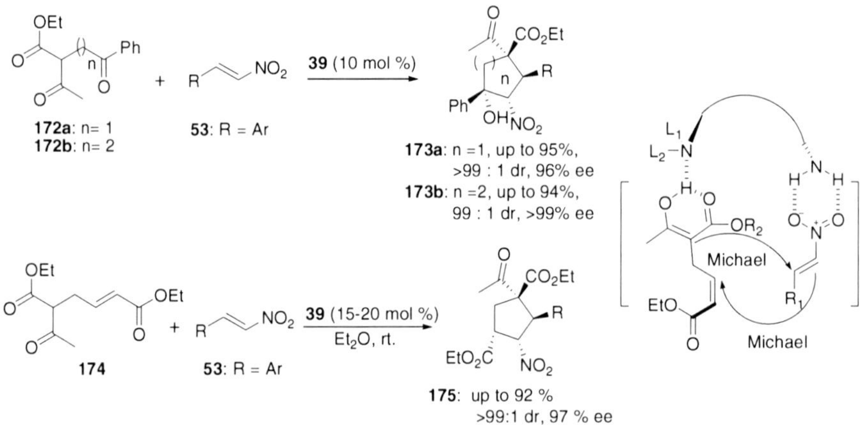

172a: n= 1
172b: n= 2
53: R = Ar
39 (10 mol %)
173a: n =1, up to 95%, >99 : 1 dr, 96% ee
173b: n =2, up to 94%, 99 : 1 dr, >99% ee

174
53: R = Ar
39 (15-20 mol %)
Et_2O, rt.
175: up to 92 % >99:1 dr, 97 % ee

Michael
Michael

Scheme 5.49 Asymmetric Michael-Henry reaction

Lu and coworkers reported that Ts-DPEN **177** (Scheme 5.50) was a good H-bond catalyst for Michael reactions of the 1, 3-dicarbonyl compounds and nitroolefines [78] in a manner similar to cinchona alkaloid catalyst **39** (see Scheme 5.49).

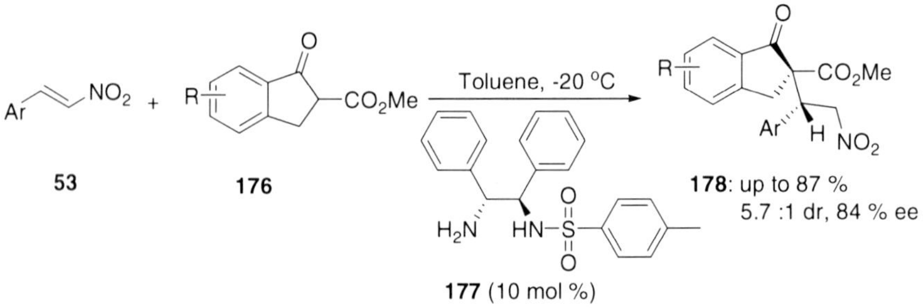

53
176
Toluene, -20 °C
177 (10 mol %)
178: up to 87 % 5.7 :1 dr, 84 % ee

Scheme 5.50 (DPEN: (R,R)-1,2-diphenylethylenediamine)

5.6 Supported Chiral Primary Amine Catalysis

Immobilization is one of the proved strategies to address some limitations of organocatalysts such as high catalyst loading and difficulties in catalyst separating and recycling. Based on their chiral primary-tertiary diamine catalysts, Luo group further developed two types of supported primary amine catalysts. In one example, the native chiral primary amine catalyst such as **DMDA** (*N,N*-dimethyl-cyclohexanediamine) can be non-covalently supported onto polystyrene (PS)-sulfonic acid via simple acid-base interactions. No derivatization was needed in this process and the obtained supported catalyst such as **179** maintains similar or even improved selectivity compared with its homogeneous counterpart [79]. This catalyst could be recycled via filtration and reused for six cycles without loss in enantiocontrol (Scheme 5.51). Notably, the deactivated catalyst can be regenerated by washing with HCl and recharging with fresh chiral diamine (Scheme 5.51).

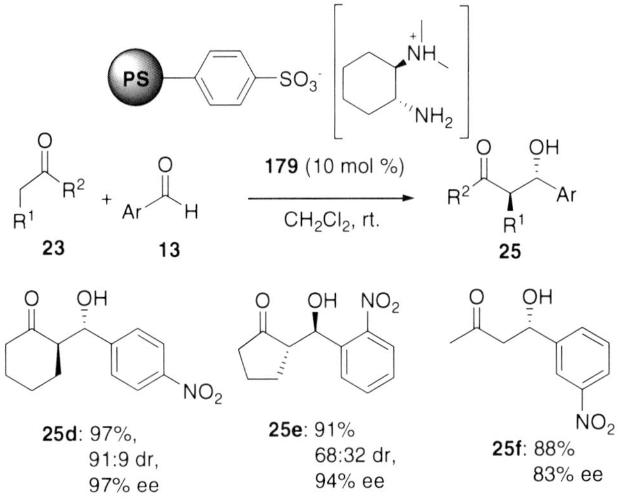

Scheme 5.51 Aldol additions

Recently, a non-covalently supported iminium-type primary amine catalyst **181** has also been developed using polyoxametalate as the solid acid support (Scheme 5.52) [80]. The catalyst was applied to asymmetric Diels-Alder reactions of α-substituted acroleins with good activity and stereoselectivity and could be recycled and reused for six runs.

In a covalent immobilization strategy, magnetic nanoparticle (MNP), was selected as the support due to its easy separation via magnetic force, large surface area, easy preparation and low cost. The desired chiral primary-tertiary diamine is

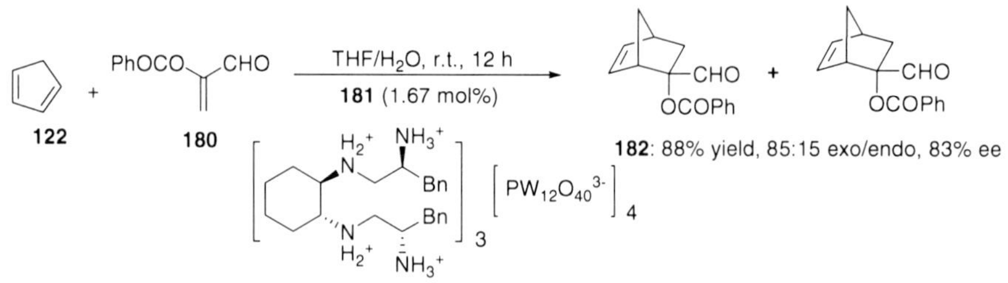

Scheme 5.52 Asymmetric Diels-Alder reaction

covalently connected via Si-O-Si bond. The obtained MNP-supported catalyst **183** showed good activity and stereoselectivity in the asymmetric direct aldol reactions of cyclohexanone [81]. The catalyst can be recycled and reused up to 11 cycles with similar activity and stereoselectivity (Scheme 5.53).

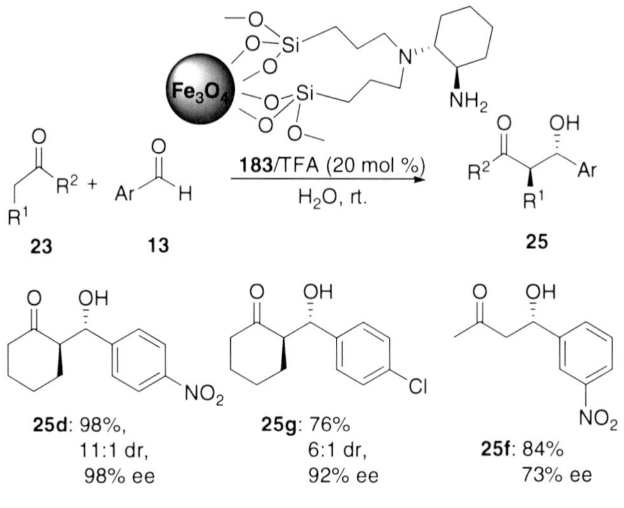

Scheme 5.53 Asymmetric direct aldol reactions

5.7 Summary and Outlook

The past 5 years has witnessed extraordinary progresses in developing chiral primary aminocatalysis. The vast resources of chiral primary amines have indeed facilitated the rapid screening and identification of the desired catalyst

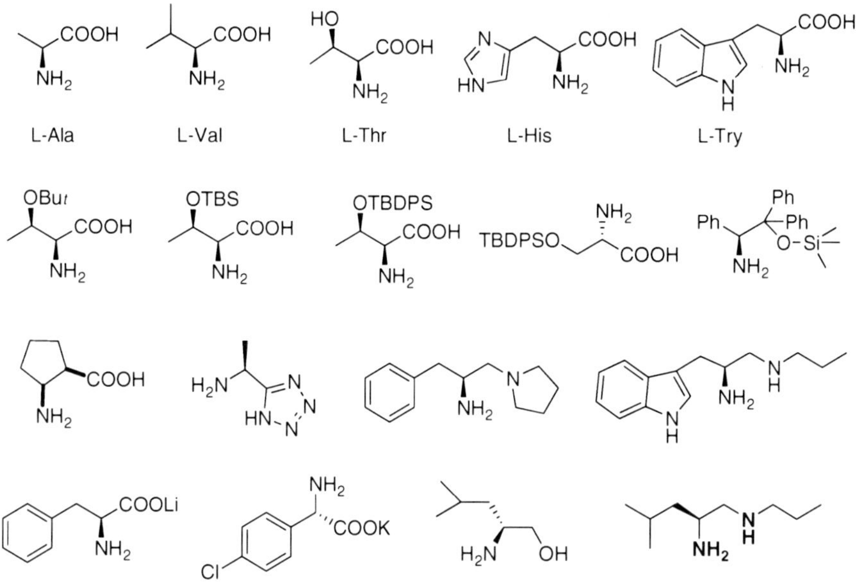

Fig. 5.2 Amino acids and derivatives

for a targeted reaction. In this regard, natural amino acids, cinchona alkaloids and commercial available chiral vicinal diamines represent three of the most frequently utilized chiral primary amine precursors. Of all the current chiral primary amine catalysts, vicinal diamine and primary amine-amide/thiourea conjugate appear as the two "privileged" catalytic motifs for a number of transformations. Collectively, several features of these primary aminocatalysis can be summarized: (1) Primary amines enable catalysis beyond the reach of secondary amines in the reactions of steric hindered substrates such as α-substituted carbonyls and α,β-unsaturated carbonyl compounds; (2) In covalent catalysis with primary amines, acidic additives are essential for both catalytic activity and stereocontrol in most cases; (3) novel catalytic modes such as concurrent cascade catalysis or non-covalent hydrogen-bonding catalysis are possible with primary amines and represent a promising direction for further development. Despite of the explosive progresses, mechanistic understanding of primary aminocatalysis has been largely lagged behind. Fundamental mechanism studies as well as the design and invent of new catalysts and new reactions serve as two research focus in the near future in the field of primary aminocatalysis Figs. 5.2–5.6.

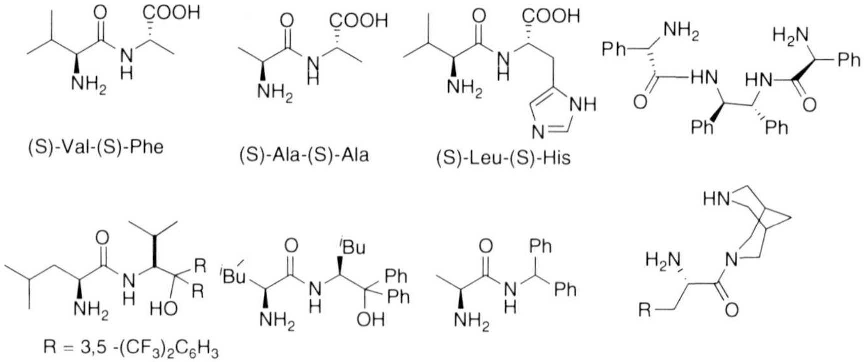

(S)-Val-(S)-Phe (S)-Ala-(S)-Ala (S)-Leu-(S)-His

R = 3,5 -$(CF_3)_2C_6H_3$

Fig. 5.3 Peptides derivatives

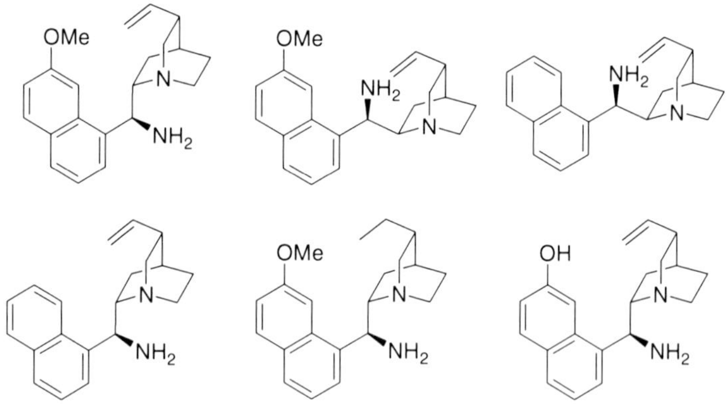

Fig. 5.4 Cinchona alkaloid derivatives

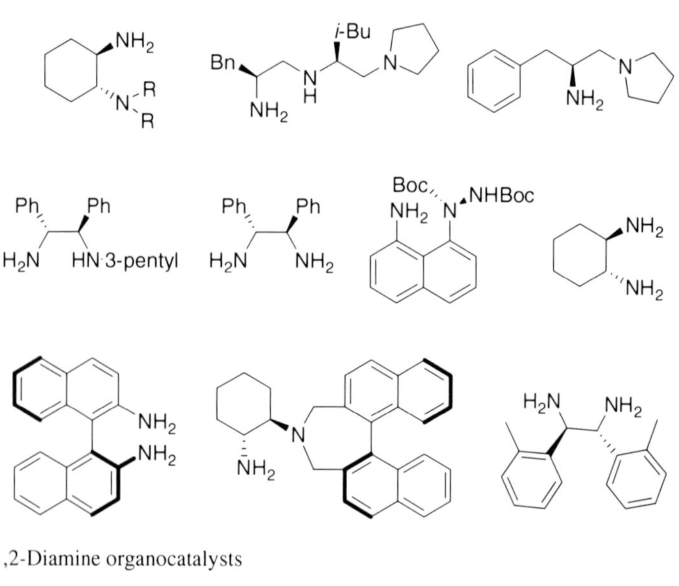

Fig. 5.5 1,2-Diamine organocatalysts

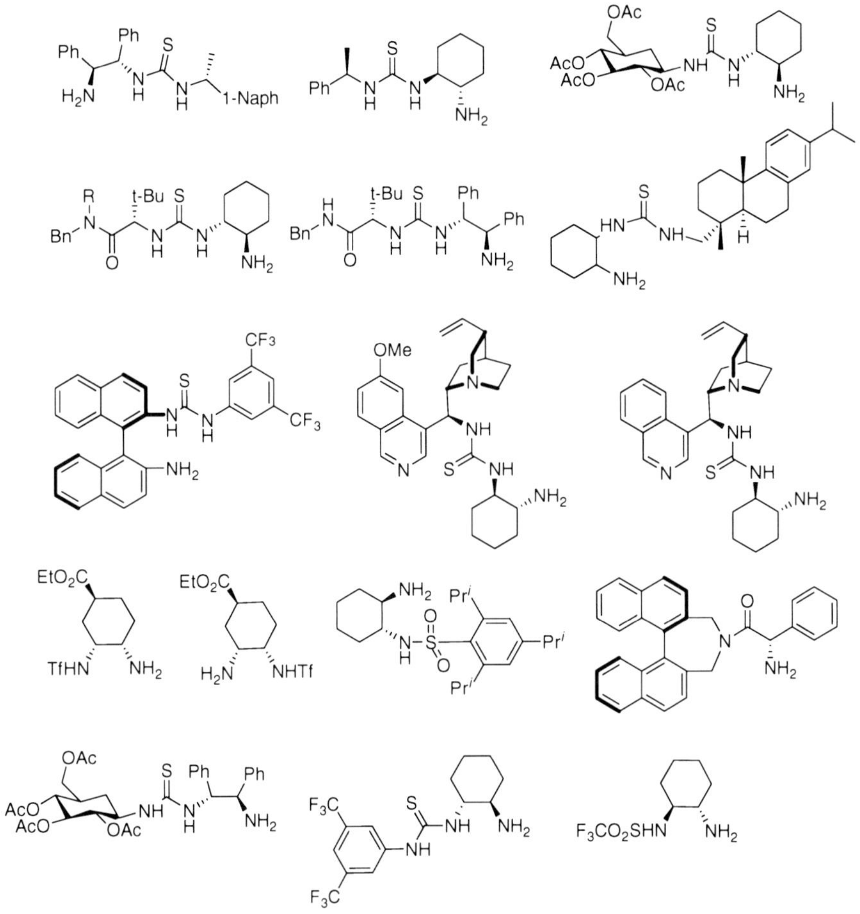

Fig. 5.6 Primary amine amide and primary amine thiourea catalysts

Acknowledgements This project was supported by the Natural Science Foundation (20972163, 21025208), MOST (2011CB808600) and the Chinese Academy of Sciences.

References

1. Reviews see: (a) Dalko PI, Moisan L (2004) Angew Chem Int Ed 43:5138; (b) Dalko PI, Moisan L (2001) Angew Chem Int Ed 40:3726
2. (a) Caro GMM, Meierhenrich UJ, Schutte WA, Barbier B, Segovia AA, Rosenbauer H, Thiemann WH-P, Brack A, Greenberg JM (2002) Nature 416:403; (b) Pizzarello S, Weber AL (2004) Science 303:1151; (c) Movassaghi M, Jacobsen EN (2002) Science 298:1904; (d) Klussmann M, Iwamura1 H, Mathew1 SP, Wells Jr DH, Pandya U, Armstrong A, Blackmond DG (2006) Nature 441:621; (e) Klussmann M, Mathew SP, Iwamura H, Wells Jr DH, Armstrong

A, Blackmond DG (2006) Angew Chem Int Ed 45:7989; (f) Mathew SP, Iwamura H, Blackmond DG (2004) Angew Chem Int Ed 43:3317; (g) Hayashi Y, Matsuzawa M, Yamaguchi J, Yonehara S, Matsumoto Y, Shoji M, Hashizume D, Koshino H (2006) Angew Chem Int Ed 45:4593
3. (a) Gefflaut T, Blonski C, Perie J, Wilson M (1995) Prog Biophys Mol Biol 63:301; (b) Barbas III CF, Heine A, Zhong G, Hoffmann T, Gramatikova S, Bjornestedt R, List B, Anderson J, Stura EA, Wilson IA, Lerner RA (1997) Science 298:2085; (c) Jean MS, Vanasse JL, Liotard B, Sygusch J (2005) J Biol Chem 280:27262; (d) Allard J, Grochulski P, Sygusch J (2001) Proc Natl Acad Sci 98:3679; (e) Heine A, DeSantis G, Luz JG, Mitchell M, Wong C-H, Wilson IA (2001) Science 294:369; (f) Fullerton SW, Griffiths JS, Merkel AB, Cheriyan M, Wymer NJ, Hutchins MJ, Fierke CA, Tooneb EJ, Naismith JH (2006) Bioorg Med Chem 14:3002; (g) Karlstrom A, Zhong G, Rader C, Larsen NA, Heine A, Fuller R, List B, Tanaka F, Wilson IA, Barbas III CF, Lerner RA (2000) Proc Natl Acad Sci 97:3878
4. Westheimer FH, Cohen H (1938) J Am Chem Soc 60:90
5. (a) Pedersen KJ (1929) J Am Chem Soc 51:2098; (b) Pedersen KJ (1934) J Phys Chem 38:559; (c) Pedersen KJ (1938) J Am Chem Soc 60:595
6. Hine J (1978) Acc Chem Res 11:1
7. (a) Hajos ZG, Parrish DR (1973) J Org Chem 38:3239; (b) Hajos ZG, Parrish DR (1974) J Org Chem 39:1615; (c) Eder U, Sauer G, Wiechert R (1971) Angew Chem Int Ed 10:496; (d) Hagiwara H, Uda H (1988) J Org Chem 53:2308; (e) Danishefsky S, Cain P (1976) J Am Chem Soc 98:4975; (f) Danishefsky S, Cain P (1975) J Am Chem Soc 97:5282
8. Reymond J-L, Chen Y (1995) J Org Chem 60:6970
9. Clemente FR, Houk KN (2005) J Am Chem Soc 127:11294
10. (a) Bassan A, Zou W, Reyes E, Himo F, Cordova A (2005) Angew Chem Int Ed 44:7028; (b) Cordova A, Zou W, Ibrahem I, Reyes E, Engqvist M, Liao W (2005) Chem Commun 3586; (c) Cordova A, Zou W, Dziedzic P, Ibrahem I, Reyes E, Xu Y (2006) Chem Eur J 12:5383; (d) Amedjkouh M (2005) Tetrahedron Asymm 16:1411
11. Mahrwald R (ed.) (2004) Modern aldol reactions. Wiley-VCH, Weinheim
12. Pizzarello S, Weber AL (2004) Science 303:1151
13. (a) Jiang Z, Liang Z, Wu X, Lu Y (2006) Chem Commun 2801; (b) Jiang Z, Yang Hui, Han X, Luo J, Wong MW, Lu Y (2010) Org Biomol Chem 8:1368; other papers of amine acid derivatives catalyzed aldol reaction, see: (c) Lombardo M, Easwar S, Pasi F, Trombini C, Dhavale DD (2008) Tetrahedron 64:9203; (d) Teo Y-C, Chua G-L, Ong C-Y, Poh C-Y (2009) Tetrahedron Lett 50:4854; (e) Malkov AV, Kabeshov MA, Bella M, Kysilka O, Malyshev DA, Pluhackova K, Kocovsky P (2007) Org Lett 9:5473; (f) Davies SG, Sheppard RL, Smith AD (2005) Chem Commun 3802
14. (a) Cordova A, Zou W, Dziedzic P, Ibrahem I, Reyes E, Xu Y (2006) Chem Eur J 12:5383; (b) Dziedzic P, Zou W, Hafren J, Cordova A (2006) Org Biomol Chem 4:38; (c) Zou W, Ibrahem I, Dziedzic P, Sunden H, Cordova A (2005) Chem Commun 4946; (d) Tsogoeva SB, Wei S (2005) Tetrahedron Asymm 16:1947; (e) Wu F-C, Da C-S, Du Z-X, Guo Q-P, Li W-P, Yi L, Jia Y-N, Ma X (2009) J Org Chem 74:4812
15. (a) Tanaka F, Barbas III CF (2007) Angew Chem Int Ed 46:5572; (b) Ramasastry SSV, Zhang H, Tanaka F, Barbas III CF (2007) J Am Chem Soc 129:288; (c) Utsumi N, Imai M, Tanaka F, Ramasastry SSV, Barbas III CF (2007) Org Lett 9:3445
16. Markert M, Scheffler U, Mahrwald R (2009) J Am Chem Soc 131:16642
17. (a) Luo S, Xu H, Li J, Zhang L, Cheng J-P (2007) J Am Chem Soc 129:3074; (b) Luo S, Xu H, Zhang L, Li J, Cheng J-P (2008) Org Lett 10:653
18. Li J, Luo S, Cheng J-P (2009) J Org Chem 74:1747
19. Luo S, Xu H, Chen L, Cheng J-P (2008) Org Lett 10:1775
20. Luo S, Qiao Y, Zhang L, Li J, Li X, Cheng J-P (2009) J Org Chem 74:9521
21. (a) Peng F-Z, Shao Z-H, Pu X-W, Zhang H-B (2008) Adv Synth Catal 350:2199; (b) Liu Q-Z, Wang X-L, Luo S-W, Zheng B-L, Qin D-B (2008) Tetrahedron Lett 49:7434; for spiro diamine catalysis, see: (c) Jiang M, Zhu S, Yang Y, Gong L, Zhou X, Zhou Q (2006) Tetrahedron Asymm 17:384

22. Nakayama K, Maruoka K (2008) J Am Chem Soc 130:17666
23. Zhou J, Wakchaure V, Kraft P, List B (2008) Angew Chem Int Ed 47:7656
24. Xu X, Wang Y, Gong L (2007) Org Lett 9:4247
25. (a) Wu X, Ma Z, Ye Z, Qian S, Zhao G (2009) Adv Synth Catal 351:158; (b) Da C-S, Che L-P, Guo Q-P, Wu F-C, Ma X, Jia Y-N (2009) J Org Chem 74:2541
26. Liu J, Yang Z, Wang Z, Wang F, Chen X, Liu X, Feng X, Su Z, Hu C (2008) J Am Chem Soc 130:5654
27. Luo S, Zhou P, Li J, Cheng J-P (2010) Chem Eur J 16:4457
28. Ibrahem I, Zou W, Engqvist M, Xu Y, Cordova A (2005) Chem Eur J 11:7024
29. Zhang H, Ramasastry SSV, Tanaka F, Barbas CF III (2008) Adv Synth Catal 350:791
30. (a) Cheng L, Han X, Huang H, Wong MW (2007) Chem Comm 40:4143; (b) Cheng L, Wu X, Lu Y (2007) Org Biomol Chem 5:1018
31. Review see: Mukherjee S, Yang JW, Hoffmann S, List B (2007) Chem Rev 107:5471
32. (a) Xu Y, Zou W, Sunden H, Ibrahem I, Cordova A (2006) Adv Synth Catal 348:418; (b) Xu Y, Co´rdova A (2006) Chem Comm 2006:460
33. (a)Xiong Y, Wen Y, Wang F, Gao B, Liu X, Huang X, Feng X (2007) Adv Synth Catal 349:2156; (b) Yang Z, Liu J, Liu X, Wang Z, Feng X, Su Z, Hua C (2008) Adv Synth Catal 350:2001; (c) Liu J, Yang Z, Liu X, Wang Z, Liu Y, Bai S, Lin L, Feng X (2009) Org Biomol Chem 7:4120
34. Sato A, Yoshida M, Hara S (2008) Chem Commun 46:6242
35. (a) Tsogoeva SB, Wei S, (2006) Chem Commun 1451; (b) Wei S, Yalalov DA, Tsogoeva SB, Schmatz S (2007) Catal Today 121:151; (c) Yalalov DA, Tsogoeva SB, Schmatz S (2006) Adv Synth Catal 348:826
36. (a) Lalonde MP, Chen Y, Jacobsen EN (2006) Angew Chem Int Ed 45:6366; (b) Huang H, Jacobsen EN (2006) J Am Chem Soc 128:7170
37. (a) Uehara H, Barbas III CF (2009) Angew Chem Int Ed 48:9848; examples of H-bond catalysis in Michael addition of nitroolefins and ketones, see: (b) Li B-L, Wang Y-F, Luo S-P, Zhong A-G, Li Z-B, Du X-H, Xu D-Q (2010) Eur J Org Chem 656; (c) Peng L, Xu X-Y, Wang L-L, Huang J, Bai J-F, Huang Q-C, Wang L-X (2010) Eur J Org Chem 2010:1849
38. (a) Liu K, Cui H, Nie J, Dong K, Li X, Ma J (2007) Org Lett 9:923; (b) M Hai, Liu K, ZhanG F-G, Zhu C-L Zhu, Nie J, Ma J-A (2010) J Org Chem 75:1402; (c) Jiang X, Zhang Y, Chan ASC, Wang R (2009) Org Lett 11:153; an example of biginelli reaction with catalyst **25,** see: (d) Wang Y, Yang H, Yu J, Miao Z, Chen R (2009) Adv Synth Catal 351:3057
39. (a) Xue F, Zhang S, Duan W, Wang W (2008) Adv Synth Catal 350:2194; (b) Rasappan R, Reiser O (2009) Eur J Org Chem 9:1305
40. McCooey SH, Connon SJ (2007) Org Lett 9:599
41. Zhu Q, Cheng L, Lu Y (2008) Chem Commun 47:6315
42. Li X, Cun L, Lian C, Zhong L, Chen Y, Liao J, Zhu J, Deng J (2008) Org Biomol Chem 6:349
43. Liu T, Cui H, Zhang Y, Jiang K, Du W, He Z, Chen Y (2007) Org Lett 9:3671
44. Brandes S, Niess B, Bella M, Prieto A, Overgaard J, Jogensen KA (2006) Chem Eur J 12:6039
45. Review see: Erkkil A, Majander I, Pihko PM (2007) Chem Rev 107:5416
46. (a) Ahrendt KA, Borths CJ, MacMillan DWC (2000) J Am Chem Soc 122:4243; (b) North AB, MacMillan DWC (2002) J Am Chem Soc 124:2458
47. Austin JF, MacMillan DWC (2002) J Am Chem Soc 124:1172
48. Chen W, Du W, Yue L, Li R, Wu Y, Ding L, Chen Y (2007) Org Biomol Chem 5:816
49. (a) Bartoli G, Bosco M, Carlone A, Pesciaioli F, Sambri L, Melchiorre P (2007) Org Lett 9:1403; recent example of similar rection , see: (b) Hong L, Sun W, Liu C, Wang L, Wong K, Wang R (2009) Chem Eur J 15:11105
50. (a) Kim H, Yen C, Preston P, Chin J (2006) Org Lett 8:5239–5241; (b) Kristensen TE, Vestli K, Hansen FK, Hansen T (2009) Eur J Org Chem 2009:5185
51. Xie J, Yue L, Chen W, Du W, Zhu J, Deng J, Chen Y (2007) Org Lett 9:413
52. (a) Aleman J, Jacobsen CB, Frisch K, Overgaard J, Jørgensen KA (2008) Chem Commun 5:632; (b) Niess B, Jørgensen KA (2007) Chem Commun 16:1620; (c) Cui HL, Chen Y C (2009) Chem Commun 30:4479

53. (a) Xie B, Chen W, Li R, Zeng M, Du W, Yue L, Chen Y, Wu Y, Zhu J, Deng J (2007) Angew Chem Int Ed 46:389; (b) Kang T, Xie J, Du W, Feng X, Chen Y (2008) Org Biomol Chem 6:2673
54. (a) Yang Y-Q, Zhao G (2008) Chem Eur J 14:10888; For other catalytic applications see: (b) Cui H-F, Yang Y-Q, Chai Z, Li P, Zheng C-W, Zhu S-Z, Zhao G (2010) J Org Chem 75:117; (c) Yang Y-Q, Chai Z, Wang H-F, Chen X-K, Cui H-F, Zheng C-W, Xiao H , Li P, Zhao G (2009) Chem Eur J 15:13295
55. (a) Li P, Wen S, Yu F, Liu Q, Li W, Wang Y, Liang X, Ye J (2009) Org Lett 11:753; (b) Li P, Wang Y, Liang X, Ye J (2008) Chem Commun 28:3302; other examples of Michael reaction of nitroalkanes and α,β-unsaturated ketones, see: (c) Mei K, Jin M, Zhang S, Li P, Liu W, Chen X, Xue F, Duan W, Wang W (2009) Org Lett 11:2864; (d) Dong L, Lu R, Du Q, Zhang J, Liu S, Xuan Y, Yan M (2009) Tetrahedron 65:4124
56. Galzerano P, Bencivenni G, Pesciaioli F, Mazzanti A, Giannichi B, Sambri L, Bartoli G, Melchiorre P (2009) Chem Eur J 15:7846
57. Vinylogous Michael reaction, see: Wang J , Chao Q, Ge Z, Cheng T, Li R (2010) Chem Commun 46:2124
58. (a) Ricci P, Carlone A, Bartoli G, Bosco M, Sambri L, Melchiorre P (2008) Adv Synth Catal 350:49; (b) Carlone A, Bartoli G, Bosco M, Pesciaioli F, Ricci P, Sambri L, Melchiorre P (2007) Eur J Org Chem 5492
59. (a) Lu X, Deng L (2008) Angew Chem Int Ed 47:7710; a recent example of aza-Michael reaction, see: (b) G Sanjib, Zhao C-G, Ding D (2009) Org Lett (2009) 11:2249
60. (a) Kim H, Yen C, Preston P, Chin J (2006) Org Lett 8:5239; (b) Corey E J (2002) Angew Chem Int Ed 41:1650
61. Kim KH, Lee S, Lee D, Ko D, Ha D (2005) Tetrahedron Lett 46:5991
62. (a) Ishihara K, Nakano K (2005) J Am Chem Soc 127:10504; (b) Ishihara K, Nakano K, Akakura M (2008) Org Lett 10:2893
63. (a) Sakakura A, Suzuki K, Nakano K, Ishihara K (2006) Org Lett 8:2229; (b) Sakakura A, Suzuki K, Ishihara K (2006) Adv Synth Catal 348:2457
64. Kano T, Tanaka Y, Osawa K, Yurino T, Maruoka K (2009) Chem Commun 15:1956
65. Singh RP, Bartelson K, Wang Y, Su H, Lu X, Deng L (2008) J Am Chem Soc 130:2422
66. Ishihara K, Nakano K (2007) J Am Chem Soc 129:8930
67. Chen W, Du W, Duan Y, Wu Y, Yang S, Chen Y (2007) Angew Chem Int Ed 46:7667
68. Pesciaioli F, Vincentiis FD, Galzerano P, Bencivenni G, Bartoli G, Mazzanti A, Melchiorre P (2008) Angew Chem Int Ed 47:8703
69. Wang X, Reisinger CM, List B (2008) J Am Chem Soc 130:6070
70. Lu X, Liu Y, Sun B, Cindric B, Deng L (2008) J Am Chem Soc 130:8134
71. Reisinger CM, Wang X, List B (2008) Angew Chem Int Ed 47:8112
72. (a) Yang JW, Fonseca MTH, Vignola N, List B (2005) Angew Chem Int Ed 44:108; (b) Mayer S, List B (2006) Angew Chem Int Ed 45:4193; (c) Ouellet SG, Tuttle JB, MacMillan DWC (2005) J Am Chem Soc 127:32; (d) Martin NJA, List B (2006) J Am Chem Soc 128: 13368
73. Zhang E, Fan C-A, Tu Y-Q, Zhang F-M, Song Y-L (2009) J Am Chem Soc 131:14626
74. (a) Bow WF, Basak A , Jolit A, Vicic DA, Tius MA (2010) Org Lett 12:440; (b) Wang J, Wang X, Ge Z, Cheng T, Li T (2010) Chem Commun 46:1751
75. Inokoishi Y, Sasakura N, Nakano K, Ichikawa Y, Kotsuki HA (2010) Org Lett 12:ASAP
76. Yalalov DA, Tsogoeva SB, Shubina TE, Martynova IM, Clark T (2008) Angew Chem Int Ed 47:6624
77. (a) Tan B, Chua PJ, Li Y, Zhong G (2008) Org Lett 10:2437; (b) Tan B, Chua PJ, Zeng X, Lu M, Zhong G (2008) Org Lett 10:3489; (c) Tan B, Shi Z, Chua PJ, Zhong G (2008) Org Lett 10:3425
78. Ju Y-D, Xu L-W, Li L, Lai G-Q, Qiu H-Y, Jiang J-X, Lu Y (2008) Tetrahedron Lett 49:6773
79. Luo S, Li J, Zhang L, Xu H, Cheng J-P (2008) Chem Eur J 14:1273
80. Li J, Li X, Zhou P, Zhang L, Luo S, Cheng J-P (2009) Eur J Org Chem 2009:4486
81. Luo S, Zheng X, Cheng J-P (2008) Chem Commun 44:5719

Chapter 6
Bifunctional Acid-Base Catalysis

Petri M. Pihko and Hasibur Rahaman

Abstract Acid-base catalysis with bifunctional catalysts is a very prominent catalytic strategy in both small-molecule organocatalysts as well as enzyme catalysis. In both worlds, small-molecule catalysts and enzymatic catalysis, a variety of different general acids or hydrogen bond donors are used. In this chapter, important parallels between small molecule catalysts and enzymes are discussed, and a comparison is also made to the emerging field of frustrated Lewis pair catalysis.

6.1 Introduction

Appreciation of the role of noncovalent interactions in catalysis has grown significantly during the past few years [1]. Although a single noncovalent interaction, such as a hydrogen bond, may not be always enough to achieve significant transition state stabilization, a combination of hydrogen bond donor/Brønsted acidic functionality and a basic functionality has been shown to constitute a privileged design strategy for enantioselective catalysis.

Typically, a combination of a strong (Brønsted) acidic catalyst and a strong basic catalyst leads to the formation of a salt and subsequent neutralization. The primary problem in the design of bifunctional acid-base catalysts lies in the prevention of this unwanted acid-base reaction. This can be achieved in two ways: (1) The use of a readily dissociating pair, such as a combination of a weak acid and strong base, or vice versa, or (2) physical separation of the acid-base pair by steric hindrance and/or a rigid molecular framework. The first approach is exemplified by the success of

P.M. Pihko (✉) • H. Rahaman
Department of Chemistry, University of Jyväskylä, P.O. Box 35, FI-40014 Jyväskylä, Finland
e-mail: Petri.Pihko@jyu.fi; hasibur.h.rahaman@jyu.fi

R. Mahrwald (ed.), *Enantioselective Organocatalyzed Reactions I: Enantioselective Oxidation, Reduction, Functionalization and Desymmetrization*, DOI 10.1007/978-90-481-3865-4_6,

catalyst scaffolds bearing a moderately basic (tertiary) amine group and a weakly acidic, hydrogen bond donor (thio)urea group. The second class is exemplified by catalysts bearing "frustrated Lewis acid-base pairs". In this chapter, we focus mostly on the structure and the design of the catalysts rather than individual reactions [2].

6.2 Urea, Thiourea and Squaramide Catalysts – Double Hydrogen Bonding Catalysis

The prototypical chiral bifunctional acid-base catalyst is quinine. Quinine, quinidine, cinchonine and cinchonidine all bear a basic tertiary amine group encased in a quinuclidine skeleton and an adjacent OH group capable of hydrogen bonding (Fig. 6.1).

In 1981, Wynberg and Hiemstra already identified the unmodified cinchona alkaloids as chiral bifunctional catalysts for enantioselective conjugate additions to cycloalkenones [3]. They proposed that the OH group of cinchonine would act as a hydrogen bond donor site and stabilize the enolate-like transition state of the conjugate addition reaction (Scheme 6.1).

Scheme 6.1 The Hiemstra-Wynberg mechanism for cinchona alkaloid catalyzed conjugate additions

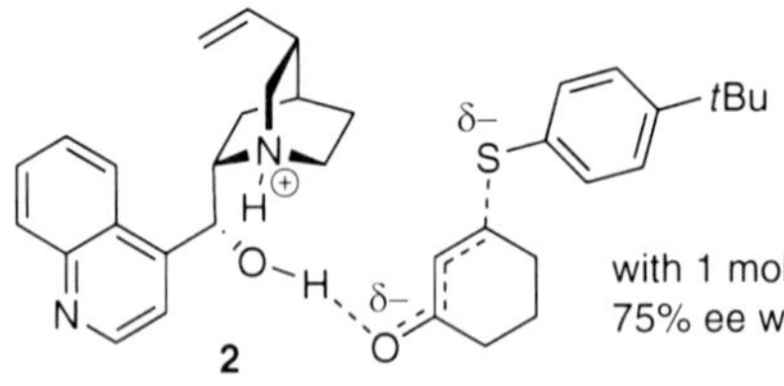

2

with 1 mol% of **2**, 75% ee was obtained

Improved catalytic activity can be expected if the hydroxyl group of the cinchona alkaloids is replaced with a better hydrogen bond donor group. The most popular variants of this type are the thiourea and the squaramide groups (Fig. 6.2). Before

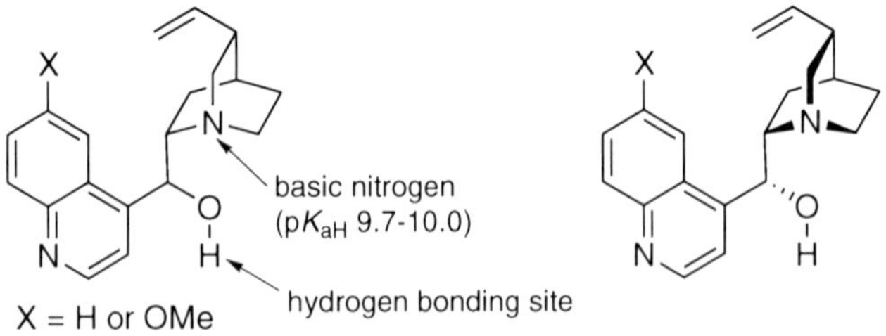

quinine/quinidine (X = OMe) or
cinchonidine/cinchonine (X = H)

quinine **1** (X = OMe)
and cinchonidine **2** (X = H)

Fig. 6.1 Structural features of bifunctional cinchona alkaloid catalysts

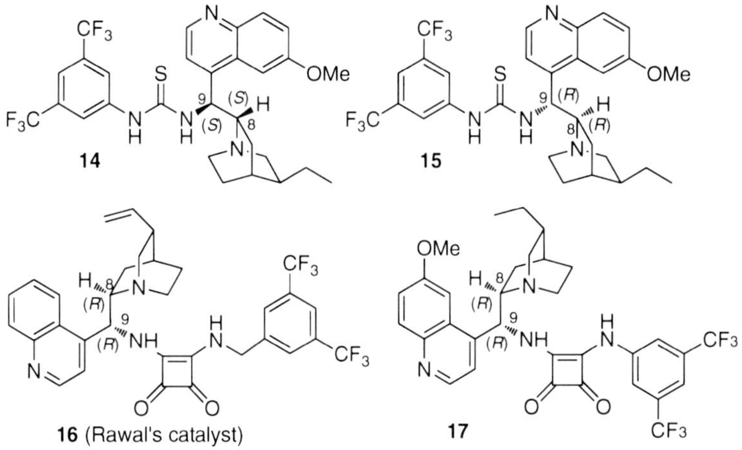

Fig. 6.2 Prototypical bifunctional catalysts derived from dihydroquinine and cinchonine

the advent of cinchona alkaloid catalysts based on these better hydrogen bond donor groups, however, Takemoto and co-workers demonstrated in 2003 that a simple combination of a tertiary amine base and thiourea in a rigid cyclohexane framework possesses remarkable catalytic activities (Scheme 6.2). This catalyst was first used to promote enantioselective conjugate additions of dialkyl malonates to *trans*-β-nitrostyrenes [4] (Scheme 6.2), and has subsequently been used for a range of reactions, including, among many other examples, e.g. stereoselective Michael addition of a prochiral α-substituted β-ketoester to *trans*-β-nitrostyrene [5], and a double Michael reaction leading to the construction of cyclic intermediate with the generation of three contiguous stereogenic centers. [6] The synthetic utility of the Takemoto catalyst **3** was also demonstrated by the enantioselective total syntheses of (*R*)-(−)-baclofen [5] and (−)-epibatidine [6, 7].

Although the design of these catalysts is relatively simple, the actual mechanisms how they operate have been subject to some debate. In the case of addition of malonates to *trans*-β-nitrostyrenes, at least two alternative mechanisms [8] have been suggested based on computational studies (Scheme 6.3). The alternatives differ mainly in how the substrates are hydrogen bonded to the catalyst. Route A, suggested by Takemoto, involves double hydrogen bonding to the nitrostyrene and a singly hydrogen bonded (possibly via a bifurcated hydrogen bond) malonate, whereas route B involves a doubly hydrogen bonded malonate and a singly hydrogen bonded nitrostyrene.

The extension of the thiourea-tertiary amine concept to the cinchona alkaloids was reported by several groups independently in 2005. The general usefulness of cinchona alkaloids as bifunctional catalysts was recognized by several groups even before this date [9]. The focus herein is on (thio)urea and squaramide catalysts since they are prototypes of more efficient catalysts. The Chen and the Dixon groups screened cinchonine and cinchonidine-derived thiourea catalysts for conjugate

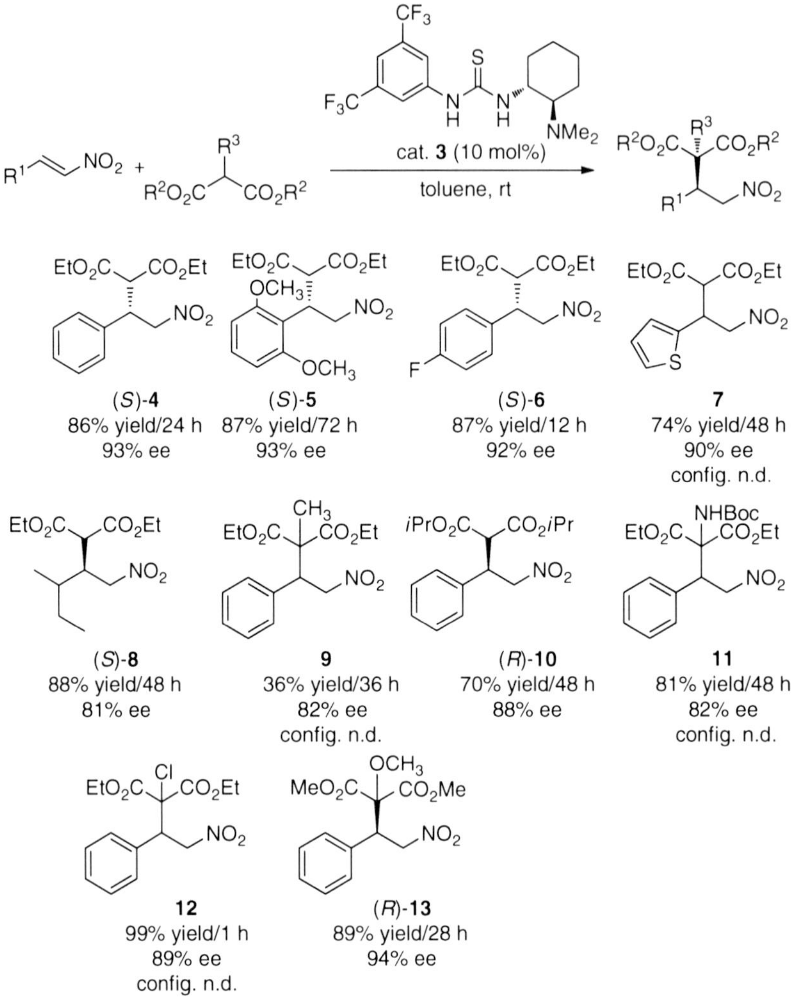

cat. **3** (10 mol%)
toluene, rt

(*S*)-**4** 86% yield/24 h 93% ee

(*S*)-**5** 87% yield/72 h 93% ee

(*S*)-**6** 87% yield/12 h 92% ee

7 74% yield/48 h 90% ee config. n.d.

(*S*)-**8** 88% yield/48 h 81% ee

9 36% yield/36 h 82% ee config. n.d.

(*R*)-**10** 70% yield/48 h 88% ee

11 81% yield/48 h 82% ee config. n.d.

12 99% yield/1 h 89% ee config. n.d.

(*R*)-**13** 89% yield/28 h 94% ee

Scheme 6.2 Bifunctional thiourea **3** catalyzes the asymmetric Michael addition of various dialkyl malonate to *trans*-α-nitrostyrene Michael acceptors

additions of thiophenols to α,β-unsaturated imines [10] and conjugate additions of malonates to *trans*-β-nitrostyrenes [11], respectively. The quinine/quinidine-derived catalysts (Fig. 6.2), first pioneered by Soós in 2005, however, have turned out to be even more popular than the cinchonine/cinchonidine-based catalysts. The squaramide catalyst structure **16** was first disclosed by the Rawal group in 2008, and in comparison with the Takemoto catalyst **3**, afforded faster rates and higher enantioselectivities in the conjugate addition of β-diketones to nitrostyrenes [12].

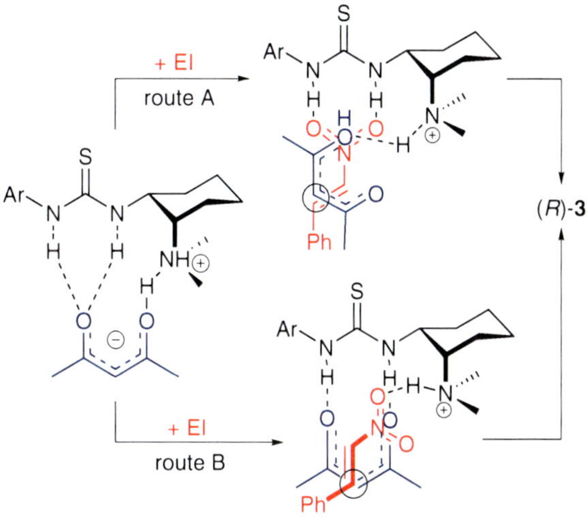

Scheme 6.3 Alternative mechanisms for the activation of a *trans*-β-nitrostyrene and acetoacetate anion with the Takemoto catalyst[8]

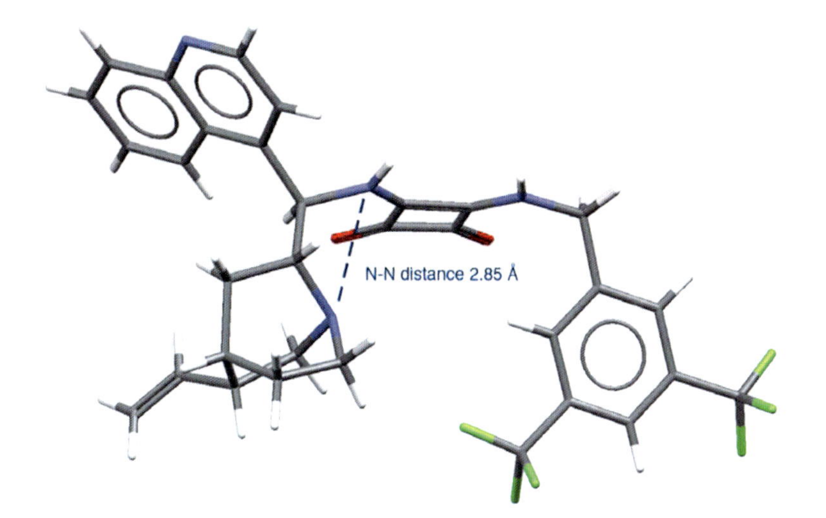

Fig. 6.3 X-ray structure of a cinchonine-derived squaramide catalyst **16** (Fig. 6.2) (CSD: NOLRIQ) [12]

In nearly all cases, catalysts bearing the unnatural C9-epi configuration (e.g. 8*S*, 9*S* for the quinine derivatives or 8*R*,9*R* for the quinidine-derived catalysts) have turned out to be more enantioselective [2]. A thorough review by Schreiner and Kotke discusses the developments in this area until 2009 [2, 13]. In spite of recent developments in the area, only a handful of X-ray crystal structures of the active catalysts have been reported. The X-ray structure of a cinchonine-derived (*8R, 9R* configuration) squaramide catalyst is displayed in Fig. 6.3.

The Soós group first applied catalyst **14** (DHQ-cat) in the addition of nitromethane to various chalcone derivatives (Scheme 6.4). The adducts were produced in excellent enantioselectivity and good yields [14].

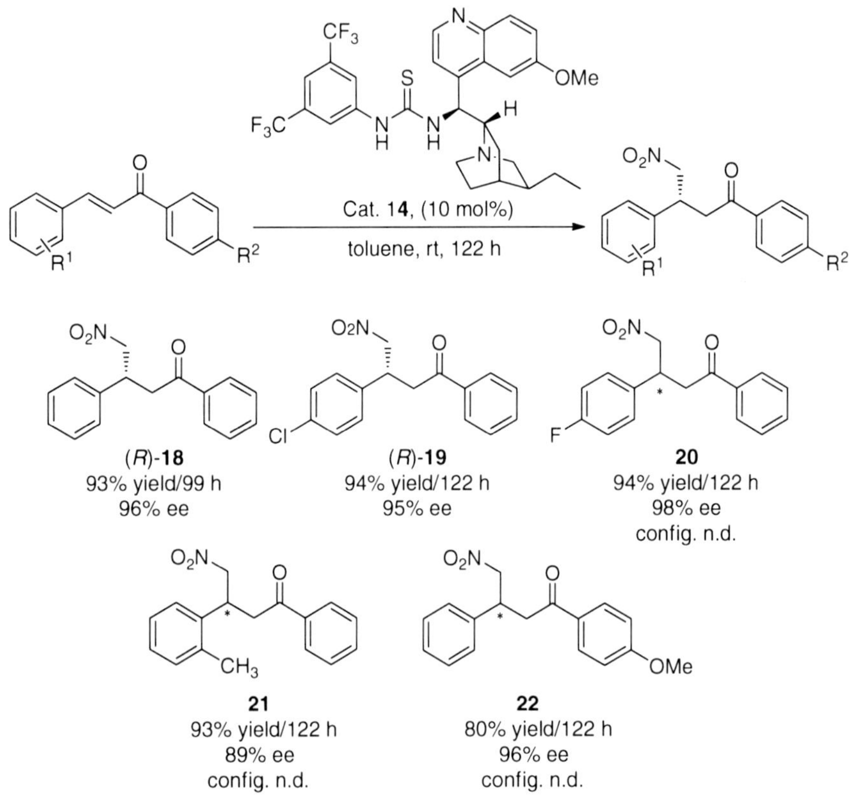

Cat. **14**, (10 mol%)
toluene, rt, 122 h

(*R*)-**18**
93% yield/99 h
96% ee

(*R*)-**19**
94% yield/122 h
95% ee

20
94% yield/122 h
98% ee
config. n.d.

21
93% yield/122 h
89% ee
config. n.d.

22
80% yield/122 h
96% ee
config. n.d.

Scheme 6.4 The conjugate addition of nitromethane to chalcones was one of the first examples of an enantioselective reactions catalyzed by cinchona alkaloid-derived thioureas (*nd*: not determined)

In collaboration with the Pápai group [8], the Soós group have also proposed a mechanistic scenario based on computational studies. The mechanism involves the binding of nitromethane anion to the thiourea catalyst by triple hydrogen bonding (Fig. 6.4).

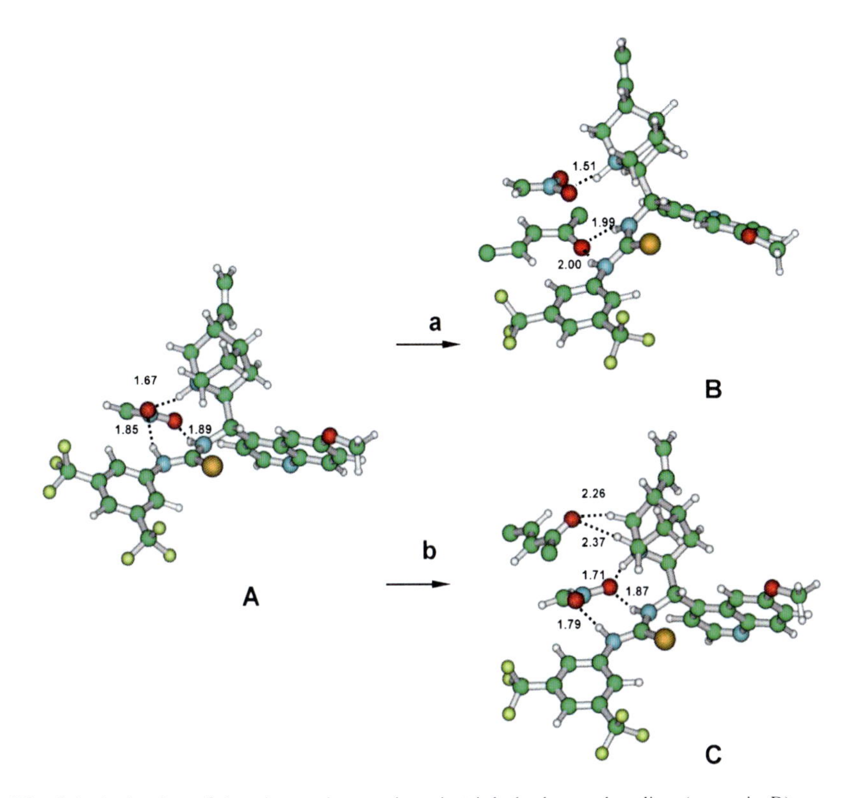

Fig. 6.4 Activation of the nitromethane anion via triple hydrogen bonding (scenario B) as proposed by the Pápai group

Also in 2005, the Connon group published [15] an independent study where they used the C9-epi-aminodihydrocinchonidine-derived thiourea catalyst **23** for the conjugate addition of malonates to β-nitrostyrenes (Scheme 6.5). Excellent enantioselectivities and good yields were obtained.

Another pioneering study [16] by the Dixon group, published in 2006, involved the addition of malonates to carbamate-protected imines (Scheme 6.6). Using the cinchonine-derived thiourea catalyst **32**, the Mannich-type products were obtained in good to excellent enantioselectivity and good yields.

In general, the best results reported in the literature so far have been obtained with substrates bearing multiple hydrogen bond acceptor sites either in the nucleophile or the electrophile. Activation of simple carbonyl compounds has been less successful, although examples are beginning to emerge. As a representative example, the Coltart group in 2010 disclosed an interesting Mannich reaction with

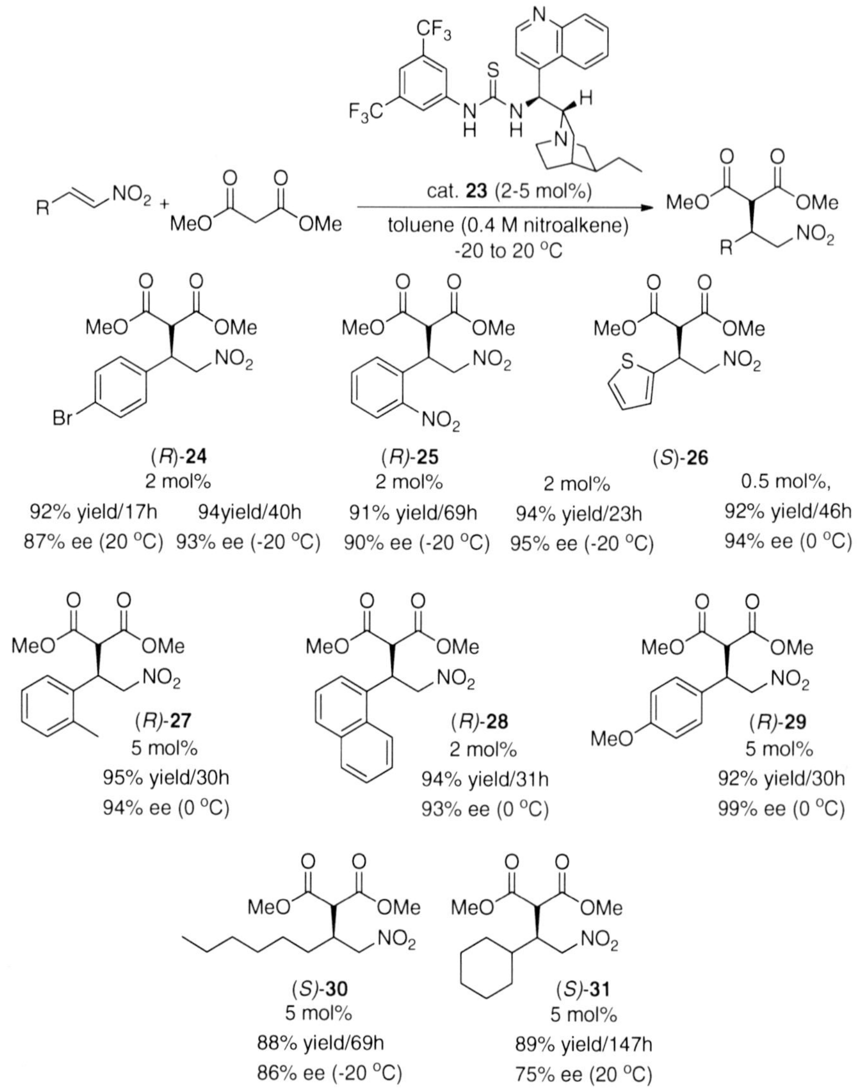

Scheme 6.5 Dihydrocinchonidine-derived catalyst **23** provides access to conjugate addition products in good to excellent enantioselectivity

sulfonylimines using relatively acidic α-aryl thioesters as nucleophiles with a dihydroquinine-derived catalyst **42** (Scheme 6.7) [17]. Although enantioselectivities were still only moderate, this study demonstrated that thioesters are also viable substrates for bifunctional acid-base catalysis.

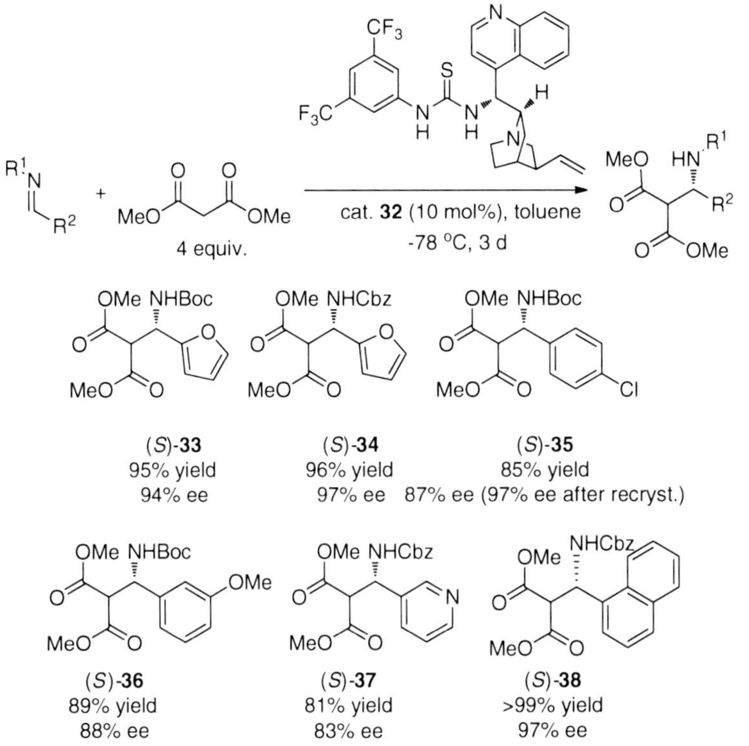

cat. **32** (10 mol%), toluene
-78 °C, 3 d

4 equiv.

(*S*)-**33** 95% yield 94% ee
(*S*)-**34** 96% yield 97% ee
(*S*)-**35** 85% yield 87% ee (97% ee after recryst.)
(*S*)-**36** 89% yield 88% ee
(*S*)-**37** 81% yield 83% ee
(*S*)-**38** >99% yield 97% ee

Scheme 6.6 Chiral Mannich adducts are obtained with the cinchonine-derived thiourea catalyst **32** from dimethyl malonate and *N*-Boc as well as *N*-Cbz aldimines

An alternative position for the (thio)urea moiety in a cinchona alkaloid-derived catalyst is the quinoline ring. This system (catalyst **55**, Scheme 6.8) provides a much larger distance for the thiourea and the basic tertiary amino group, and the Hiemstra group used this design successfully in an enantioselective Henry reaction [18] (Scheme 6.8). Many other catalyst designs failed to give good enantioselectivities in this transformation, including the 9-epi-aminocinchona alkaloid derived thiourea catalysts [18].

Interestingly, nearly all highly enantioselective bifunctional hydrogen bond/base catalysts involve 1,2-diamino substructures in their design. The success behind this type of design lies in the fact that the acid and the base units cannot associate intramolecularly due to strain (in principle, a 5-membered ring bearing a bent N-H---NR_2 hydrogen bond would be possible, but would involve a very strained bent hydrogen bond and possibly also some torsional strain in the N-C-C-N unit). Examples of 1,2-diamino bifunctional organocatalysts (**62**, [19] **63**, [20] **64–67** [21]) are described in Scheme 6.9. The Hiemstra catalyst **55** (Scheme 6.8) is an important exception to this rule and constitutes one of the very rare examples where the hydrogen bond donor and the basic unit have been further separated [22].

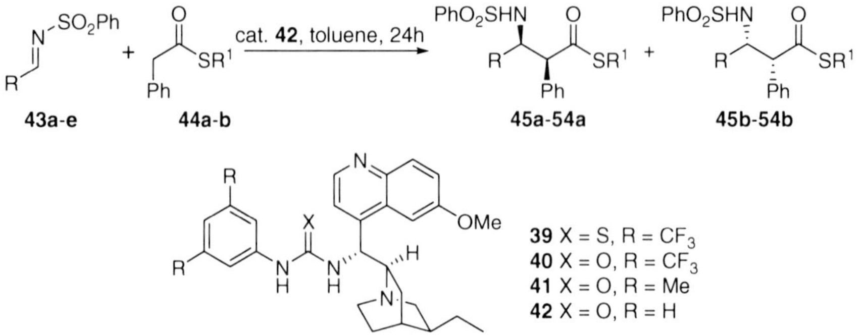

39 X = S, R = CF_3
40 X = O, R = CF_3
41 X = O, R = Me
42 X = O, R = H

Entry	Sulfonylimine (R)	Thioester (R[1])	**42** (mol%)	Product	*Syn*: *Anti*	er (a:b)	Yield (%)
1	**43a** (Ph)	**44a** (Et)	20	**45**	95:5	87:13	77
2	**43a**	**44b** (CH_2CF_3)	5	**46**	93:7	81:19	95
3	**43b** (furanyl)	**44a**	20	**47**	92:8	76:24	66
4	**43b**	**44b**	5	**48**	92:8	73:27	94
5	**43c** (4-Cl-C_6H_4)	**44a**	20	**49**	93:7	85:15	73
6	**43c**	**44b**	5	**50**	91:9	74:26	90
7	**43d** (4-OMe-C_6H_4)	**44a**	20	**51**	93:7	84:16	41
8	**43d**	**44b**	5	**52**	83:17	76:24	50
9	**43e** (4-Me-C_6H_4)	**44a**	20	**53**	98:2	88:12	78
10	**43e**	**44b**	5	**54**	97:3	66:34	85

Scheme 6.7 Mannich reaction with various sulfonylimines, promoted by dihydroquinine-derived catalyst **42**

However, an issue that has not been typically addressed in the design of bifunctional organocatalysts bearing hydrogen bond donors and a basic amino group is the possibility for intermolecular catalyst self-association in the solution (i.e. via the formation of dimers or oligomers). This might lead to reduced activity, or, if the dimers or oligomers are still active catalysts, could also lead to reduced enantioselectivity.

The Song group [23] at SKKU, Korea, have addressed this very problem by demonstrating that catalysts such as **73** do indeed give enantioselectivities that vary with the concentration of the catalyst in the enantioselective ring opening of azlactones (drop of ee from 87% to 84% in the range 0.4–1.0 M of substrate with 10 mol% of catalyst). However, the bis-cinchona alkaloid catalysts such as **69** afforded a constant enantioselectivity (94% ee) over the range of 0.4–1.0 M (Scheme 6.10). The lack of self-aggregation was further demonstrated by DOSY studies on the catalyst system.

The Song group has also developed monobasic cinchona alkaloid catalysts [24] such as the bulky sulfonamide **77** that do not display self-aggregation behavior. These catalysts are highly useful for the enantioselective alcoholysis of meso anhydrides [25], as demonstrated in Scheme 6.11.

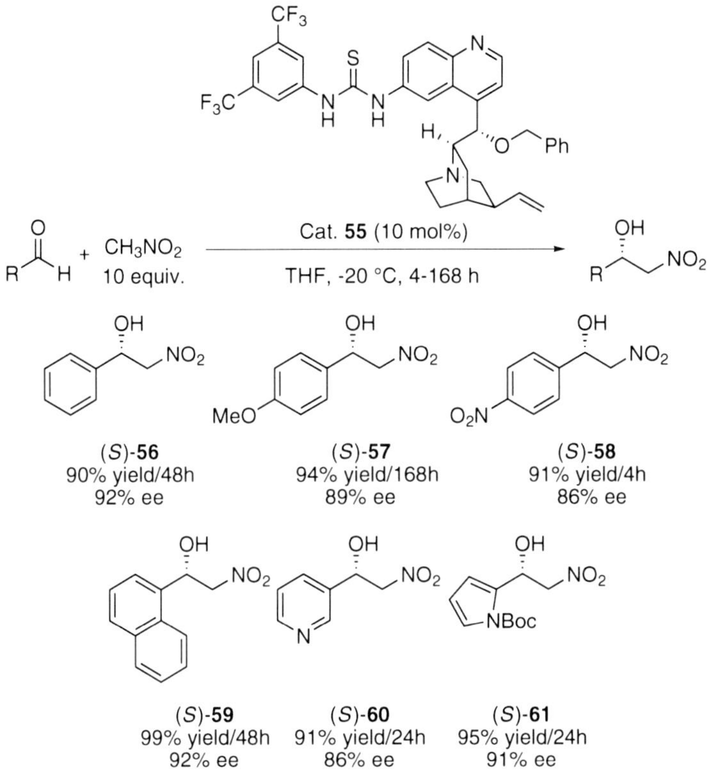

Cat. **55** (10 mol%)

THF, -20 °C, 4-168 h

CH_3NO_2 10 equiv.

(*S*)-**56** 90% yield/48h 92% ee

(*S*)-**57** 94% yield/168h 89% ee

(*S*)-**58** 91% yield/4h 86% ee

(*S*)-**59** 99% yield/48h 92% ee

(*S*)-**60** 91% yield/24h 86% ee

(*S*)-**61** 95% yield/24h 91% ee

Scheme 6.8 Representative adducts obtained from the asymmetric Henry reaction between nitromethane and (hetero) aromatic aldehydes under bifunctional catalysis of C6′-thiourea-functionalized cinchona alkaloid **55**

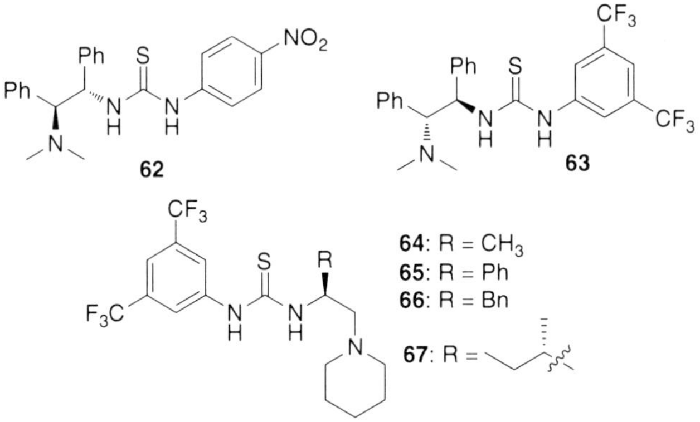

64: R = CH_3
65: R = Ph
66: R = Bn
67: R =

Scheme 6.9 Examples of 1,2-diamino bifunctional organocatalysts

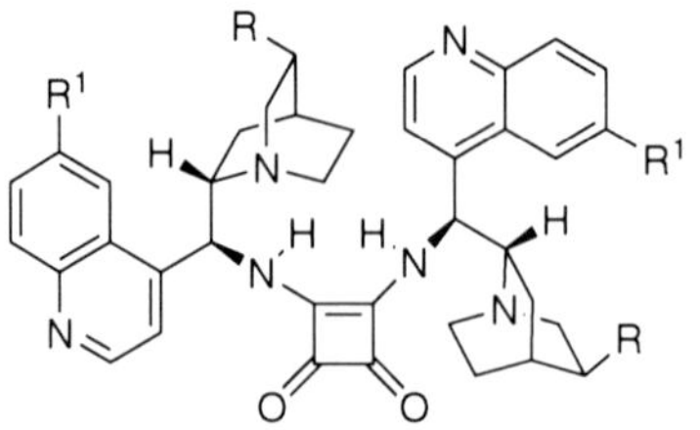

68, **Bis-QN-SQA**, R = vinyl, R^1 = OMe
69, **Bis-HQN-SQA**, R = ethyl, R^1 = OMe
70, **Bis-CD-SQA**, R = vinyl, R^1 = H
71, **Bis-GCD-SQA**, R = ethyl, R^1 = H

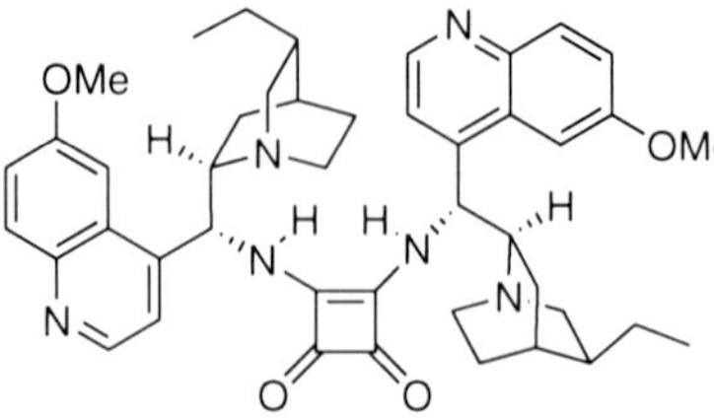

72, **Bis-HQD-SQA**

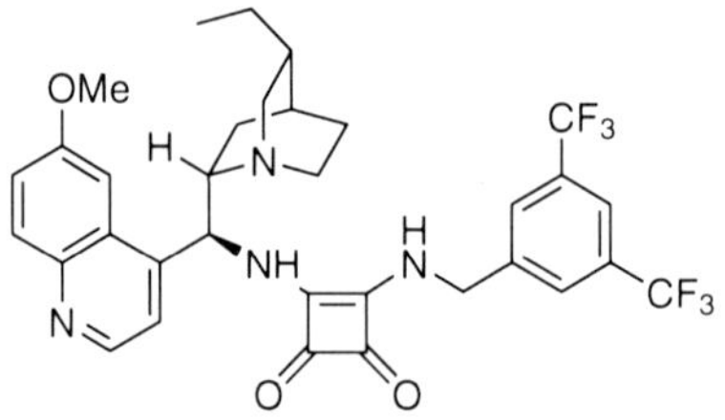

73, **HQN-SQA**

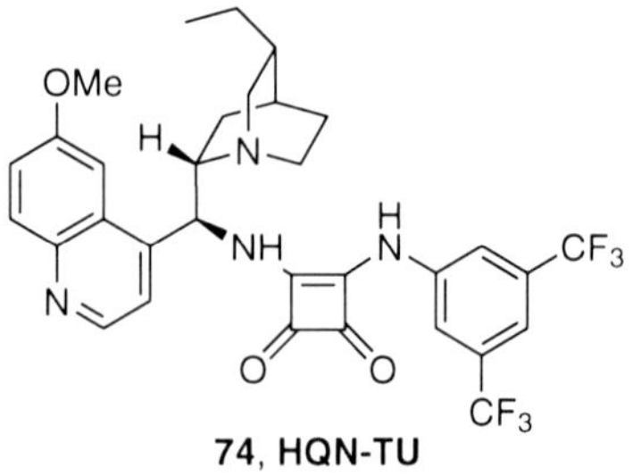

74, **HQN-TU**

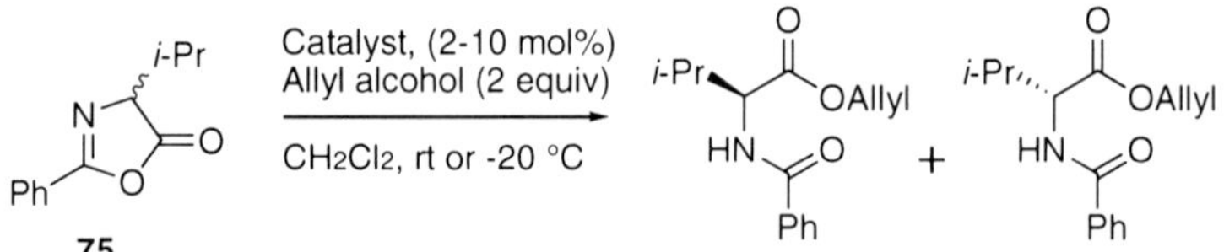

75 → Catalyst, (2-10 mol%), Allyl alcohol (2 equiv), CH_2Cl_2, rt or -20 °C → (*S*)-76 + (*R*)-76

Entry	Catalyst (mol%)	T(°C)	Time	Yield (%)	ee (%)	Main product
1	**68** (10)	rt	6 h	99	93	**(*S*)-76**
2	**69** (10)	rt	3 h	95	94	**(*S*)-76**
3	**69** (5)	rt	6 h	92	94	**(*S*)-76**
4	**69** (2)	rt	24 h	98	94	**(*S*)-76**
5	**69** (10)	−20 °C	8 h	98	97	**(*S*)-76**
6	**70** (10)	rt	6 h	96	93	**(*S*)-76**
7	**71** (10)	rt	3 h	98	94	**(*S*)-76**
8	**72** (20)	rt	48 h	97	91	**(*R*)-76**
9	**73**(10)	rt	24 h	86	86	**(*S*)-76**
10	**74** (10)	rt	12 h	95	75	**(*S*)-76**

Scheme 6.10 Catalytic DKR of the racemic valine-derived azalactone **75** with allyl alcohol

6.2.1 Enzymatic Bifunctional Acid-Base Catalysts

Most enzymes utilizing oxyanion holes [26] are typical examples of bifunctional acid-base catalysts. Two representative examples are discussed in detail to illustrate the similarity of enzymatic mechanisms to designed organocatalysts.

77

78	79	80	81
92% yield/1.5 h	90% yield/4.5 h	90% yield/5h	88% yield/6h
96% ee/rt	96% ee/rt	94% ee/rt	95% ee/rt

82	83	84
95% yield/5 h	95% yield/4 h	97% yield/4 h
98% ee/rt	91% ee/rt	94% ee/rt

Scheme 6.11 Enantioselective ring opening of *meso* anhydrides affords chiral monoesters of valuable dicarboxylic acids in excellent enantioselectivity using the sulfonamide catalyst **77**

Enzymes with oxyanion holes are now known to catalyze a wide range of reactions with substrates bearing a carbonyl moiety. Examples include thioesters, oxygen esters, peptides and ketones (Scheme 6.12). Two classes of high-energy intermediates with oxyanions are generated in these reactions, a tetrahedral intermediate and an enolate. A significant fraction of these enzymes are metal ion independent (i.e., organocatalytic), and no other cofactors are required either.

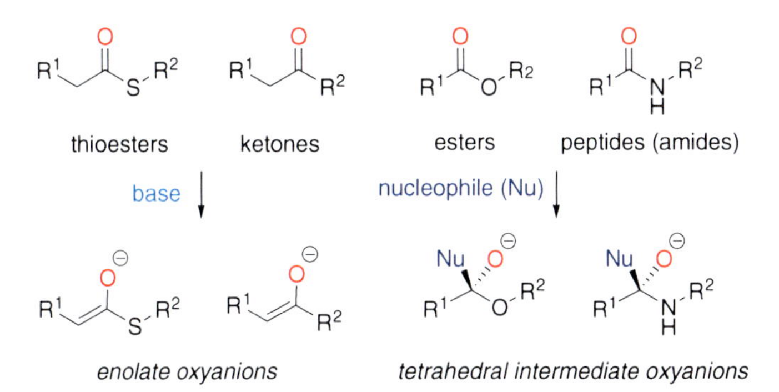

Scheme 6.12 High-energy oxyanion intermediates are formed by e.g. enolization of thioesters and ketones/aldehydes, or by addition of nucleophiles to esters and amides

A very well studied example is the oxyanion hole in serine proteases. In chymotrypsin the oxyanion hole is formed by main chain NH groups of two peptide units, Gly193 and Ser195 [27] (Scheme 6.13). More generally, the classical oxyanion hole stabilizes the negatively charged oxyanion of a tetrahedral intermediate and exists of two main chain peptide NH groups [28].

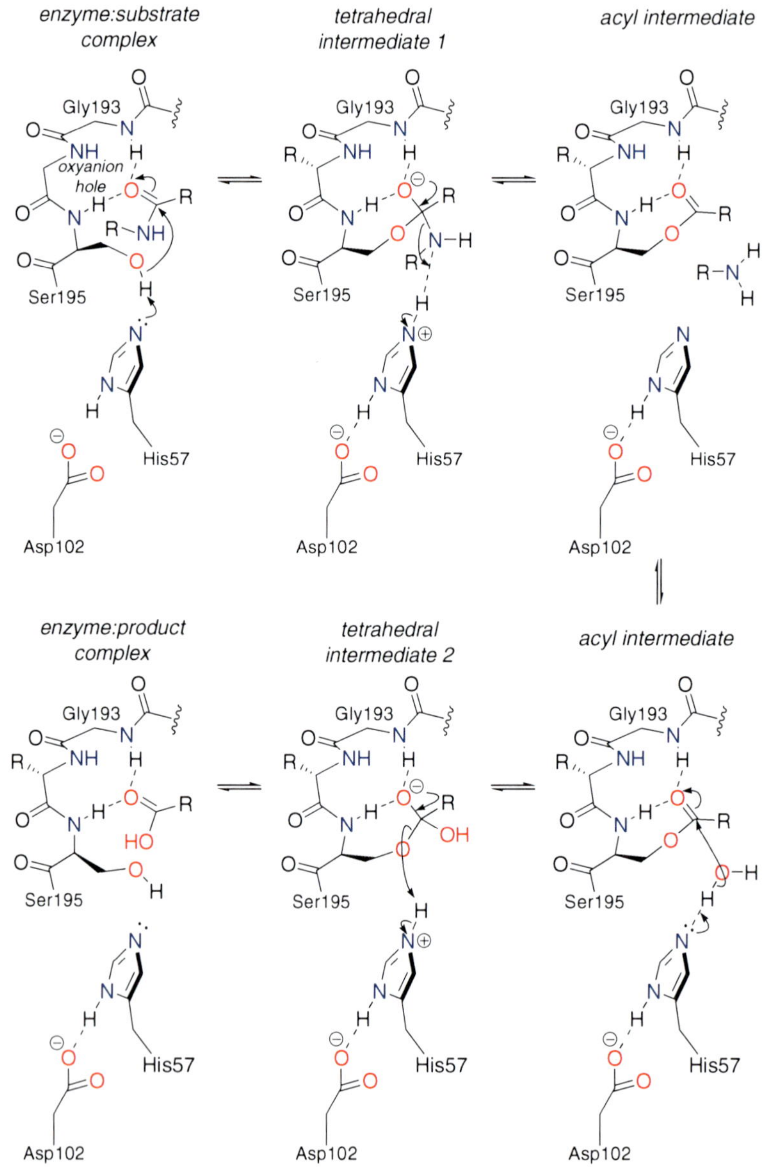

enzyme:substrate complex · *tetrahedral intermediate 1* · *acyl intermediate*

enzyme:product complex · *tetrahedral intermediate 2* · *acyl intermediate*

Gly193, Ser195, His57, Asp102, oxyanion hole

Scheme 6.13 The reaction mechanism of chymotrypsin

The tetrahedral intermediate is generated by a nucleophilic attack on the carbonyl carbon atom of the activated nucleophile, which in the case of chymotrypsin and trypsin is the catalytic Ser195. The mechanism of chymotrypsin can be compared to that of Song catalyst **77** that catalyzes the nucleophilic ring opening of acid anhydrides. The role of the sulfonamide group is played by the oxyanion hole amide NH groups, and the role of the cinchona alkaloid-derived base is played by the Asp-His-(Ser) dyad/triad.

Another enzyme that bears a close resemblance to bifunctional acid-base catalysts is enoyl-CoA hydratase (ECH). This enzyme is discussed here as an example of an enolate-stabilizing enzyme. The mechanism of ECH involves an enolate intermediate that is formed upon addition of a nucleophile, water molecule, to an α,β-unsaturated thioester. This reaction is the second step in the metabolic oxidation of fatty acids. The X-ray structures of enoyl hydratase enzymes complexed with various substrates analogs are available [29, 30]. A simplified view of the mechanism of ECH is presented in Scheme 6.14. The lifetime of the enolate intermediate is not

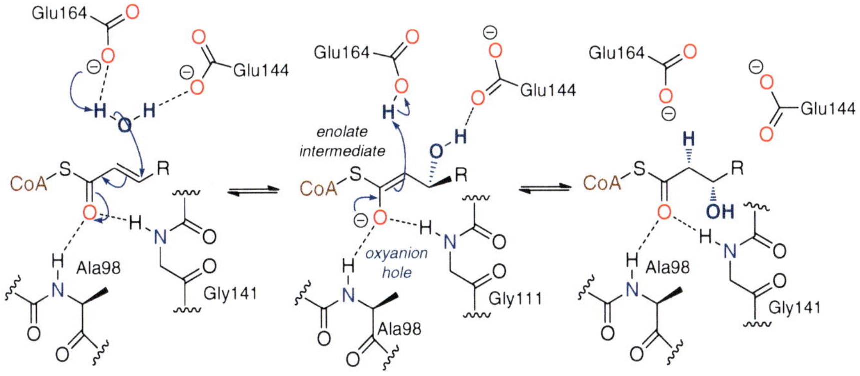

Scheme 6.14 The reaction mechanism and mode of catalysis by enoyl-CoA hydratase (ECH)

known [29] but it is likely to be short; nevertheless, the mechanism depicted below is consistent with the available evidence that the reaction proceeds via a *syn*-addition of H and OH [31].

A three-dimensional picture of the active site of ECH complexed with an inactive substrate 4-(dimethylamino)cinnamoyl-CoA is presented in Fig. 6.5.

The active site of ECH is an example of an active site built on the framework of the crotonase fold. It is now well established that this crotonase fold provides an active site framework that has been used by Nature for a wide range of different chemical reactions. These reactions have been reviewed recently [32].

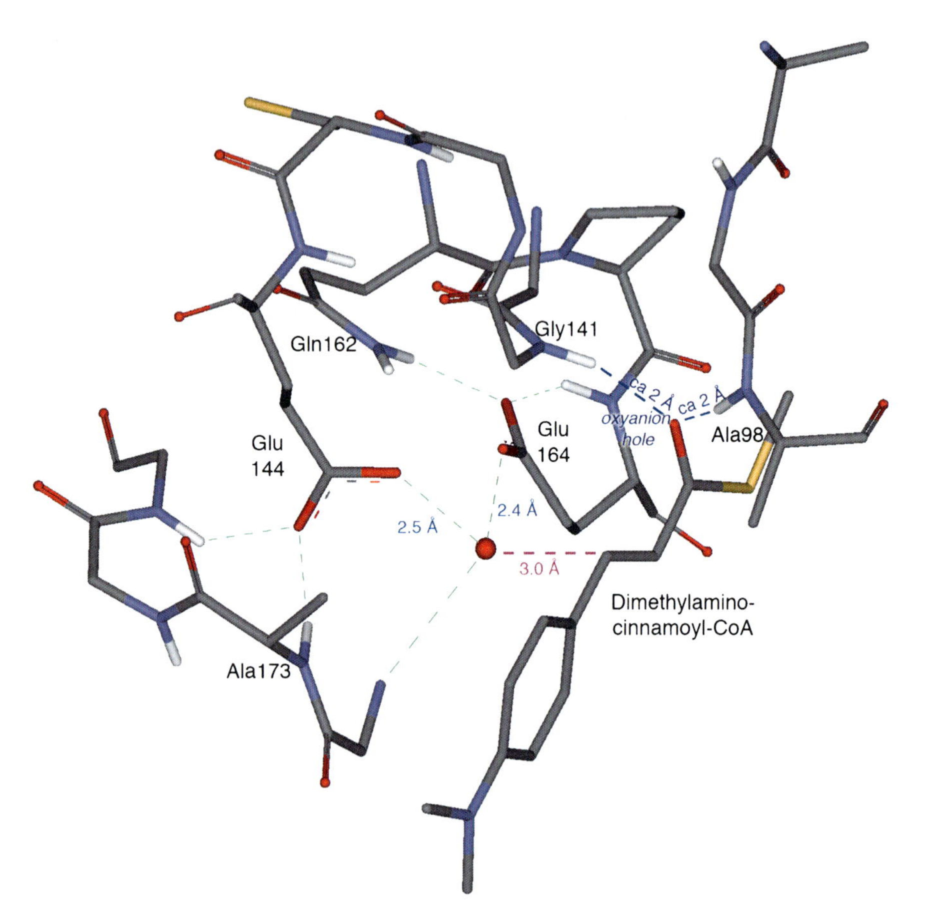

Fig. 6.5 The active site of enoyl-CoA hydratase (ECH) (PDB: 1EY3) [29]. The attacking water nucleophile (red sphere, hydrogen atoms are not shown) is activated by two glutamate moieties, Glu144 and Glu164. The glutamate bases are anchored by hydrogen bonds to Gln162 and Ala173. The thioester carbonyl, in turn, is activated by hydrogen bonds from Ala98 and Gly141

6.3 Frustrated Lewis Acid–Lewis Base Catalysis

The bifunctional acid-base catalysts discussed above are all Brønsted acid/Brønsted base catalysts. However, it is also possible to design related catalysts that are based on Lewis acid/Lewis base pairs. To prevent the self-association and irreversible intermolecular adduct formation, sterically encumbered Lewis acids and Lewis bases are typically used. Such hindered pairs are often called "frustrated Lewis pairs" (FLPs). This field has been reviewed very thoroughly recently [33]. Herein, we focus on the catalytic aspects of FLPs, especially their role as bifunctional hydrogen activators.

In pioneering work, the reversible action of FLPs was demonstrated by Stephan and co-workers [34]. Starting from a zwitterionic salt **87**, reduction with Me_2SiHCl provided another zwitterion, an internal phosphonium borohydride **88**. This species turned out to be relatively robust, but heating to 150°C forced the loss of H_2 to generate the phosphino-borane **86** as an orange-red solid. **86** reacted with H_2 at 25°C to regenerate the zwitterion **88**, thus completing the cycle. The phosphino-borane **86** turned out to be monomeric in solution (Scheme 6.15).

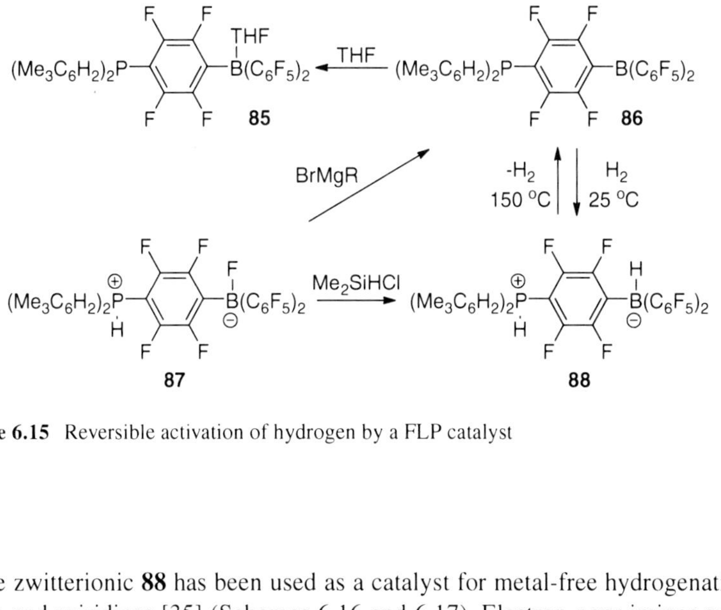

Scheme 6.15 Reversible activation of hydrogen by a FLP catalyst

The zwitterionic **88** has been used as a catalyst for metal-free hydrogenation of imines and aziridines [35] (Schemes 6.16 and 6.17). Electron-poor imines such as the sulfonimide **90** required longer reaction times, possibly due to rate-determining protonation of the imine nitrogen. Imines bearing less sterically demanding substituents could be reduced only with a stoichiometric amounts of the zwitterion, suggesting that the product amines might bind too tightly to the Lewis acidic boron center. The mechanism of the hydrogenation reaction appears to involve a sequence of imine protonation, followed by reduction of the iminium salt with a borohydride [34]. A computational study by the Pápai group agrees with the proposed mechanism [36] (Scheme 6.16).

A more reactive catalyst pair is provided by placing the Lewis acid and Lewis basic phosphorus and boron moieties in vicinal carbon atoms, such as in catalyst **97** (Scheme 6.18) [37]. This design is very similar to that of (thio)urea-tertiary amine catalysts discussed in the preceding section. Using this catalyst **97**, enamines can also be reduced by the FLP catalyst. These reactions likely proceed via the intermediacy of the corresponding iminium ions.

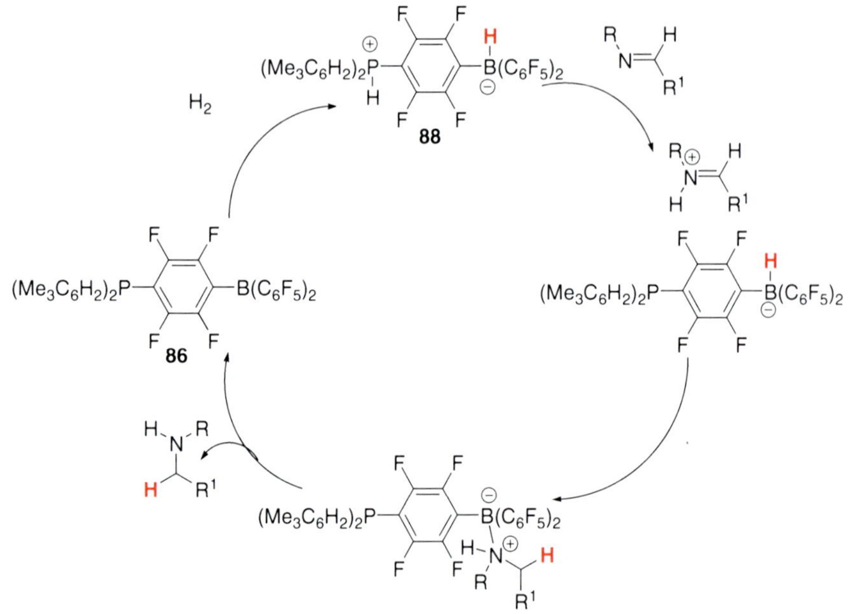

H_2

$(Me_3C_6H_2)_2\overset{\oplus}{P}(H)$–$C_6F_4$–$\overset{\ominus}{B}(H)(C_6F_5)_2$

88

R–N=CH–R^1

$R\overset{\oplus}{N}(H)$=CH–R^1

$(Me_3C_6H_2)_2P$–C_6F_4–$\overset{\ominus}{B}(H)(C_6F_5)_2$

$(Me_3C_6H_2)_2P$–C_6F_4–$B(C_6F_5)_2$

86

H–N(R)–CH(H)–R^1

$(Me_3C_6H_2)_2P$–C_6F_4–$\overset{\ominus}{B}(C_6F_5)_2$–$\overset{\oplus}{N}H(R)$–CH(H)–$R^1$

Scheme 6.16 Postulated catalytic cycle for FLP-catalyzed hydrogenation of imines

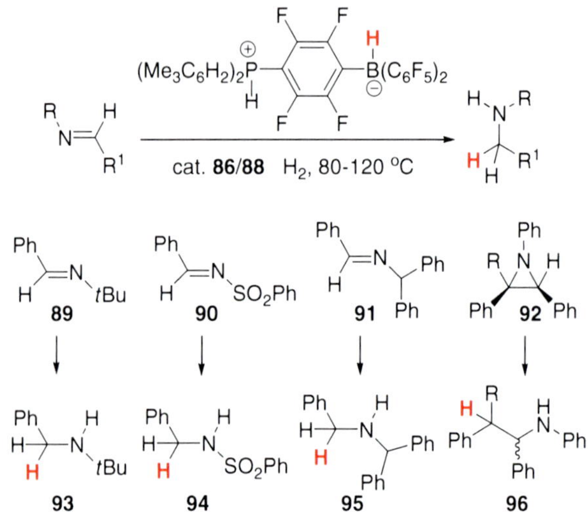

Scheme 6.17 Reduction of imines and aziridines with a FLP catalyst

$(Me_3C_6H_2)_2\overset{\oplus}{P}(H)$–$C_6F_4$–$\overset{\ominus}{B}(H)(C_6F_5)_2$

R–N=CH–R^1 → H–N(R)–CH_2–R^1

cat. **86/88** H_2, 80-120 °C

Ph–CH=N–tBu **89** → Ph–CH_2–NH–tBu **93**

Ph–CH=N–SO_2Ph **90** → Ph–CH_2–NH–SO_2Ph **94**

Ph–CH=N–CHPh_2 **91** → Ph–CH_2–NH–CHPh_2 **95**

R, Ph, H, Ph, N–Ph **92** → H–C(R)(Ph)–CH(Ph)–NH–Ph **96**

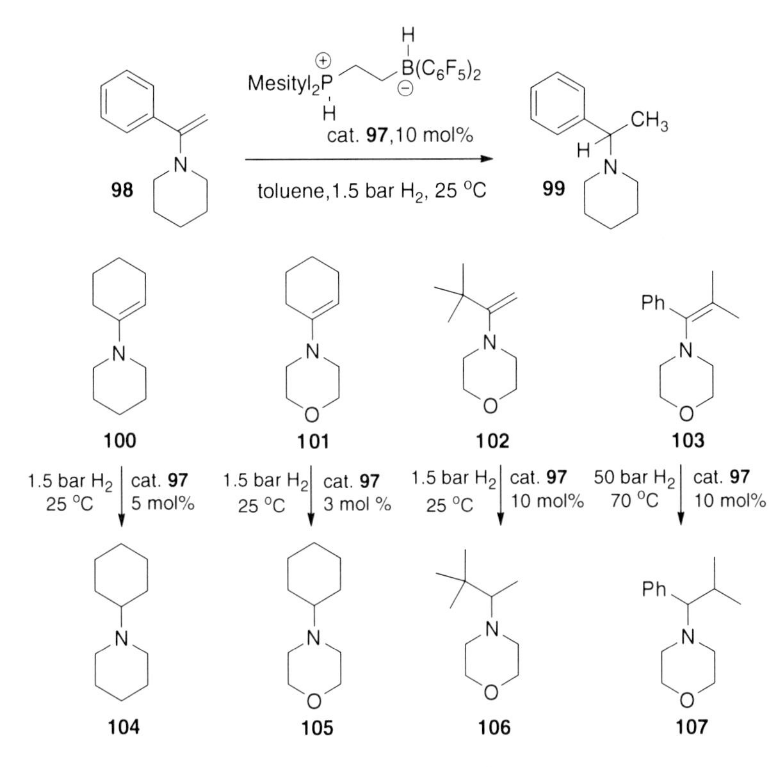

Scheme 6.18 Hydrogenation of enamines with a FLP catalyst

A very interesting FLP catalyst design was recently disclosed by Repo, Rieger and co-workers [38]. Coined the *ansa-aminoborane* [39] FLP catalyst, it is based on the use of amines instead of phosphines as the Lewis basic component (Scheme 6.19). This catalyst is also active in 10 mol% quantities for the reduction of imines and enamines. The X-ray structure of the ansa-catalyst is illustrated below (Fig. 6.6).

Very recently, Klankermayer and co-workers described the first enantioselective hydrogenation with chiral FLP catalyst that afforded enantioselectivities better than 80% ee [40]. Although this catalyst was not strictly a bifunctional FLP catalyst, the result is mentioned here since the covalent, enantioselective FLP catalysts are very likely the next step in FLP catalysis. The Klankermayer catalyst **115** was able to reduce a number of imines enantioselectively to amines (Scheme 6.20).

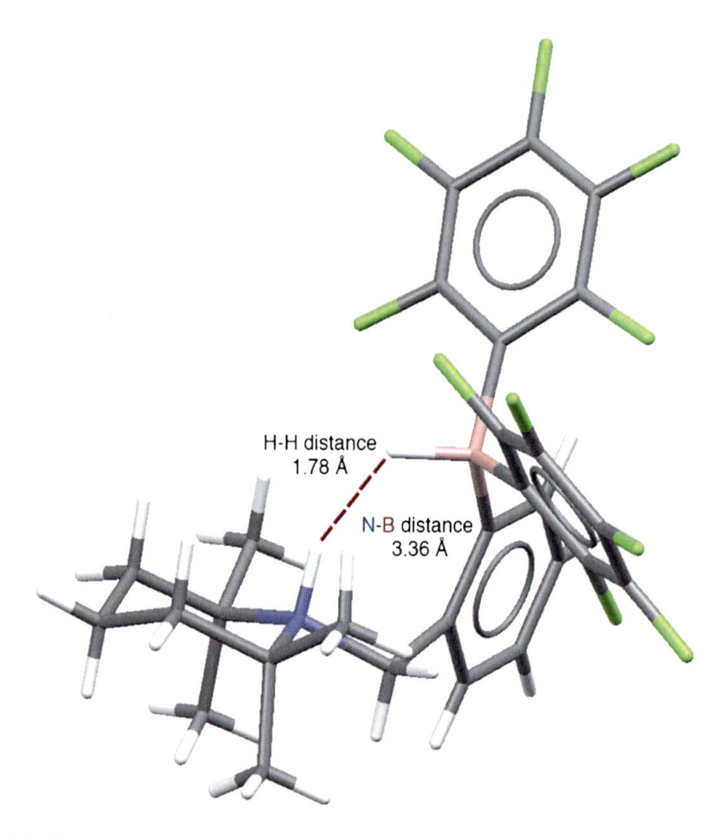

Fig. 6.6 X-ray crystal structure of the ansa-ammonium-borohydride catalyst **108** (CCDC: ROHSAJ)

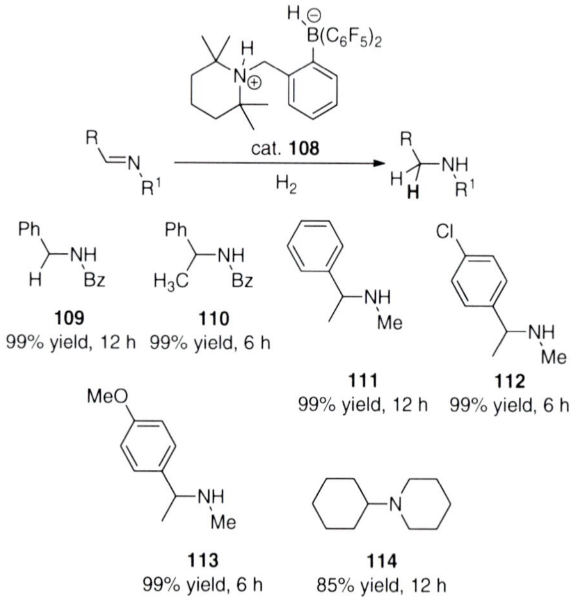

$B(C_6F_5)_2$

cat. **108**

H_2

109
99% yield, 12 h

110
99% yield, 6 h

111
99% yield, 12 h

112
99% yield, 6 h

113
99% yield, 6 h

114
85% yield, 12 h

Scheme 6.19 Metal-free hydrogenation of imines and enamines using the ansa-catalyst **108**

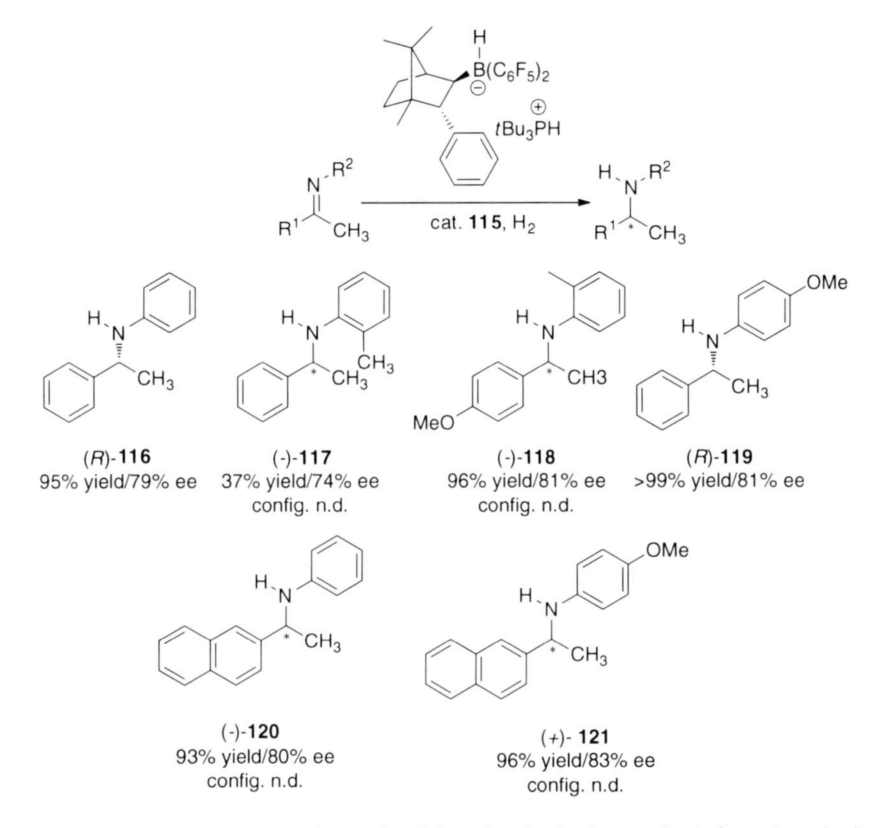

Scheme 6.20 The first highly enantioselective FLP-catalyzed reduction reaction (*nd*: not determined)

6.4 Conclusions

The use of noncovalent interactions, such as hydrogen bonding and ion pair (Coulombic) interactions, offers significant advantages in terms of flexibility and rapid turnovers. A large portion of the research has focused on establishing new enantioselective transformations based on bifunctional catalyst designs. However, less attention has been paid to understanding the structure-activity relationships of these catalysts. We believe this work will be essential in developing future catalysts with higher activities. The lessons learned from enzymes show that suitably positioned hydrogen bond donors and basic groups can achieve wonders in catalysis.

The key question is how to emulate this with small molecules with their much more limited range of binding capacity? Perhaps future catalysts will require a more widespread use of other secondary interactions, such as cation-π interactions, to achieve higher selectitivities [1].

References

1. For a recent overview, see: Knowles RR, Jacobsen EN (2010) Proc Natl Acad Sci, doi: 10.1073/pnas.1006402107
2. For a comprehensive review of thiourea and urea catalysts, see: Kotke M, Schreiner P (2009) '(Thio)urea Organocatalysts.' In: Pihko PM (ed) Hydrogen Bonding in Organic Synthesis, Wiley-VCH, Weinherm, Germany p 141
3. Hiemstra H, Wynberg H (1981) J Am Chem Soc 103:417
4. Okino T, Hoashi Y, Takemoto Y (2003) J Am Chem Soc 125:12672
5. Okino T, Hoashi Y, Furukawa T, Xu X, Takemoto Y (2005) J Am Chem Soc 127:119
6. Hoashi Y, Yabuta T, Takemoto Y (2004) Tetrahedron Lett 45:9185
7. Hoashi Y, Yabuta T, Yuan P, Miyabe H, Takemoto Y (2006) Tetrahedron 62:365
8. Hamza A, Schubert G, Soos T, Papai I (2006) J Am Chem Soc 128:13151
9. (a)Tian SK, Chen YG, Hang JF, Tang L, McDaid P, Deng L (2004) Acc Chem Res 37:621; (b) Doyle AG, Jacobsen EN (2007) Chem Rev 107:5713
10. Li BJ, Jiang L, Liu M, Chen YC, Ding LS, Wu Y (2005) Synlett 603
11. Ye J X, Dixon D J, Hynes P S (2005) Chem Commun 4481
12. Malerich JP, Hagihara K, Rawal VH (2008) J Am Chem Soc 130:14416
13. For other reviews of this area, see: (a) Takemoto Y (2005) Org Biomol Chem 3:4299; (b) Taylor MS, Jacobsen EN (2006) Angew Chem Int Ed 45:1520; (c) Connon SJ (2006) Chem Eur J 12:5418; (d) Miyabe H, Takemoto Y (2008) Bull Chem Soc Jpn 81:785. (e) Connon SJ (2009) Synlett 354. (See also ref 9b)
14. Vakulya B, Varga S, Csampai A, Soos T (2005) Org Lett 7:1967
15. McCooey SH, Connon SJ (2005) Angew Chem Int Ed 44:6367
16. (a) Tillman AL, Ye JX, Dixon DJ (2006) Chem Commun 1191. For a review, see (b) Tin A, Schaus SE (2007) Eur J Org Chem 5797–5815
17. Kohler M, Yost JM, Garnsey MR, Coltart DM (2010) Org Lett 12:3376–3379
18. (a) Marcelli T, van Maarseveen JH, Hiemstra H (2006) Angew Chem Int Ed 45:7496–7504; (b) Palomo C, Oiarbide M, Laso A (2007) Eur J Org Chem 2007:2561–2574; (c) Marcelli T, van der Haas RNS, van Maarseveen JH, Hiemstra H (2006) Angew Chem Int Ed 45:929–931; (d) Marcelli T, van der Haas RNS, van Maarseveen JH, Hiemstra H (2005) Synlett 18:2817–2819
19. Liao Y-H, Liu X-L, Wu Z-J, Cun L-F, Zhang X-M, Yuan W-C (2010) Org Lett 12:2896
20. Liu TY, Long J, Li BJ, Jiang L, Li R, Wu Y, Ding LS, Chen YC (2006) Org Biomol Chem 4:2097
21. Gao Y, Ren Q, Wang L, Wang J (2010) Chem Eur J 16:13068
22. For another interesting example of a multifunctional Brønsted acid catalyst, see: (a) Nugent BM, Yoder RA, Johnston JN (2004) J Am Chem Soc 126:3418; (b) Davis TA, Wilt JC, Johnston JN (2010) J Am Chem Soc 132:2880
23. Lee JW, Ryu TH, Oh JS, Bae HY, Jang HB, Song CE (2009) Chem Commun 7224
24. (a) Oh SH, Rho HS, Lee JH, Lee JE, Youk SH, Chin J, Song CE (2008) Angew Chem Int Ed 47:7872; (b) Park SE, Nam EH, Jang HB, Oh JS, Some S, Lee YS, Song CE (2010) Adv Synth Catal 352:2211
25. It should be noted that the presence of a hydrogen bond donor/ Brønsted acid is not required for high enantioselectivities in alcoholysis of meso anhydrides. For leading references to other cinchona- alka loid based catalysts for this transformation, see: Li H, Liu X, Wu F, Tang L, Deng L (2010) Proc Nat Acad Sci, doi: 10.1073/pnas.1004439107
26. For a comprehensive review of oxyanion holes in catalysis, see: Pihko P, Rapakko S, Wierenga RK (2009) 'Oxyanion Holes and Their Mimics' In: Pihko PM (ed) Hydrogen Bonding in Organic Synthesis, Wiley-VCH, Weinherm, Germany p 43
27. (a) For reviews, see: Frey PA, (2004) J Phys Org Chem 17:511; (b) Hedstrom L (2002) Chem Rev 102:4501; (c) Sedolisins (serine-carboxyl peptidases), such as kumamolisin-As, offer an interesting counterexample. In this carboxypeptidase, the oxyanion hole includes an aspartate group which appears to protonate the developing oxyanion. For a computational study of this system, see: Guo H, Wlodawer A, Guo H (2005) J Am Chem Soc 127:15662

28. (a) Blow D (2000) Structure 8:77; (b) Henderson R (1970) J Mol Biol 54:341
29. Bahnson BJ, Anderson VE, Petsko GA (2002) Biochemistry 41:2621
30. Engel CK, Mathieu M, Zeelen JP, Hiltunen JK, Wierenga RK (1996) EMBO J 15:5135
31. Willadsen P, Eggerer H (1975) Eur J Biochem 54:247
32. (a) Hamed RB, Batchelar ET, Clifton IJ, Schofield CJ (2008) Cell Mol Life Sci 65:2507; (b) Holden HM, Benning MM, Haller T, Gerlt JA (2001) Acc Chem Res 34:145
33. For recent reviews of frustrated Lewis pairs, see: (a) Stephan DW, Erker G (2010) Angew Chem Int Ed 49:46; (b) Stephan DW (2009) Dalton Trans 3129
34. (a) Welch GC, Stephan DW (2007) J Am Chem Soc 129:1880; (b) Welch GC, Juan RRS, Masuda JD, Stephan DW (2006) Science 314:1124
35. Chase PA, Welch GC, Jurca T, Stephan DW (2007) Angew Chem 119:8196; (2007) Angew Chem Int Ed 46: 8050
36. For hydrogenation with $P(tBu)_3/B(C_6F_5)_3$ pair see Rokob TA, Hamza A, Stirling A, Pápai I (2009) J Am Chem Soc 131:2029
37. (a) Spies P, Schwendemann S, Lange S, Kehr G, Frohlich R, Erker G (2008) Angew Chem 120:7654; (2008) Angew Chem Int Ed 47:7543; (b) Axenov KV, Kehr G, Frohlich R, Erker G (2009) J Am Chem Soc 131:3454; (c) Schwendemann S, Tumay TA, Axenov KV, Peuser I, Kehr G, Frohlich R, Erker G (2010) Organometallics 29:1067
38. (a) Sumerin V, Schulz F, Atsumi M, Wang C, Nieger M, Leskela M, Repo T, Pyykko P, Rieger B (2008) J Am Chem Soc 130:14117; (b) Sumerin V, Schulz F, Nieger M, Atsumi M, Wang C, Leskela M, Pyykko P, Repo T, Rieger B (2009) J Organomet Chem 694:2654
39. The terms ansa-aminoborane [ansa (lat.) = "handle"] refer to the use of the successful concept of ansa- metallocenes where a bridge between two Cp-ligands forces them into a distinct geometry and hence nfluences the specific reactivity of these compounds.
40. Chen C, Wang Y, Klankermayer J (2010) Angew Chem Int Ed 49:9475

Chapter 7
Chalcogen-Based Organocatalysis

Ludger A. Wessjohann, Martin C. Nin Brauer, and Kristin Brand

Abstract Most current organocatalysts are based on nitrogen (or phosphorus) as reactive atom, including also most processes depending on proton acidity and/or Lewis basicity. Only few organocatalytic systems use organochalcogens, although such reactions are of great importance in nature, especially evident in hydrolases with serine or cysteine as catalytic hotspot, or in oxidoreductases with cysteine or selenocysteine as key players. Catalytic processes in nature commonly rely on the nucleophilic or redox properties of chalcogen atoms.

Accordingly early attempts in chemical catalysis using organochalcogens concentrate either on systems reminiscent of catalytic diads and triads of enzymes with catalysts consisting of a hydroxyl or sulfhydryl group that is activated as nucleophile by a neighboring base (catalytic diads and triads). Other "traditional" uses of chalcogen-based catalysts comprise chiral dioxiranes and oxaziranes for epoxidations, and sulfur redox catalysts, the latter especially in the application of sulfur ylides covered by the predominant work of Aggarwal et al. Since the advent of "Organocatalysis" as a distinct subfield of catalysis, not only these traditional organochalcogen catalyst systems excelled; also new applications are more systematically studied now, including not only oxygen and sulfur but increasingly selenium – and to a smaller extent – even tellurium based catalysis [372]*. If nature and its several thousand years of selection of catalysis modes serve as a reference, group VI-based catalysis is yet very much below its real potential in chemical organocatalysis. This contribution thus aims at giving the reader an entry into this so much underutilized field, which offers ample room especially for those who like to try new paths and who not only wish expand on existing processes of well established nitrogen-based catalysts.

*Added in proof.

L.A. Wessjohann (✉) • M.C.N. Brauer • K. Brand
Leibniz Institute of Plant Biochemistry, Weinberg 3, D-06120 Halle (Saale), Germany
e-mail: wessjohann@ipb-halle.de; MartinClaudio.NinBrauer@ipb-halle.de; Kristin.Brand@ipb-halle.de

R. Mahrwald (ed.), *Enantioselective Organocatalyzed Reactions I: Enantioselective Oxidation, Reduction, Functionalization and Desymmetrization*, DOI 10.1007/978-90-481-3865-4_7,

7.1 Transacylation

Ester hydrolysis, esterification or acylation, and transesterification reactions play an important role in nature and organic synthesis including the synthesis of natural products [1, 2]. Applications of hydrolytic or esterification reactions range from laboratory syntheses (e.g. with the acyl moiety as key intermediate or as protecting group) to industrial scale production of bulk chemicals, biodiesel or food (e.g. ester units as building blocks for polymerisation and polycondensation reactions, transesterification of triglycerides) [3, 4]. Transesterification reactions in nature include for example enzymatic reactions of lipases, esterases and other hydrolases that rely on the catalytic triad of serine proteases, which will be discussed in the first part of this section [5–8].

The most important technical applications of catalytic hydrolysis and acylation involve technical enzymes, as used in food processing, washing powders, or deracemisations. Especially the latter application has also found significant application in chemical synthesis. The kinetic resolution of chiral, racemic esters, anhydrides, or alcohols relies on the faster conversion of only one substrate enantiomer by the chiral catalyst, whereas the other enantiomer ideally remains unchanged. A special case within kinetic resolutions is the desymmetrization of "prochiral" *meso*-compounds like *meso*-anhydrides (**2**) or *meso*-diols (**5**) that requires a selective conversion of one of the two enantiotopic functional groups (carbonyl or OH-group, Scheme 7.1).

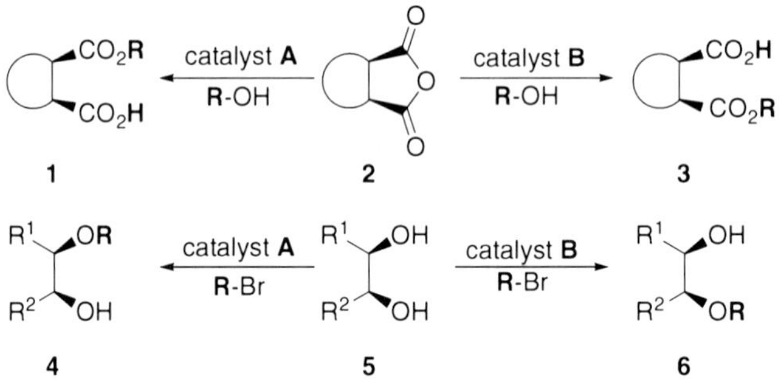

Scheme 7.1 Desymmetrization of *meso*-anhydrides **2** and *meso*-diols **5**

For acylation reactions, a variety of acyl-transfer catalysts are well-known like tertiary amines (e.g. (−)-quinine), *N*-heteroaromatic compounds (e.g. the Steglich catalyst DMAP), or phosphines which act as general base or general nucleophile catalysts. Many such catalysts are bifunctional, i.e. they possess two functional groups, one being the catalytic site (*N*-atom or chalcogen), or a co-catalytic activator (*N*-atom), and others may serve as binding site, e.g. a carbonyl group (**7**) or hydroxyl group (**8**, Scheme 7.2). However, most current organocatalysts use group-V elements (N, P) as catalytic center, in contrast to nature that uses chalcogens (O, S, Se) in its hydrolases.

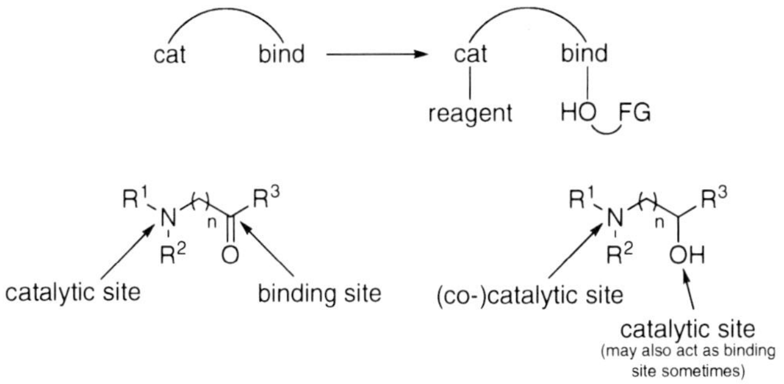

cat bind → cat bind
reagent HO FG

catalytic site, binding site — **7**
(co-)catalytic site, catalytic site (may also act as binding site sometimes) — **8**

Scheme 7.2 Bifunctional acyl-transfer catalysts

7.1.1 Serine Protease Mimetics – Histidine-Serine/Threonine/ Cysteine-Carboxylate – Systems (Catalytic Triads: N→OH/SH→CO_2^-)

Peptidases or proteases are enzymes that hydrolyse peptide bonds [9]. Proteolytic enzymes can be classified in five classes on basis of their catalytic mechanism: aspartic, metallo-, cysteine, threonine and serine peptidases, whereby the latter three follow the same basic mechanism (Scheme 7.3) [10]. Another classification of peptidases on the basis of statistically significant similarities in amino acid sequences was presented by Rawlings et al. (*MEROPS* database) [11]. Serine proteases (SP) alone cover approximately one-third of all known proteases, and can accelerate the peptide hydrolysis very efficiently ($k_{cat}/k_{uncat} = 10^{10}$ fold) [6, 11, 12]. SPs also hydrolyse other acyl compounds such amides, anilides, esters, and thioesters [11].

The most accepted mechanism for chymotrypsin-like SPs is detailed in Scheme 7.3. In the acylation step (cf. **11**) the hydroxyl of a serine (Ser195 in chymotrypsin) is activated by deprotonation with histidine (His57), which in turn is activated by a proton relay to a carboxylate of an aspartate (Asp102), i. e. the Ser—His—Asp catalytic triad produces a serine anion. This undergoes a nucleophilic attack to the carbonyl of the substrate **10** to form a tetrahedral intermediate **12**. The oxyanion is stabilized by hydrogen bonding in the so-called oxyanion hole (NHs of Ser195 and Gly193, cf. **12**). The liberation of amine **13** as leaving group generates the acylenzyme **14**. The deacylation step starts with the attack of water to the acylenzyme, assisted again by His57/Asp102, generating a second tetrahedral intermediate **15**. The liberation of carboxylic acid product **16** and regeneration of catalytic site complete the cycle [11]. Such catalytic triads are varied in nature by changing serine for threonine, cysteine or even selenocysteine [13]. Aspartate can be substituted by glutamate in some instances.

The great efficiency combined with selectivity of SPs in cleaving esters or amides is both inspiration and goal for synthetic chemists who want to design organocatalysts

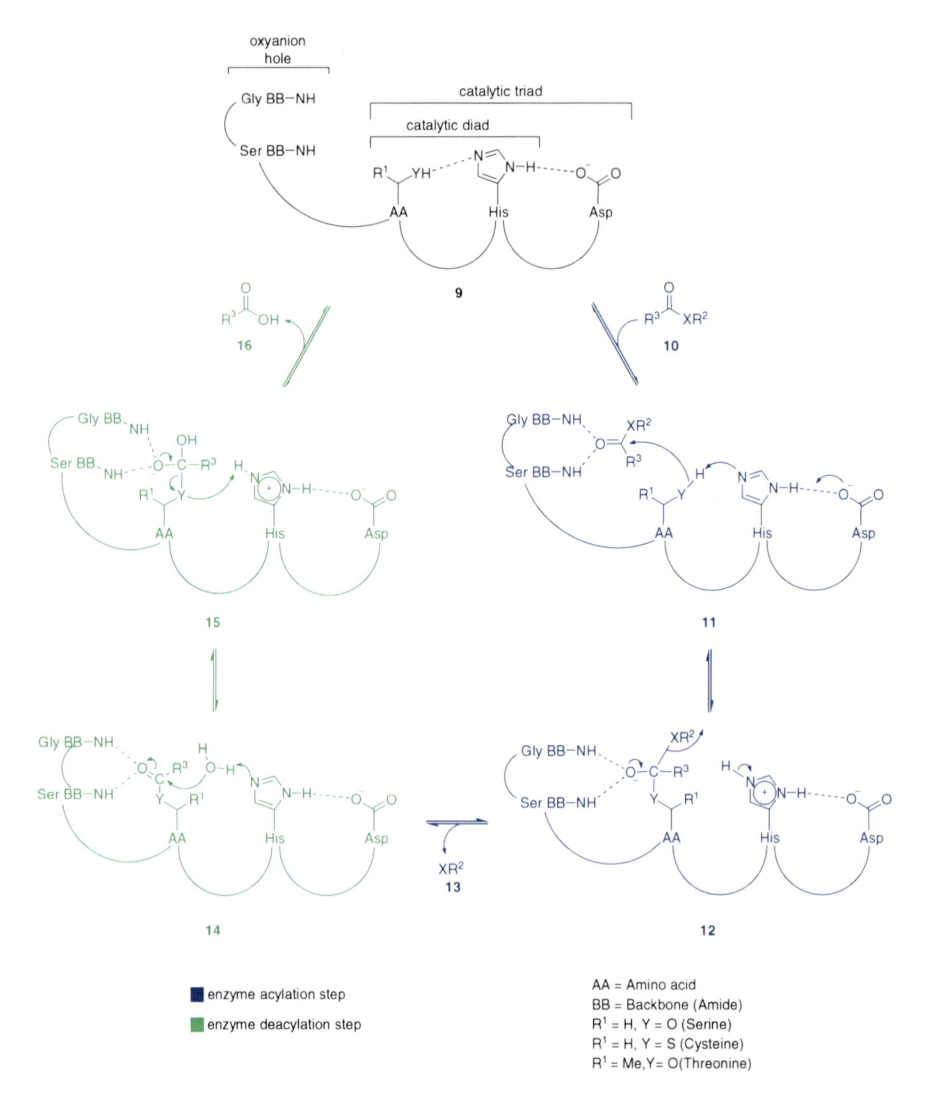

Scheme 7.3 General hydrolysis mechanism of proteolytic enzymes

that mimic the catalytic power of these enzymes. Of the many groups working on this exciting theme, some selected examples are presented below.

Breslow and collaborators used β-cyclodextrine as secondary binding (host) moiety to select substrates like an enzyme. They show in detailed studies that β-cyclodextrine and modified β-cyclodextrine catalyze the cleavage of esters, mimicking serine proteases in a very rudimentary way [14–20]. Reaction conditions e.g. buffer, solvent, influence of the structure of the ester, and pressure effect the performance of the catalyst. In his reports he shows that β-cyclodextrine can enhance the rate of the acylation reactions by 10^2–10^6 or better, compared to the uncatalyzed reactions [16]. The catalytic mechanism proposed by Breslow is shown in simplified form in Scheme 7.4. However, in reality, only an acylation occurs and the catalyst is not regenerated.

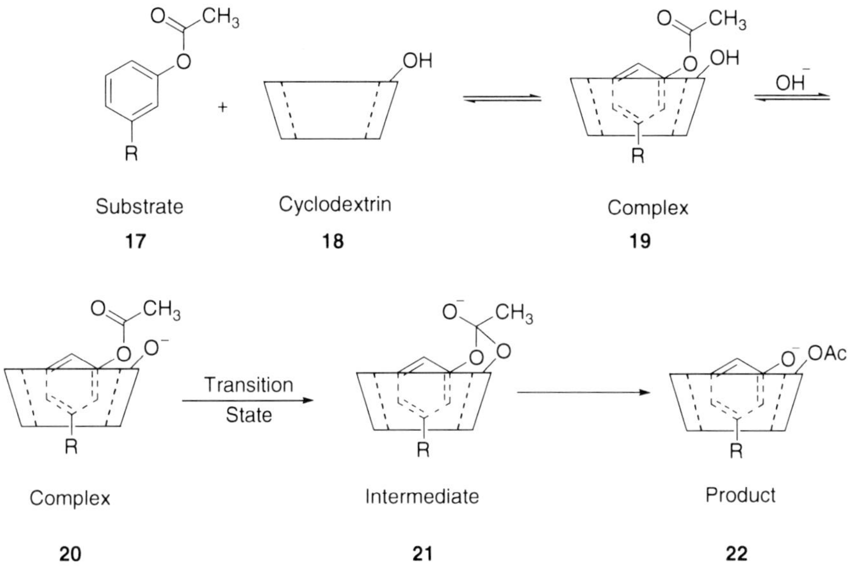

Substrate **17** + Cyclodextrin **18** ⇌ Complex **19** ⇌ (OH⁻) Complex **20** → (Transition State) Intermediate **21** → Product **22**

Scheme 7.4 Simplified catalytic mechanism of β-CD promoted ester hydrolysis as proposed by Breslow

Using modified β-cyclodextrine **23**, Bender et al. report a β-cyclodextrine based artificial enzyme mimic with a catalytic triad that has catalyzed hydrolysis rates of 10^5–10^6 faster for norbornyl cinnamate ester [21]. They show that 1 mol of **23** could be used in the hydrolysis of more than 10 mol of substrate, which proves a multiple turnover of the catalyst. The proposed mechanism is shown in Fig. 7.1. This mechanism was evaluated and questioned by Zimmerman [22].

Rao et al. showed that β-cyclodextrine substituted with an imidazole in position 2 (**25**) is 70 times more effective in the hydrolysis of *p*-nitrophenylacetate compared to a β-cyclodextrine substituted at position 6 (Scheme 7.5) [23].

Liu et al. report that 6,6'-telluroxy-bis(6-deoxy-*β*-cyclodextrin) (**26**) acts as an enhancer for the hydrolysis of 4,4'-dinitrophenyl carbonate (**28**) with a significant rate acceleration. The proposed mechanism is shown in Scheme 7.6 but appears to be unlikely as the substrate is suggested to enter the cyclodextrin at the narrow end twice [24].

A different approach was tried by Cram et al. They designed the artificial host **31** with an arrangement of functional groups that partially mimics a serine protease. In sequential reports Cram demonstrated the synthesis of hosts **32**, **33** and **34** [25–32]. The reactivity for acylation of the imidazolyl group of the host **34** (via formation of an intermediate amino ester salt host-guest complex) was higher by a factor of 10^{10}

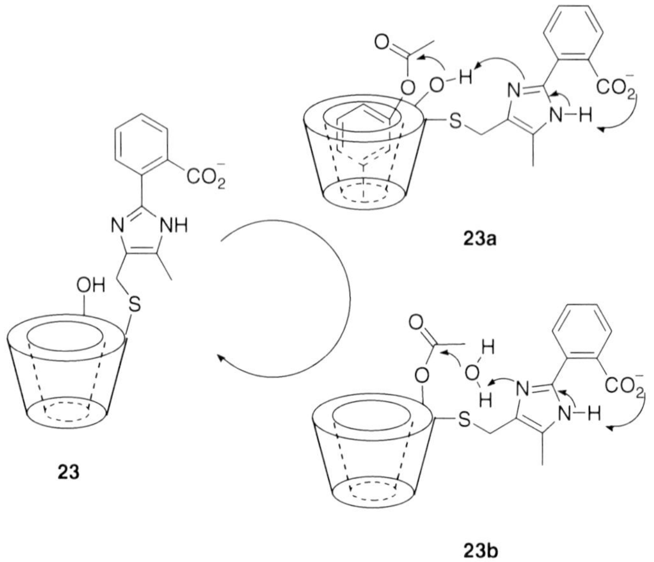

Fig. 7.1 Bender´s hydrolytic triad β–cyclodextrine: **23a** proposed acylation step; **23b** proposed deacylation step

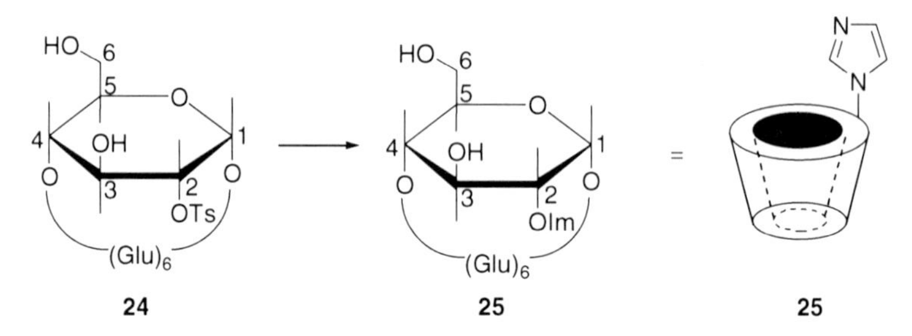

Scheme 7.5 Synthesis of 2-imidazolyl *β*-cyclodextrine **25**

when compared with 4-phenylimidazole. The originally designed mimic **31** is not published to date (Fig. 7.2).

Reymond et al. report the synthesis of peptide dendrimers displaying multiple serine-histidine (diad) residues to catalyze ester hydrolysis reactions [33]. Best results where obtained with the fourth generation dendrimer **36**. The rate acceleration of the hydrolysis of a fluorescent nonanoyl ester (**35**) was 140,000-fold more efficient than with 4-methyl-imidazole as reference catalyst (Scheme 7.7).

The same group also reported that dendrimers built by interaction of Dap-His-Ser display a strong effect for the catalytic hydrolysis with enzyme-like kinetics in aqueous medium [34]. The rate enhancement of the catalyzed reaction versus the uncatalyzed one is 90,000-fold employing 2.5 mol% of catalyst **40** (Scheme 7.8) which displays His-Thr diad elements.

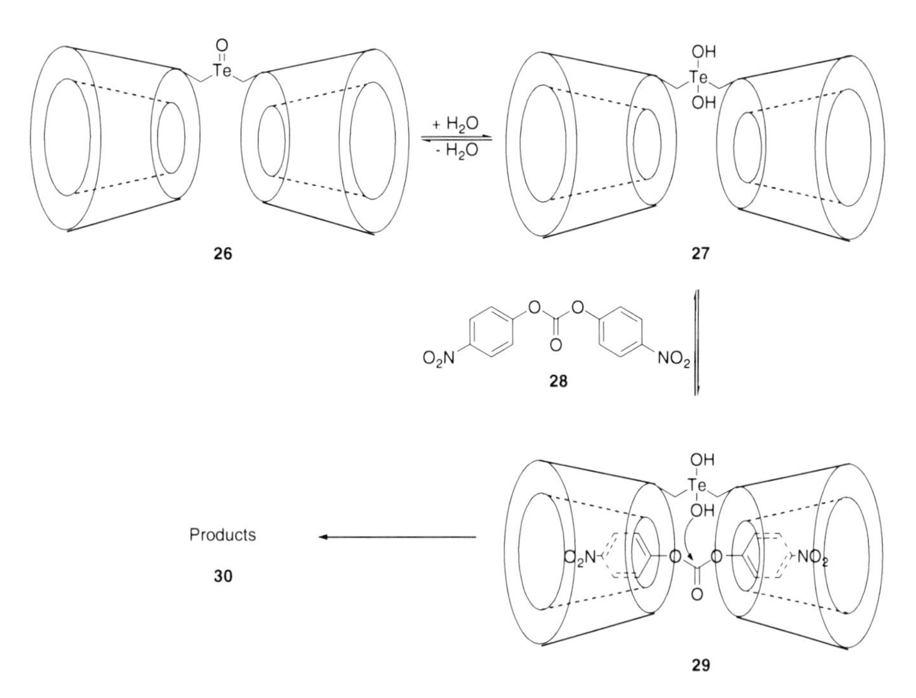

Scheme 7.6 Telluroxy bis-(β-cyclodextrine) **26** catalyzes the hydrolysis of 4,4′-dinitrophenyl carbonate

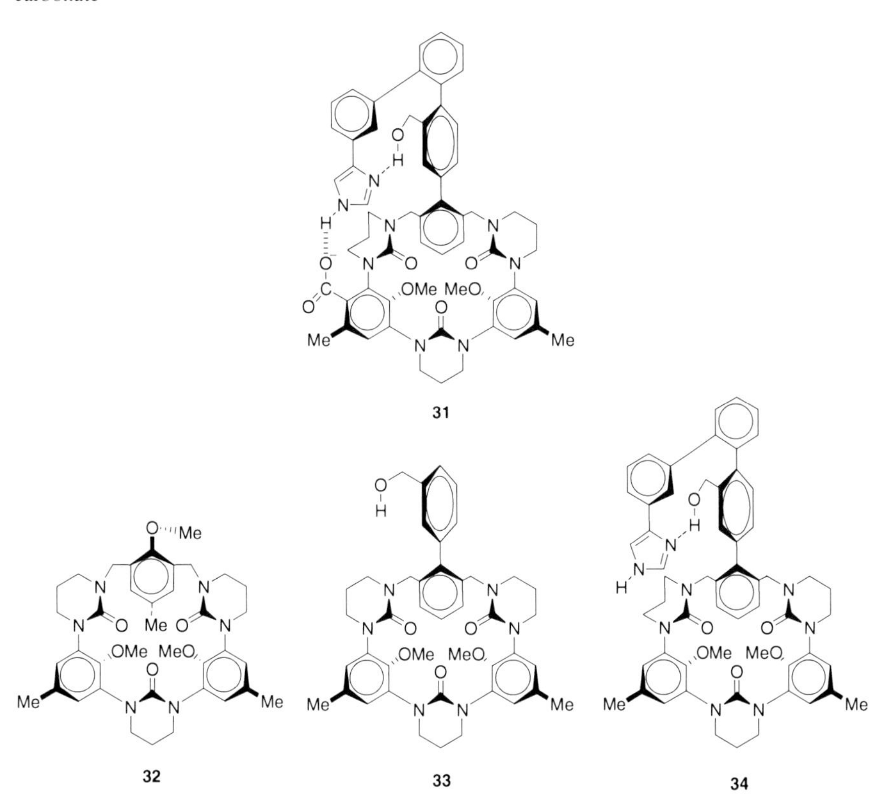

Fig. 7.2 Partial serine protease mimics: theoretical design catalyst **31** by Cram and realized hosts **32, 33** and **34**

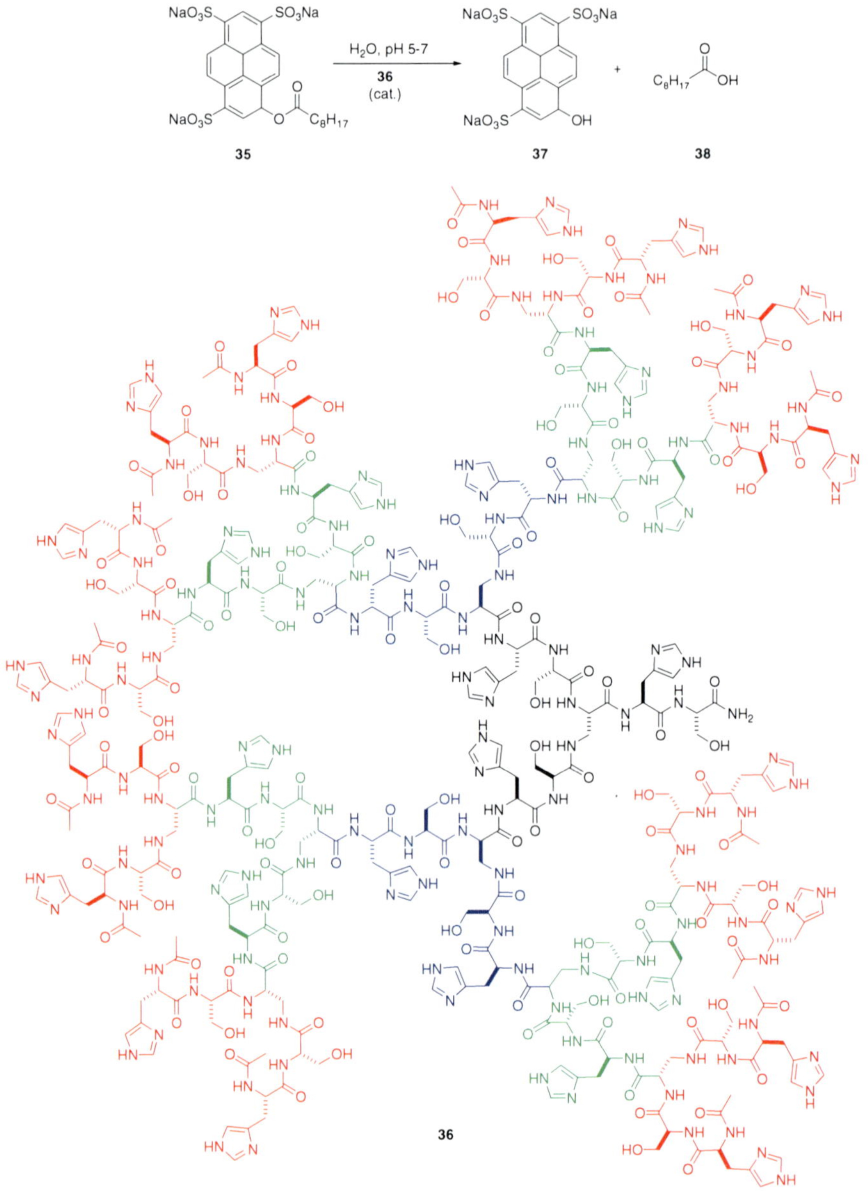

Scheme 7.7 Hydrolysis of nonanoyl pyrene trisulfonate (**35**) catalyzed by dendrimer **36**

The proposed mechanism for the hydrolysis reaction is shown in Scheme 7.9 [35]. Interestingly it does not involve the chalcogen amino acids Ser or Thr, which where shown to be irrelevant in this highly flexible setup. This may be taken as a hint that SP-triad catalysis is strongly dependent on the correct spatial arrangement of the triad components.

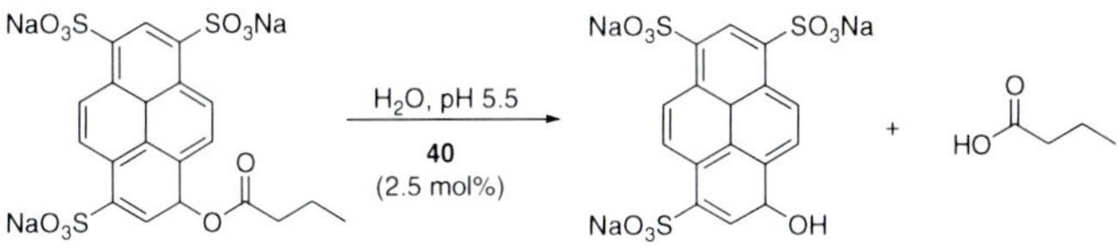

Scheme 7.8 Structure of catalytic peptide third generation dendrimer **40**

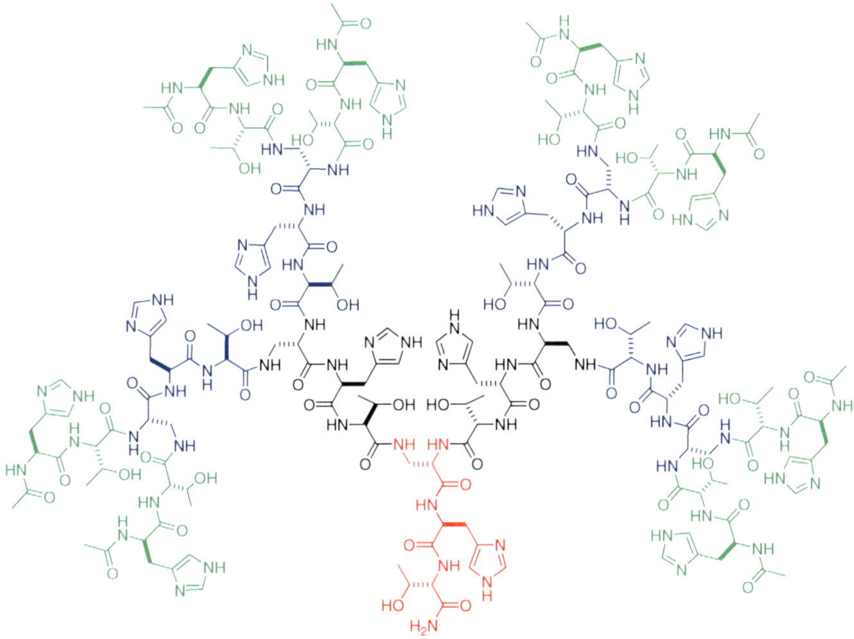

Later Reymonds group reported the synthesis of dendrimer **45** (Fig. 7.3) that catalyzes ester hydrolysis with a single catalytic histidine residue in the core [36]. According to the author, a pair of arginine residues in the first-generation branch assists substrate binding.

7.1.2 Amino Alcohols and Analogues (Catalytic Diads: N→OH/SH)

Early studies describe the transesterification between an active ester and a nucleophile like water catalyzed by amino alcohols [37–46]. For the hydrolysis of active *p*-nitrophenyl acetates, amino alcohol catalysts served as models for the catalytic diad of serine proteases. Kinetic studies show that the hydrolysis catalyzed by an amino alcohol proceeds via an acylation-deacylation mechanism.

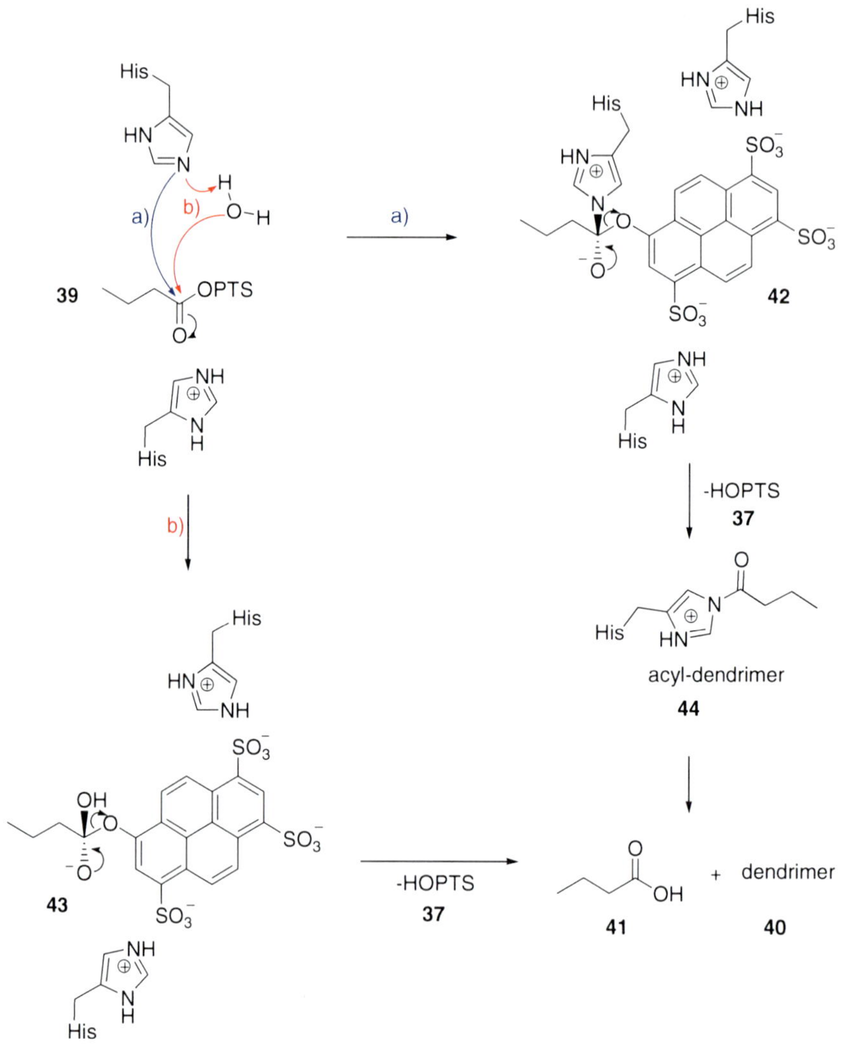

Scheme 7.9 Proposed mechanism for the His-dendrimer-catalyzed ester hydrolysis: (**a**) nucleophilic attack of histidine generates an acyl-dendrimer intermediate; (**b**) general base catalyzed nucleophilic attack of a water molecule

The catalysts operate by a mechanism in which the chalcogen functional group (carbonyl/nucleophile or hydroxyl group) of the catalyst attacks the carbonyl of an active ester to form the acyl catalyst intermediate in a first step. This is supported by the nitrogen of the catalyst which acts as a general base. Subsequently, the intermediate undergoes hydrolysis or alcoholysis, to provide the acid or ester, respectively, and to regenerate the catalyst. This step is also base catalyzed by the proximal nitrogen of the catalyst (Scheme 7.10) [47–52].

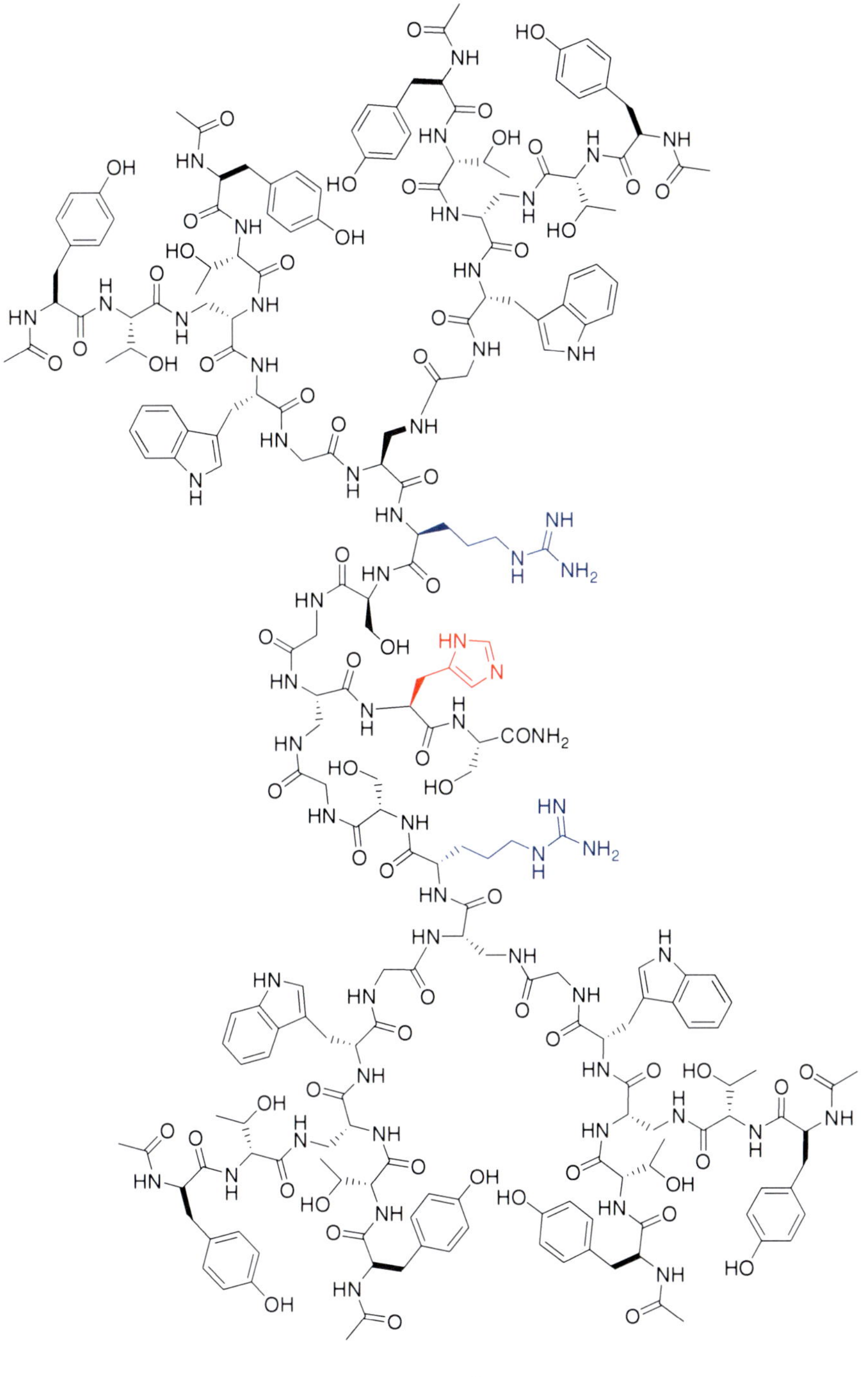

45

Fig. 7.3 Peptide dendrimer **45** with a single catalytic site

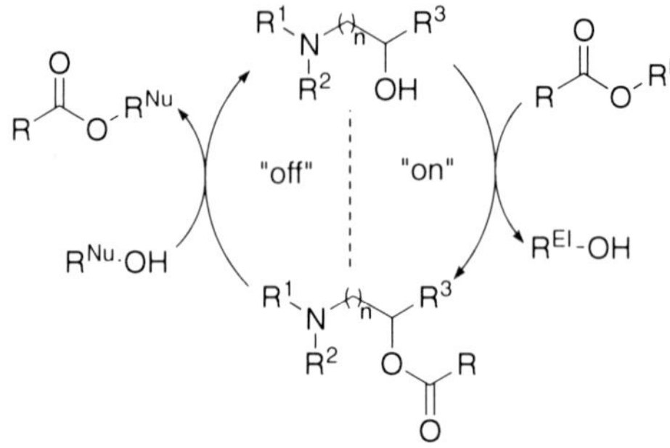

Scheme 7.10 Acylation-deacylation mechanism for amino-alcohols

In 1996, Sammakia et al. reported a new catalyst, 2-formyl-4-pyrrolidinopyridine (FPP, **46**), whose mechanism of action for the hydroxyl-directed methanolysis of α-hydroxy esters was studied [53, 54]. In contrast to the amino alcohol catalysts, FPP possesses a carbonyl group as binding site. With the help of kinetic ^{1}H-NMR-studies they showed that FPP catalyzes the methanolysis of *p*-nitrophenyl esters of glycolic acid via hemiacetal **47** and dioxolanone intermediate **48** (Scheme 7.11).

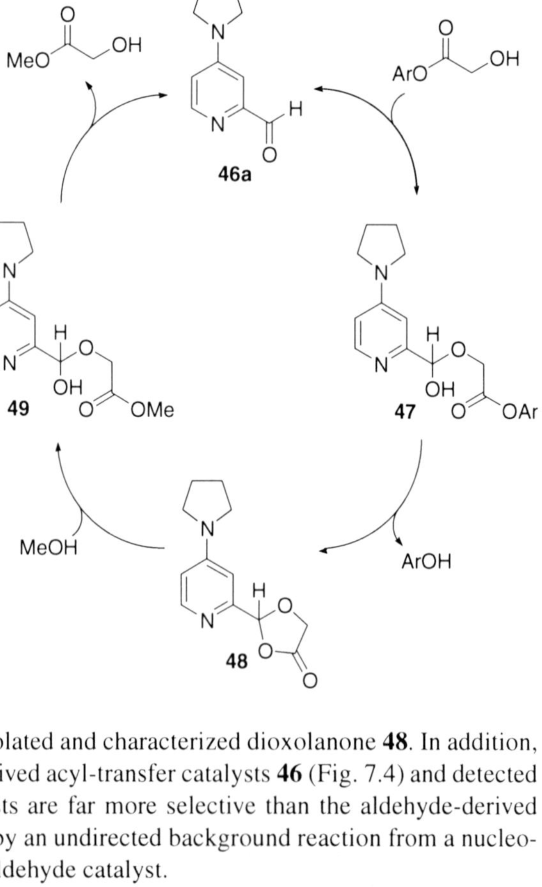

Scheme 7.11 Mechanism of action for 2-formyl-4-pyrrolidinopyridine (FPP) (**46a**)

As further evidence, they isolated and characterized dioxolanone **48**. In addition, they tested further pyridine-derived acyl-transfer catalysts **46** (Fig. 7.4) and detected that the ketone-derived catalysts are far more selective than the aldehyde-derived ones [55]. This was explained by an undirected background reaction from a nucleophilic attack of MeOH to the aldehyde catalyst.

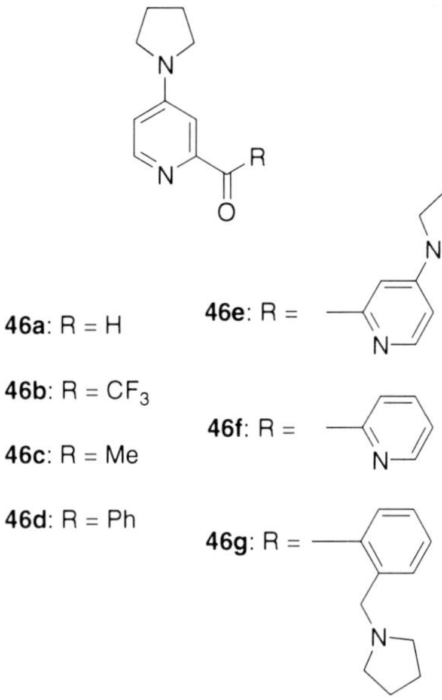

Fig. 7.4 Pyridine-derived acyl-transfer catalysts

46a: R = H
46b: R = CF_3
46c: R = Me
46d: R = Ph
46e: R =
46f: R =
46g: R =

In 2003, Sammakia et al. described two additional types of amino alcohol-derived acyl-transfer catalysts **51**, bearing electron-withdrawing groups in close proximity to the hydroxyl group (Scheme 7.12) [56].

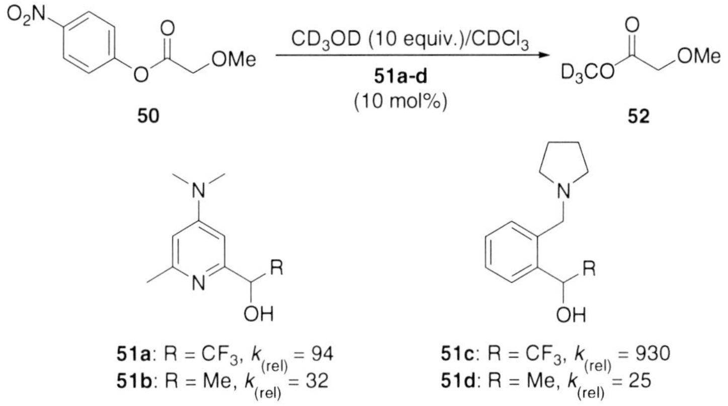

CD_3OD (10 equiv.)/$CDCl_3$, **51a-d** (10 mol%): **50** → **52**

51a: R = CF_3, $k_{(rel)} = 94$
51b: R = Me, $k_{(rel)} = 32$
51c: R = CF_3, $k_{(rel)} = 930$
51d: R = Me, $k_{(rel)} = 25$

Scheme 7.12 Methanolysis of *p*-nitrophenyl methoxyacetate (**50**) catalyzed by **51**

Catalyst **51c** proved to be the most active one due to the higher acidity of the OH-group. With non-racemic versions of these catalysts, Sammakia et al. achieved the kinetic resolution of protected α-amino acid derivatives in high *ee* (94–99%) and with acceptable selectivities ($E = S = 20$–86). [57, 58]

In 2008, the transacylation properties of a series of dialkyl amino alcohol catalysts **53** and **54** was studied in more detail by the Wessjohann group (Fig. 7.5) [52].

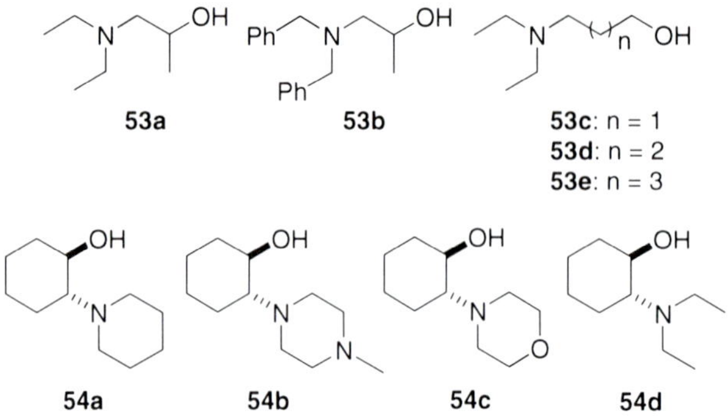

53a 53b 53c: n = 1 53d: n = 2 53e: n = 3

54a 54b 54c 54d

Fig. 7.5 Series of dialkyl amino alcohols **53** and **54**. The catalysts **54a** and **54d** are the most active ones [52]

The influence of structural and electronic parameters on the acylation and deacylation rate were separately studied (catalyst-on half cycle vs. catalyst-off half cycle; cf. scheme 7.10). Through kinetic ^{1}H-NMR-studies it was proven that the acylation rate of the dialkyl amino alcohol catalysts and an acyl donor (butyric anhydride) depends on the number of (carbon) spacer atoms between hydroxyl and tertiary amine, the flexibility of the molecule and the presence and position of further heteroatoms. Besides, it could be detected that the methanolysis (off-half cycle) of the formed β-amino ester intermediate follows a similar trend as the acylation reaction, but appeared to be rate limiting in this studies setup. The information was used for the selective auto-catalytic acylation and deacylation of complex natural antibiotics.

Sakai and co-workers, reported the synthesis of a biomimetic trifunctional organocatalyst **56** that mimics the active site of serine proteases. It shows a high acceleration for the acyl transfer reaction with highly active, irreversible vinyl trifluoroacetate (**55**) up to almost $4*10^6$-fold vs. background (Scheme 7.13) [59]. The proposed mechanism is shown in Scheme 7.14.

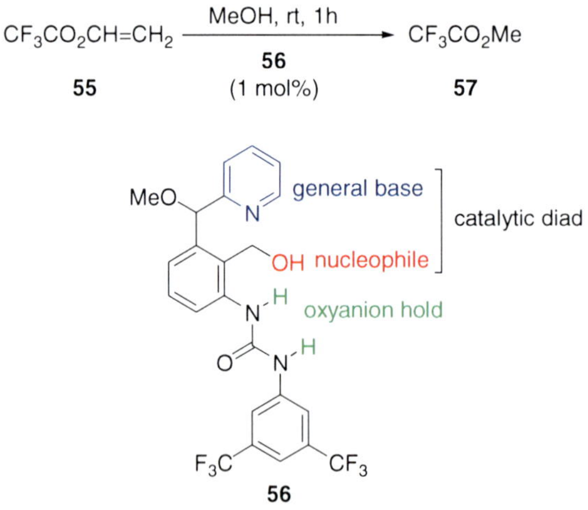

$CF_3CO_2CH{=}CH_2$ (**55**) → [MeOH, rt, 1h; **56** (1 mol%)] → CF_3CO_2Me (**57**)

general base
catalytic diad
nucleophile
oxyanion hold
56

Scheme 7.13 Methanolysis of vinyl trifluoroacetate catalyzed by Sakai's trifunctional catalyst **56**

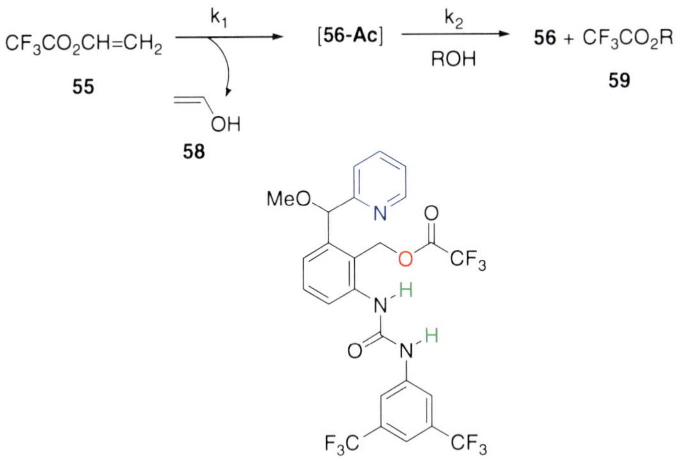

Scheme 7.14 Proposed mechanism for the acyl-transfer reaction catalyzed by **56**

7.1.3 *Kinetic Resolution and Desymmetrization of Cyclic Anhydrides*

The stereoselective ring opening of cyclic anhydrides gives optically active hemiesters which are versatile intermediates for the synthesis of, e. g., bioactive compounds [60–62]. A variety of bifunctional catalysts like derivatives of cinchona alkaloids and β-amino alcohols have been developed for the stereoselective alcoholysis of anhydrides. The mechanism for the desymmetrization and kinetic resolution of cyclic anhydrides conform to the catalytic cycle for transacylations (see Sect. 7.1.2, Scheme 7.10) [63]. The hydroxyl group of, e.g., a (-)-quinine catalyst activates the electrophilic anhydrides while the quinuclidine group of the catalyst activates the nucleophilic alcohol via a hydrogen bonding interaction.

In 1985, Oda et al. reported the first metal-free enantioselective anhydride opening catalyzed by various cinchona alkaloids **61** in excellent yields (95%) and with moderate *ee*'s (70% *ee*) using 10 mol% catalyst (Scheme 7.15) [64, 65]. Oda investigated the opening of glutaric and succinic anhydrides and observed that stereochemistry and enantioselectivities are highly dependent on a specific substrate/catalyst combination. Related studies were published in 1988 and 1990 by Aitken et al. (Scheme 7.15) [66, 67]. These authors improved the *ee* slightly (76% *ee*) by increasing the catalyst loading (50 mol%).

Later, Bolm and coworkers reported excellent enantioselectivities (85–99%) for the desymmetrization of *meso*-anhydrides by an optimized reaction protocol

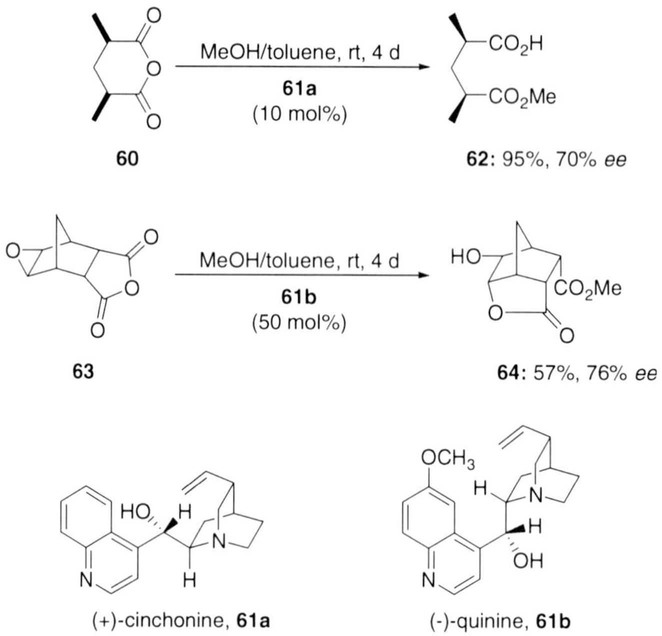

MeOH/toluene, rt, 4 d
61a
(10 mol%)
60
CO_2H
CO_2Me
62: 95%, 70% *ee*

MeOH/toluene, rt, 4 d
61b
(50 mol%)
63
HO
CO_2Me
64: 57%, 76% *ee*

(+)-cinchonine, **61a**
OCH_3
(-)-quinine, **61b**

Scheme 7.15 Anhydride opening catalyzed by cinchona alkaloids **61**

with low temperature, an excess of methanol, toluene as solvent and stoichiometric amounts of quinine and quinidine [68, 69]. Bolm ascertained that quinine catalyzes the nucleophilic attack to the *pro*-(*R*) carbonyl group while quinidine always exhibits the opposite selectivity [68–72]. Selected results are presented in Table 7.1.

In parallel, Deng et al. investigated a variety of mono- and bis-cinchona alkaloid derivatives **68** in the desymmetrization of cyclic anhydrides by alcoholysis [73–75]. In Fig. 7.6 some examples of the alkaloids are depicted.

These modified cinchona alkaloids (**68**) have usually been used as ligands in the Sharpless asymmetric dihydroxylation [76, 77]. Deng developed a protocol based on a catalytic amount of the alkaloid derivatives (5–30 mol%) acting at room temperature in ether as solvent. Under these conditions, methanolysis of several monocyclic, bicyclic, and tricyclic succinic or glutaric anhydrides was achieved in good yields and excellent *ee* (Table 7.2). The catalysts can be recovered by a simple extractive procedure and reused without any loss in yield and enantioselectivities [73–75].

The desymmetrization method was applied for the asymmetric synthesis of (+)-biotin [78] and the γ-amino acid baclofen [79]. Furthermore, Deng et al. reported the kinetic resolution of cyclic anhydrides of β,γ-unsaturated α-amino acids in the presence of the dimeric cinchona alkaloid catalyst $(DHQD)_2AQN$ **68a** (10 mol%)

Table 7.1 Desymmetrization of *meso*-anhydrides catalyzed by **61b–c**

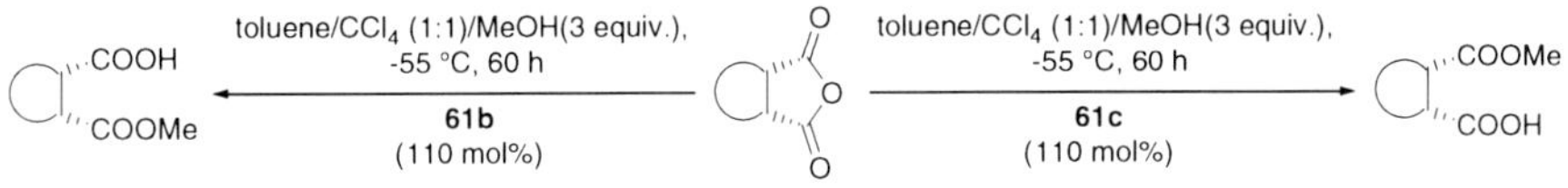

66 **65** **67**

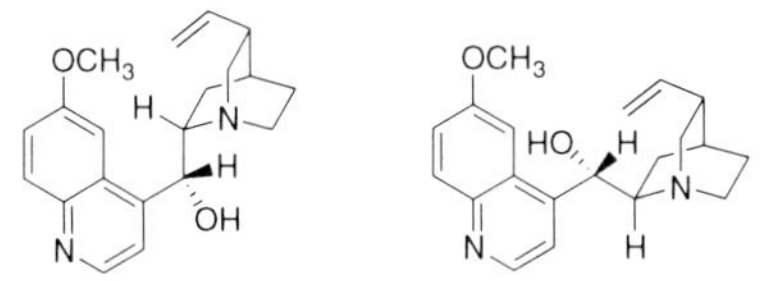

(-)-quinine, **61b** (+)-quinidine, **61c**

Entry	*Meso*-anhydrides **65**	(−)-Quinine, **61b** products **66** Yield (%)	*ee* (%)	(+)-Quinidine, **61c** products **67** Yield (%)	*ee* (%)
1	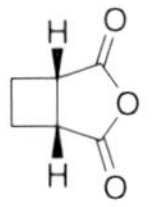	99	94	93	87
2	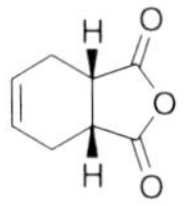	93	95	99	93
3	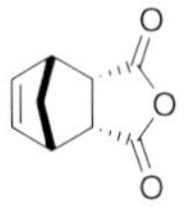	98	99	94	93
4	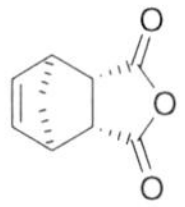	96	96	91	87

with selectivity (E- or S-) factors up to 170 affording the ethyl esters (*R*)-**72** and the corresponding amino acid (*S*)-**74** (Table 7.3) [80].

In 2007, Furukawa et al. described large-scale applications of the alkaloid derived catalysts **75** (Fig. 7.7) for the kinetic resolution of urethane-protected α-amino acid *N*-carboxyanhydrides and the desymmetrization of cyclic *meso*-anhydrides [81]. The *O*-propargylquinidine (OPQD) **75b** and quinine (OPQ) catalyst **75a** were described as highly enantioselective (70–97%) and practical, e.g. in the synthesis of (*S*)-*N*-Boc-propargylglycine, a key intermediate for a commercial β-amino acid (BAY 10-8888/PLD-118).

In 2009, Hameršak et al. reported an unexpected inversion of the induced configuration, dependent on the degree of quinine loading [82]. They observed that during the desymmetrization of glutaric *meso*-anhydrides the product inverted from about 40% *ee* of (*R*)-configuration to about 40% *ee* of the opposite enantiomer by decreasing the quinine loading.

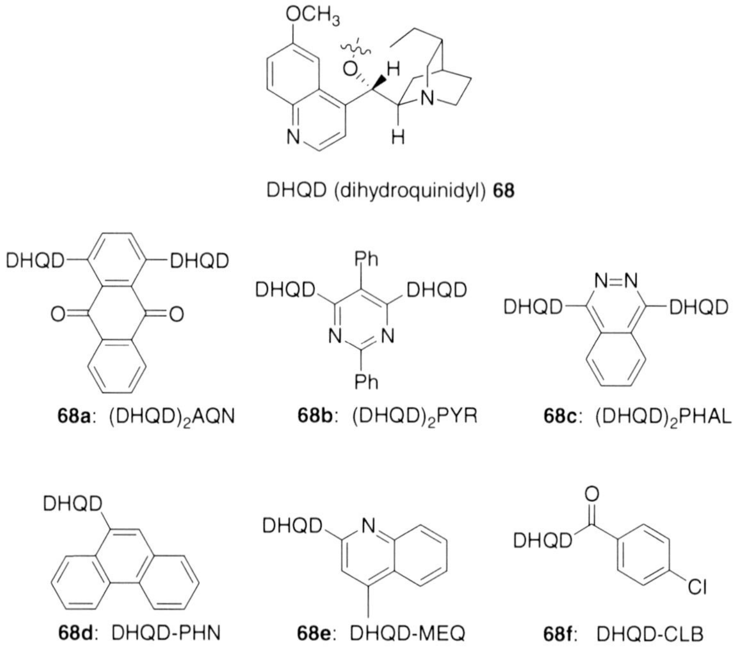

DHQD (dihydroquinidyl) **68**

68a: $(DHQD)_2AQN$ **68b**: $(DHQD)_2PYR$ **68c**: $(DHQD)_2PHAL$

68d: DHQD-PHN **68e**: DHQD-MEQ **68f**: DHQD-CLB

Fig. 7.6 Mono- and bis-cinchona alkaloid derivatives

Table 7.2 Opening of anhydrides catalyzed by cinchona alkaloid derived **68a**

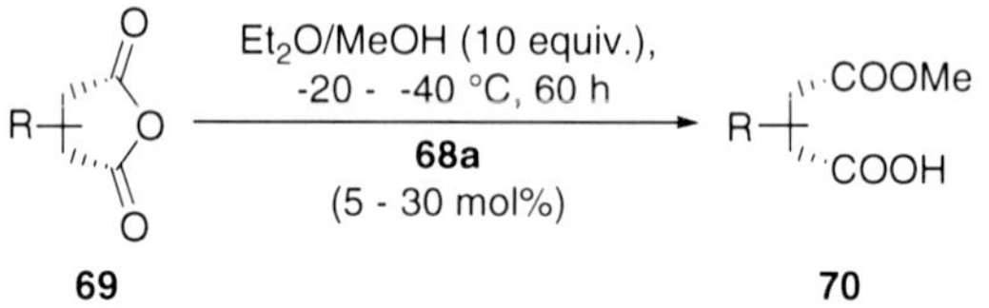

69 → **70**; Et_2O/MeOH (10 equiv.), -20 - -40 °C, 60 h, **68a** (5 - 30 mol%)

Entry	*meso*-anhydrides **69**	Yield (%) **70**	*ee* (%) **70**
1	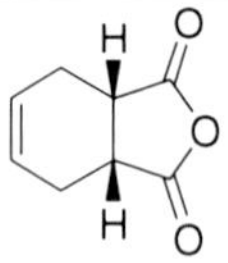	99	95
2	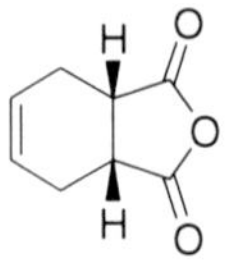	95	98
3	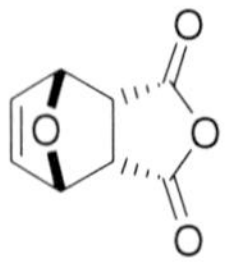	74	92

(continued)

Table 7.2 (continued)

Entry	*meso*-anhydrides **69**	Yield (%) **70**	*ee* (%) **70**
4	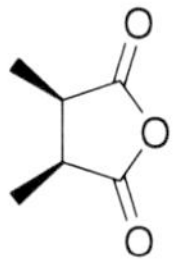	93	98
5	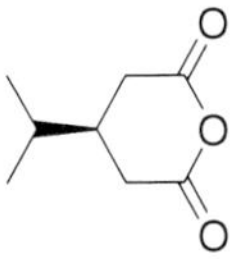	72	90

Table 7.3 Kinetic resolution of cyclic anhydrides catalyzed by **68a**

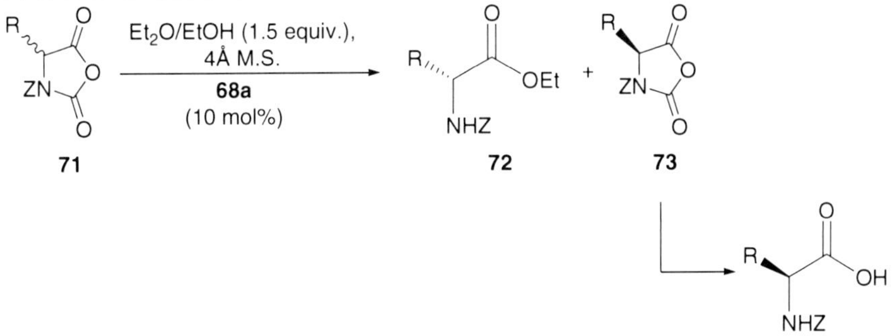

Et_2O/EtOH (1.5 equiv.), 4Å M.S., **68a** (10 mol%)

71 → **72** + **73**

Entry	R	Temperature (°C)	Time (h)	Conv. (%)	*ee* (yield)(%) (*R*)-**72**	*ee* (yield)(%) (*S*)-**74**	*S*
1		−78	14	51	93 (49)	95 (42)	114
2	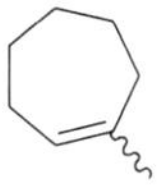	−40	7.5	57	75 (55)	96 (42)	39
3	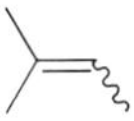	−52	19	52	93 (50)	99 (44)	149
4		−52	18	52	85 (50)	95 (44)	40
5		−40	13	57	74 (57)	96 (38)	26
6	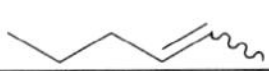	−73	42	49	95 (48)	91 (40)	125

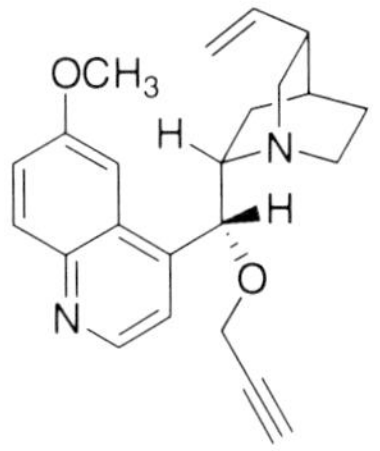

QPQ **75a** 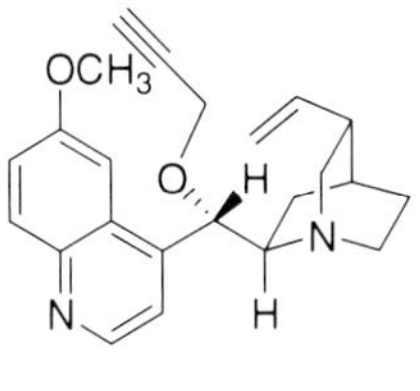 QPQD **75b**

Fig. 7.7 Alkaloid derived catalyst for enantioselective kinetic resolution of α-amino acids and desymmetrization of *meso*-anhydrides

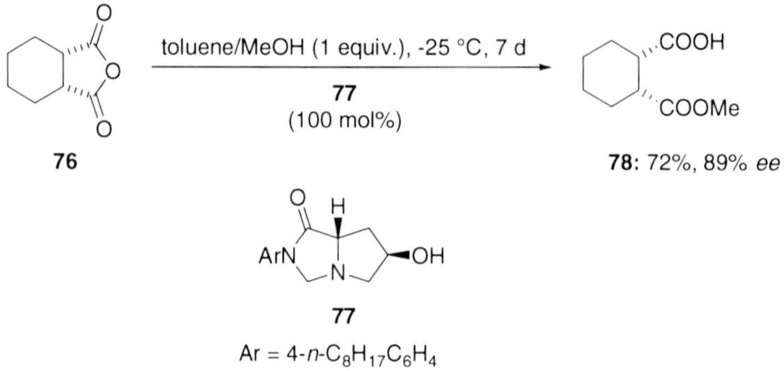

76

toluene/MeOH (1 equiv.), -25 °C, 7 d

77 (100 mol%)

COOH

COOMe

78: 72%, 89% *ee*

ArN

OH

77

Ar = 4-*n*-$C_8H_{17}C_6H_4$

Scheme 7.16 Enantioselective methanolysis of *meso*-anhydride catalyzed by bicyclic tertiary amines **77**

Table 7.4 Enantioselective alcoholysis of anhydrides catalyzed by β-amino alcohols **80a–c**

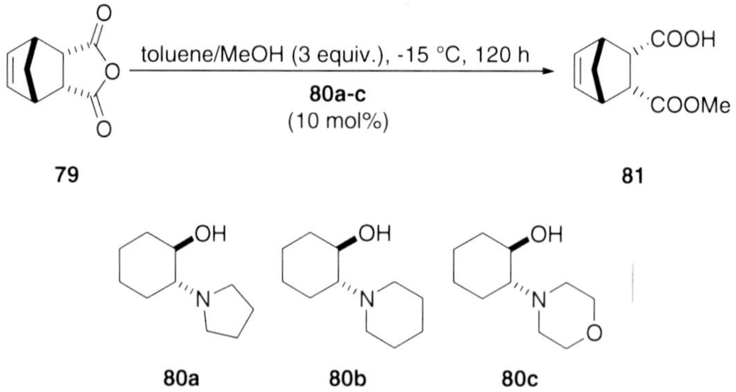

toluene/MeOH (3 equiv.), -15 °C, 120 h

80a-c (10 mol%)

79

81

80a 80b 80c

Entry	Catalyst	Yield (%) **81**	*ee* (%) **81**
1	**80a**	85	56
2	**80b**	83	82
3	**80c**	81	76

In contrast to the alkaloid derived catalyst, some work groups focused their attention on synthetic amino alcohols as desymmetrization catalyst. In 2001, Uozomi et al. reported a series of *N*-chiral bicyclic tertiary amines **77** for the methanolysis of cyclohexane-1,2-dicarboxylic anhydride [83, 84]. In the presence of a stoichiometric amount of catalyst **77**, the desymmetrization of the anhydride proceeds with up to 89% *ee* (Scheme 7.16).

Later, Bolm et al. presented ordinary β-amino alcohols **80** for the enantioselective alcoholysis of various anhydrides. Up to 82% *ee* was obtained (Table 7.4) [62, 85].

In 2007, Irie et al. reported asymmetric methanolysis of cyclic *meso*-anhydrides with tripodal 2,6-*trans*-1,2,6-trisubstituted piperidine **83** as chiral amine catalyst (Table 7.5) [86]. A good level of enantioselectivity (up to 81% *ee*) was achieved in the presence of a catalytic amount of 1-5 mol% catalyst.

Table 7.5 Tripodal piperidine **83** catalyzed opening of anhydrides

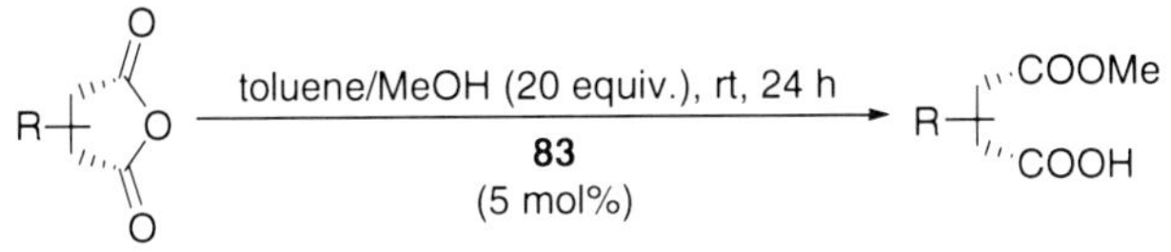

Entry	*Meso*-anhydride **82**	Hemiester **84**	Yield (%) **84**	*ee* (%) **84**
1	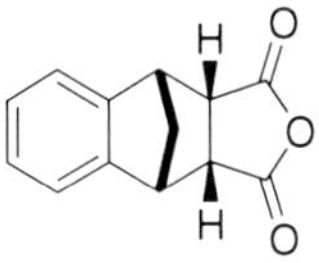	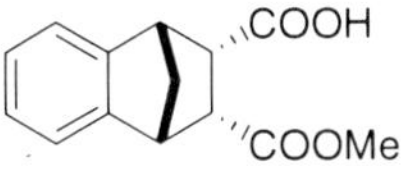 COOH / COOMe	71	81[a]
2	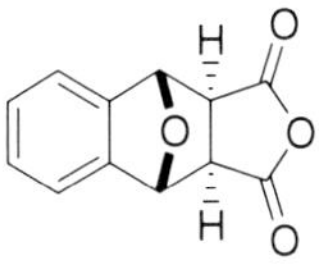	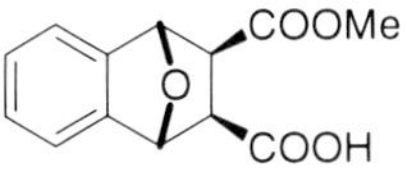 COOMe / COOH	99	65
3	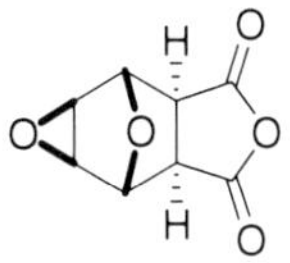	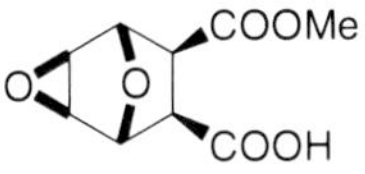 COOMe / COOH	99	68
4	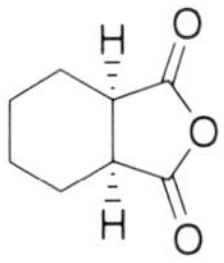	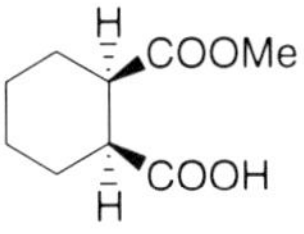 COOMe / COOH	99	52

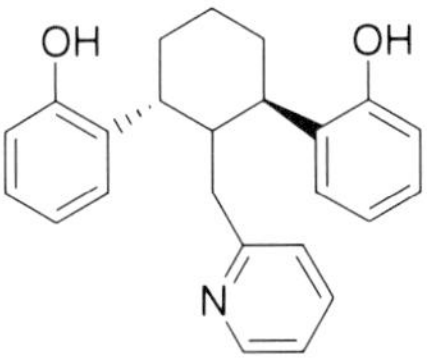

[a]Reaction proceed at 0°C

7.1.4 *Kinetic Resolution of Racemic Alcohols/Desymmetrization of Meso-Diols*

In the previous chapters, achiral substrates or substrates with a chiral acid moiety have been considered. But the kinetic resolution of racemic alcohols and the desymmetrization of *meso*-diols are equally important for the synthesis of pharmaceutical and natural substances [87]. Enantioselective acylation of racemic

Fig. 7.8 Bifunctional quinidine phosphinite catalyst

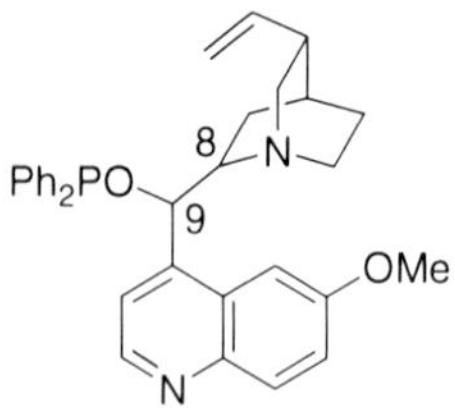

85d

Fig. 7.9 *O*-Acylquinidine catalysts

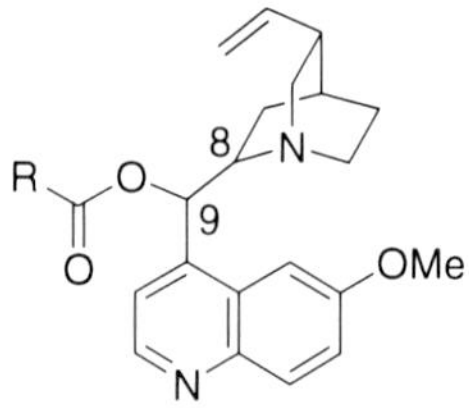

86a: R = Ph (quinine 8*S*, 9*R*)
86b: R = Ph (quinidine 8*R*, 9*S*)
86c: R = *t*-Bu (quinidine 8*R*, 9*S*)

alcohols or *meso*-diols has been extensively studied using enzymatic processes [88] or non-enzymatic catalysts [89–107]. The proposed catalytic mechanism for the acylative kinetic resolution and desymmetrization of alcohols conform to the catalytic cycle of the trans-acylation as demonstrated before. For a bifunctional catalyst like a quinidine phosphinite **85d** (Fig. 7.8), it has been postulated that the two functional groups – the Lewis-basic trivalent phosphorus center and a Brønsted-basic tertiary amino group – act cooperatively [108–111]. The phosphinite moiety activates an acylating reagent (e.g. acyl chloride) and the nitrogen atom of the quinuclidine traps a proton from the hydroxyl group of the alcohol. Since in the phosphinites the involvement of phosphorus or oxygen is both important, it is incorporated here. However, compounds **85** are not organochalcogen-catalysts in *sensu strictu*.

In 1985, Duhamel reported *O*-acylquinidines **86** (Fig. 7.9) as catalysts for the kinetic benzoylation of 2-cyclopenten-1,4-diol with 47% enantioselectivity, albeit in low yield (8%) [108].

In 1997, Kawabata et al. described the kinetic resolution of amino alcohol derivatives with a chiral nucleophilic catalyst **87** (Fig. 7.10) [90]. The compound could effectively help to acylate cylic *cis*-amino alcohol derivatives whose amino groups are protected as *p*-(dimethylamino)benzoates ($S = 10$–21). Up to 69% *ee* were achieved for the monoacylated product [112].

In 2003, Fujimoto et al. reported the asymmetric desymmetrization of *meso*-1,2-diols with benzoylbromide catalyzed by phosphinite derivatives of cinchona alkaloids **85** (Table 7.6) [109]. The phosphinite derivative **85a** of cinchonidine which was generated *in situ* by using Hünig's base.

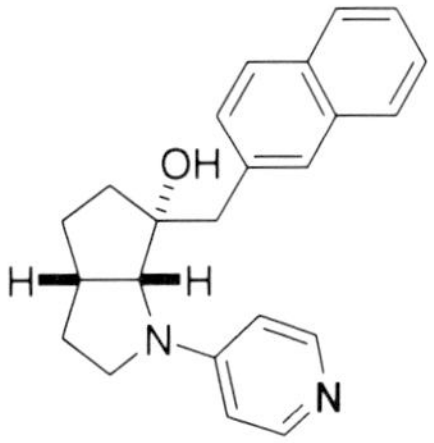
OH
H
H
N
N

87

Fig. 7.10 Kawabata's chiral nucleophilic catalyst **87**

Table 7.6 Desymmetrization of *meso*-diol **88** catalyzed by phosphinite derivatives of cinchona alkaloids **85** and **86**

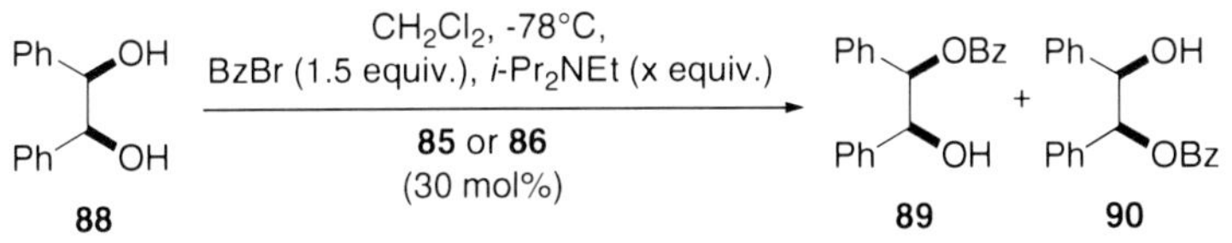
Ph OH
Ph OH
88
CH2Cl2, -78°C,
BzBr (1.5 equiv.), i-Pr2NEt (x equiv.)
85 or 86
(30 mol%)
Ph OBz
Ph OH
89
+
Ph OH
Ph OBz
90

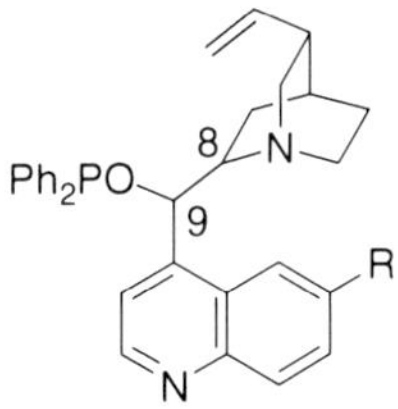
Ph2PO
8
9
N
R
N

85a: R = H (cinchonidine 8*S*,9*R*)
85b: R = H (cinchonine 8*R*,9*S*)
85c: R = OMe (quinine 8*S*,9*R*)
85d: R = OMe (quinidine 8*R*,9*S*)

Entry	Catalyst	Time (h)	Major product	Yield (%)	*ee* (%)
1	**85a**	24	**89**	83	78
2	**86b**	1.5	**90**	99	82
3	**86c**	3.5	**89**	58	22
4	**86d**	3.5	**90**	59	29

The enantioselective acylation proceeded with 34% yield of the monobenzoylated diol in 74% *ee*. The diastereomeric cinchonine phosphinite **85b** gave nearly quantitative yield and 82% *ee*. Further studies with other *meso*-1,2-diols proceeded also in good *ee* (86–94%).

In 2006, Fujimoto et al. reported the kinetic resolution of *d,l*-1,2-diols which was also catalyzed by **85** [111]. The racemic C_2-symmetric 1,2-diols were kinetically resolved by the acylation reaction to afford the monoacylated product and unreacted diol with good to excellent selectivities (Table 7.7).

Table 7.7 Kinetic resolution of *d,l*-1,2-diols catalyzed by **85d**

EtCN, -78°C; *p*-$CF_3C_6H_4COCl$ (0.65 equiv.), *i*-Pr_2NEt (0.5 equiv.); **85d** (30 mol%): **91** → **92** + **93**

Entry	Diol	*ee* of **92** (%)	*ee* of **93** (%)	Conv. (%)	*S*
1		98	99	50	525
2		89	15	14	20
3		81	63	44	18
4		74	52	41	11
5		68	69	50	11

7.2 Addition to C=X Bonds

7.2.1 Addition of Alkyl Groups to C=X Bonds – Allylation of Benzylhydrazones

Khiar et al. reported the allylation of benzylhydrazones using non-racemic monosulfoxides, methylene-bridged C_2-symmetric bissulfoxides, and ethylene-bridged C_2-symmetric bissulfoxides. The reactions have moderate yields but good enantioselectivity. However, the yields hardly exceed the catalyst loading of 50 mol%. The enantioselectivity of the process is highly dependent on the spacer between the two sulfinyl groups and the concentration of the catalyst in the reaction (Scheme 7.17) [113].

7.2.2 Addition of Cyanide to C=X Bonds – Hydrocyanation

The addition of cyanide to a carbonyl group results in the formation of a α-hydroxy nitrile (a cyanohydrin) which can act as intermediate in the synthesis of

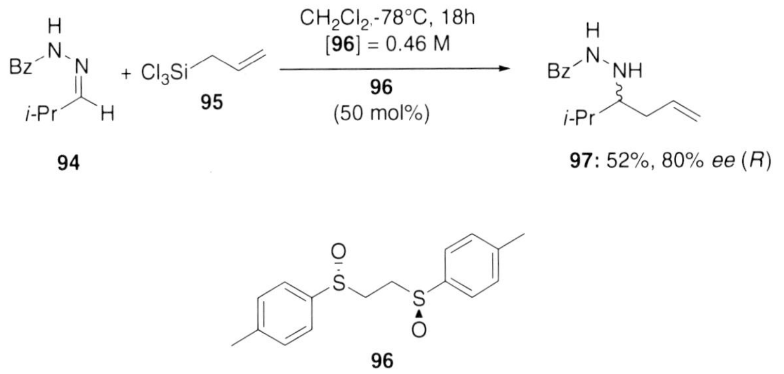

Scheme 7.17 C_2-symmetric bis-sulfoxide **96** catalyzes the enantioselective allylation of benzylhydrazones

α-hydroxy acids, α-hydroxy aldehydes or α-hydroxy ketones, and β-amino alcohols [114–117]. In this section, the quite extensively studied catalytic hydrocyanation of aldehydes and imines (Strecker reaction) will be covered only shortly. Methods which rely exclusively on Lewis acidic metal ion complexes (e.g. of peptides) are omitted. In all hydrocyanations, severe background of uncatalyzed reaction is a common problem.

7.2.2.1 Hydrocyanation of Aldehydes

In 2000, Kagan et al. presented the addition of trimethylsilylcyanide (TMSCN) (**99**) to aldehydes catalyzed by the monolithium salt of (*S*)-(−)-BINOL (**106**) [118]. The proposed mechanism includes a hypervalent silicon intermediate. In an initial attack of the anionic chiral catalyst, a pentavalent silicon complex is formed (**100**). The complex acts as Lewis acid and is supposed to coordinate the carbonyl group of the aldehyde (**101**) to a hexavalent species (**102**). The next step is the enantioselective transfer of cyanide (**103**), followed by elimination of the anionic catalyst. Consequently, the TMS cyanohydrin (**104**) is formed (Scheme 7.18).

Kagan et al. studied a number of chiral bisnaphtholates **106** and **107** which formed a number of highly enantiomerically enriched cyanohydrins in up to 97% *ee* (Table 7.8) [118, 119].

7.2.2.2 Hydrocyanation of Imines (Strecker Reaction)

The Strecker reaction is an efficient method for the preparation of α-amino acids which was first reported by Adolf Strecker over 100 years ago [120]. Now, several methods for asymmetric cyanations catalyzed by metal-complexes or organocatalysts have been developed [121–124].

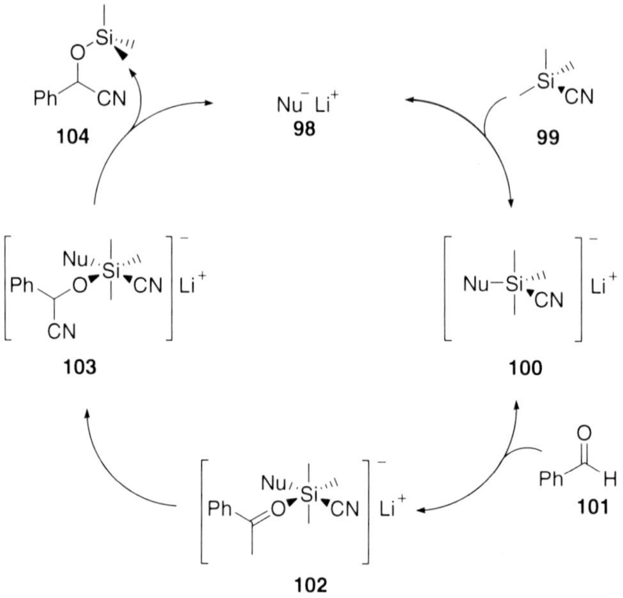

Scheme 7.18 Proposed mechanism for the addition of trimethylsilylcyanide to aldehydes (Nu = catalyst)

Table 7.8 Enantioselective hydrocyanation of aldehydes catalyzed by **106** and **107**

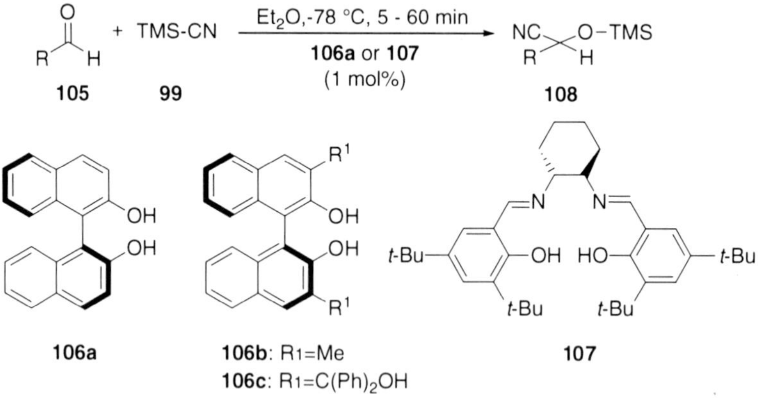

Et_2O, -78 °C, 5 - 60 min; **106a** or **107** (1 mol%)

106b: R1=Me
106c: R1=C(Ph)$_2$OH

Entry	R	Catalyst	Yield (%) **108**	*ee* (%) **108**
1	Ph	**106a**	96	56
2	Ph	**107**	98	86
3	4-Me-Ph	**106a**	95	59
4	4-Me-Ph	**107**	96	93

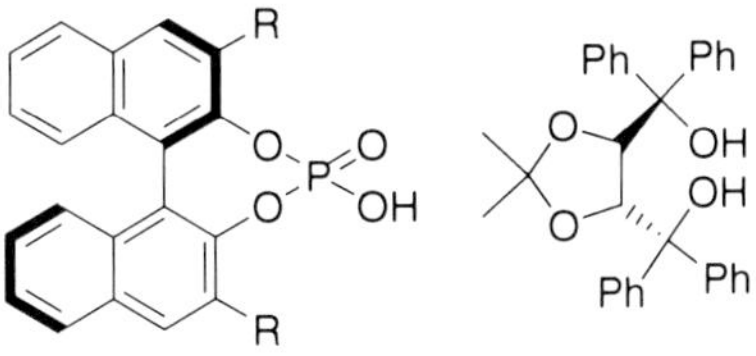

Fig. 7.11 Ruepings chiral BINOL phosphate catalysts

In 2001, Feng et al. reported chiral *N*-oxides **110–112** for the enantioselective cyanation of aldimines [125–127]. The yields (63–93%) are good, but enantioselectivities (37–73%) are still modest [125]. After optimization of the reaction conditions, however, up to 97% *ee* could be achieved (Scheme 7.19) [126, 127].

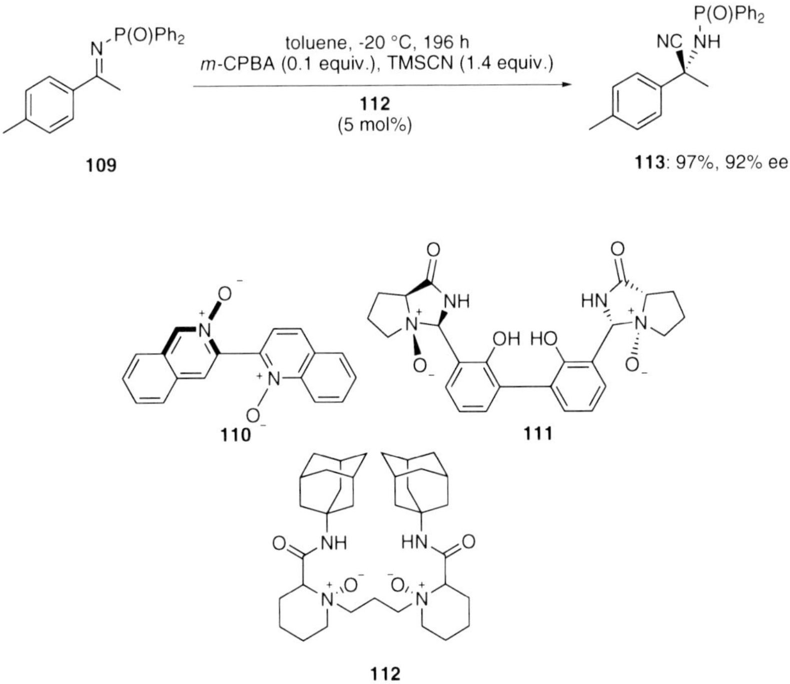

Scheme 7.19 Enantioselective cyanation of aldimines catalyzed by chiral *N*-oxides **110–112**

In 2007, Rueping et al. reported an enantioselective Strecker reaction catalyzed by chiral BINOL phosphate **114** and TADDOL **115** catalysts (Fig. 7.11) [128]. The corresponding amino nitriles were obtained in good isolated yields and enantioselectivities (up to 97% *ee*).

In 2008, Sun et al. described the reduction of ketimine catalyzed by *S*-chiral bissulfinamides **117** (Table 7.9) [129]. These catalysts proved to be very efficient (up to 98% yield, up to 93% *ee*).

Table 7.9 Reduction of ketimines catalyzed by *S*-chiral bissulfinamides **117**

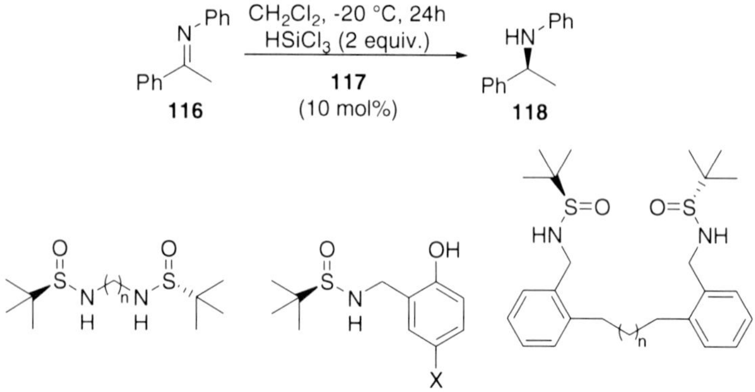

Entry	Catalyst	Yield (%) **118**	*ee* (%) **118**
1	**117a** (n=5)	93	91
2	**117b** (X=H)	84	88
3	**117c** (n=5)	78	91

7.2.3 *Addition of Enolates to C=X Bonds*

7.2.3.1 Intermolecular Aldol Reactions

The asymmetric catalysis of aldol reactions with chiral Lewis bases is an important method to form C-C bonds [130–139]. The emergence of Lewis base-activated Lewis acid catalysis is pioneered by the Denmark group [140–142]. They used a variety of chiral phosphoramides as catalysts for the enantioselective intermolecular aldol reactions (Scheme 7.20).

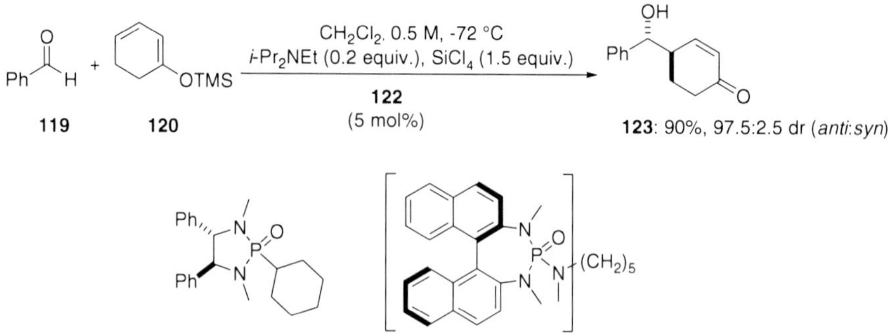

Scheme 7.20 Enantioselective intramolecular aldol reaction catalyzed by chiral phosphoramides **121** and **122**

The aldol products can be obtained in high yield and with excellent enantioselectivities (<90% *ee*). Although the oxygen of the phosphoramides is crucial as a ligating atom during catalysis, the intermolecular aldol reaction itself proceeds via a trichlorosilyl enolate. Accordingly, this type of reaction is not further discussed in this chapter. The chiral phosphoramide base serves only as activator and steric modifier of tetrachlorosilane (Scheme 7.21) [143–145].

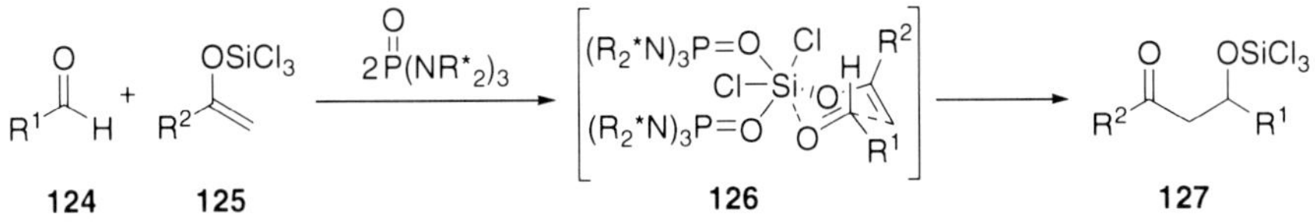

Scheme 7.21 Lewis-base catalyzed aldol addition

7.2.3.2 Morita-Baylis-Hillman Reactions

In the classical Morita-Baylis-Hillman (MBH) reaction an α,β-unsaturated ester (electrophilically activated alkene), is activated by the reversible Michael-addition of a tertiary amine catalyst (e.g. DABCO), producing a zwitterion intermediate, the enolate moiety of which can react with an aldehyde to form an aldolate zwitterion. Retro-Michael-addition then regenerates the catalyst and the MBH-product (Scheme 7.22). The catalyst is sometimes used in high amounts (over stoichiometric) and often the reaction is very sensitive to the Michael acceptor used.

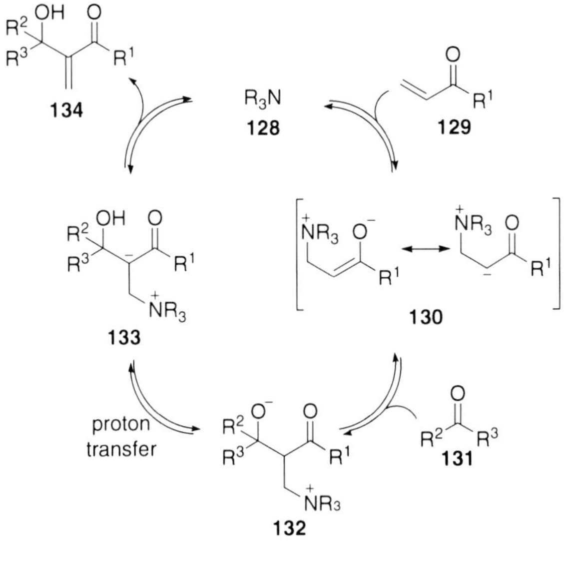

Scheme 7.22 Mechanism of classical Morita-Baylis-Hillman reaction

Table 7.10 Sulfide and selenide catalyzed MBH reaction in the presence of $TiCl_4$ for additional carbonyl activation

Entry	Chalcogenide	Yield (%) **149**
1	**137**	60
2	**138**	67
3	**139**	85
4	**140**	69
5	**141**	71
6	**142**	71
7	**143**	70
8	**144**	74
9	**145**	78
10	**146**	71
11	**147**	76
12	**148**	69

Several chalcogenide catalyzed MBH-type reactions are reported [146]. Instead of the common *tert*-amines or phosphanes, also higher organochalcogenides can act as nucleophilic activator. Such Morita-Baylis-Hillman reactions catalyzed by sub-stoichiometric amounts of sulfides and selenides in the presence of Lewis acid to activate the carbonyl group were described by Kataoka and co-workers [147, 148]. The reaction of *p*-nitrobenzaldehyde and 2-cyclohexenone has been used for screening a series of chalcogenide catalysts in dichloromethane at room temperature. The best result was found when 10 mol% of chalcogenide where employed with a stoichiometric amount of $TiCl_4$ in the presence of excess 2-cyclohexenone as Michael acceptor (3 equiv., Table 7.10).

The proposed catalytic cycle is shown in the Scheme 7.23.

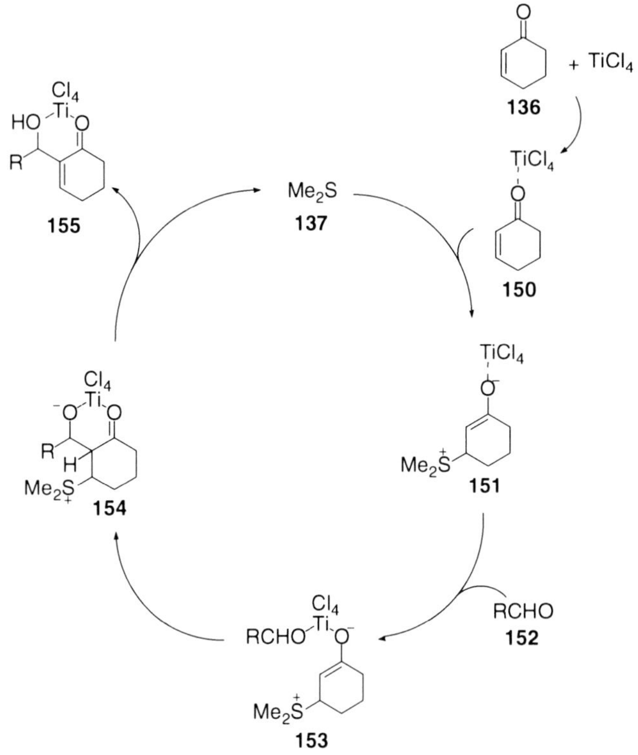

Scheme 7.23 Proposed mechanism for a MBH reaction catalyzed by sulfides and selenides in the presence of a Lewis acid

The reaction was expanded to a number of different Michael acceptors and aldehydes [148]. Some results for the reaction catalyzed with chalcogenides **137** and **139** are summarized in Table 7.11.

The method was extended to α,β-unsaturated thioesters [149]. The catalysts **137**, **160** and **161** were employed in 10 mol% (Scheme 7.24) [150]. In this reaction some hydrogen chloride adduct was formed as byproduct or intermediate.

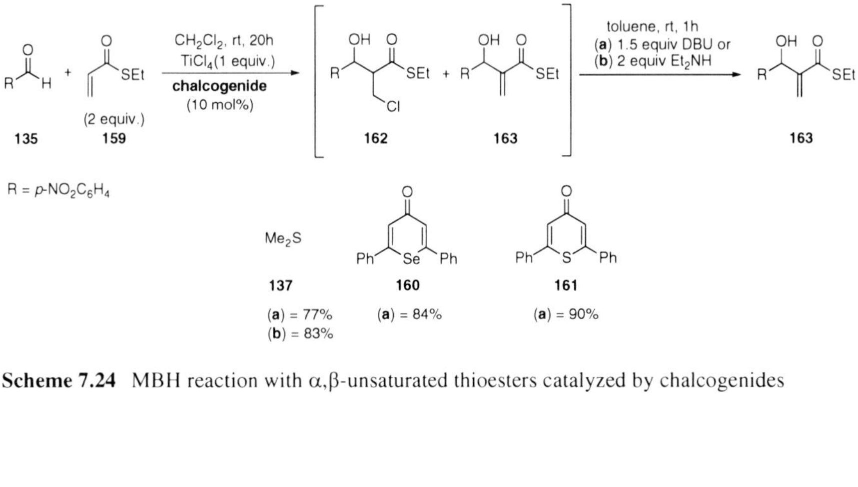

Scheme 7.24 MBH reaction with α,β-unsaturated thioesters catalyzed by chalcogenides

Table 7.11 MBH reaction catalyzed by sulfide **137** and diselenide **139**

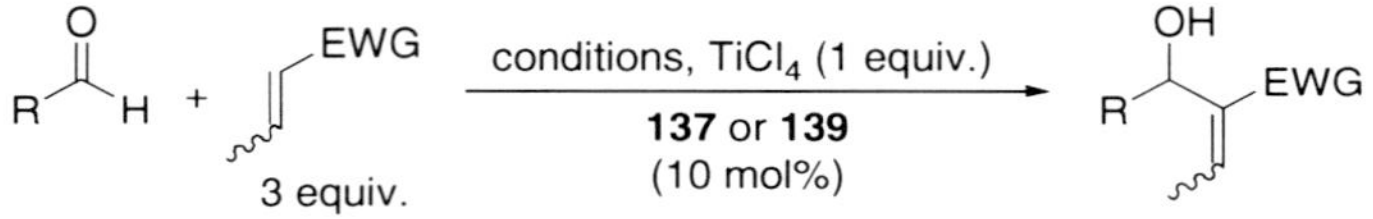

156 + **157** (3 equiv.) → conditions, $TiCl_4$ (1 equiv.), **137** or **139** (10 mol%) → **158**

Entry	Chalcogenide	Aldehyde	Michael acceptors	Conditions	Product	Yield (%)
1	**139**	*p*-$NO_2C_6H_4CHO$	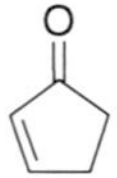	rt, 1 h	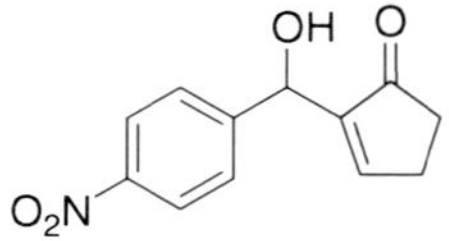	70
2	**139**	*p*-$NO_2C_6H_4CHO$	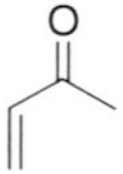	rt, 1 h	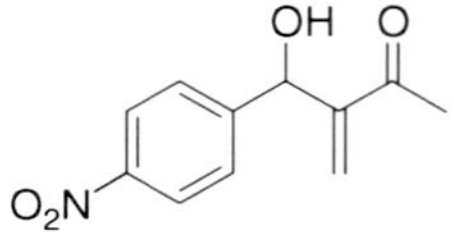	67
3	**139**	PhCHO	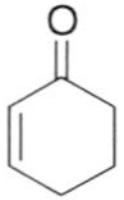	rt, 1 h	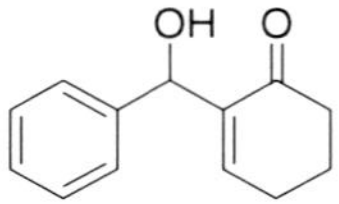	52
4	**137**	*p*-$NO_2C_6H_4CHO$	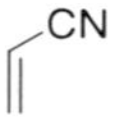 CN	Reflux, 24 h	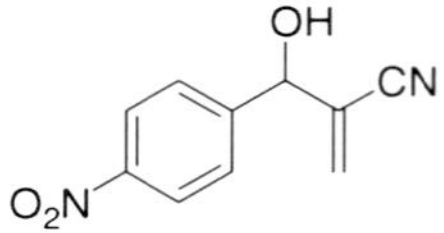	88
5	**139**	*p*-$NO_2C_6H_4CHO$	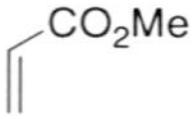 CO_2Me	rt, 2 min	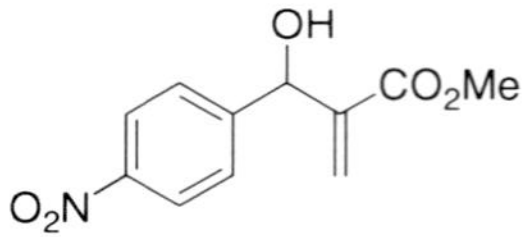	49

Table 7.12 MBH with alkynoates catalyzed by sulfide **137**

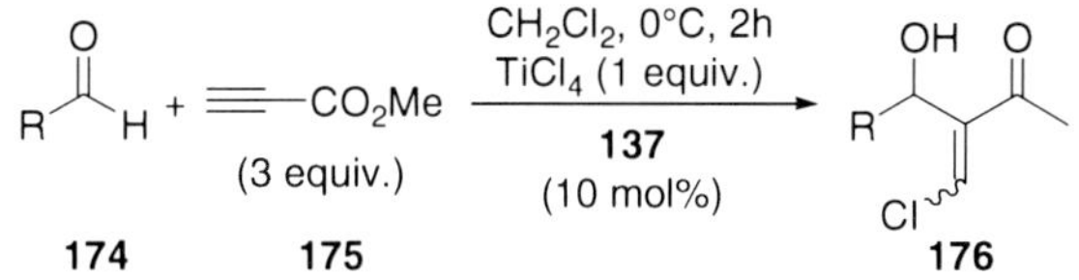

Entry	R	Yield (%) **176**	*E*:*Z* ratio
1	*p*-$NO_2C_6H_4$	86	*E* only
2	*p*-$CF_3C_6H_4$	73	*E* only
3	*p*-ClC_6H_4	73	*E* only
4	*p*-FC_6H_4	89	*E* only

The raw product showed a ratio of **163**: ***syn*-162**: ***anti*-162** = 5 : 65 : 30, i.e. with **162** as major initial product. A reformulated mechanism was proposed (Scheme 7.25) [151].

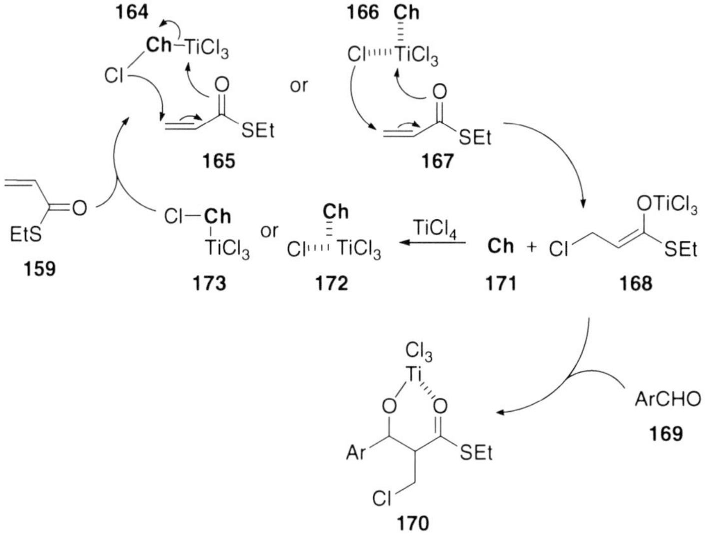

Scheme 7.25 Reformulated mechanism for MBH reaction catalyzed by chalcogenides (**Ch**) in the presence of Lewis acid

A further expansion utilized alkynoates (Table 7.12). Here HCl-addition is also observed, with a strong preference for the *E*-isomer (Scheme 7.26). Dimethylsulfide (**137**) was the most active catalyst [152, 153].

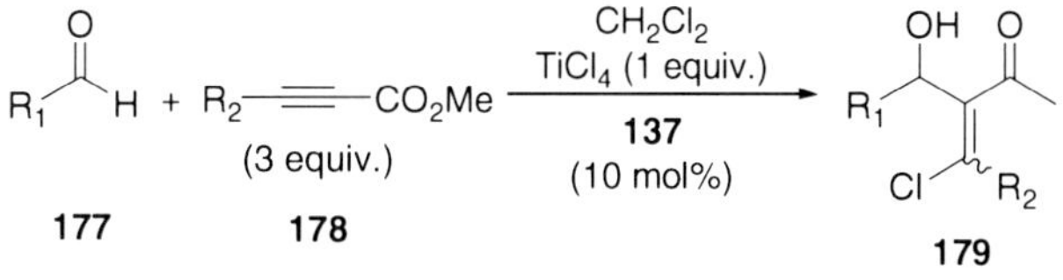

Scheme 7.26 MBH with alkynoates catalyzed by sulfide **137**

Kataoka's method was further explored by Basavaiah et al. with nonenolizable α-keto esters. The best result is shown in Scheme 7.27 [154].

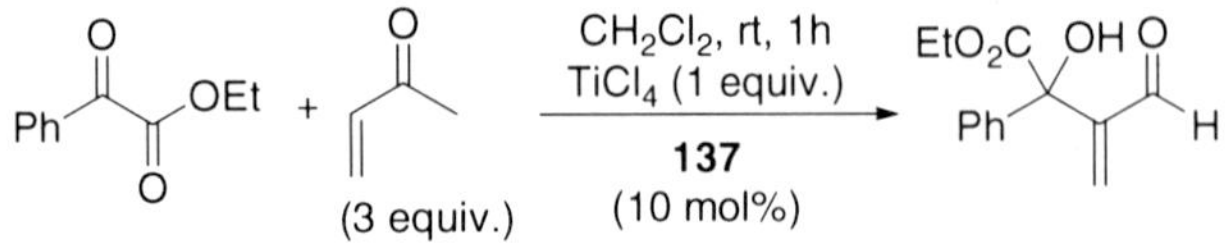

CH_2Cl_2, rt, 1h
$TiCl_4$ (1 equiv.)
137 (10 mol%)
180 + **181** (3 equiv.) → **182**: 73%

Scheme 7.27 Basavaiah's best result with nonenolizable α-keto esters

Asymmetric MBH were reported with (−)-menthyl and (−)-8-phenylmenthyl glyoxylates [155]. Only one of the possible diastereoisomers was observed in ^{1}H NMR (Scheme 7.28).

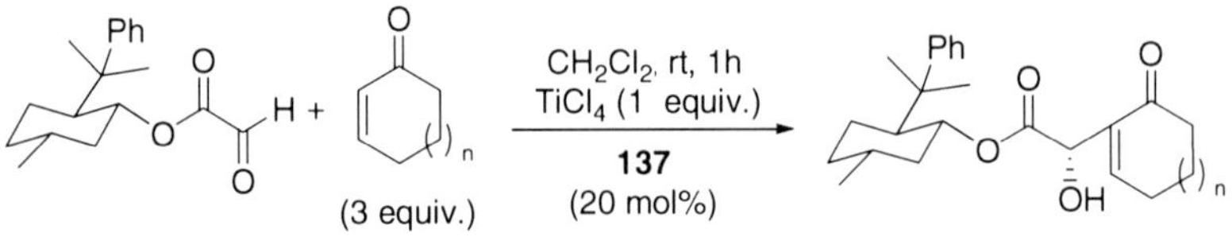

CH_2Cl_2, rt, 1h
$TiCl_4$ (1 equiv.)
137 (20 mol%)
183 + (3 equiv.) **184**: n=0; **136**: n=1 →
185a: n=0, 76%, >95% *dr*
185b: n=1, 78%, >95% *dr*

Scheme 7.28 Diastereoselective MBH catalyzed by **137**

Another enantioselective version used the optically active sulfides **186a–b**. These gave excellent yields but showed no enantioselectivity (Scheme 7.29) [156].

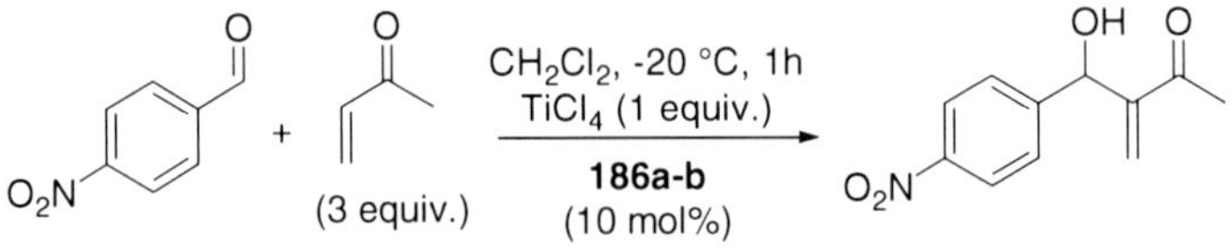

CH_2Cl_2, -20 °C, 1h
$TiCl_4$ (1 equiv.)
186a-b (10 mol%)
135 + **181** (3 equiv.) → **187**

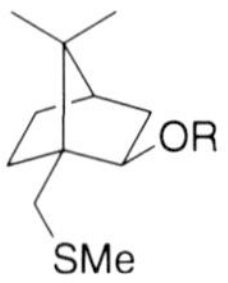

186a: R = H, 95%, 2% *ee* (*R*)
186b: R = Me, 97%, 1% *ee* (*R*)

Scheme 7.29 MBH catalyzed by chiral sulfides **186**

Metzner et al. reported an acetal (Mukaiyama-type) MHB reaction with a combination of a sulfide catalyst and TBDMSOTf under basic conditions, i.e. without Lewis metal ion co-activation of the electrophilic enone. A substoichiometric amount of **189** (20 mol%) at −40°C gave 42% of **190** (Scheme 7.30) [157].

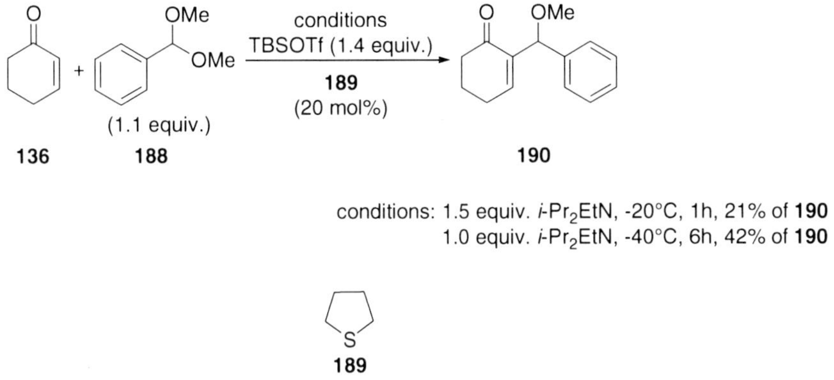

conditions: 1.5 equiv. i-Pr_2EtN, -20°C, 1h, 21% of **190**
1.0 equiv. i-Pr_2EtN, -40°C, 6h, 42% of **190**

Scheme 7.30 Mukaiyama-type MBH reaction catalyzed by sulfide **189**

The proposed mechanism for the reaction is shown in Scheme 7.31.

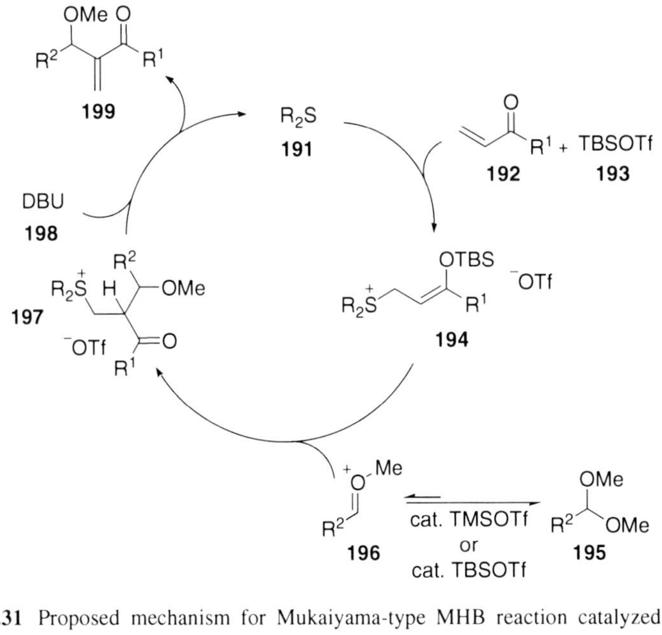

Scheme 7.31 Proposed mechanism for Mukaiyama-type MHB reaction catalyzed by sulfide under basic conditions

7.3 Redox and Ylide Reactions

7.3.1 *Reductions*

7.3.1.1 Telluride Catalyzed Dehalogenation and Deselenation

In 1985 Suzuki et al. reported that vicinal dibromides can be debrominated to olefins by potassium disulfite in the presence of catalytic amounts of bis(4-methoxyphenyl) telluride (**201**) in a two-phase system [158]. The yields are good to excellent with high stereoselectivity (Table 7.13).

The proposed catalytic cycle is shown in Scheme 7.32.

Detty et al. demonstrated that more electron rich diorganotellurides (**207**–**209**) (Fig. 7.12) provide terminal alkenes and *cis* and *trans* 1,2-disubstituted olefins [159]. They use sodium ascorbate or glutathione as reducing agents.

Comasseto et al. reported the smooth reductive deselenation of α-phenylseleno esters in the presence of 2.5 mol% of **211** (Scheme 7.33) [160]. In this highly unusual reaction, a putative mechanistic pathway involves a tellurolate (from $NaBH_4$-reduction) that attacks at selenium in a nucleophilic substitution reaction with the ester enolate as leaving group.

Table 7.13 Dehalogenation reaction catalyzed by telluride **201**

Entry	R^1	R^2	Relative config.	*trans*:*cis*	Yield (%)
1	Ph	Ph	*erythro*	100:0	92
2	Ph	Ph	*threo*	6:94	88
3	Ph	4-pyridyl	*erythro*	100:0	quant.
4	Ph	COPh	–	100:0	61
5	Ph	CO_2Et	–	100:0	57

Scheme 7.32 Proposed catalytic cycle for the dehalogenation reaction catalyzed by tellurides

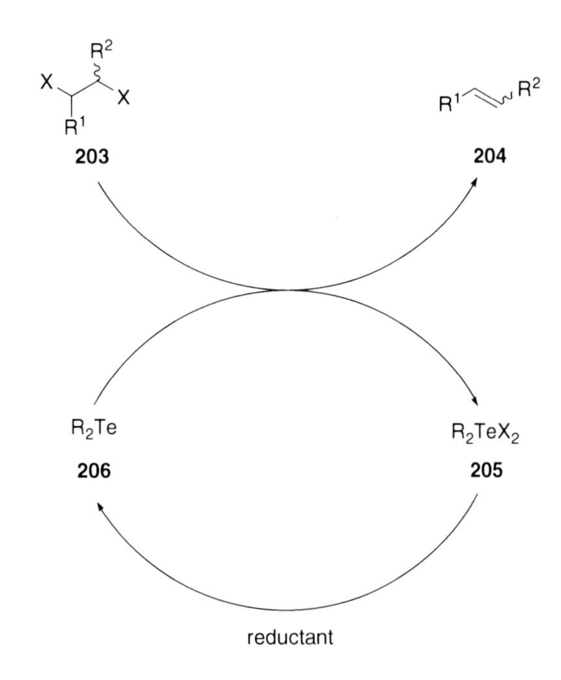

Fig. 7.12 Diorganotelluride dehalogenation catalysts

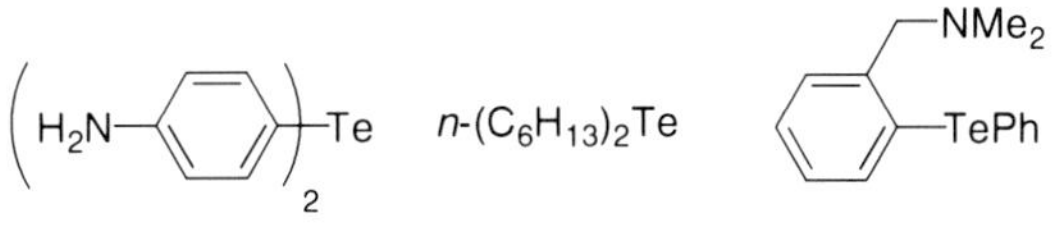

$(H_2N-C_6H_4)_2Te$ **207** $n\text{-}(C_6H_{13})_2Te$ **208** $2\text{-}(Me_2NCH_2)C_6H_4TePh$ **209**

Scheme 7.33 Deselenation of carbonylic compounds catalyzed by telluride **211**

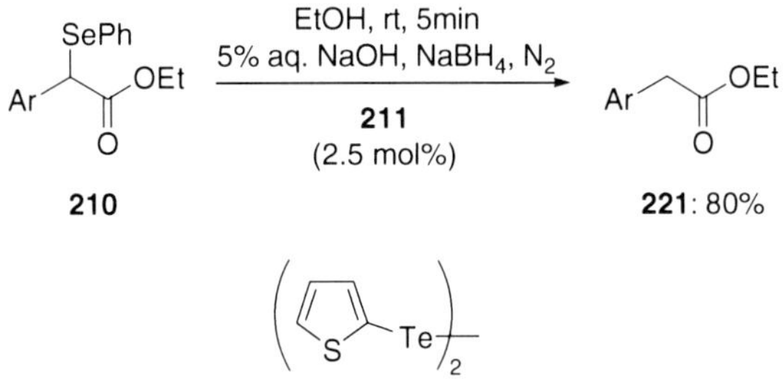

ArCH(SePh)C(O)OEt **210** → [EtOH, rt, 5min; 5% aq. NaOH, $NaBH_4$, N_2; **211** (2.5 mol%)] → $ArCH_2C(O)OEt$ **221**: 80%

$(2\text{-thienyl-Te})_2$ **211**

Table 7.14 Olefination of α-bromoester or -ketone catalyzed by telluride **215**

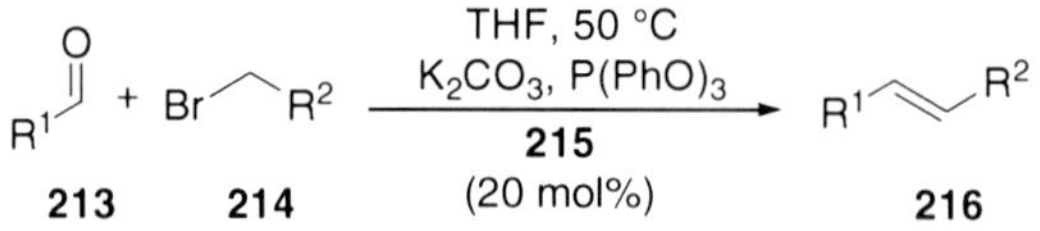

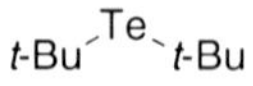

Entry	R^1	R^2	Time(h)	Yield (%)
1	p-ClC_6H_4	CO_2Me	13	89
2	Ph	CO_2Me	13	98
3	PhCH=CH	CO_2Me	17	80
4	p-MeC_6H_4	CO_2Me	18	95
5	$Me(CH_2)_8$	CO_2Me	17	83
6	p-ClC_6H_4	Bz	17	89
7	2-Furyl	Bz	16	91
8	Ph	CO*i*-Pr	14,5	98
9	2-pyridil	CO*i*-Pr	14,5	98

7.3.2 *Reactions Involving Chalcogenide Ylides*

7.3.2.1 Telluride Catalyzed Olefinations (Wittig Type Te-Ylide Reaction)

Huang et al. developed a reductive one-pot olefination method with telluride as catalyst. A α-bromoester or -ketone reacts with an aldehyde in the presence of a catalytic amount (20 mol%) dibutyl telluride (**215**); potassium carbonate serves as base, and triphenyl phosphite as reductant. The products are formed via an intermediate tellurium ylide in good to excellent yields with *E* stereoselectivity (Table 7.14, Scheme 7.34) [161].

Tang and co-workers synthesized a soluble polyethyleneglycol (PEG) supported telluride **226** of which just 2 mol% are sufficient for a quantitative conversion. A further simplification is the use of hydrogensulfite as reductant. The catalyst could be recycled, but with partial loss of activity. Selected results are presented in Table 7.15 [162].

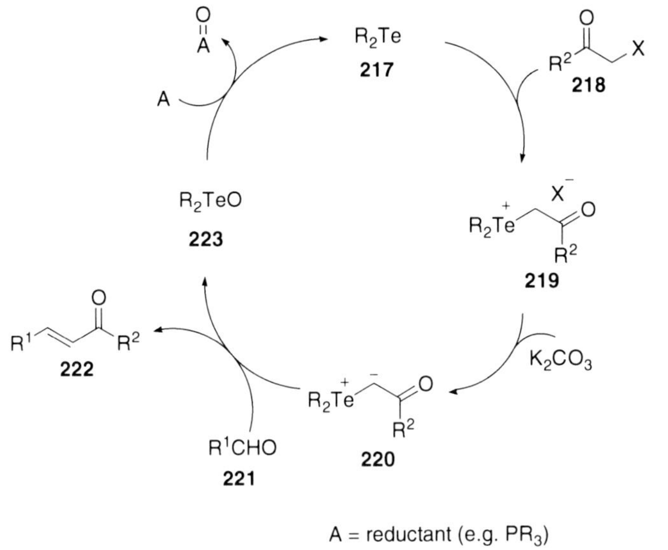

A = reductant (e.g. PR_3)

Scheme 7.34 Proposed catalytic cycle for the dialkyltelluride catalyzed olefination (reductive aldol condensation)

Table 7.15 Olefination catalyzed by PEG-supported telluride **226**

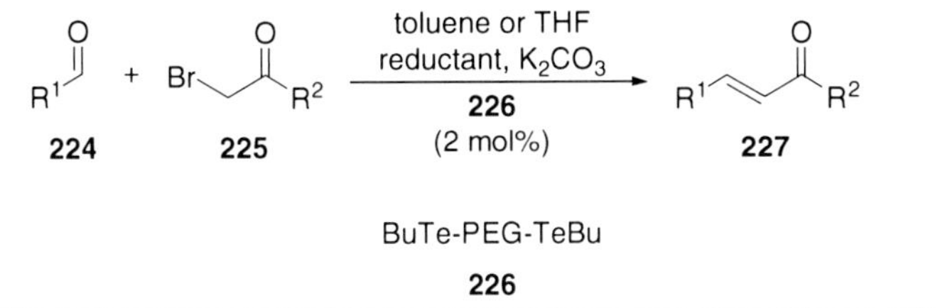

toluene or THF, reductant, K_2CO_3, **226** (2 mol%): **224** + **225** → **227**

BuTe-PEG-TeBu

226

Entry	R^1	R^2	t (h)	Reductant	Yield	*E/Z*
1	p-$CH_3OC_6H_4$	OEt	43	$P(OPh)_3$	94	>99:1
2	p-$CH_3OC_6H_4$	OtBu	43	$NaHSO_3$	87	>99:1
3	p-$CH_3C_6H_4$	OEt	23	$P(OPh)_3$	93	90:10
4	p-$CH_3C_6H_4$	OtBu	24	$NaHSO_3$	96	94:6
5	2-furyl	OEt	12	$P(OPh)_3$	96	>99:1
6	2-furyl	OtBu	11	$NaHSO_3$	88	>99:1
7	cyclohexyl	OEt	48	$P(OPh)_3$	70	>99:1
8	Cyclohexyl	OtBu	48	$NaHSO_3$	76	>99:1

7.3.2.2 Cyclopropane Formation

Cyclopropane Formation via Chalcogen-Ylides by Alkylation/Deprotonation

Tang and co-workers report a catalytic asymmetric cyclopropanation method where the sulfide reacts with an activated alkyl bromide to afford the sulfonium intermediate [163]. This is deprotonated by base to generate the sulfur ylide that reacts with electrondeficient alkenes via Michael-addition and S_N-attack of the enolate formed to give the cyclopropane and to regenerate the sulfide (alkylation-deprotonation, Scheme 7.35).

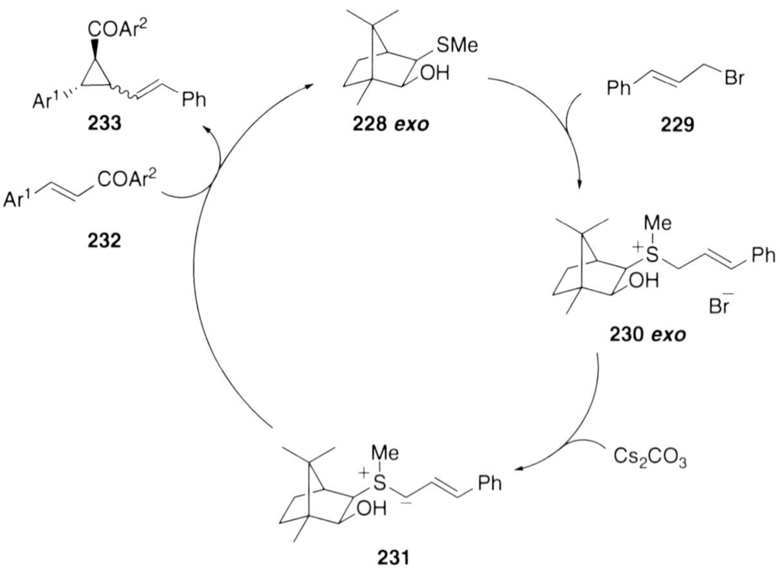

Scheme 7.35 Catalytic cycle for the cyclopropane formation catalyzed by a chalcogenide organocatalyst

The best results were obtained with 20 mol% camphor-derived sulfonium salt **230**, with 1.5 equivalents of 3-phenylallyl bromide and one equivalent of chalcone at 0°C in a mixture of *tert*-butyl alcohol and acetonitrile (2.5:1). The cyclopropanation product was obtained in moderate diastereoselectivity (86/14), good enantioselectivity (82% *ee*) and in high yield (92%). The method could be extended to a variety of α,β-unsaturated carbonyl compounds (Table 7.16). The chiral sulfide **228** could also be employed directly without preformation of the sulfonium salt **230** with similar results.

For the *endo* sulfonium salt from sulfide **228** (Fig. 7.13), the authors report formation of the opposite enantiomer of **235**.

Table 7.16 Cyclopropane formation catalyzed by *exo* sulfonium salt **230**

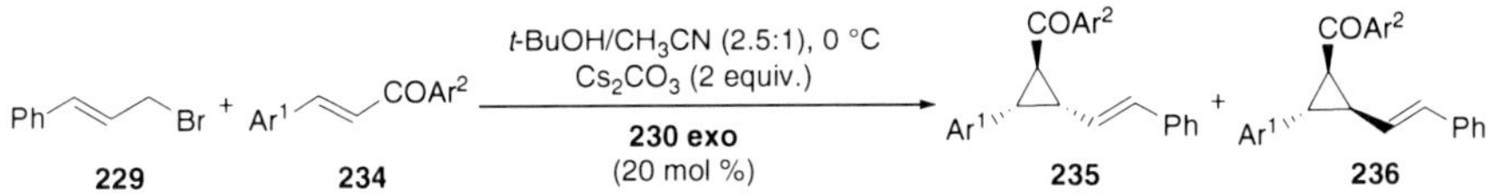

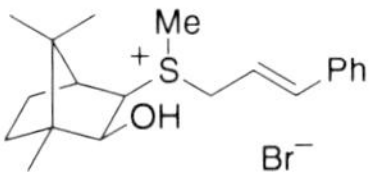

Ph‍CH=CHCH₂Br (229) + Ar¹CH=CHCOAr² (234) → t-BuOH/CH₃CN (2.5:1), 0 °C, Cs_2CO_3 (2 equiv.), **230 exo** (20 mol %) → **235** + **236**

230 exo

Entry	Ar^1	Ar^2	Time(h)	Yield(%)	**235/236**	*ee*(%)
1	Ph	Ph	36	92	86/14	82
2	p-ClC_6H_4	Ph	30	86	76/24	78
3	p-BrC_6H_4	Ph	30	89	75/25	77
4	p-$MeOC_6H_4$	Ph	54	80	86/14	81
5	o-BrC_6H_4	Ph	20	90	77/23	88
6	Ph	*p*-MeC_6H_4	57	87	87/13	78

Fig. 7.13 *Endo* sulfonium salt **230** derived from **228**

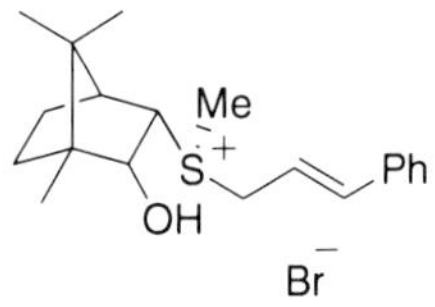

230 *endo*

According to the author, the neighbouring hydroxyl group plays a crucial role in the reaction [163]. If it is methylated, the cyclopropanation does not take place and only rearrangement products are isolated. The free hydroxyl group guides the substrate to the right face to ensure enantioselectivity and was proposed to form hydrogen bonds to the substrate.

Huang and co-workers reported the catalytic cyclopropanation of a series of β-aryl enones with an excess of trimethylsilylallyl bromide using *i*-Bu_2Te (**237**) as catalyst (20 mol%) in the presence of solid Cs_2CO_3 (2 equiv.) in THF/trace H_2O at 50°C. The proposed mechanism is shown in Scheme 7.36 [164]. In this reaction, high yields and excellent diastereoselectivities were achieved (Table 7.17).

Tang et al. reported an asymmetric variation of Huang's method with chiral telluronium salt **247** (Scheme 7.37) [165].

Sulfide Catalyzed Intramolecular Cyclopropane Formation and Synthesis of Chromenes

Tang et al. describe a catalytic intramolecular cyclopropanation for the preparation of benzobicyclic compounds with the [*n*.1.0] unit in moderate to good yields with

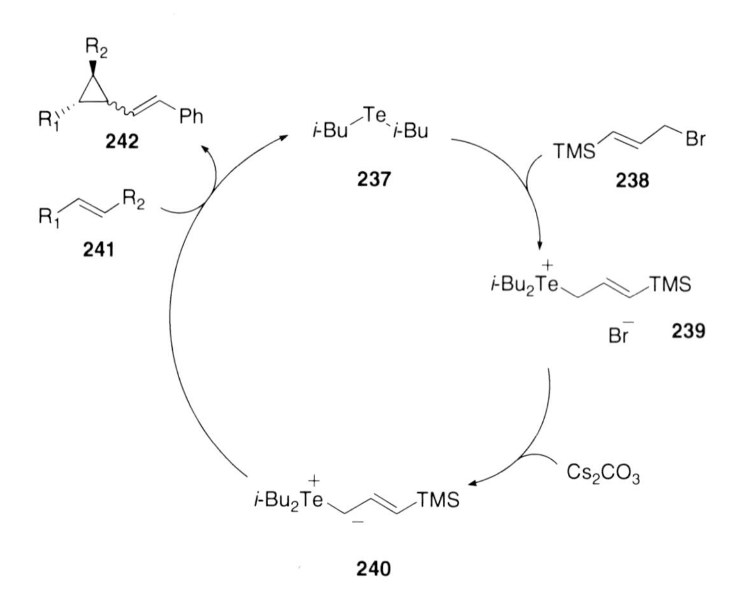

Scheme 7.36 Putative mechanism for the cyclopropane formation catalyzed by telluride **237** via an alkylation-deprotonation cycle

Table 7.17 Cyclopropanation of β-aryl enones catalyzed by diisobutyltelluride (**237**)

243 + Br–CH₂CH=CH–TMS **238** → (THF/traces H_2O, 50 °C, Cs_2CO_3, **237** (20 mol%)) → **244** + **245**

237: *i*-Bu–Te–*i*-Bu

Entry	R^1	R^2	**244/245**	Yield (%)
1	Ph	COPh	>98:2	70
2	p-ClC_6H_4	COPh	>99:1	64
3	p-$NO_2C_6H_4$	COPh	>99:1	83
4	p-MeC_6H_4	COPh	>98:2	70
5	Ph	COCH=CHPh	>99:1	58

246 + Br–CH₂CH=CH–TMS **238** → (THF, Cs_3CO_2, **247** (20 mol%)) → **248** + **249**

$R^1 = R^2 =$ Ph: 91%, 86% *ee*, **248/249** = 90:10

$R^1 =$ *p*-Cl-C_6H_4, $R^2 =$ Ph: 94%, 89% *ee*, **248/249** = 91:9

247: telluronium salt, BPh_4^-

Scheme 7.37 Asymmetric cyclopropane formation catalyzed with telluronium salt **247**

tetrahydrothiophene (**189**) as catalyst (Scheme 7.38) [166]. The proposed mechanism is shown in Scheme 7.39.

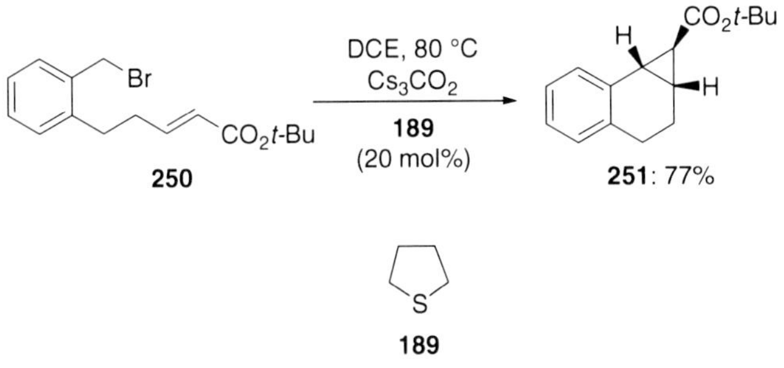

Br
CO_2t-Bu
250
DCE, 80 °C
Cs_3CO_2
189
(20 mol%)
H
CO_2t-Bu
H
251: 77%
S
189

Scheme 7.38 Synthesis of benzobicyclic compounds by intramolecular cyclopropanation catalyzed by tetrahydrothiophene

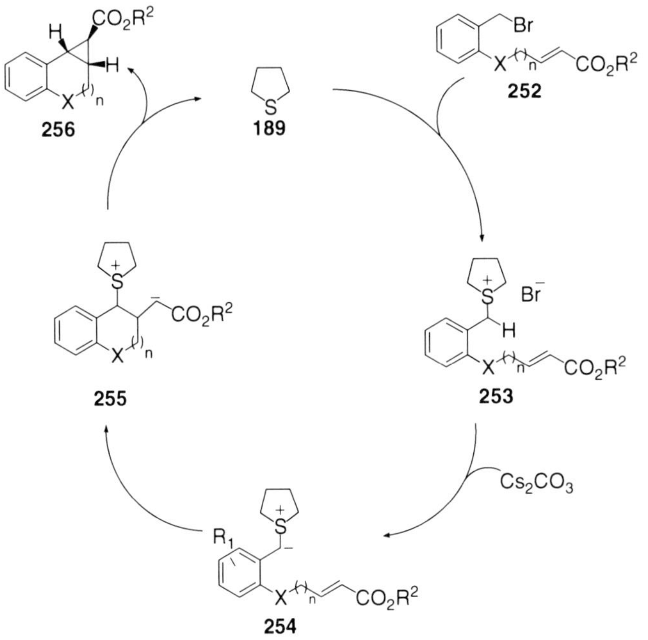

H
CO_2R^2
H
X
n
256
S
189
Br
X
n
CO_2R^2
252
S
+
CO_2R^2
X
n
255
S
+
Br^-
H
X
n
CO_2R^2
253
Cs_2CO_3
S
+
R_1
X
n
CO_2R^2
254

Scheme 7.39 Proposed catalytic cycle for the intermolecular cyclopropanation catalyzed by chalcogenides

Tang et al. reported an unexpected tetrahydrothiophene-catalyzed rearrangement/annulation reaction via a tandem Michael addition/elimination/substitution for the synthesis of chromenes. The products are 2*H*-chromene- or 4*H*-chromene-2-carboxylates, depending on the choice of the base. With K_2CO_3 2*H*-chromene is formed, with Cs_2CO_3, 4*H*-chromene is the major product (Scheme 7.40) [167].

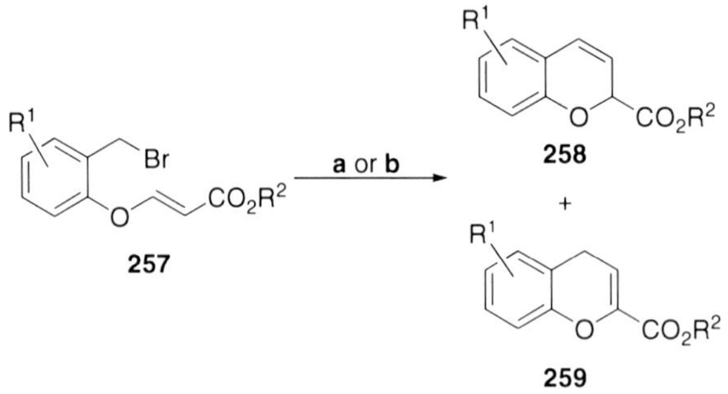

$R^1 = 6$-*t*-Bu, $R^2 = CO_2Et$; **a** = 10 mol% **189**, K_2CO_3, DCE, 80 °C; 99%, **258/259**: >99:1
$R^1 = 1$-naphthyl, $R^2 = CO_2Et$; **b** = 100 mol% **189**, Cs_2CO_3, DCE, 80 °C; 85%, **258/259**: 1:>99

Scheme 7.40 Synthesis of 2*H*-chromene and 4*H*-chromene derivatives

The proposed mechanism is shown in Scheme 7.41. Tetrahydrothiophene reacts with the benzyl bromide moiety to form a sulfonium salt, which is deprotonated by K_2CO_3 to generate the corresponding sulfonium ylide *in situ*. An intramolecular Michael addition of the ylide to the acrylate moiety, followed by a β-elimination and a cyclative S_N2'-substitution to re-liberate the sulfide completes the catalytic cycle. The stronger base Cs_2CO_3 can isomerize the 2*H*-chromene to 4*H*-chromene.

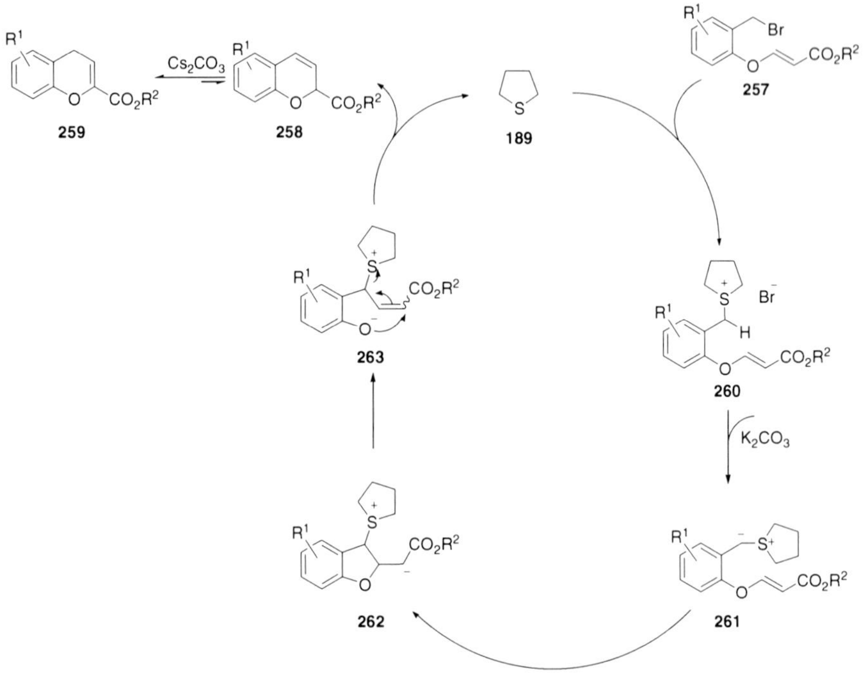

Scheme 7.41 Proposed mechanism for tandem Michael addition/elimination/substitution synthesis of 2*H*-chromene and 4*H*-chromene derivatives catalyzed by sulfides

Cyclopropane Formation via Chalcogen-Ylides Generated from Carbene Sources

Aggarwal and co-workers reported a method for the *in situ* chalcogen ylide generation via sulfide ↔ metal carbene interchange which is applicable to cyclopropanations (this chapter), aziridine and epoxide formations (following chapters) [168–170]. The general schemes for the generation of chalcogen-ylides from diazo compounds or tosylhydrazones via metal carbenes is given in Schemes 7.42 and 7.43, respectively [171].

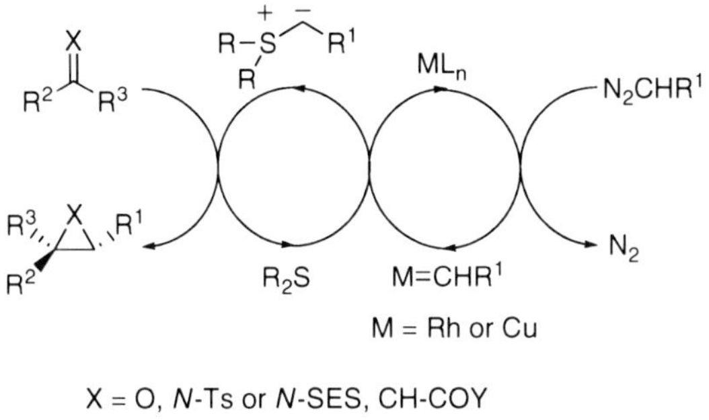

Scheme 7.42 Catalytic epoxide (X = O), aziridine (X = *N*-Ts or *N*-SES) and cyclopropane (X = CH-COY) formation mediated by chalcogen-ylides generated from diazo compounds via metal carbenes (start at upper right corner)

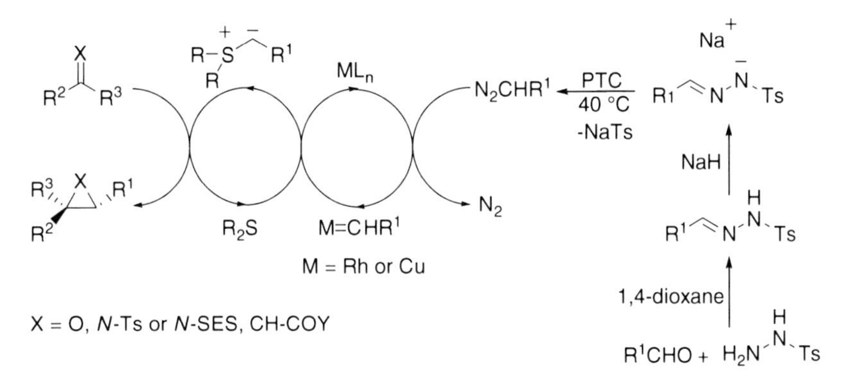

Scheme 7.43 Catalytic epoxide (X = O), aziridine (X = *N*-Ts or *N*-SES), and cyclopropane (X = CH-COY) formation mediated by chalcogen-ylides generated from metal carbenes using diazo compounds generated *in situ* from tosylhydrazones (start at lower right)

Aggarwal's group studied a variety of sulfides as catalysts for cyclopropanations. They observed that thianes (six-membered-ring sulfides) are better catalysts compared to the thiolanes (five-membered-rings). The slower reaction of the ylide form of the latter ones with enones provides time for an alternative reaction. Ylide anion equilibration eventually enables a Sommelet-Hauser rearrangement (Scheme 7.44) [169]. The propensity for this alternate reaction is lower in six-membered-ring ylides.

Table 7.18 Cyclopropanation of enones catalyzed by cyclic pentamethylene sulfide with ylide formation *via* metal-catalyzed decomposition of diazo compounds

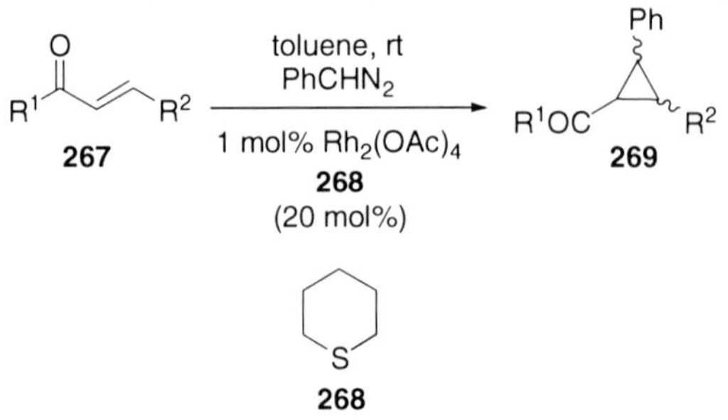

Entry	Enone (R^1, R^2)	Time (h)[a]	Yield (%)[b]	Isomeric ratio[c]
1	Ph, Ph	12	70	4:1
2	Me, Ph	12	82	4:1
3	Ph, Me	12	50	1:1:1
4	Me, Me	12	30	1:1:1

[a]Addition of $PhCHN_2$ using syringe pump
[b]Yield of *cis* and *trans* isomers
[c]Ratio determined by ^1H NMR spectroscopy

Thus increased yields, diastereo- and enantioselectivities are observed with the thianes. The ylides can be formed from halides, diazo compounds, or from tosylhydrazone salts via diazo compounds as shown above [168–170].

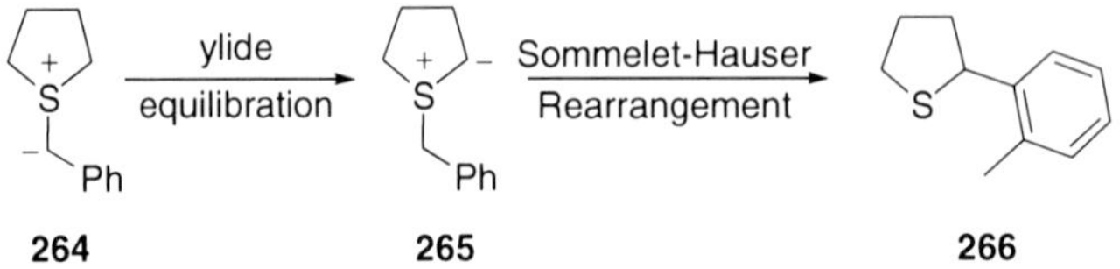

Scheme 7.44 Sommelet-Hauser rearrangement prevalent in five-membered sulfur ylides

Cyclic and acyclic enones can be used with varied diastereoselectivity. The best results are summarized in Table 7.18–7.20 [168, 169]. However, some data given in the papers are ambiguous with respect to the stereochemical assignment.

7.3.2.3 Aziridine Formation

Aziridines can be formed by nitrene (–like) additions to double bonds, or by carbene (–like) additions to imines. Both processes especially the latter one can also occur by a stepwise addition-elimination process [172]. Similar to cyclopropanations, aziridinations of imines can follow two protocols: alkylation-deprotonation-substitution and sulfide reactions with metallocarbenes.

The best results starting from tosylhydrazone salts are summarized in Table 7.21 [170]

Table 7.19 Asymmetric cyclopropanation of enones with an asymmetric cyclic sulfide

Entry	Enone (R^1, R^2)	t (h)[a]	Yield (%)[b]	Isomeric ratio[c]	*ee* (%)[d]
1	Ph, Ph	12	38	4:1	97
2	Me, Ph	12	14	4:1	>98

[a]Addition of $PhCHN_2$ using syringe pump
[b]Yield of *cis* and *trans* isomers
[c]Ratio determined by ^1H NMR spectroscopy
[d]Determined using chiral HPLC

Table 7.20 Cyclopropanation of enones with ethyl diazoacetate catalyzed by pentamethylene sulfide **268**

Entry	Enone (R^1, R^2)	Yield (%)[a]	Isomeric ratio[b]
1	Ph, Ph	77	4:2:1
2	Me, H	43	>95:5
3	OEt, CO_2Et	68	>95:5

[a]Combined yield of all isomers
[b]Ratio determined by ^1H NMR spectroscopy

Aziridine Formation with Chalcogen-Ylides by Alkylation/Deprotonation

The synthesis of β-phenylvinylaziridines from *N*-sulfonylimines and cinnamyl bromide mediated by a catalytic amount of dimethyl sulfide in presence of solid K_2CO_3 was the first organocatalytic aziridination reported (Scheme 7.45) [173]. The proposed mechanism is shown in Scheme 7.46.

Table 7.21 Asymmetric cyclopropanation of electron deficient alkenes using tosylhydrazone salts catalyzed with chiral sulfide **278**

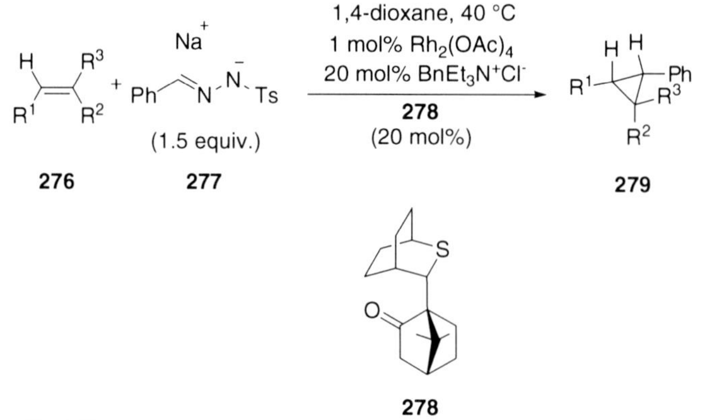

1,4-dioxane, 40 °C
1 mol% $Rh_2(OAc)_4$
20 mol% $BnEt_3N^+Cl^-$
278
(20 mol%)

(1.5 equiv.)

276 **277** **279**

278

Entry	R[1]	R[2]	R[3]	Yield (%)[a]	Ratio (*trans*:*cis*)[b]	*ee* (%)[c]
1	Ph	H	COPh	73	4:1	91
2	Me	H	COPh	50	4:1	90
3	H	*N*-succinimide	CO_2Et	55	1:7	91
4	H	$N(Boc)_2$	CO_2Me	72	1:6	92

[a]Combined yield of all isomers
[b]Ratio of *trans* : *cis* isomers was determined by ^1H NMR spectroscopy
[c]Determined using chiral HPLC

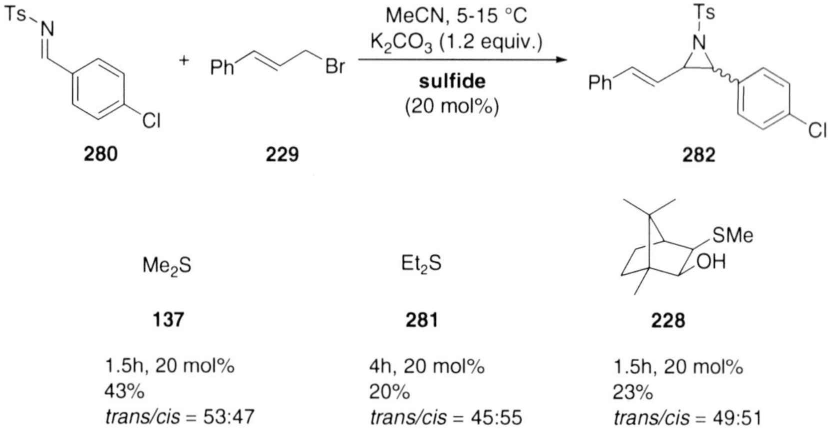

MeCN, 5-15 °C
K_2CO_3 (1.2 equiv.)
sulfide
(20 mol%)

280 **229** **282**

Me_2S **137**: 1.5h, 20 mol%; 43%; *trans/cis* = 53:47

Et_2S **281**: 4h, 20 mol%; 20%; *trans/cis* = 45:55

228: 1.5h, 20 mol%; 23%; *trans/cis* = 49:51

Scheme 7.45 Aziridination of imines catalyzed by chalcogenides

A change of the sulfide loading from stoichiometric to catalytic amounts (20 mol%) reduced yields and increased reaction times. The diastereoselectivity was very poor, enantioselectivities for the chiral organocatalyst **228** were not reported. Attempts to extend the reaction to other bromides were unsuccessful. Saito et al. utilized the same one pot protocol with another chiral catalyst (**289**) in the presence of an excess of aryl halides and base. He achieved moderate to excellent conversions. Use of dry acetonitrile prevented imine hydrolysis and enantioselectivities became good to excellent (Scheme 7.47) [172].

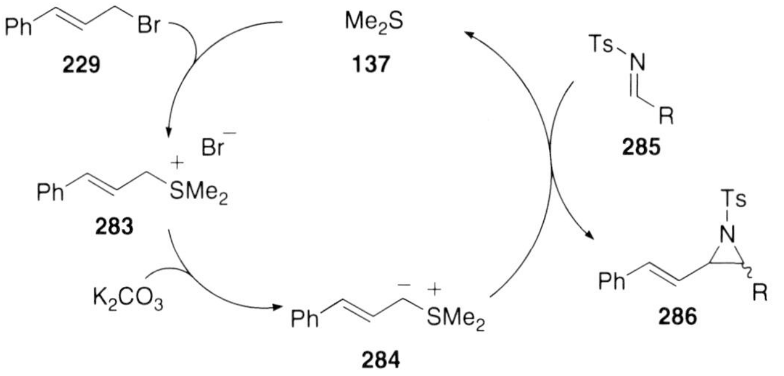

Scheme 7.46 Proposed mechanism for the aziridine formation catalyzed by sulfides

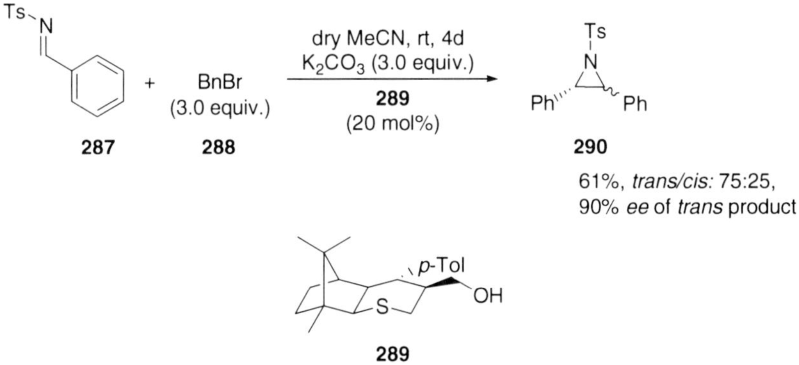

Scheme 7.47 Asymmetric aziridination catalyzed by sulfide **289**

Aziridine Formation with Chalcogen-Ylides from Carbene Sources

Aggarwal et al. [174] reported a novel organochalcogenid mediated aziridination where the sulfur ylides are generated by the reaction of metallocarbenes formed with diazo compounds (see section "Cyclopropane Formation via Chalcogen-Ylides Generated from Carbene Sources", Scheme 7.42). Furthermore, the diazo compounds can be generated *in situ* from tosylhydrazones (see section "Cyclopropane Formation via Chalcogen-Ylides Generated from Carbene Sources", Scheme 7.43) [175, 176]. In detailed studies Aggarwal et al. [175, 177] show that the process can be applied to electron deficient benzaldimines. With 20 mol% dimethyl sulfide (**137**), high yields of aziridines were obtained with a preference for *trans*–diastereoisomers (Table 7.22) [175].

Also *N,N*-diethyl diazoacetamide and diazoacetate could be used, but higher temperatures were required for the decomposition of the diazo compounds (60°C).

Table 7.22 Diastereoselective aziridination of imines catalyzed by dimethylsulfide (**137**)

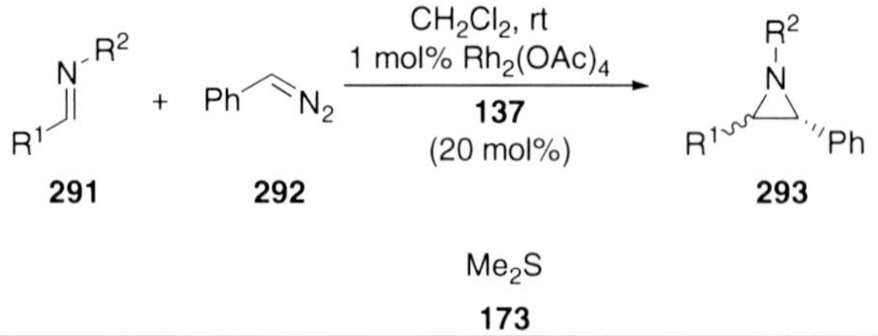

291 + 292 → (CH_2Cl_2, rt, 1 mol% $Rh_2(OAc)_4$, **137** (20 mol%)) → 293

Me_2S
173

Entry	R^1	R^2	Yield (%)[a]	Ratio (*trans:cis*)
1	Ph	Ts	91	4:1
2	Ph	DPP	83	3:1
3	Ph	SES	92	3:1
4	p-ClC_6H_4	SES	88	3:1
5	p-MeC_6H_4	SES	96	3:1

[a]The yield refers to the total yield of *trans* and *cis* isomers

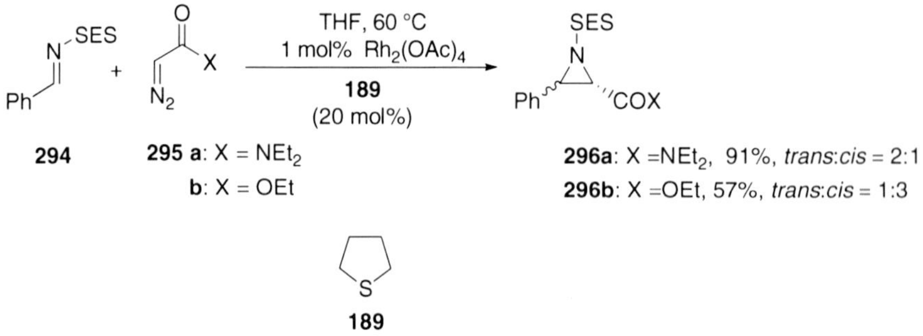

294 + **295 a**: X = NEt_2; **b**: X = OEt → (THF, 60 °C, 1 mol% $Rh_2(OAc)_4$, **189** (20 mol%)) →

296a: X =NEt_2, 91%, *trans:cis* = 2:1
296b: X =OEt, 57%, *trans:cis* = 1:3

189

Scheme 7.48 Aziridination of *N*-2-(trimethylsilyl)ethanesulfonyl-imine (*N*-SES imine) with diazoacetamides and diazoacetate catalyzed by sulfide **189**

In this case tetrahydrothiophene (**189**) in THF was employed instead of dimethylsulfide in CH_2Cl_2 (Scheme 7.48) [175].

In an asymmetric version, sulfide **271** shows the best results (Table 7.23). However, any reduction of the amount of catalyst is negatively reflected in the yield of the reaction [175, 177]. With ester diazoacetates the formation of *cis*-product is favoured (Scheme 7.49) [177].

With *N*-tosyl-hydrazone salts as diazo precursors the best results were achieved using sulfide **302** as catalyst, with loadings of 5 mol%. The formation of *trans*-aziridines is favoured, and the diastereoselectivity is moderate to good, the enantioselectivity excellent. The process can be applied to a broad range of electrophiles and diazoprecursors, examples are given in the Table 7.24 [170, 171].

Table 7.23 Enantioselective aziridination of *N*-SES imines

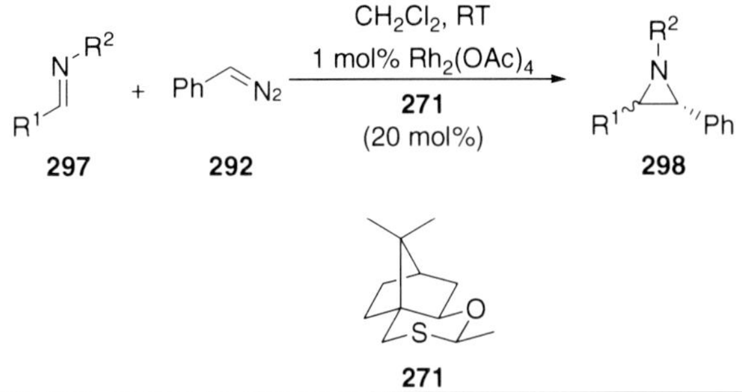

Entry	R^1	R^2	Yield (%)[a]	Ratio (*trans:cis*)	*ee*[a] (%)
1	Ph	SES	47	3:1	95
2	p-MeC_6H_4	SES	91	3:1	93
3	p-ClC_6H_4	SES	58	3:1	88
4	(E)-PhCHCH	SES	62	5:1	93

[a]*ee* of *trans* product

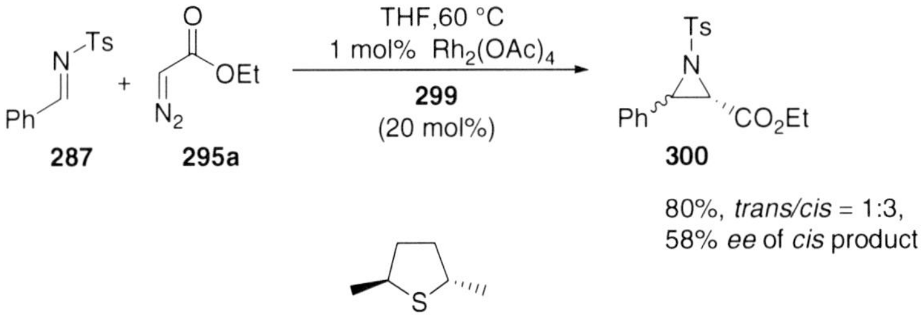

80%, *trans/cis* = 1:3, 58% *ee* of *cis* product

Scheme 7.49 Enantioselective aziridine formation from *N*-Ts imines with diazoacetate catalyzed by **299**

7.3.2.4 Epoxide Formation

Chiral epoxides can be prepared (a) by a stereoselective oxidation of prochiral alkenes (e.g. by Sharpless and Jacobsen epoxidation), or (b) by an enantioselective alkylidenation of prochiral oxo compounds with ylides, carbenes, or through the Darzens reaction.

Epoxide Formation with Chalcogen-Ylides by Alkylation/Deprotonation

Furukawa et al. introduced the catalytic ylide mediated epoxidation by using a reaction system that is composed of alkyl sulfide, alkyl halide and aldehyde in the

Table 7.24 Asymmetric aziridination of imines with hydrazone salts as carbene precursors catalyzed by **302**

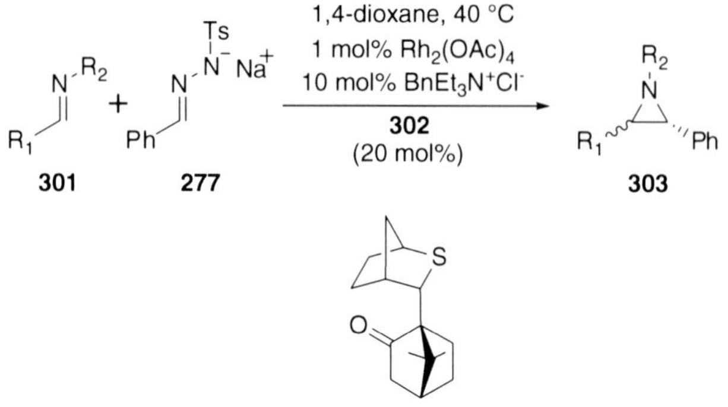

Entry	R^1	R^2	Yield (%)[a]	Ratio (*trans:cis*)	*ee*[a] (%)
1	Ph	SES	75	2.5:1	94
2[b]	Ph	SES	66	2.5:1	95
3	p-ClC$_6$H$_4$	SES	82	2:1	98 (81)
4	3-furfuryl	Ts	72	8:1	95
5	Ph	Cl$_3$COCO-	71	6:1	90

[a] *ee* of *trans* products (*ee* of *cis* products)
[b] 5 mol% of the sulfide used

presence of base under phase-transfer conditions. In this reaction, a sulfur ylide is generated from an alkylsulfonium salt that is deprotonated by potassium carbonate *in situ*. The ylide transfers an alkylidene group to the aldehyde under epoxide formation, thereby regenerating the sulfide (alkylation-deprotonation-substitution: Scheme 7.50) [178].

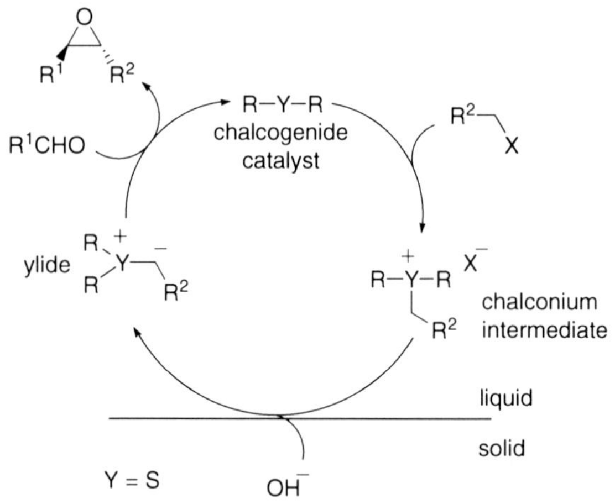

Scheme 7.50 Proposed catalytic cycle for epoxide formation with chalcogen-ylides *via* alkylation / deprotonation /substitution mechanism in a phase transfer system

This approach was extended to an enantioselective version using optically active sulfides **186**, **306** and **307** derived from (+)-camphorsulfonic acids. The best result is shown in Scheme 7.51 [179].

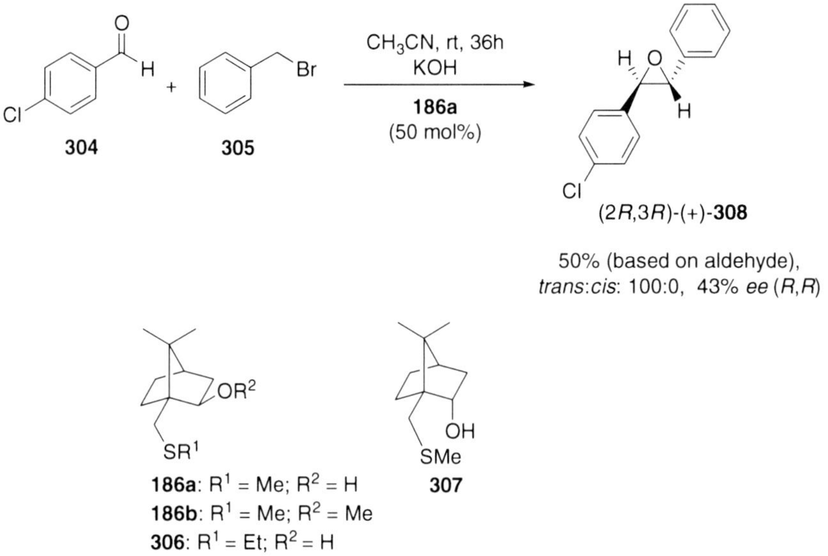

CH_3CN, rt, 36h
KOH
186a
(50 mol%)

304 **305**

(2*R*,3*R*)-(+)-**308**

50% (based on aldehyde), *trans*:*cis*: 100:0, 43% *ee* (*R*,*R*)

186a: R^1 = Me; R^2 = H
186b: R^1 = Me; R^2 = Me
306: R^1 = Et; R^2 = H
307

Scheme 7.51 Epoxidation of aldehydes employing (+)-camphorsulfonic acid derived chiral sulfides

Since Furukawa's work, new chalcogen catalysts with great structural variation were synthesised (Scheme 7.52) [180–202]. The best results were reported by Metzner [181–184, 194], Aggarwal [195] and Huang [196].

In 1990, Huang et al. reported the synthesis of vinyl epoxides from aldehydes catalyzed by diisobutyl telluride (**237**). The ditelluride works with a variety of different aldehydes. An example is shown below (Scheme 7.53) [198].

Metzner at al. used selenide **323** and telluride **325** as catalysts (Scheme 7.54) [199, 200]. In the case of the selenide, the enantioselectivity was high, but there was no diastereoselectivity. The telluride had higher diastereo- and enantioselectivities, but gave lower yields.

Most reports remark that the best reaction environment is a (sometimes biphasic) mixture of polar solvents, like *t*-BuOH/H_2O or CH_3CN/H_2O in proportions of 9:1, in presence of the bases like NaOH or KOH at room temperature [146]. These conditions reduce the possibilities of side reactions (e.g. Cannizzaro reaction, Williamson alkylation, reaction with solvent) [181]. Kinetic studies demonstrated that this reaction has two slow steps: the alkylation (formation of sulfonium salt from the catalytic sulfide) and the addition of the ylide to the carbonyl compound [184].

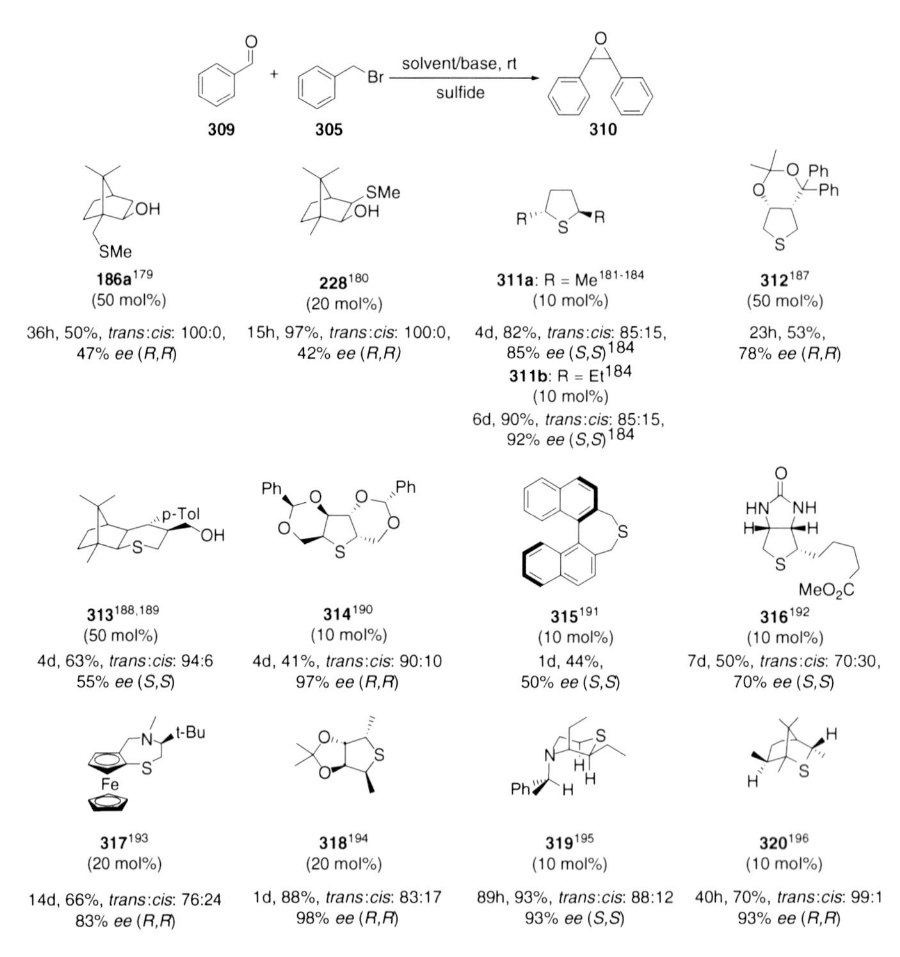

Scheme 7.52 Organosulfide catalysts and their best results obtained in the stilbene epoxide formation from benzaldehyde with benzyl bromide

PhCHO + Br (allyl bromide) → $THF/Et_2O/H_2O$, 50 °C, 19h, Cs_2CO_3, **237** (10 mol%)

309 **321**

322: 75%, *cis/trans*: 71:29

i-Bu–Te–*i*-Bu

237

Scheme 7.53 Synthesis of vinyl epoxides catalyzed by telluride **237**

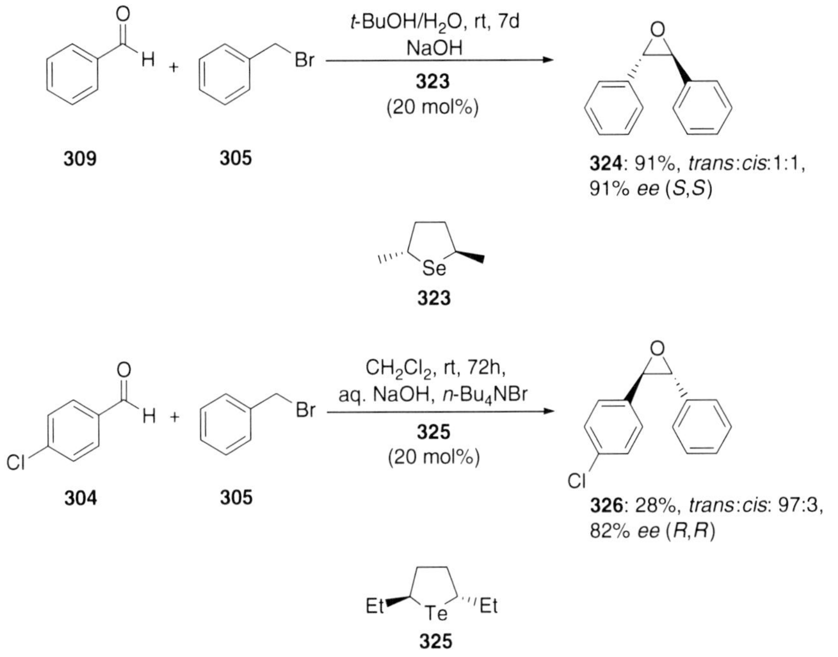

t-BuOH/H_2O, rt, 7d
NaOH
323
(20 mol%)

309 **305**

324: 91%, *trans*:*cis*:1:1,
91% *ee* (*S*,*S*)

Se
323

CH_2Cl_2, rt, 72h,
aq. NaOH, *n*-Bu_4NBr
325
(20 mol%)

304 **305**

326: 28%, *trans*:*cis*: 97:3,
82% *ee* (*R*,*R*)

Et Te Et
325

Scheme 7.54 Asymmetric epoxidation catalyzed by selenide **323** and telluride **325**

The formation of sulfonium salt is a slow and reversible reaction. The formation of the ylide is rapid and reversible [182]. The protic conditions keep the ylide at low concentration. The reversibility avoids ring cleavage of the cyclic catalyst and at the same time rapid ylide (re-)generation enables epoxidation [182]. Additives (e.g. Bu_4NI, NaI, Bu_4NHSO_4, catechol) have been used to activate the alkylhalide in order to increase the rate of the formation of sulfonium salt. It is supposed that the additives can also work as phase transfer catalysts in the reaction, supporting the extraction of the hydroxide anion into the organic phase and/or the deprotonation of the sulfonium salt [184, 194].

Because of the basic conditions of the reaction it is difficult to extend the use to enolizable aldehydes. Another limiting factor is the possibility of a [2,3]-sigmatropic shift if allyl halides are used (Scheme 7.55). This problem can be overcome with two strategies: (a) the design of a sulfide that favours α-deprotonation, and/or (b) introduction of a higher grade of hindrance in the α'-positions, so that the alternative deprotonation in this position has a kinetic disadvantage [201].

Only low diastereoselectivities were observed in the product when vinyl oxiranes are prepared from non-β-substituted allyl halides (Scheme 7.56a) [202]. Improvements of the diastereo- and enantioselectivities can be achieved when β-substituted allyl halides are used in combination with chiral sulfides (Scheme 7.56b) [201].

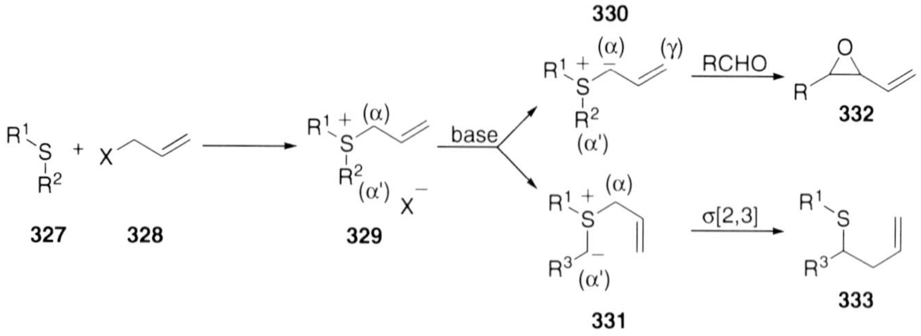

Scheme 7.55 Possible alternative reaction for epoxide formation with allyl and related halides by [2,3]-sigmatropic shift

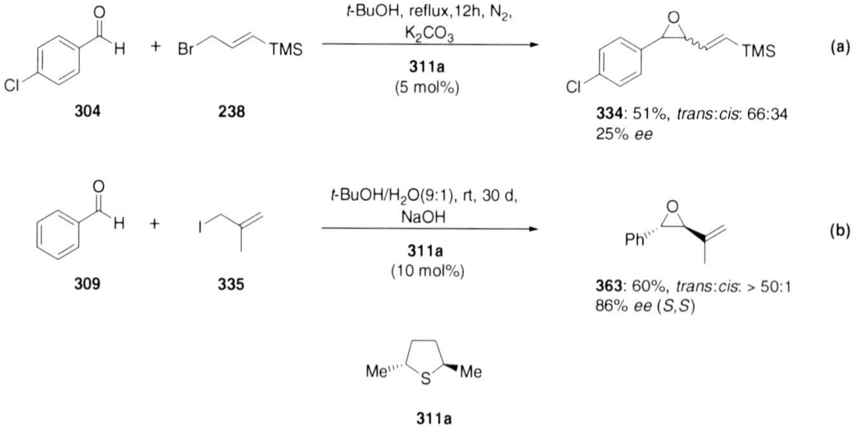

Scheme 7.56 Synthesis of vinyl oxiranes catalyzed by chiral sulfide **311a** with (**a**) non-β-substituted vinyl halide; (**b**) β-substituted allyl halides used

Some authors report recovery of the catalyst ranging between 70% to quantitative [188, 190]. A sulfur-functionalized chiral ionic liquid derived from (+)-isoborneol (**337**) was prepared by Huang's group (Scheme 7.57) [203]. The epoxidation was conducted in pure water without any organic solvent. The catalyst could be reused five times without any remarkable decrease in yields and enantioselectivities.

Epoxide Formation with Chalcogen-Ylides from Carbene Sources

In 1994 Aggarwal et al. were the first who reported a catalytic cycle in which sulfur ylides were generated in the presence of carbonyl compounds by metal-catalyzed decomposition of diazo compounds [204]. The proposed mechanism is shown in Scheme 7.58.

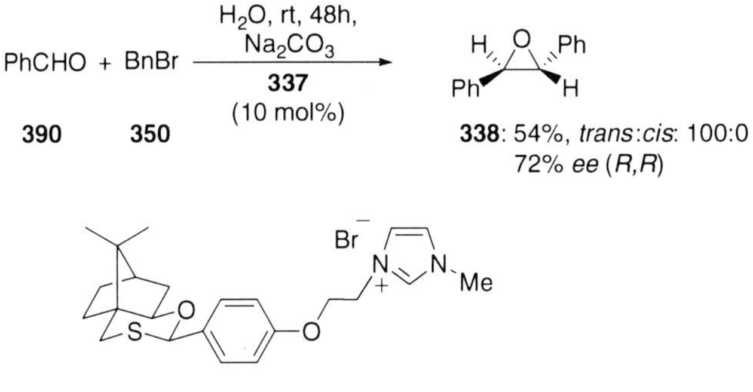

Scheme 7.57 Epoxidation of aldehydes catalyzed with sulfur-functionalized chiral ionic liquid

Scheme 7.58 Catalytic cycle proposed for the epoxide formation catalyzed by chalcogen-ylides with diazo-derived carbene as ylide source

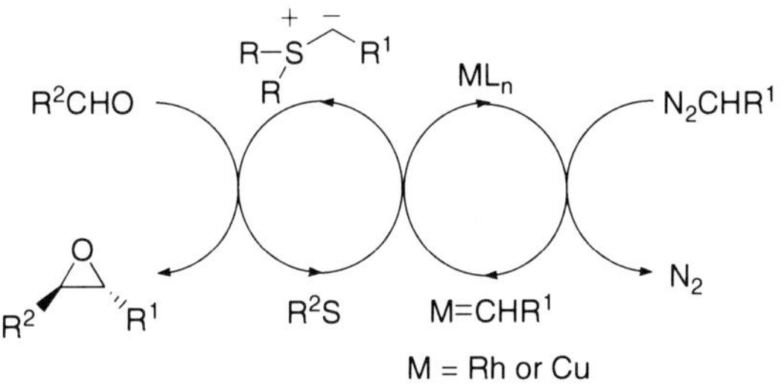

Two main obstacles were well-known: (a) diazo compounds can react directly with carbonyl compounds generating homologated products (Arndt-Eistert reaction, Scheme 7.59a), (b) diazo compounds can dimerize in the presence of metal salts and form alkenes (Scheme 7.59b) [204]. Other reactions involve carbene insertions or acid/water addition.

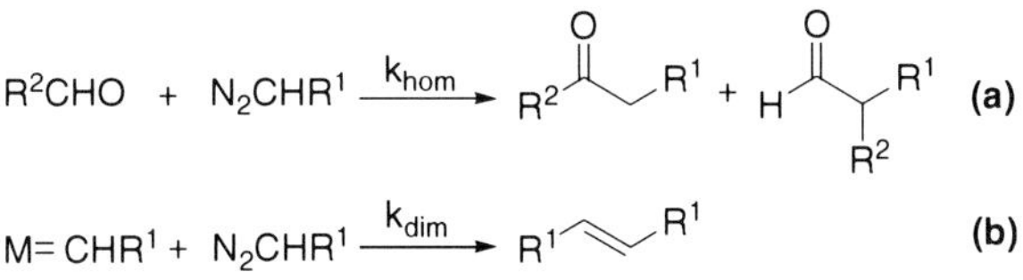

Scheme 7.59 Most relevant side reactions of diazo compounds as carbene precursors. (**a**) Arndt-Eistert reaction; (**b**) carbene dimerization

The problems with side reactions were avoided when the diazo compound was maintained at low concentration. This could be reached by its slow addition to the reaction mixture [204]. Aggarwal's method has the disadvantage of additional metal catalysis, but possesses two important advantages compared to Furukawa's: the formation of the ylide occurs in only one step and the reaction is conducted under neutral conditions. This extends the applicability from aromatic to aliphatic aldehydes as well as to base sensitive ones [205–207]. Moreover achiral and chiral sulfide catalysts could be used in substoichiometric quantities. A crucial point is the nature of the catalyst, because the metallocarbene can follow alternate paths (see above). So the sulfide needs to react faster [204]. Therefore the reaction parameters (e.g. solvent effects, variation in the nature and amount of catalyst, scope of substrates, effect of the metal catalyst, diastereo and enantioselectivities) have been studied in detail [208, 209].

The reaction works well in most solvents, but diastereoselectivity control is better in polar solvents due a higher reversibility of betaine formation. The best chiral sulfide organocatalysts are those with the sulfur moiety in a cyclic structure. Compound **271** produces the best enantio- and diastereoselectivities with metal complex $Cu(acac)_2$ (Scheme 7.60). It enables the use of substoichiometric amounts of chiral sulfide keeping yields acceptable with excellent diastereo and enantioselectivities. The reaction is very sensitive to the concentration of the sulfide [206].

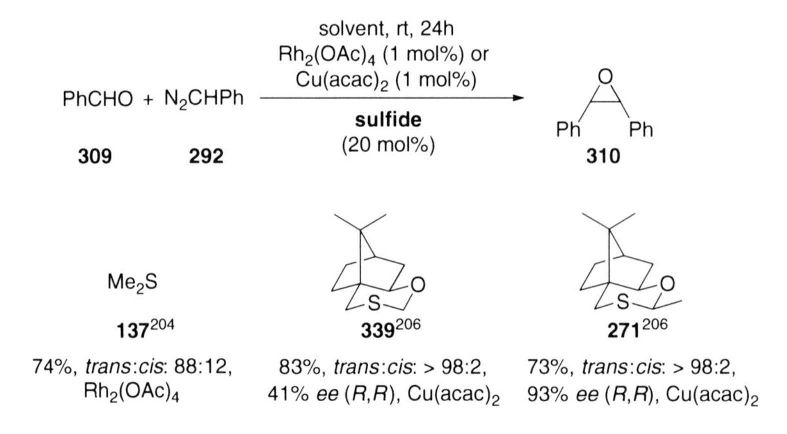

Scheme 7.60 Epoxide formation catalyzed by **137, 339** and **271**

An improvement of Aggarwal's method is the use of tosylhydrazone salts as substitute for diazo compounds. The tosylhydrazone salts are generated *in situ*, and decompose to the diazo compound in the presence of base and a phase transfer catalyst (Scheme 7.61). Aromatic, heteroaromatic, and unsaturated aldehydes furnish the corresponding epoxides in high yields and high *trans* selectivities. Aliphatic aldehydes give lower yields and diastereoselectivities [171, 210–212].

Detailed studies were carried out and the best combination of yield, diastereo- and enantioselection were achieved when sodium tosylhydrazone salts, 1 mol% $Rh_2(OAc)_4$, 5–10 mol% $BnEt_3N^+Cl^-$, and sulfide **302** (Fig. 7.14) were used in toluene, acetonitrile, 1,4-dioxane or trifluorotoluene at temperatures between 30°C and 40°C [211, 212].

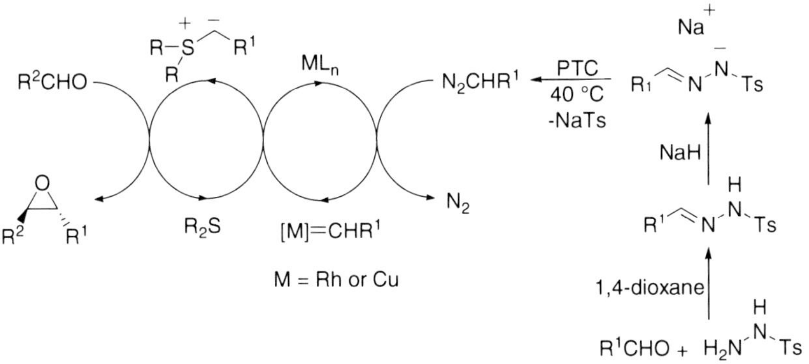

R^2CHO — $R{-}S^{+}(R){-}CH^{-}R^1$ — ML_n — N_2CHR^1 ← PTC, 40 °C, -NaTs — $R^1CH{=}N{-}N^{-}Ts\ Na^+$

R_2S — [M]=CHR[1] — N_2

M = Rh or Cu

NaH; 1,4-dioxane; $R^1CHO + H_2N{-}NH{-}Ts$

Scheme 7.61 Catalytic cycle proposed for the epoxide formation catalyzed by chalcogen-ylides with tosylhydrazone salts as ylide source

Fig. 7.14 The best sulfide organocatalyst for the epoxide formation in Aggarwals systems

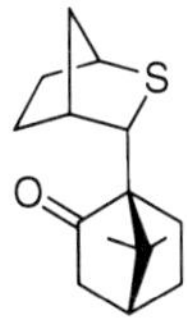

302

Origin of Diastereo- and Enantioselectivities in Chalcogen Ylid Additions to Polar Double Bonds

The diastereoselectivity of the ylide mediated oxirane formation is connected to the degree of reversibility of the betaine intermediate formed by the reaction of the chalcogen ylide with the carbonyl compound (Scheme 7.62). When the

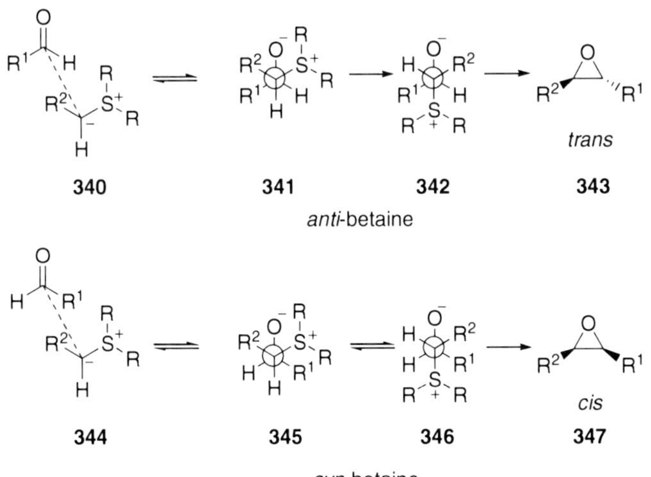

340 ⇌ 341 → 342 → 343 (trans)

anti-betaine

344 ⇌ 345 ⇌ 346 → 347 (cis)

syn-betaine

Scheme 7.62 *Syn* and *anti-* betaine intermediates

formation of both *trans* and *syn* betaine intermediates is irreversible, a low grade of diastereoselection is observed. If one of the betaines is formed reversibly and can return to the starting material while the other one can not (easily), a high diastereoselectivity results, as is the case with stabilized ylides (e.g. aryl stabilized ylides).

Factors that affect the *syn*-betaine reversibility are: (a) thermodynamic stability of starting materials; (b) steric hindrance of the ylide and/or the aldehyde; (c) effect of the solvents and if applicable metal ions. With chiral sulfides (R = non racemic) additionally enantioselection is possible [212].

7.3.3 Oxidation

7.3.3.1 The Catalytic Epoxidation of Electron Rich Olefins

Optically active epoxides are found in many natural products [213] and are highly versatile intermediates and building blocks. Asymmetric epoxidations of double bonds have long been employed using metal-mediated methods such as the Sharpless [214] and Jacobsen-Katsuki [215, 216] epoxidation. Metal-free asymmetric epoxidations are mostly mediated by chiral dioxiranes and oxaziridines. Dioxiranes or their respective precursor ketones represent some of the oldest and most versatile organocatalysts for the asymmetric epoxidation of olefins. They are particularly useful for unfunctionalized *trans*-, disubstituted, trisubstituted and terminal olefins [217–224].

Dioxiranes can be generated *in situ* from chiral ketones and oxone ($KHSO_5$) [225–238]. Besides oxone, a combination of H_2O_2 and organic nitrile can be used as oxidant [239]. In principle, if the dioxirane is formed *in situ*, the ketone can be used as catalyst. According to the commonly accepted mechanism, the oxidant (oxone or H_2O_2/RCN) initially adds to the ketone which generates a peroxide intermediate. This undergoes a base-catalyzed intramolecular ring-closing reaction to generate the dioxirane. The dioxirane transfers an oxygen atom to the olefin to form the corresponding epoxide, and to regenerate the ketone (Scheme 7.63). Vidal-Ferran et al. could confirm the mechanism with isotope labelling studies [239].

The first asymmetric epoxidation using chiral ketones **355** and **356** was reported by Curci et al. in 1984 (Fig. 7.15) [240, 241].

Only low enantioselectivities (12%) were obtained for the epoxidation of *trans*-stilbene or *trans*-β-methylstyrene despite high catalyst loadings (20–300 mol%). During the past few years, new and better chiral ketones have been reported. In order to better describe the development of ketone catalysts for the asymmetric epoxidation, these will be divided into groups of different structural features.

C_2-Symmetric Binaphthyl-Based and Related Ketone Catalysts

In 1996, Yang et al. for the first time achieved preparatively useful enantioselectivities (e.g. 85% *ee* in the epoxidation of 4,4′-diphenyl-*trans*-stilbene) with

Scheme 7.63 Proposed mechanism for the epoxidation of eletron rich olefines catalyzed by ketones and stoichiometric oxidative agent

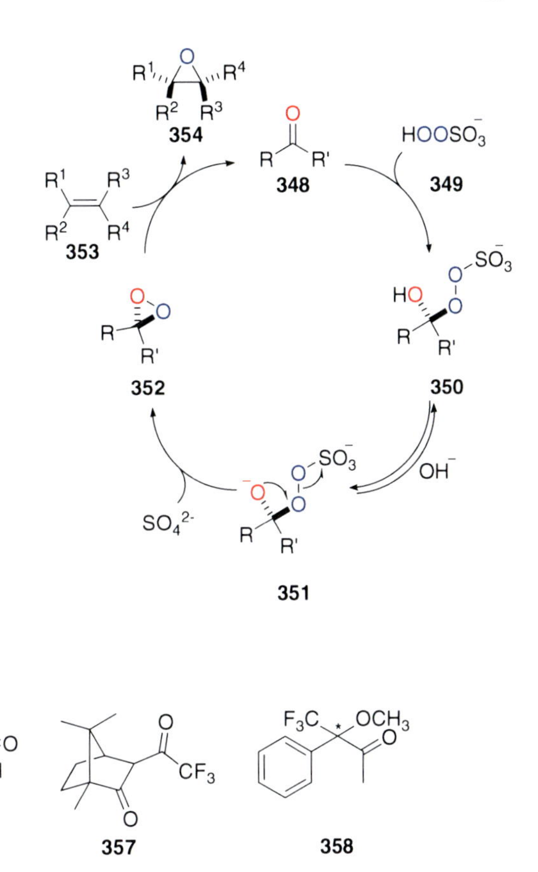

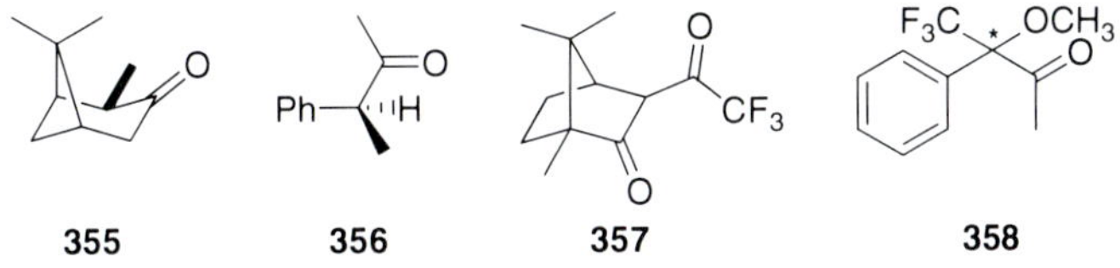

Fig. 7.15 First chiral ketone catalysts reported

binaphthylene-derived ketones **359** at 5 mol% catalyst loading [242–244]. Investigations of the influence of substituent X on the binaphtylene unit identified the ketone **359d** as highly reactive and selective catalyst (Fig. 7.16).

In 1997, Song et al. reported the ether-linked C_2-symmetric ketones **360** and **361** which showed lower reactivity and enantioselectivities (Table 7.25) than ketone **359** [245, 246].

Denmark and co-workers used 7-membered C_2-symmetric carbocyclic biaryl ketones **362** for the asymmetric epoxidation of *trans*-olefins [247]. Ketone **362c** was found to be more active than ketones **359** (Table 7.25). At the same time the structurally related fluorinated binaphthyl ketones **363** of similar catalytic profile were reported by Behar et al. in 2002 [248].

In 2001, Tomioka et al. investigated several seven-membered cycloalkanones **364** and **365** bearing 1,2-diphenylethane-1,2-diamine and cyclohexane-1,2-diamine backbones [249, 250]. However, only 30% *ee* could be achieved for the epoxidation of *trans*-stilbene with ketone **364b**, while the tricyclic ketone **366** and bicyclic ketone **367** gave 64% *ee* and 57% *ee*, repectively [251].

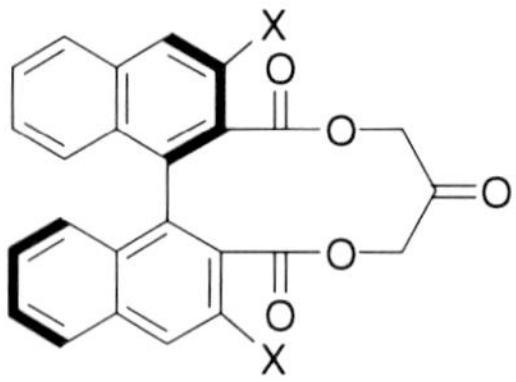

359a: X = H
359b: X = Cl
359c: X = Br
359d: X = 1,3-dioxan-2-yl

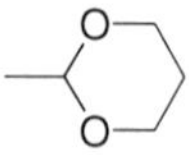

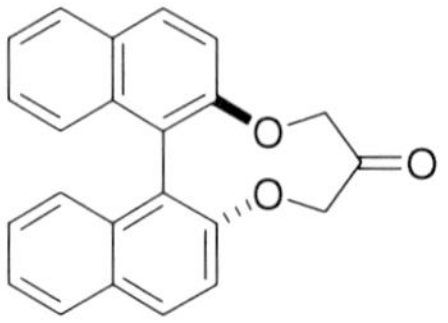

360 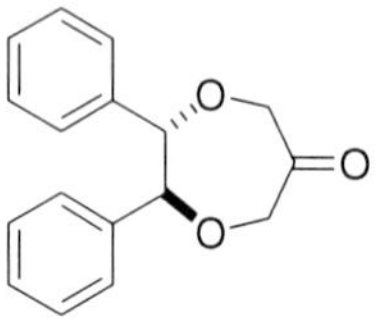 **361**

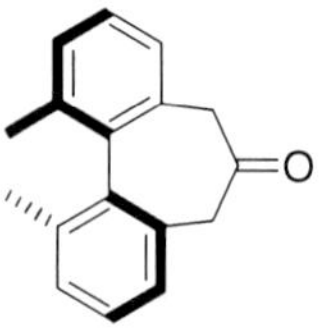

362a 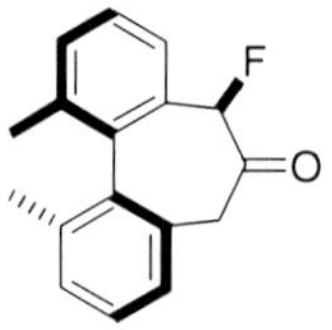 **362b** 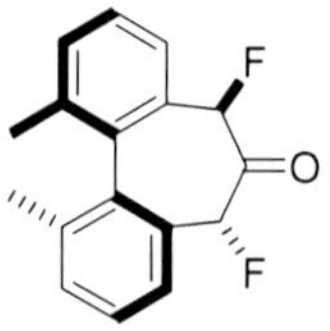 **362c** 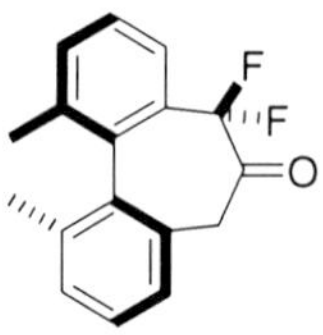 **362d**

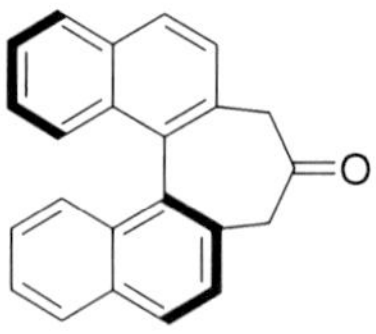

363a 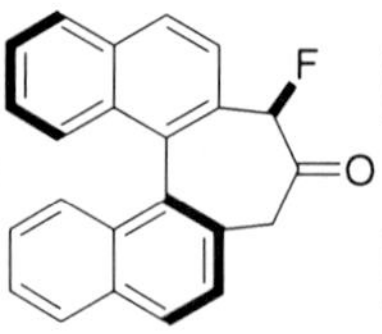 **363b** 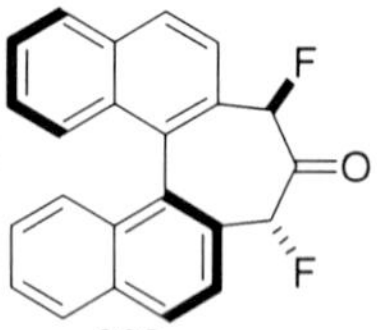 **363c**

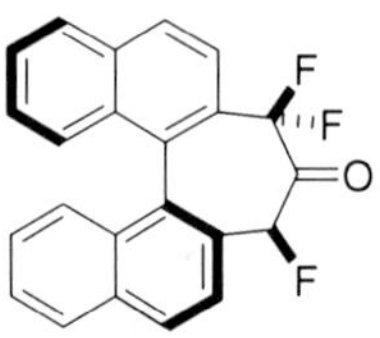

363d 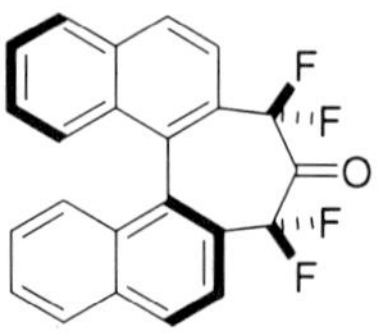 **363e**

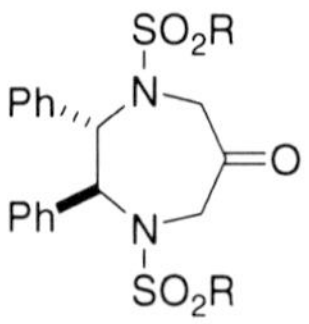

364a: R = CF_3
364b: R = Ph
364c: R = C_6H_5

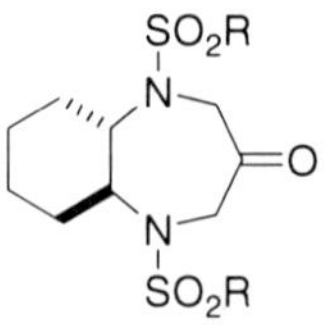

365a: R = CF_3
365b: R = Ph

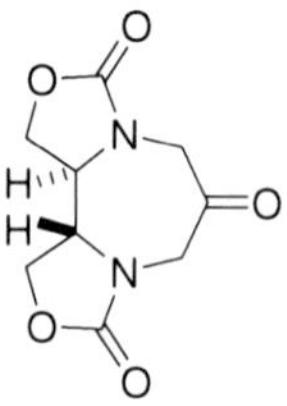

366 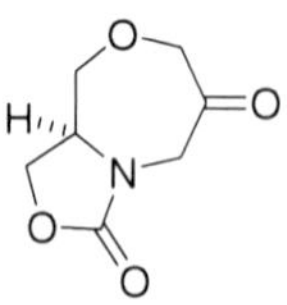 **367**

Fig. 7.16 C_2-symmetric ketone catalysts for asymmetric epoxidations

Table 7.25 Epoxidation of trans olefins catalyzed by chiral ketones

rt or 0 °C, $NaHCO_3$ (15.5 equiv.), Oxone (5 equiv.), **Ketone** (10-100 mol%)

Ph–CH=CH–Ph **368** → **369**

Entry	Catalyst	Mol%	Yield (%)	*ee* (%)
1	**359d**	10	> 90	84 (*S*,*S*)
2	**360**	100	79	26 (*S*,*S*)
3	**361**	100	72	59 (*S*,*S*)
4	**362c**	100	46	94 (*R*,*R*)

Table 7.26 Bicyclo[3.2.1]octan-3-one catalysts and selected results

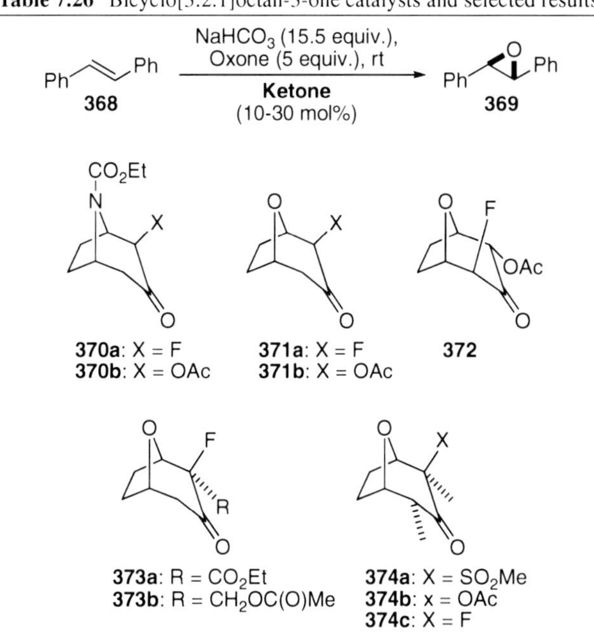

Entry	Catalyst	Mol%	Conv. (%)	*ee* (%)
1	**370a**	10	100	76 (*R*,*R*)
2	**371b**	20	85	93 (*R*,*R*)
3	**372**	10	100	64 (*S*,*S*)
4	**373a**	10	92	77 (*S*,*S*)
5	**374c**	30	67	68

Bicyclo[3.2.1]octan-3-ones and Related Ketone Catalysts

In 1998, Armstrong et al. reported fluoro ketone catalysts **370** based on the tropinone skeleton [252, 253]. Ketone **370a** epoxidized a variety of olefins quickly and in good yield with up to 83% *ee* for phenylstilbene. Other bicyclo[3.2.1]octan-3-ones **371–373** bearing electronegative substituents at the α-position are equally suitable (Table 7.26) [254].

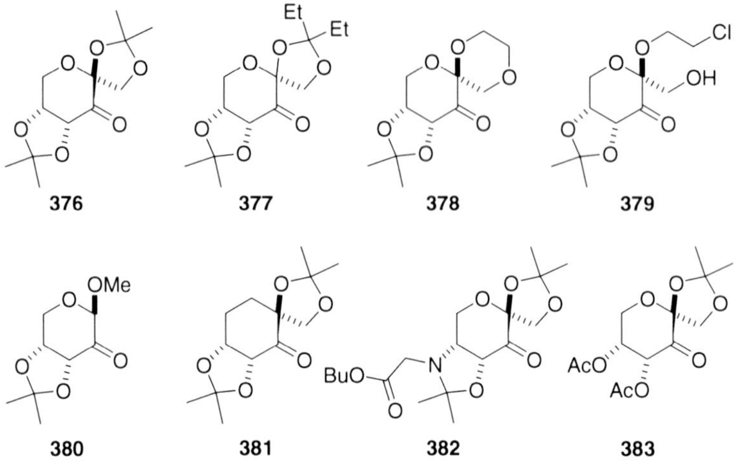

376 377 378 379

380 381 382 383

Fig. 7.17 Carbohydrate-based ketone catalysts

Klein et al. also described the epoxidation of stilbene with new 2-substituted 2,4-dimethyl-8-oxabicyclo[3.2.1]octan-3-ones **374** with similar *ee* as with catalysts **373** (Table 7.26) [255].

Carbohydrate-Based Ketone Catalysts

In 1996, Shi et al. reported fructose-derived ketone **376** as a highly reactive and enantioselective (<90% *ee*) catalyst for the epoxidation of *trans*-olefins, trisubstituted olefins, fluoroolefins, conjugated enynes, and silyl enol ethers with 20–30 mol% catalyst loading at an elevated pH of 10 [256]. This ketone can be prepared from

Scheme 7.64 Chiral fructose-derived ketone catalyst **376**

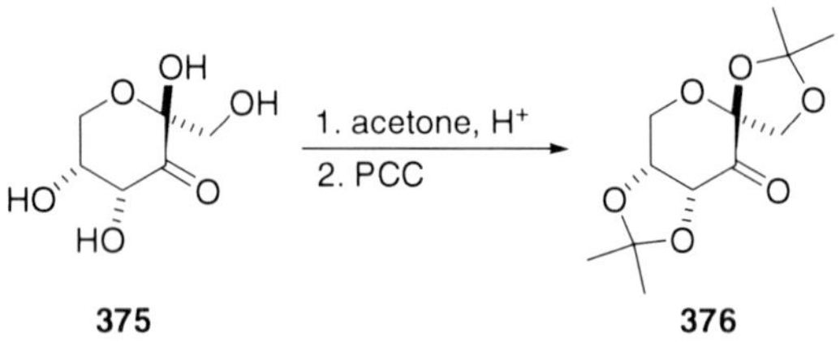

375 → 1. acetone, H+ 2. PCC → 376

inexpensive and readily available D-fructose in two simple steps (Scheme 7.64) [257–259]. Because of the important influence of the pH on the epoxidation success with ketone **376** (pH > 10 leads to autodecomposition of oxone, pH ≤ 8 to decomposition of **376** by Baeyer-Villiger oxidation) a high catalyst loading is required for a good conversion of olefin substrates [233, 237, 238].

To avoid such complications, the Shi group synthesized a variety of ketone catalysts **377–383** (Fig. 7.17). Only catalyst **382** shows comparable epoxidation results at a lower catalyst loading of 5 mol% [260–265].

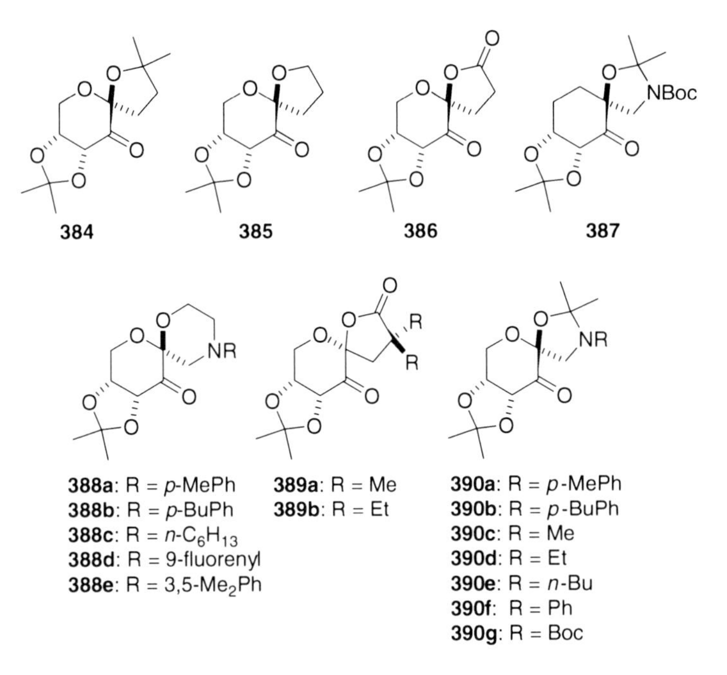

Fig. 7.18 Variations in the α-spiro moiety of the pyranone catalysts

Table 7.27 Asymmetric epoxidation of olefines catalyzed by oxazolidine derivatized ketone **390c**

$R^1R^2C{=}CR^3R^4$ → (-10 °C or 0 °C, K_2CO_3 (6.7-13.4 equiv.), Oxone (1.6-3.2 equiv.), **390c** (10-30 mol%)) → epoxide (R^1, R^2, R^3, R^4)

Entry	Substrate	Yield (%) product	*ee* (%)
1	Ph	99	84
2	Ph	93	90
3	Ph	94	84
4		54	80
5	Ph	66	85

Many other variations of the basic structure **376** have been explored (Fig. 7.18), of which the oxazolidinone-containing ketone catalysts **390** were found to be most effective (Table 7.27) [266–276].

In 2002, Shing et al. reported glucose-derived ketones **391** and **392**. Ketone **391** epoxidizes *trans*-stilbene with up to 71% *ee* [277]. A series of L-arabinose-derived ketones **393–399** followed, and up to 90% *ee* was obtained for *trans*-stilbene epoxide with ketone **396** [278]. In the same year, Zhao et al. reported three fructose-derived ketones and aldehydes **400–402** for the asymmetric epoxidation [279]. Aldehyde **402** achieved 94% *ee* for *trans*-stilbene. In 2009, Davis et al. presented a variety of conformationally restricted ketones **403**, prepared from *N*-acetyl-D-glucosamine which show useful selectivities with terminal olefins (styrene: 81% *ee*, Fig. 7.19) [280].

Other Chiral Ketone Catalysts

In 1997 Shi used a series of pseudo-C_2-symmetric ketones **404** bearing two fused rings at each side of the carbonyl group [281, 282]. Ketones **404** are characterized by good enantioselectivities (96% *ee* in the epoxidation of phenylstilbene), even for the epoxidation of electron deficient olefins. Overall, ketones **404** are less enantioselective than **376** for the epoxidation of *trans*-, disubstituted, and trisubstituted olefins. Also, the sterically relaxed ketones **405** and **406** with a CH_2-group next to the keto-group show lower enantioselectivities and reactivities for the epoxidation. Zhao et al. reported 85% *ee* for stilbene with catalyst **406** (Fig. 7.20) [283].

In 1998, Yang et al. described a series of cyclohexanones **407**, in which *ee*'s for the epoxidation vary depending on the *meta*- or *para*-phenylsubstituent of the *trans*-stilbene substrates (Fig. 7.20) [284].

In 2000, Solladié-Cavallo synthesized fluorinated ketones **408** from (+)-dihydrocarvone and investigated them in the asymmetric epoxidation of different *trans*-stilbenes and silyl enol ethers (Fig. 7.20) [285–287]. Later she reported rigid *trans*-decalones **409** which gave up to 70% *ee* (**409a**) and 20% *ee* (**409b**) in the epoxidation of *trans*-β-methylstyrene (Fig. 7.20) [288].

In 2001, Bortolini et al. described a series of keto bile acids **410** as dioxirane precursors for the asymmetric epoxidation [289, 290]. *Trans*-stilbene can be epoxidized in up to 98% *ee* (Fig. 7.21).

The readily available chiral oxazolidinone trifluoromethyl ketone **411** was reported by Armstrong et al. in 1999. Unfortunately, only to 34% *ee* was achieved for 1-phenylcyclohexene (Fig. 7.22) [291].

Low enantioselectivities were also observed for β-cyclodextrine modified ketoesters [292, 293].

7.3.3.2 Epoxidation Catalyzed by Selenides and Selenoxides

Knochel et al. reported the use of electron deficient butylselenide (**413**) as catalyst in the epoxidation of diverse alkenes [294]. The catalyst was selectively soluble in perfluorinated solvents and could easily be recovered by phase separation for reuse without a decrease in yields. An example is shown in Scheme 7.65.

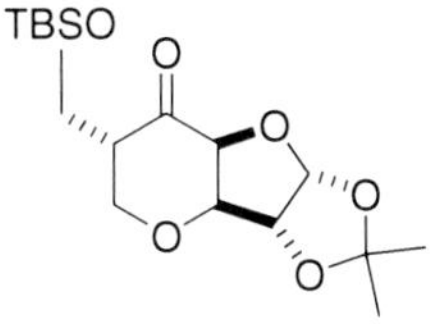

391 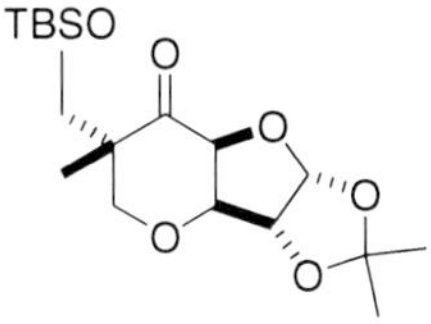 **392a** 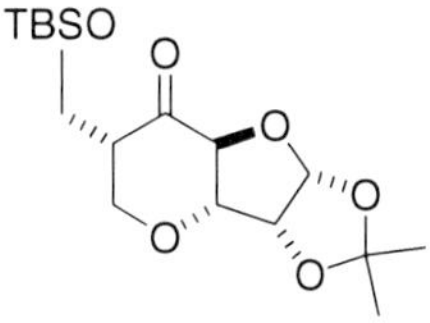 **392b**

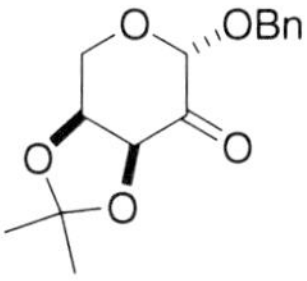

393 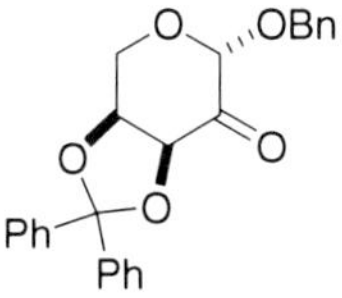 **394** 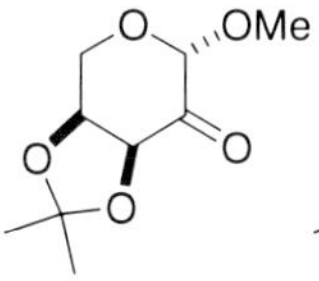 **395** 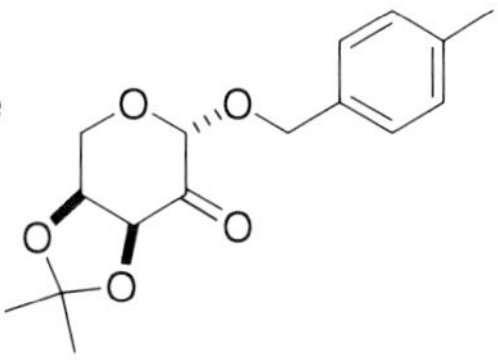 **396**

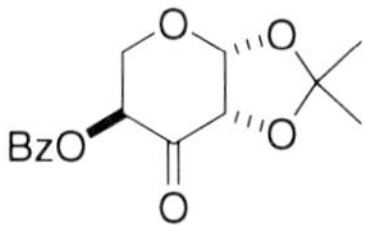

397 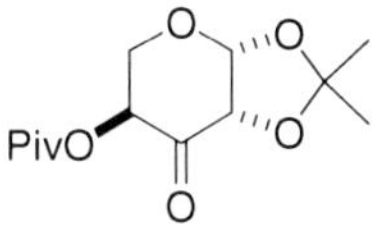 **398** 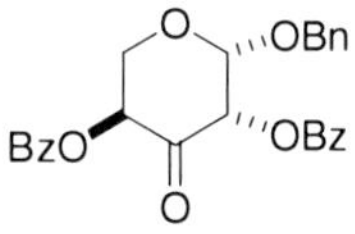 **399**

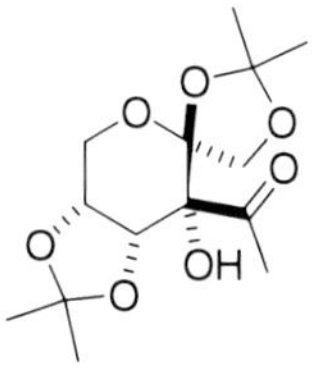

400 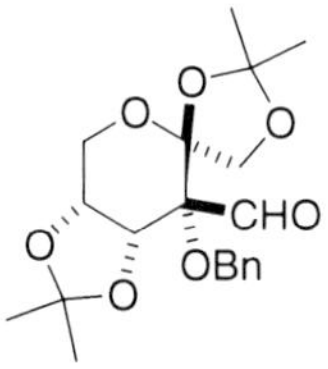 **401** 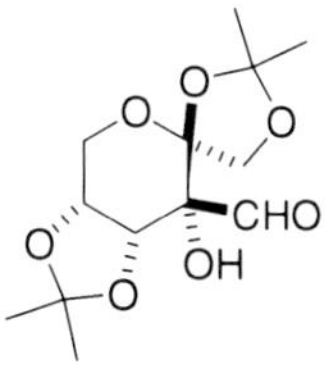 **402**

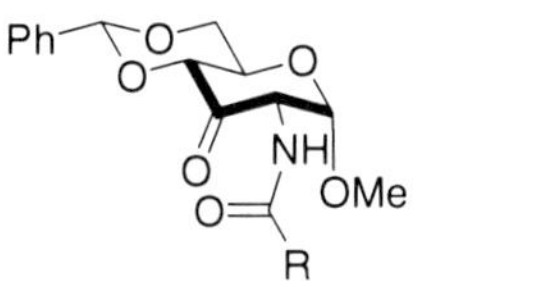

403a: R = Me
403b: R = Ph
403c: R = p-$CF_3C_6H_4$
403d: R = p-ClC_6H_4
403e: R = p-$MeOC_6H_4$
403f: R = p-$Me_2NC_6H_4$

Fig. 7.19 Further sugar derived ketone catalysts

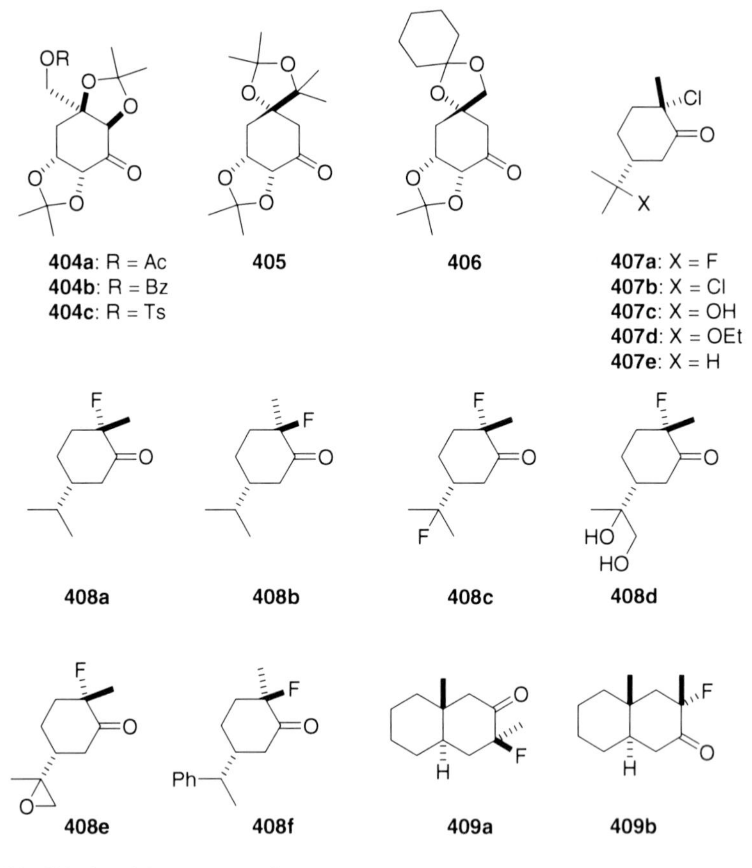

Fig. 7.20 Chiral cyclohexanone catalysts

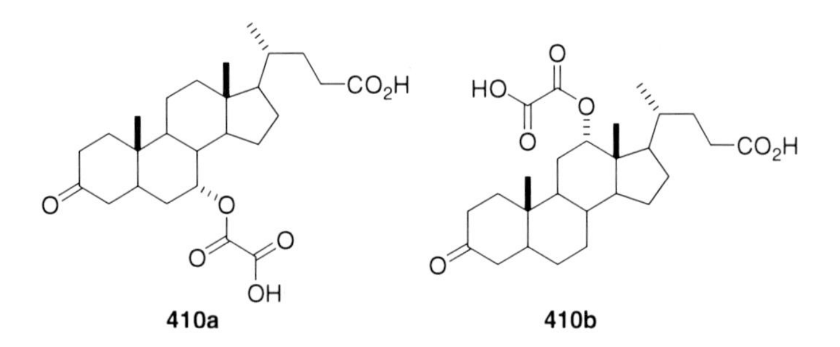

Fig. 7.21 Keto bile acid oxalates as catalysts

Fig. 7.22 Oxazolidinone ketone

O
O N CF3
O
Ph
411

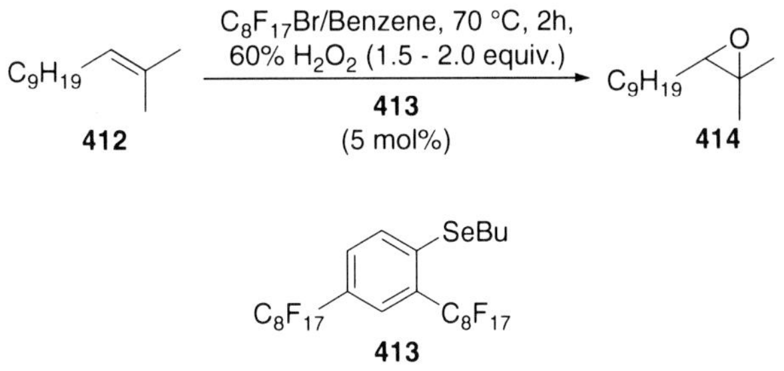

Scheme 7.65 Epoxidation of alkenes catalyzed by selenide **413**

Detty et al. reported the use of selenoxide **416** as catalyst for the epoxidation of alkenes (Scheme 7.66) [295].

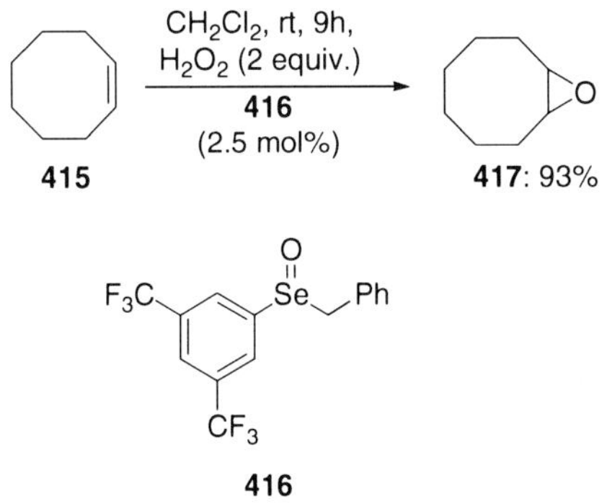

Scheme 7.66 Epoxidation of alkenes catalyzed by selenoxide **416**

7.3.3.3 The Catalytic Epoxidation of Electron Poor Alkenes

The asymmetric epoxidation of electron deficient alkenes like α,β-unsaturated esters, ketones and nitriles often is not efficient with the reagents suitable for electron rich systems. Prominent examples for the successful epoxidation of α,β-enones are the well-known Weitz-Scheffer epoxidation using alkaline hydrogen peroxide or hydroperoxides in the presence of base [296, 297].

α,β-Unsaturated acid derivatives can also be oxidized by chiral dioxiranes or by α,α-diarylprolinol and primary β-(alkyl)amino alcohols. The catalytic mechanism of dioxirane oxidations was described before. The epoxidation of enones with α,α-diarylprolinol and primary β-(alkyl)amino alcohol catalysts uses *tert*-butylhydroperoxid (TBHP) as an oxidizing agent. The usual mechanism was supplemented by the involvement of the amino alcohol, proposed by Lattanzi et al. (Scheme 7.67) [298, 299].

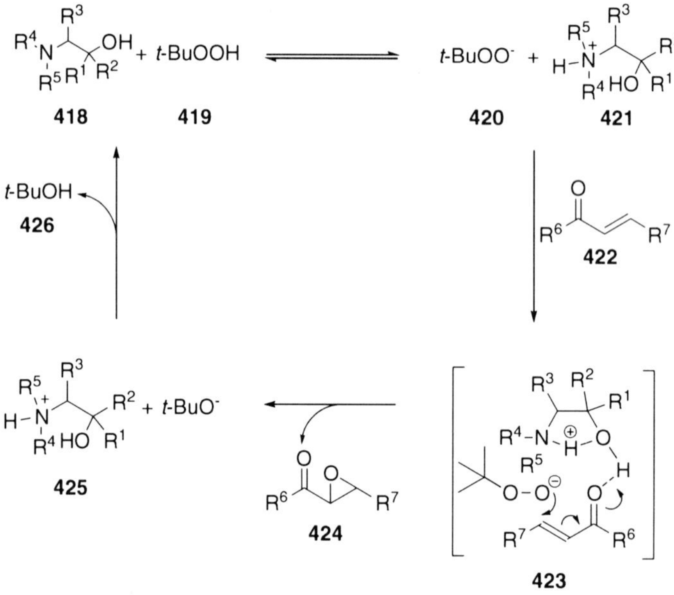

Scheme 7.67 Proposed mechanism for the epoxidation of enones catalyzed by amino alcohols

In the putative catalytic cycle, the catalyst deprotonates TBHP to form the active catalytic species, the alkyl-ammonium / *tert*-butyl peroxy ion pair. The hydroxyl group acid-activates the enone by forming a hydrogen bond with the carbonyl group while the peroxide adds in Michael-fashion. The intermediate enolate attacks the O-O bond intramolecularly and forms the epoxide and the *tert*-butoxide anion. Finally, the *tert*-butoxide acts as base to regenerate the catalyst.

Epoxidation of Enones Mediated by Chiral Dioxiranes

In 1998, Armstrong et al. reported α-fluorinated ketone **370a** which provided 64% *ee* for the epoxidation of methyl cinnamate [252]. Better enantioselectivies are achieved with from (−)-quinic acid derived ketones **404a** and **427** for the epoxidation of ethyl cinnamate (86% **404a**, 89% *ee* **427**) as reported by Wang et al. in 1999 (Fig. 7.23) [282].

In 2000, Solladié-Cavallo et al. reported the asymmetric epoxidation of *p*-methylcinnamate with fluorocyclohexanones **408** [286]. Higher conversion and *ee*' s were achieved with **408a** than with **408b**. Presumably, the axial fluorine is a more efficient electron withdrawing substituent than equatorial fluorine (Table 7.28).

In 2001, Bortolini used keto bile acids **431** which epoxidized *p*-methylcinnamic acid in good yields (45–94%) and high *ee*'s (26–95%) (Fig. 7.24) [300, 301].

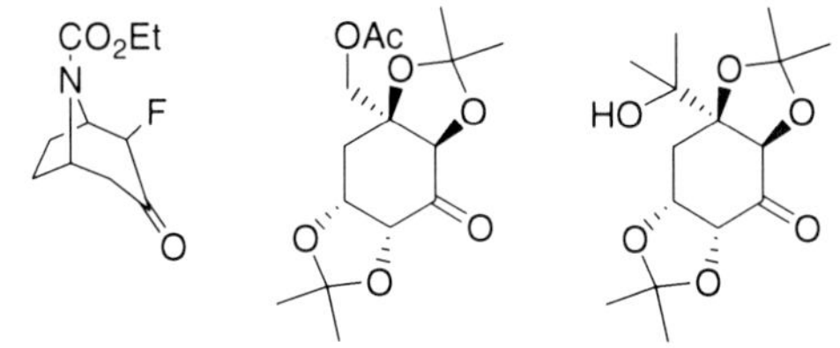

Fig. 7.23 (–)-Quinic acid derived ketones and α-fluorinated ketone catalysts

370a **404a** **427**

Table 7.28 Fluorocyclohexanones **408a–b** in the de asymmetric epoxidation of methylcinnamate

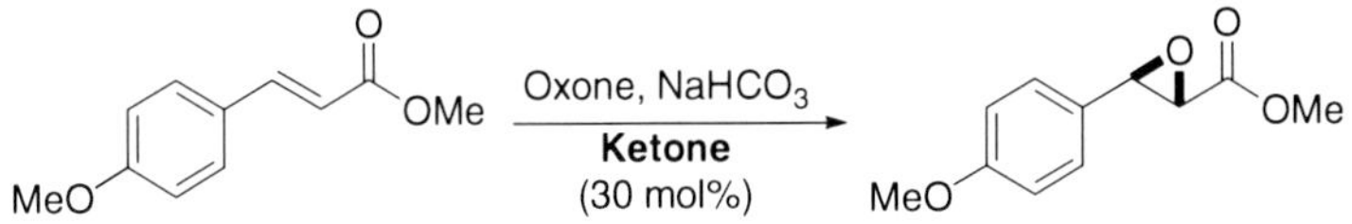

428 → Oxone, $NaHCO_3$, **Ketone** (30 mol%) → **429**

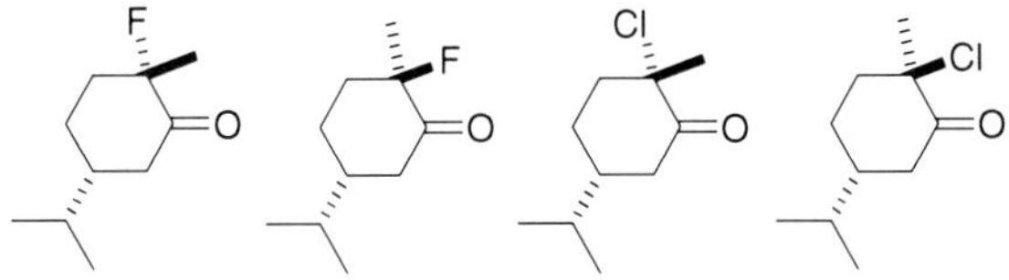

408a **408b** **430a** **430b**

Entry	Catalyst	Conv. (%)	*ee* (%)
1	**408a**	99	40 (2*R*,3*S*)
2	**408b**	43	6 (2*S*,3*R*)

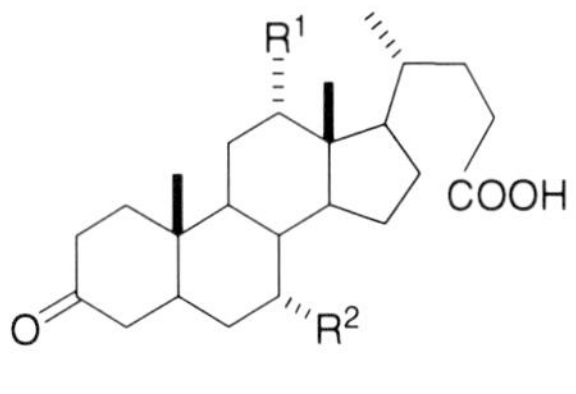

Fig. 7.24 Oxo bile acid epoxidation catalysts (R = H, OH etc)

431

In 2002, Shi et al. reported fructose-derived diacetate ketone **434** [263]. This catalyst was effective for the epoxidation of a variety of electron-deficient α,β-unsaturated esters (82–97% *ee*). Further fructose-derived diester and monoester ketones **434–438** were investigated for asymmetric epoxidation (Table 7.29). Ketone **434a** has been found to be the most effective catalyst [264].

In 2004, Seki et al. use the 11-membered C_2-symmetric biaryl ketones **359** and **439** for asymmetric epoxidation of methyl *para*-methoxycinnamate (80% *ee*) (Fig. 7.25) [302].

Table 7.29 Asymmetric epoxidation of ethyl cinnamate catalyzed by sugar ketones **434–438**

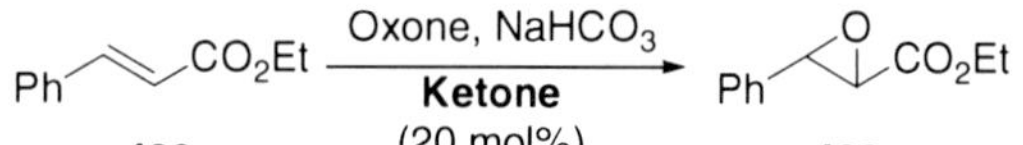

Oxone, $NaHCO_3$, **Ketone** (20 mol%): **432** → **433**

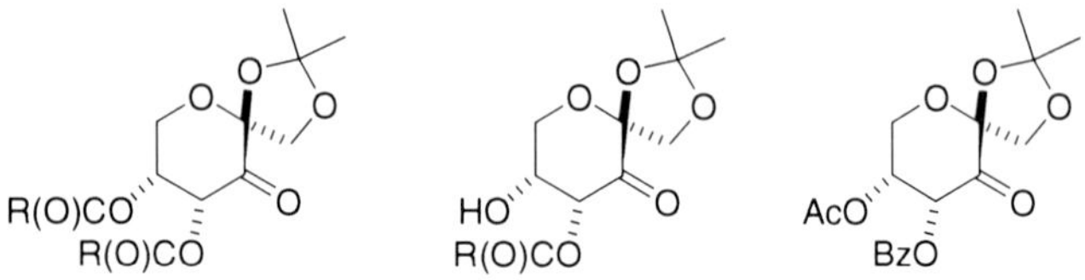

434a: R = Me
434b: R = Et
434c: R = CH_2Cl
434d: R = CH_2OMe
434e: R = *i*-Pr

435a: R = Me
435b: R = Ph

436

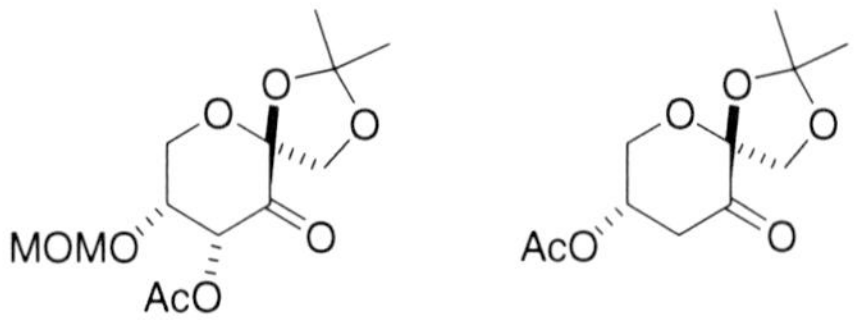

437

438

Entry	Catalyst	Conv. (%)	*ee* (%)
1	**434a**	47	95
2	**434b**	27	94
3	**434c**	16	94
4	**434d**	34	94
5	**434e**	10	90
6	**435a**	1	29
7	**436**	10	91
8	**437**	4	62
9	**438**	5	42

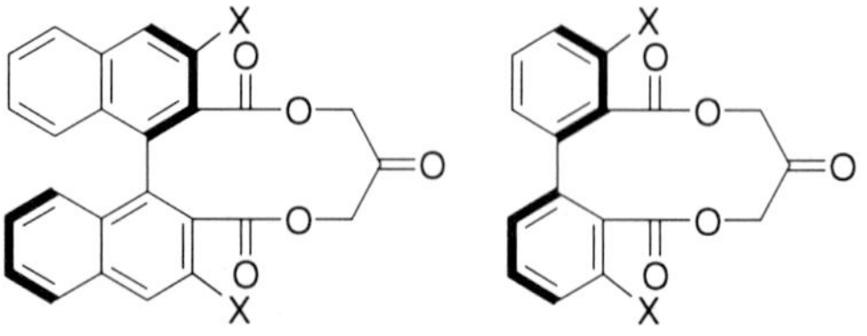

Fig. 7.25 C_2-symmetric biaryl ketones catalysts

(*R*)-**359a**: X = H
(*R*)-**359b**: X = Cl

(*R*)-**439a**: X = Cl
(*R*)-**439b**: X = NO_2

Epoxidation of Enones Mediated by α,α-Diarylprolinol and Primary β-Amino Alcohols

In 2005, Lattanzi et al. reported a series of α,α-diphenyl-L-prolinols **442** for the enantioselective epoxidation of different α,β-enones. The epoxides have been obtained in good yields and with up to 80% *ee* [298]. Further investigations were carried out in order to improve the enantioselectivities. On the one hand the substitution pattern of the phenyl residue was tuned and on the other hand the size of the cyclic amine was modified. In these studies, the catalysts **442e–f** proved to be more effective compared to other α,α-aryl-prolinols and secondary cyclic β-amino alcohols **443** and **444** (Table 7.30) [299, 303, 304].

Furthermore, different *β*-amino alcohols **445** were tested in the epoxidation of enones (Table 7.31) [305]. Simple primary *β*-amino alcohols **445a–b** afforded low levels of *ee* whereas the phenyl substituted amino alcohols **445c** and **445e–f** show moderate activity and *ee*. Catalayst **445d** and **445g** proved to be very poor catalysts.

Table 7.30 Enantioselective epoxidation of phenyl cinnamate catalyzed by α,α-diphenyl-L-prolinols **442b–444**

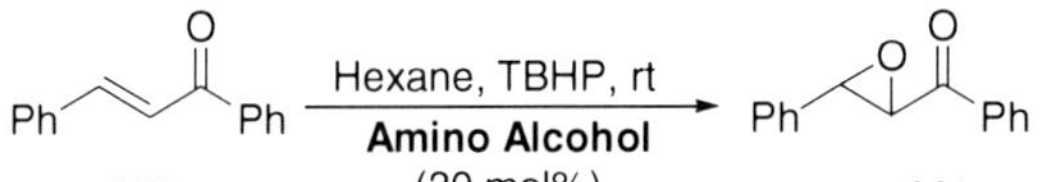

Hexane, TBHP, rt
Amino Alcohol
(30 mol%)

440 → **441**

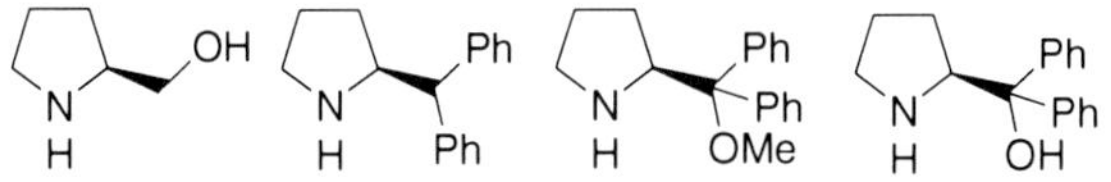

442a **442b** **442c** **442d**

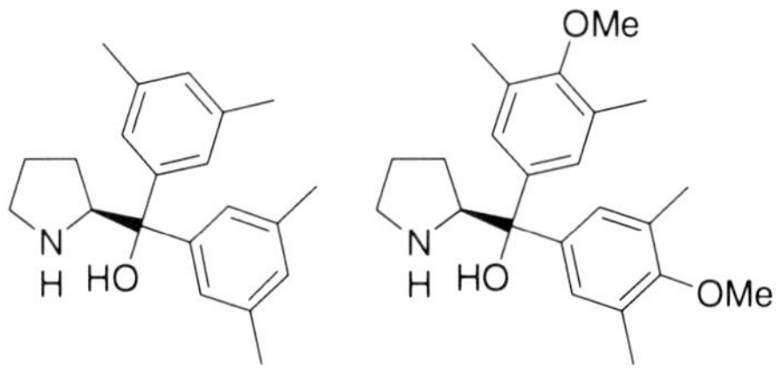

442e **442f**

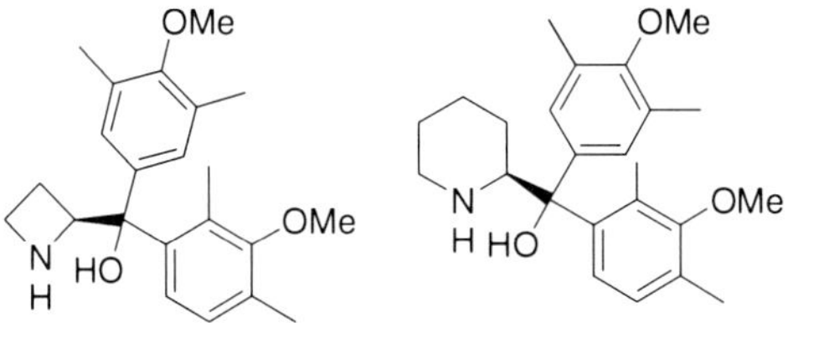

443 **444**

Table 7.30 (continued)

Entry	Catalyst	Conv. (%)	*ee* (%)
1	**442a**	82	racemic
2	**442b**	22	23
3	**442c**	14	63
4	**442d**	72	75
5	**442e**	90	91
6	**442f**	93	89
7	**443**	33	85
8	**444**	48	75

Table 7.31 Asymmetric epoxidation of phenyl cinnamate catalyzed by *β*-amino alcohols **445**

Hexane, TBHP, rt; **Amino Alcohol** (30 mol%); **440** → **441**

445a **445b** **445c** **445d** **445e** **445f** **445g**

Entry	Catalyst	Conv. (%)	*ee* (%)
1	**445a**	37	19
2	**445b**	30	34
3	**445c**	41	22
4	**445d**	18	racemic
5	**445e**	58	35
6	**445f**	21	52
7	**445 g**	5	7

In 2007, Zhao et al. reported a series of more successful 4-substituted-*α,α*-diarylprolinols **446** (Fig. 7.26), with up to 96% *ee* on enones with catalyst version **446c** [306]. A series of fluoro-substituted α,α-diarylprolinols **447** (Fig. 7.26) gave moderate yield (31–67%) and good *ee* (65–84%) [307]. Loh et al. employed (1*R*,3*S*,4*S*)-2-azanorbornyl-3-methanol **448** (Fig. 7.26) in enone epoxidation in low to good yield (25–88%) and with moderate enantioselectivity (69–88%) [308].

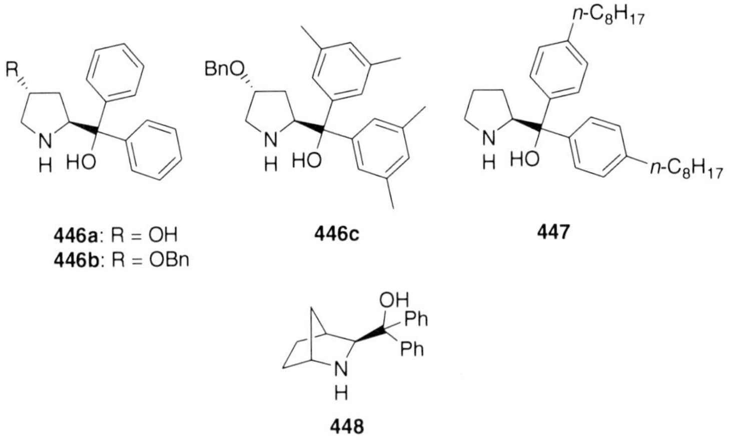

Fig. 7.26 Higher substituted α,α-diarylprolinol catalysts employed in epoxidation reactions

7.3.3.4 Allylic and Related Oxidations by Catalytic (Cyclo-) Selenenylation-Deselenenylation

In 1981 Torri et al. introduced the organocatalytic allylic oxidation by an oxyselenenylation-deselenenylation process (Scheme 7.68) [309]. The approach consists of an *in situ* electrochemical generation and recycling of the oxy-selenenylation

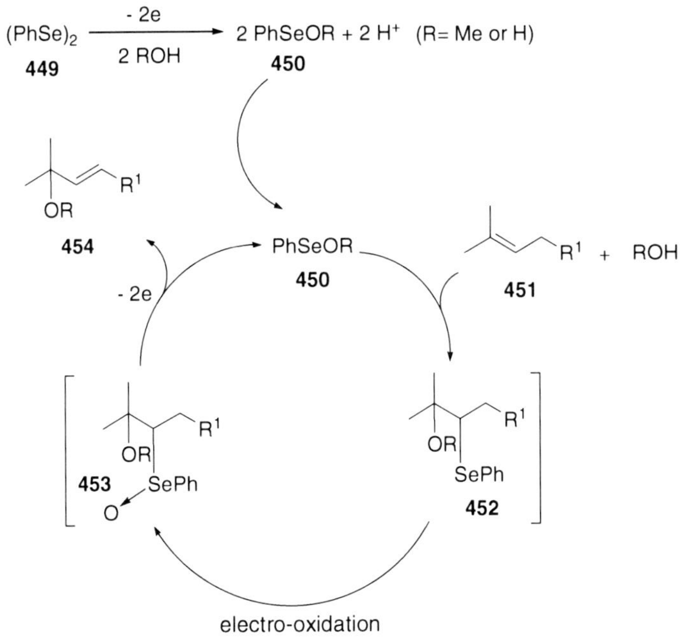

Scheme 7.68 Electrochemical oxyselenenylation-deselenenylation via selenoxide

Fig. 7.27 Diselenides with internal tertiary amines

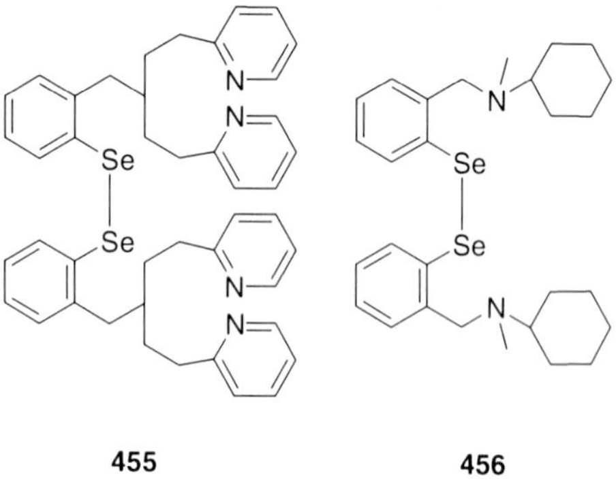

reagent **450** formed *in situ* from a catalytic amount (10 mol%) of diphenyl diselenide (**449**). Employing this protocol, allylic alcohols were produced in aqueous acetonitrile, or methyl ethers in methanol.

The reaction of the electrophilic selenium **450** with an olefin **451** and water or methanol produces the oxyselenide **452**. Subsequent electrochemical oxidation of **452** provides the corresponding selenoxide **453** that undergoes a *syn* elimination to produce the desired *trans* product **454** and regenerates the selenenic acid **450**. The addition of sulfate is crucial because it prevents the conversion of phenylselenenic acid into inert phenylseleninic acid by disproportionation or electro oxidation.

In 1992, Tomoda and Iwaoka reported the first catalytic chemical conversion of alkenes into allylic ethers and esters [310]. The reactions were performed employing 10 mol% of diselenides **455** or **456** (Fig. 7.27) and copper (II) nitrate trihydrate in methanol or acetic acid in the presence of excess of sodium persulfate as an oxidizing agent. The best results were achieved when using the diselenide **456** in the presence of molecular sieves 3 Å in acetic acid (Scheme 7.69). This suggests that

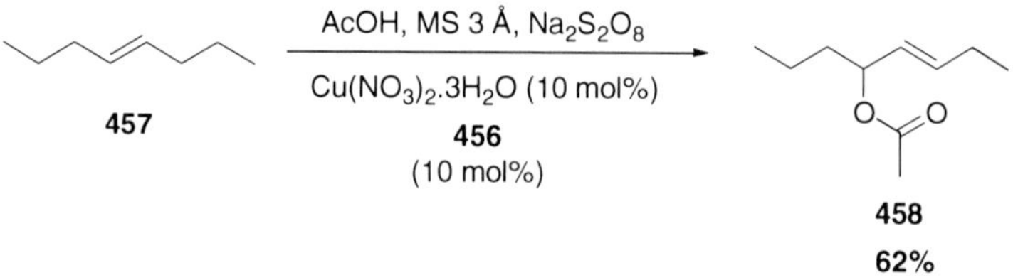

Scheme 7.69 Selenenylation-deselenenylation of olefine catalyzed by diselenide **456**

removal of the water formed during the reaction is essential for the smooth addition of selenenic acid to the alkene. With diselenide **456**, the reaction also takes place in the absence of copper(II) nitrate trihydrate. The copper nitrate, if present, may be involved chiefly in the initial step of the catalytic reaction. Moreover, the strong interaction between electrophilic selenium and an internal tertiary amine interrupts the disproportionation of a selenenic acid intermediate [310].

Table 7.32 Synthesis of γ-alkoxy and γ-hydroxy α,β-unsaturated derivatives catalyzed by diphenyl diselenide (**449**)

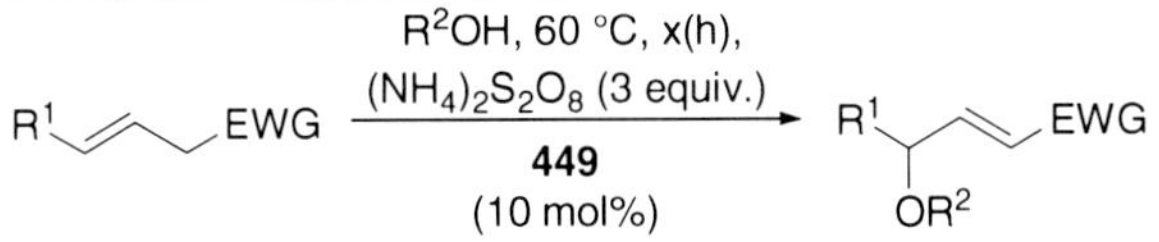

R^2OH, 60 °C, x(h), $(NH_4)_2S_2O_8$ (3 equiv.), **449** (10 mol%)
R^1–CH=CH–CH$_2$–EWG (**459**) → R^1–CH(OR^2)–CH=CH–EWG (**460**)

Entry	EWG	R^1	Solvent[a] (R^2OH)	Time (h)	Yield (%)
1	CO_2Me	Et	MeOH	2	90
2	CO_2Me	Et	H_2O[a]	5	80
3	CO_2Me	Ph	MeOH	2	83
4	CO_2Me	Ph	$(CH_2OH)_2$	2	75
5	CN	Me	MeOH	3	83

[a]Acetonitrile was added as cosolvent in the reaction run in water

In 1993, Tiecco et al. reported a similar reaction for β,γ-unsaturated esters, amides or nitriles with 10 mol% of diphenyl diselenide (**449**) and an excess of ammonium persulfate in alcohols or water to afford the γ-alkoxy or γ-hydroxy α,β-unsaturated derivatives in good yields (Table 7.32) [311].

The proposed mechanism is similar to the one discussed above (Scheme 7.70).

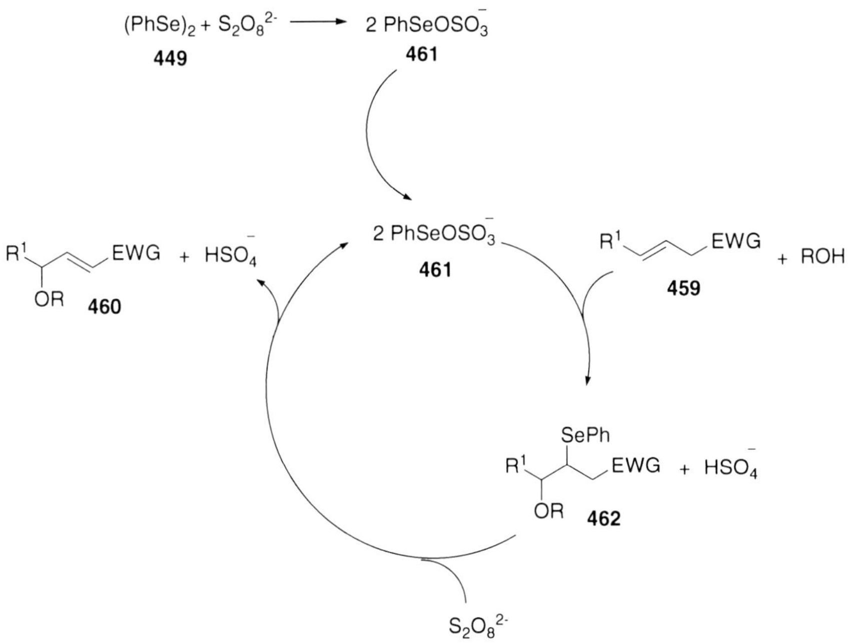

$(PhSe)_2 + S_2O_8^{2-}$ → 2 $PhSeOSO_3^-$
449 **461**

2 $PhSeOSO_3^-$ **461**
R^1–CH=CH–CH$_2$–EWG **459** + ROH
R^1–CH(OR)–CH(SePh)–CH$_2$–EWG **462** + HSO_4^-
$S_2O_8^{2-}$
R^1–CH(OR)–CH=CH–EWG **460** + HSO_4^-

Scheme 7.70 Proposed mechanism for the multi-step one-pot synthesis of γ-alkoxy or γ-hydroxy α,β-unsaturated esters and nitriles

The presence of the electron withdrawing group seems to be essential for the success of these reactions, otherwise product mixtures were observed. An extension to an intramolecular version of the method was reported for the synthesis of butenolides **464** in good yields [312]. These compounds were obtained from the reaction of β,γ-unsaturated carboxylic acids (**463**) with 10 mol% of diselenide (**449**) and an excess of ammonium persulfate in acetonitrile (Scheme 7.71). The carboxylic group acts as both the internal nucleophile and as the electron withdrawing group in the α-position.

Scheme 7.71 Synthesis of butenolides **464** catalyzed by diphenyl diselenide (**449**)

463 → 464 (21-97%): MeCN, 60 °C, x(h), $(NH4)_2S_2O_8$, **449** (10 mol%)

R = H, Me, Et, $PhCH_2$
R^1 = H, Me, Et, Ph

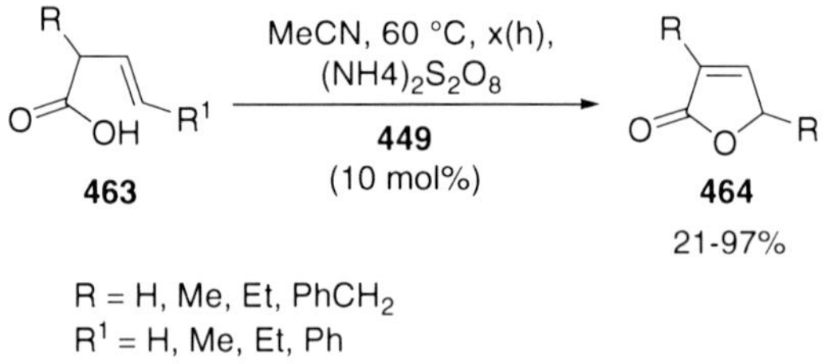

Also 2,3,5-trisubstituted furans (Scheme 7.72a) and 2,5-dihydrofurans (Scheme 7.72b) have been synthesized from their corresponding α-substituted β,γ-unsaturated ketones and 3-alkenols [313, 314].

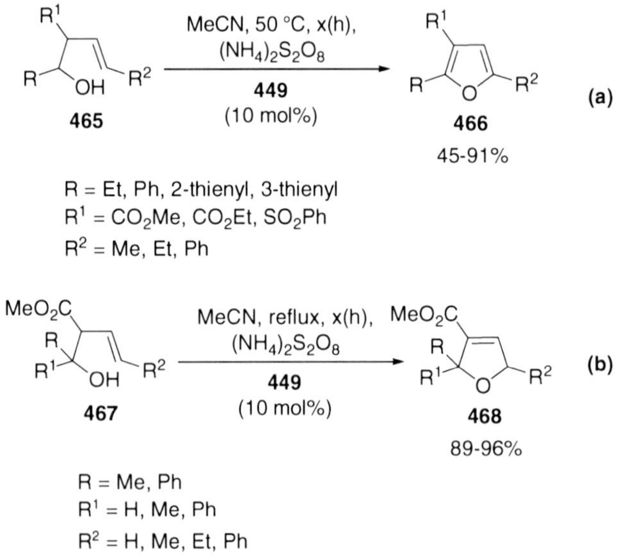

(a) **465** → **466** (45-91%): MeCN, 50 °C, x(h), $(NH_4)_2S_2O_8$, **449** (10 mol%)

R = Et, Ph, 2-thienyl, 3-thienyl
R^1 = CO_2Me, CO_2Et, SO_2Ph
R^2 = Me, Et, Ph

(b) **467** → **468** (89-96%): MeCN, reflux, x(h), $(NH_4)_2S_2O_8$, **449** (10 mol%)

R = Me, Ph
R^1 = H, Me, Ph
R^2 = H, Me, Et, Ph

Scheme 7.72 Synthesis of (**a**) 2,3,5-trisubstituted furans and (**b**) 2,5-dihydrofurans catalyzed by diphenyl diselenide (**449**)

In 1994, Tomoda et al. reported a catalytic asymmetric conversion of *trans*-*β*-methylstyrene (**469**) into its optically active allylic ether (**471**) using a C_2 symmetric chiral amino sugar-derived selenium catalyst **470** (Scheme 7.73) [315].

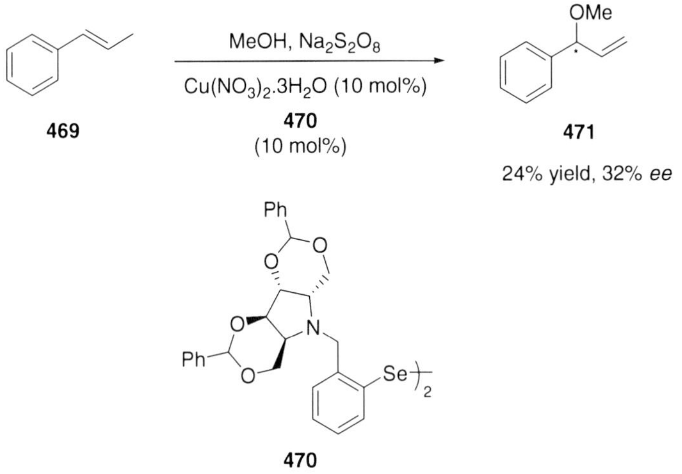

OMe
MeOH, $Na_2S_2O_8$
$Cu(NO_3)_2.3H_2O$ (10 mol%)
469
470
(10 mol%)
471
24% yield, 32% *ee*
Ph
O
O
O
N
Ph
O
Se
2
470

Scheme 7.73 Conversion *trans*-*β*-methylstyrene into its optically active allylic ether catalyzed by a chiral pyrrolidine-modified selenium catalyst **470**

In 1997, Fukuzawa et al. reported the conversion of *β,γ*-unsaturated esters **472** into the optically active *γ*-alkoxy *α,β*-unsaturated esters **474** using catalytic amounts of the chiral diferrocenyl diselenide **473** in the presence of ammonium persulfate. A selected example is presented in Scheme 7.74 [316].

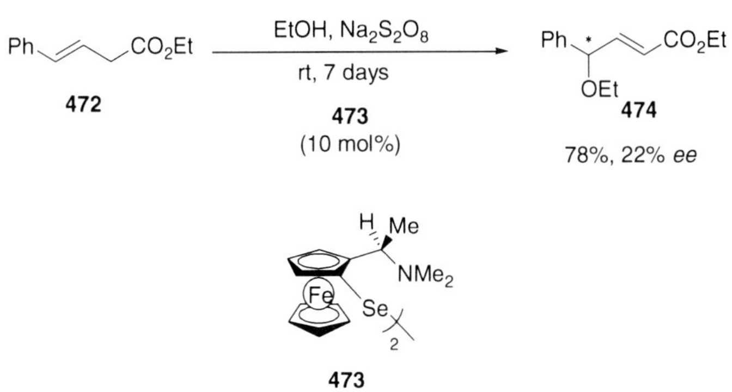

Ph
CO_2Et
EtOH, $Na_2S_2O_8$
rt, 7 days
472
473
(10 mol%)
Ph
CO_2Et
OEt
474
78%, 22% *ee*
H
Me
NMe_2
Fe
Se
2
473

Scheme 7.74 Conversion of *β,γ*-unsaturated ester **472** into the optically active *γ*-alkoxy *α,β*-unsaturated ester **474** catalyzed by chiral diferrocenyl diselenides catalyst **473**

Wirth et al. reported a catalytic asymmetric oxyselenenylation-elimination reaction using 10 mol% of chiral nitrogen-containing selenide **475**. *Trans-β*-methylstyrene (**469**) gives the allylic ether **471** with 75% *ee* but in only 23% yield (Scheme 7.75) [317].

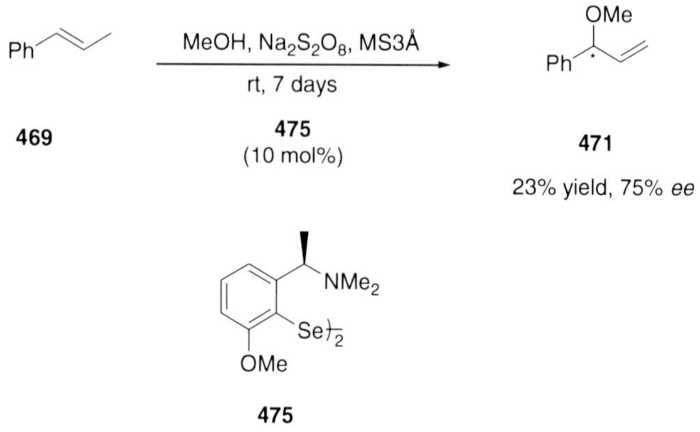

Ph
469
MeOH, $Na_2S_2O_8$, MS3Å
rt, 7 days
475
(10 mol%)
OMe
Ph
471
23% yield, 75% *ee*

NMe$_2$
Se)$_2$
OMe
475

Scheme 7.75 Oxyselenenylation-elimination using chiral nitrogen-containing selenide **475** as catalyst

Tiecco et al. have employed a sulfur analogue **476** (5 mol%) which gave the products in high yield and with good *ee* values (Table 7.33) [318].

In 2006 Wirth et al. reported an asymmetric catalytic electrochemical selenenylation-elimination sequence. A selected result is shown in Scheme 7.76 [319].

The same group has reported a convenient catalytic one-pot approach for the synthesis of butenolides **482** from butenoic acids **480** using [bis(trifluoroacetoxy) iodo]benzene (BTI, **481**) as stoichiometric oxidant in acetonitrile, employing 5 mol% of diphenyl diselenide (**449**) as catalyst. The proposed mechanism and the results are shown in Scheme 7.77 [320].

A similar route is followed for the synthesis of isocoumarin derivatives from stilbene carboxylic acids (Scheme 7.78) [321].

7.3.3.5 Olefin Dihydroxylation

Tiecco et al. reported the first general and efficient diphenyl diselenide (**449**) catalyzed dihydroxylation of olefins employing hydrogen peroxide as the oxidizing agent affording diols in good yields and high diastereoselectivity. Selected results are presented below (Table 7.34) [322].

Table 7.33 Selenenylation-deselenenylation sequences promoted by catalytic amounts (5 mol%) of diselenide **476**

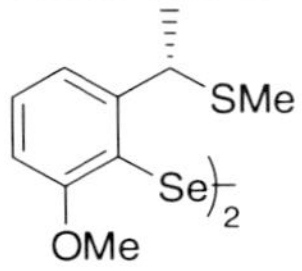

SMe
Se)$_2$
OMe
476

Alkenes	Solvent	T (°C)	Time (h)	Product	Yield (%)	*ee* (%)[a]
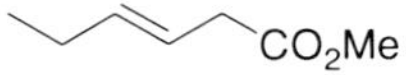 CO_2Me	MeOH	20	68	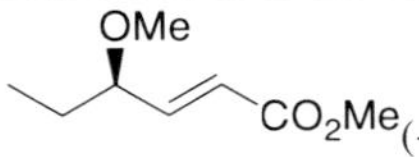 OMe CO_2Me (+)(R)	98	68
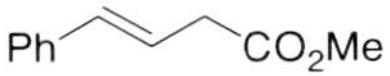 Ph CO_2Me	MeOH	20	48	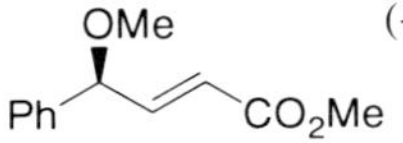 OMe Ph CO_2Me (−)-(S)	98	78
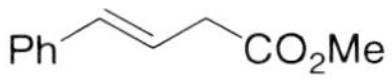 Ph CO_2Me	$MeCN/H_2O$	20	96	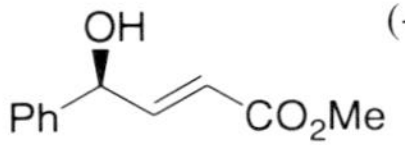 OH Ph CO_2Me (−)-(S)	98	82
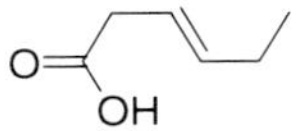 O OH	CH_2Cl_2	-30	48	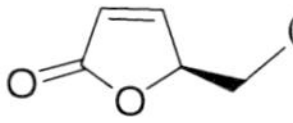 O O (−)-(R)	85	55

[a]The enantiomeric excesses were determined by recording the proton NMR spectra in the presence of (*S*)-(+)-2,2,2-trifluoro-1-(9-anthryl)ethanol

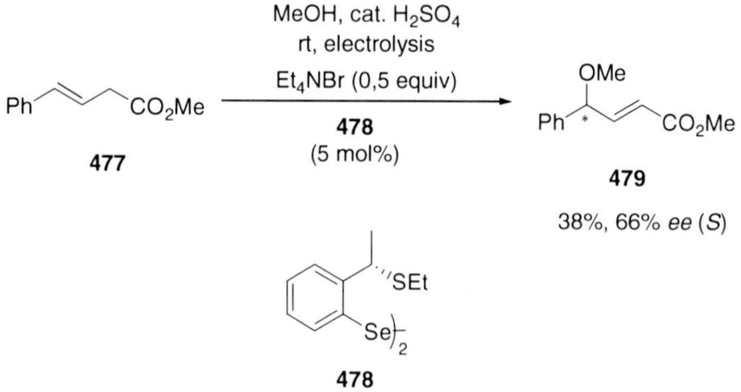

MeOH, cat. H_2SO_4
rt, electrolysis
Et_4NBr (0,5 equiv)
478
(5 mol%)

Ph CO$_2$Me **477** → OMe Ph * CO$_2$Me **479**

38%, 66% *ee* (*S*)

SEt
Se)$_2$
478

Scheme 7.76 Electrochemical selenenylation-elimination sequence catalyzed by chiral diselenide 478

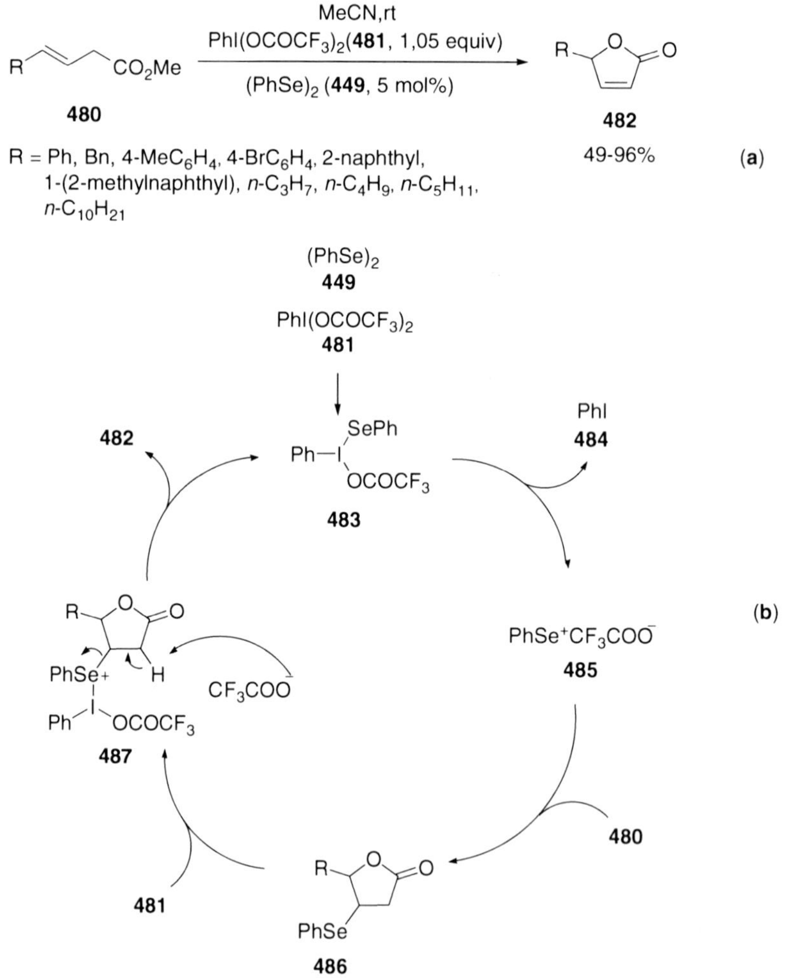

MeCN,rt
PhI(OCOCF$_3$)$_2$(**481**, 1,05 equiv)
(PhSe)$_2$ (**449**, 5 mol%)

R CO$_2$Me **480** → **482**

R = Ph, Bn, 4-MeC$_6$H$_4$, 4-BrC$_6$H$_4$, 2-naphthyl, 1-(2-methylnaphthyl), *n*-C$_3$H$_7$, *n*-C$_4$H$_9$, *n*-C$_5$H$_{11}$, *n*-C$_{10}$H$_{21}$

49-96% (**a**)

(PhSe)$_2$
449
PhI(OCOCF$_3$)$_2$
481

Ph—I(SePh)(OCOCF$_3$)
483

PhI
484

PhSe$^+$CF$_3$COO$^-$
485

480

PhSe
486

481

PhSe+ H
Ph I OCOCF$_3$
487

CF$_3$COO$^-$

482

(**b**)

Scheme 7.77 (**a**) Synthesis of butenolides from butenoic acids catalyzed by diphenyl diselenide with BTI (**481**) as oxidant; (**b**) Proposed catalytic cycle

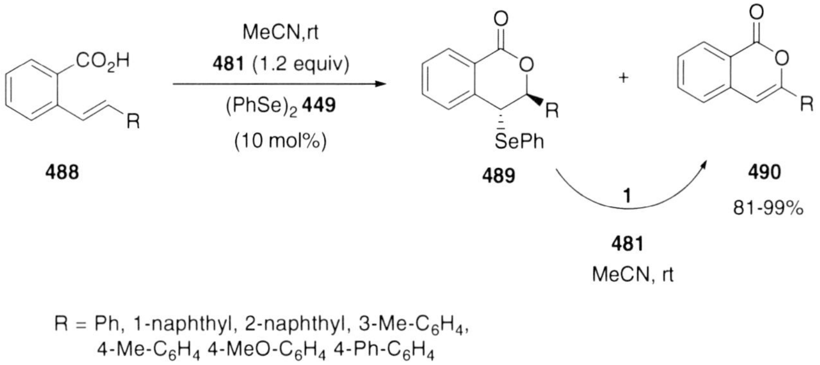

CO_2H R **488** — MeCN,rt; **481** (1.2 equiv); $(PhSe)_2$ **449** (10 mol%) → O O R SePh **489** + O O R **490** 81-99%

489 → **490**: **1**; **481**; MeCN, rt

R = Ph, 1-naphthyl, 2-naphthyl, 3-Me-C_6H_4,
4-Me-C_6H_4 4-MeO-C_6H_4 4-Ph-C_6H_4

Scheme 7.78 Synthesis of isocoumarins by the cyclization of stilbene carboxylic acids catalyzed by diphenyl diselenide

Table 7.34 Selected results of the dihydroxylation of olefins catalyzed by diphenyl diselenide **449**

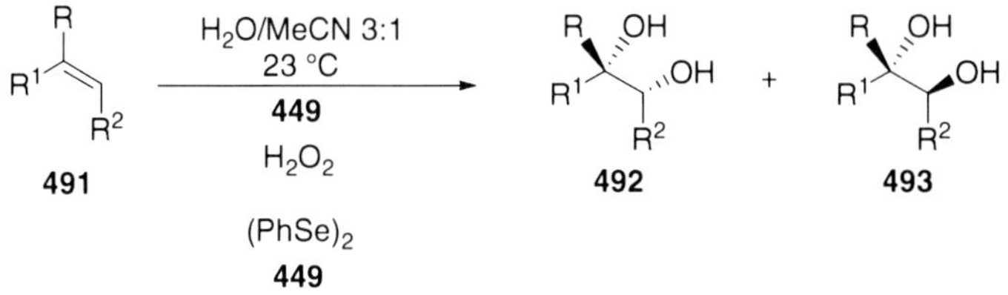

R R^1 R^2 **491** — H_2O/MeCN 3:1; 23 °C; **449**; H_2O_2 → R OH R^1 OH R^2 **492** + R OH R^1 OH R^2 **493**

$(PhSe)_2$
449

Entry	R		R^1	R^2	**449** (mol%)	T(h)	Yield (%)[a]	**492/493**[b]
1	45a	(+)-*p*-menth-1-ene			50	72	quant.	11/89
2	45b	(+)-*p*-menth-1-ene			10	72	98	11/89
3	45c	Ph	H	Me	10	72	quant.	46/54
4	45d	Me	-$(CH_2)_4$-		50	24	95	97/3
5	45e	Me	-$(CH_2)_4$-		10	24	76	99/1
6	45f	Ph	-$(CH_2)_4$-		50	30	98	80/20
7	45g	Ph	-$(CH_2)_4$-		10	96	75	80/20
8	45h	*n*-Pr	H	*n*-Pr	50	144	50	0/100
9	45i	*n*-Pr	H	*n*-Pr	10	300	50	0/100

[a]Isolated yields
[b]Deter

A reasonable mechanism is shown in Scheme 7.79. The first step is a peroxy acid-catalyzed epoxidation of the alkene (**491**). The epoxide intermediate **496** then rapidly evolves to the corresponding diol **493** either following a S_N2 mechanism (**path B**, Scheme 7.79), or through a S_N1-process via a carbocation intermediate (**497**, **path A**, Scheme 7.79), depending on the electronic and steric properties of the starting alkene.

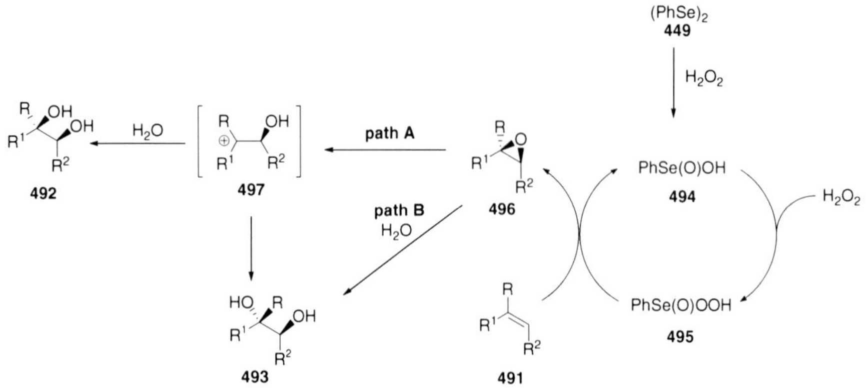

Scheme 7.79 Proposed mechanism for the dihydroxylation of olefins catalyzed by diphenyl diselenide **449**

Furthermore, the authors demonstrated that excellent enantiomeric excesses can be obtained with hydrogen peroxide at −10°C by employing the sulfur-containing chiral diselenide **499** (Scheme 7.80). ^{77}Se NMR indicated that the diselenide **499** suffered oxidation to seleninic acid or peracid. ^{1}H NMR and ^{13}C NMR signals of the aliphatic moiety suggested that the sulfur has been oxidized too, thus the corresponding sulfone may be involved.

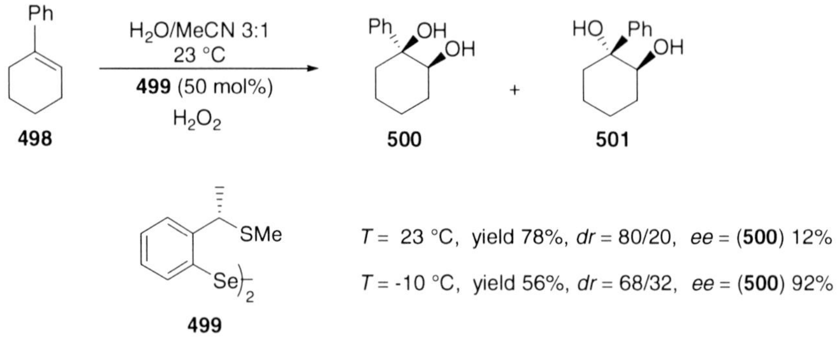

T = 23 °C, yield 78%, dr = 80/20, ee = (**500**) 12%

T = -10 °C, yield 56%, dr = 68/32, ee = (**500**) 92%

Scheme 7.80 Enantioselective dihydroxylation of olefins catalyzed by sulfur-containing chiral diselenide **499**

7.3.3.6 Oxidation of Alcohols and Benzyl Compounds

The oxidation of alcohols to carbonyl products with stable nitroxyl radicals as catalyst has been recently reviewed [323–326]. The best-known oxidation catalyst is the tetraalkylnitroxyl radical TEMPO (2,2,6,6-tetramethylpiperidine-1-oxyl).

In general, nitroxyl radicals are mild oxidation reagents containing the *N,N*-disubstituted NO-group with one unpaired electron. At strongly acidic conditions (pH < 2) the radical disproportionates to the oxoammonium cation and its hydroxylamine counterpart. Above pH 3 the reverse reaction occurs which affords two molecules of the nitroxyl radical (Scheme 7.81).

Scheme 7.81 pH dependent equilibrium of TEMPO

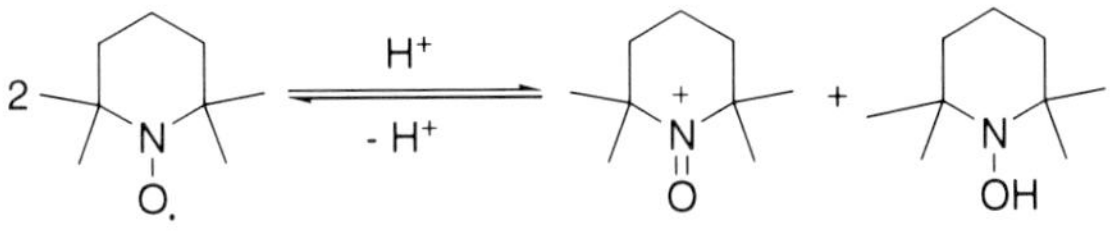

These reversible transformations are part of the catalytic cycle suggested for the oxidation of alcohols which is shown in Scheme 7.82.

Scheme 7.82 Proposed catalytic cycle for the oxidation of alcohols by nitroxides

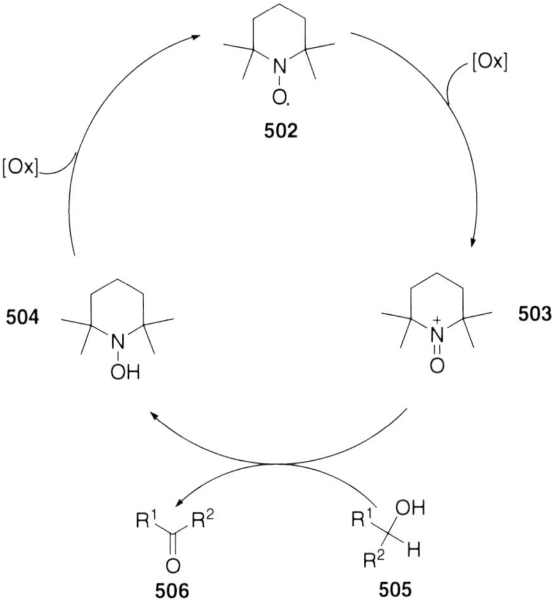

The nitroxyl radical serves as a catalyst precursor which is oxidized by using different primary oxidants (e.g. sodium chlorite [327], electrochemical oxidation [328–330] or periodic acid [331]). The resulting oxoammonium cation is considered to be the true catalytic oxidant which is reduced to the corresponding hydroxylamine during the oxidation of the alcohol. Subsequently, the hydroxylamine is re-oxidized with a suitable stoichiometric oxidant to the nitroxyl radical.

The first oxidation of primary alcohols was described by Golubev et al. in 1965 [332]. Oxoammonium salt as the oxidant was separately synthesized from 4-hydroxy-TEMPO with elemental chlorine. Later, Semmelhack et al. generated the oxoammonium cation *in situ* by electrochemical oxidation [328]. In 1987, Anelli and coworkers used 4-methoxy-TEMPO as catalyst in combination with sodium hypochlorite at 0°C under slightly basic conditions [333–335]. Thus, it was possible to regenerate the oxoammonium cation from the nitroxyl radical continuously. The TEMPO/hypochlorite protocol (Anelli-Montanari process, Scheme 7.83) has been widely applied in organic synthesis.

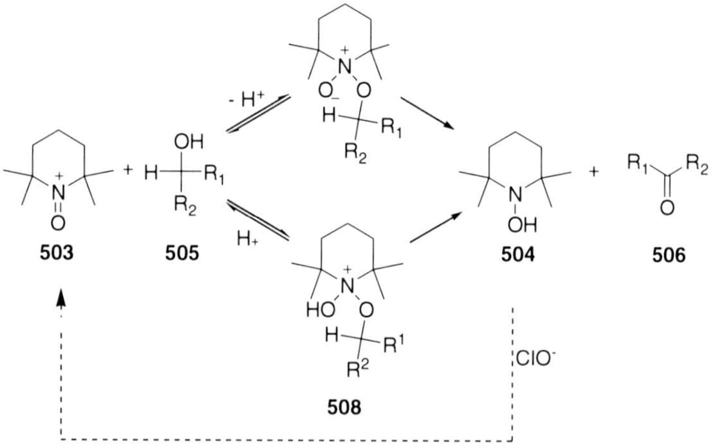

507

- H+

503 505 504 506

H+

ClO-

508

Scheme 7.83 Anelli-Montanari process

Several oxidations of simple alcohols with immobilized symmetric or racemic nitroxyl radicals have been reported since 1985 [336, 337]. Advantages of heterogeneous catalysts are the milder reaction conditions, simpler application and better recovery after the reaction. In addition, they show good selectivities (≥ 99) at short reaction times (0.5–6 h) and low catalyst loadings (1 mol%). Several groups designed variants of immobilized TEMPO, e.g. on a polymer or inorganic support [338, 339], such as silica [340], or mesoporous silica [341, 342], on a silica matrix by the sol-gel method [343, 344], or on magnetic C/Co nanoparticles [345]. In Table 7.35 some catalysts (**509–516**) and their selectivities are depicted.

In 1996, Rychnovsky et al. used optically active binaphthyl-based nitroxyl radicals as catalysts for the kinetic resolution of racemic secondary alcohols (Table 7.36) [346]. They achieved a fair selectivity factor ($S = 7.1$) for the preferred oxidation of one enantiomer.

Table 7.35 Some examples of nitroxyl radicals catalyst

Entry	Catalyst	Mol %	Alcohol	Time (h)	Product	Conv. (%)	Sel. (%)	Ref.
1[b]	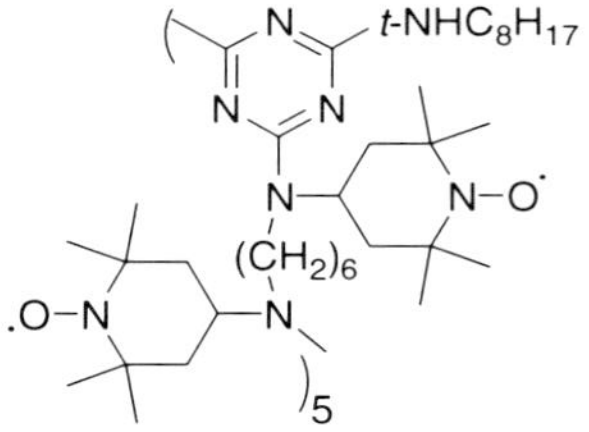 t-NHC$_8$H$_{17}$; N; N–O·; (CH$_2$)$_6$; .O–N; 5 **509**	1	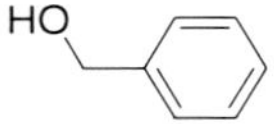 HO	0.5	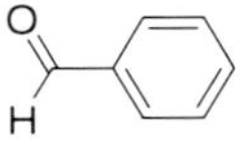 O; H	100	>99	321
2[c]	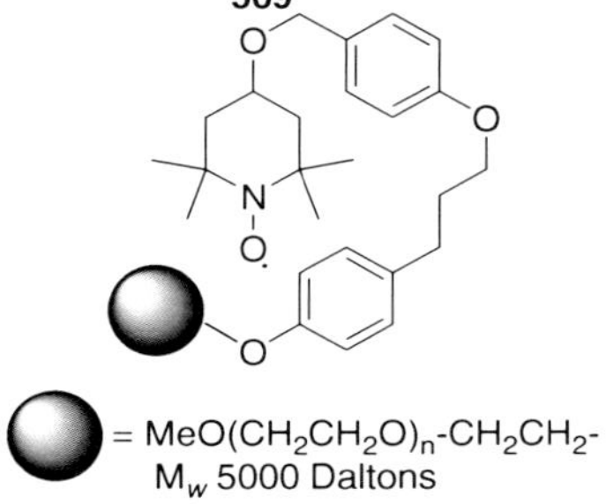 O; N; O·; = MeO(CH$_2$CH$_2$O)$_n$-CH$_2$CH$_2$- M$_w$ 5000 Daltons **510**	1	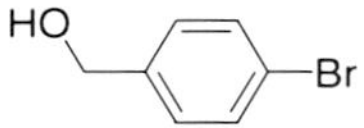 HO; Br	5	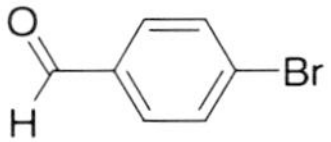 O; H; Br	90	>99	322
3[c]	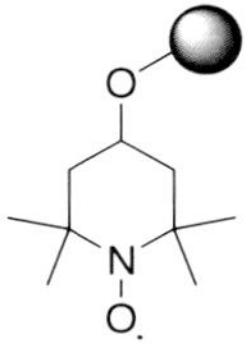 O; N; O· **511**	1	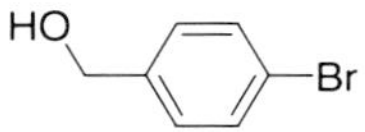 HO; Br	5	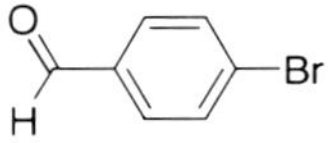 O; H; Br	72	>99	322

(continued)

Table 7.35 (continued)

Entry	Catalyst	Mol %	Alcohol	Time (h)	Product	Conv. (%)	Sel. (%)	Ref.
4[b]	N–O˙, N–H = silica **512**	1	HO	0.5	O, H	> 98	[a]	323
5[b]	MCM-41, O, O–Si, O, O, N–O˙ **513**	23	HOH_2C, O, OCH_3, HO, OH, OH	[a]	$NaCO_2$, O, OCH_3, HO, OH, OH	[a]	[a]	324
6[c]	SBA-15, O, O–Si, OH, OEt, N–H, N–O˙ **514**	1	HO	0.9	O, H	100	>99	325
7	O–Si–O, O, O, n, N–H, N–O˙ **515**	20	HOH_2C, O, OCH, HO, OH, OH	4	$NaCO_2$, O, OCH_3, HO, OH, OH	> 98	[a]	326
8[b]	Co, N, N, N, O, N–O˙ **516**	2.5	HO	1	O, H	> 98	[a]	327

[a]Is not reported; sel. = selectivity (chemoselectivity)
[b]Oxidant is NaOCl
[c]Oxidant is O_2 (1 atm)

Table 7.36 Resolution of racemic secondary alcohols catalyzed by the chiral binaphthyl-based nitroxyl radical **518**

CH_2Cl_2/H_2O, 0 °C,30 min
NaOCl (0.6-0.7 equiv.), KBr (0.1 equiv.)
S-(-) **518**
(0.5-1 mol%)

517 → **519** + **520**

S-(-) **518**

Entry	R¹	R²	*ee* (%)	Conversion (%)	*S*
1	Ph	Me	98	87	7.1
2	Np	Me	57	56	4.5
3	2-Me-Ph	Me	73	58	6.8
4	2-Cl-Ph	Me	89	70	6.0
5	4-Me-Ph	Me	64	58	5.1
6	Ph	Et	57	59	3.9

Furthermore there are many other achiral and chiral TEMPO derivatives, a variety of which is summarized in Table 7.37 [335, 347–349].

Iwabuchi et al. use chirally modified 2-azaadamantane *N*-Oxyls (AZADOs) **525–533** for the enantioselective oxidative kinetic resolution of racemic secondary alcohols instead of TEMPO (Table 7.38). TEMPO is inefficient in the oxidation of structurally hindered secondary alcohols. The modified AZADOs exihibit superior catalytic activity and high enantioselectivities for the oxidation of structurally hindered secondary alcohols. A further alternative to TEMPO is the readily available 9-azabicyclo[3.3.1]nonane *N*-oxyl (ABNO, **534**) which is kinetically more efficient than AZADO (Table 7.38) [350–352].

Onomura et al. reported several azabicyclo-*N*-oxyls which are applicable as catalysts in electrooxidations of sterically hindered secondary alcohols to the corresponding ketones (Table 7.39) [353, 354]. The *N*-oxyls **537–540** proved to be much more effective than TEMPO. Ishii and coworkers used *N*-hydroxyphthalimides (NHPI) with molecular oxygen to oxidize benzyl groups to benzoyl groups [371].

7.3.3.7 Baeyer-Villiger Oxidation

In 1977, Grieco et al. employed benzeneperoxyseleninic acid (**495**) generated *in situ* from benzeneseleninic acid (**494**) and hydrogen peroxide (**541**). The peroxy acid was employed in overstoichiometric amounts as reagent for the Baeyer-Villiger reaction (Scheme 7.84) [355].

Table 7.37 Selected further TEMPO derived catalysts

521

Entry	X	R^1	R^2	R^3	R^4	R^5	ref.
1	C	OCH^3	H	Me	Me	Me	335
2	C	$OC(O)R_F$; R_F = n-C_7F_{15} = CH_2CH_2(n-C_8F_{17})	H	Me	Me	Me	347
3	C	$HNC(O)R_F$; R_F = n-C_7F_{15} = CH_2CH_2(n-C_8F_{17})	H	Me	Me	Me	347
4	C	$-N(R_F)_2$; R_F = $CH_2CH_2CH_2$(n-C_8F_{17})	H	Me	Me	Me	347
5	C	O–triazine bearing two $N(R_F)_2$ groups; R_F = $CH_2CH_2CH_2$(n-C_8F_{17})	H	Me	Me	Me	347
6	C	OH	H	Me	Me	Me	348
7	C	OAc	H	Me	Me	Me	348
8	O	–	H	Ph	Me	Me	349
9	O	–	H	Ph	Ph	Me	349
10	O	–	H	Ph	Me	Ph	349
11	O	–	=O	Me	Me	Me	349
12	N	CH(Ph)Me	=O	Cy	Cy	Me	349
13	O	–		Np	Np	Me	349

Table 7.38 Kinetic resolution of 2-phenyl-cyclohexanol catalyzed by AZADOs and ABNO

522 → conditions A or B, Nitroxyl Catalyst (1-2 mol%) → **523** + **524**

conditions A: CH_2Cl_2 (0.2 M), -40 °C, TCCA (0.2 equiv), $NaHCO_3$ (2 equiv.)
conditions B: CH_2Cl_2/aq. $NaHCO_3$, 0 °C, NaOCl (1.5 equiv.), KBr (0.1 equiv.), n-Bu_4NBr 0.05 equiv.)

525: R = H; **526**: R = Me; **527**: R = *n*-Bu; **528**: R = *i*-Pr; **529**: R = Me; **530**: R = *i*-Pr; **531**; **532**: R = H; **533**: R = Me; **534** (ABNO)

Entry	Catalyst	Mol%	Method	Time (h)	Yield (%)	*ee* (%)
1	**525**	2	A	3	52	8
2	**526**	2	A	3	55	96
3	**527**	2	A	3	52	98
4	**528**	2	A	3	38	41
5	**529**	2	A	24	50	−70
6	**530**	2	A	24	29	−23
7	**531**	2	A	3	55	98
8	**533**	1	B	0.3	99	[a]
9	**534**	1	B	0.3	100	[a]
10	**TEMPO**	1	B	0.3	16	[a]

TCCA trichloroisocyanuric acid
[a] Not reported

A few years later, Taylor and Flood report the Baeyer-Villiger oxidation at room temperature catalyzed by 1.5 mol% of polystyrene-bound phenylseleninic acid **544** with H_2O_2 in dichloromethane [356]. Selected results are shown in Scheme 7.85b. The catalyst was synthesized from mercurated polystyrene and selenium dioxide (Scheme 7.85a).

In 1987, Syper studied the oxidation of α,β-unsaturated aldehydes with hydrogen peroxide, catalyzed by benzeneseleninic acids and their precursors. Bis-2-nitrophenyl diselenide **546** was the best catalyst. Selected results are shown in Scheme 7.86 [357].

The same author investigated a series of organoselenium compounds for the Baeyer-Villiger oxidation of aromatic aldehydes and ketones, transformed to phenols via aryl formates. The oxidizing agent in this system is organoperoxyseleninic acid (cf. Scheme 7.84) [358]. Again electronpoor diselenides **546** and **549** were the best catalysts in these reactions (Scheme 7.87).

Table 7.39 Eletrooxidations of (neo)menthol catalyzed by nitroxyl catalysts

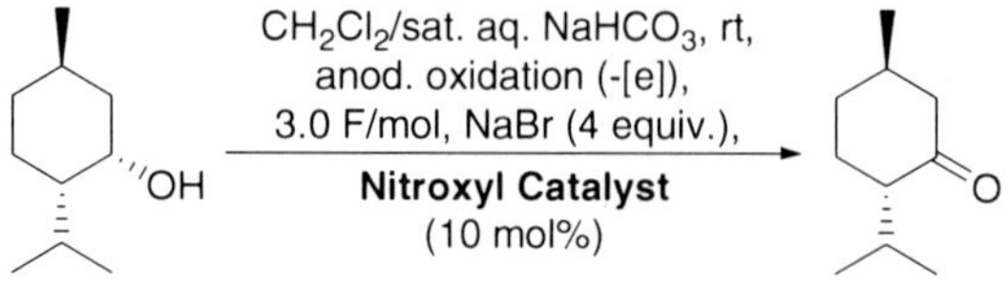

CH_2Cl_2/sat. aq. $NaHCO_3$, rt, anod. oxidation (-[e]), 3.0 F/mol, NaBr (4 equiv.), **Nitroxyl Catalyst** (10 mol%)

535 → **536**

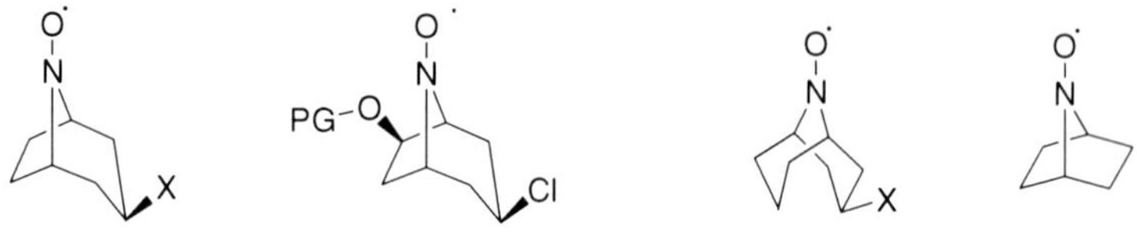

537a: X=H
537b: X=Cl

538a: PG=H
538b: PG=Acetyl
538c: PG=Benzoyl
538d: PG=Pivaloyl
538e: PG=1-Naphthoyl
538f: PG=2-Naphthoyl
538g: PG=Tosyl
538i: PG=2-Phenylbenzoyl
538j: PG=3,5-Dimethylbenzoyl
538k: PG=1-(2-Methylnaphtoyl)
538l: PG=Phenylcarbamoyl

539a: X=H
539b: X=Cl

540

Entry	Catalyst	Yield (%)
1	**537a**	86
2	**537b**	99
3	**539a**	82
4	**539b**	92
5	**540**	76
6	**TEMPO**	23

PhSe(O)OH + H_2O_2 ⇌ PhSe(O)OOH

494 **541** **495**

Scheme 7.84 *In Situ* generation of benzeneperoxyseleninic acid **495**

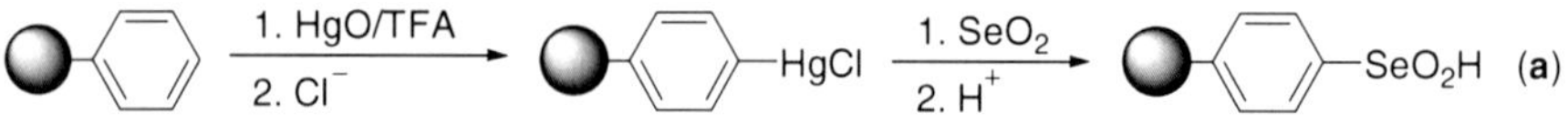

542 → (1. HgO/TFA; 2. Cl^-) → **543** (–HgCl) → (1. SeO_2; 2. H^+) → **544** (–SeO_2H) **(a)**

CH_2Cl_2, rt; H_2O_2 (1.8 equiv); **544** (1.5 mol%) **(b)**

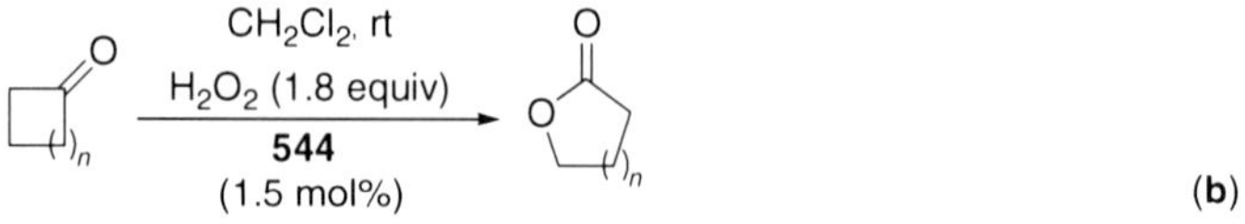

n = 1,2

n = 1, 3h, 96%
n = 2, 72h, 98%

Scheme 7.85 (**a**) Synthesis of polystyrene-bound phenylseleninic acid **544**; (**b**) Baeyer-Villiger oxidation of cyclic ketones catalyzed by **544**

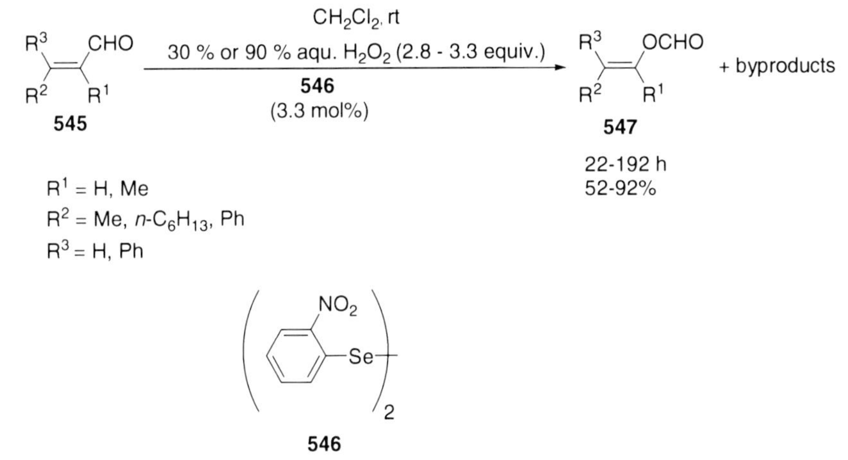

Scheme 7.86 Baeyer-Villiger oxidation of α,β-unsaturated aldehydes with hydrogen peroxide catalyzed by **546**

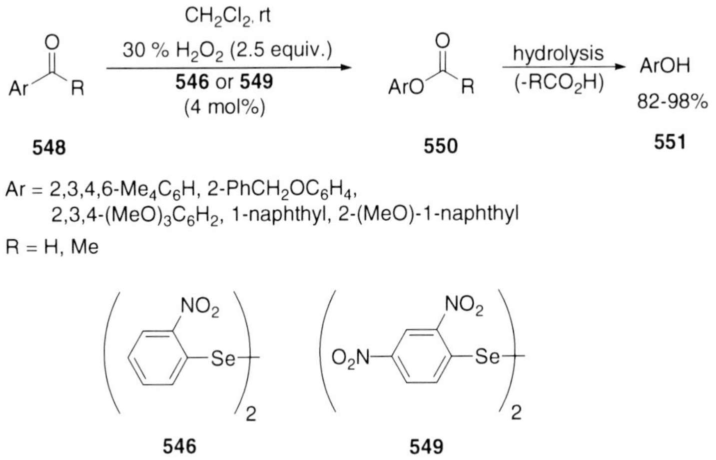

Scheme 7.87 Synthesis of phenols from aromatic aldehydes and ketones catalyzed by diselenides **546** or **549**

Ten Brink et al. reported the Baeyer-Villiger oxidation in homogeneous solution catalyzed by diselenide **553** and hydrogen peroxide as oxidant at room temperature [359]. They achieve faster and excellent conversion and high selectivity employing only 1 mol% catalyst. They obtained lactones from cyclic ketones, and phenols and acids from aromatic and aliphatic aldehydes. Selected results and the proposed mechanism are shown in Scheme 7.88.

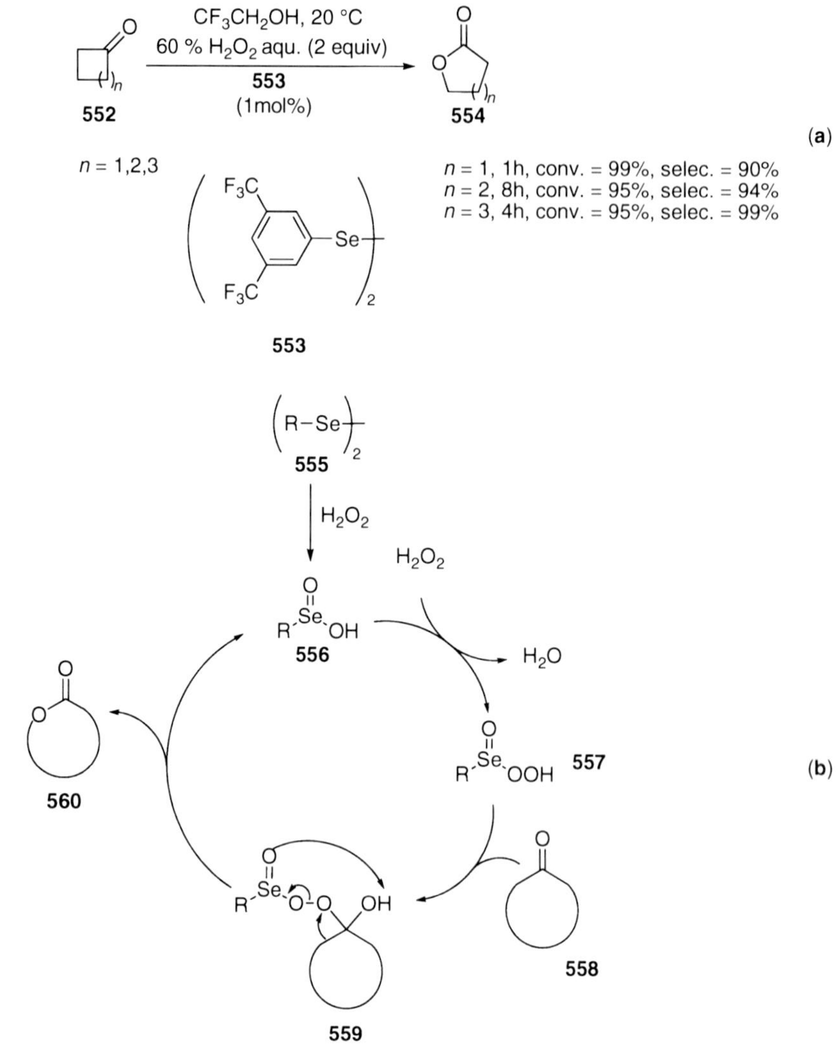

Scheme 7.88 (**a**) Baeyer-Villiger oxidation catalyzed by diselenide **553**; (**b**) Proposed mechanism for the reaction of electron poor diselenides

Other perfluoroalkyl-phenyl analogues **561** proved difficult to prepare and handle, but 3,5-bis(perfluorooctyl)phenyl butyl selenide **562** was accessible as precursor for the identical active species **563**. The catalyst **562** showed higher activity (turn over frequency) compared to diselenide **553** in the oxidation of aldehydes and ketones with aqueous hydrogen peroxide. Furthermore, **562** could be used to oxidize substrates under fluorous monophasic, biphasic, and triphasic conditions (Fig. 7.29) [360].

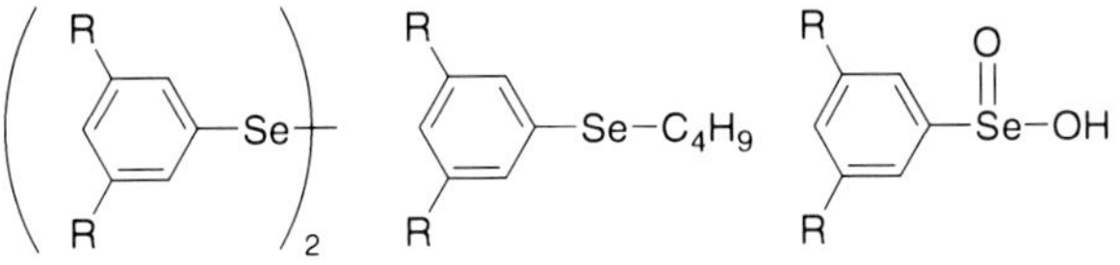
R
Se
2
Se−C4H9
O
Se−OH

561 562 563

R = C_8F_{17}

Fig. 7.29 Perfluoro (di-) selenide catalysts

Ichikawa et al. synthesized the organoselenium catalyst **565**, which shows good yields for the Baeyer-Villiger oxidation without the use of fluorous solvent [361]. A selected example is shown in Scheme 7.89.

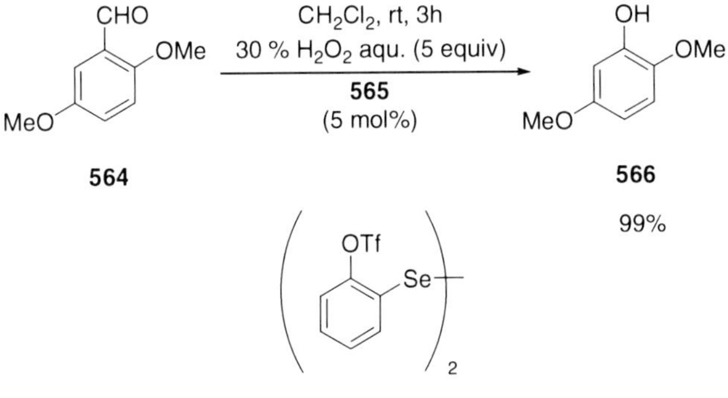
CHO
OMe
MeO
CH_2Cl_2, rt, 3h
30 % H_2O_2 aqu. (5 equiv)
565
(5 mol%)
OH
OMe
MeO
564
566
99%
OTf
Se
2

565

Scheme 7.89 Baeyer-Villiger oxidation catalyzed by diselenide **565**

Detty et al. reported the use of selenoxide **416** as catalyst in the Baeyer-Villiger oxidation of aldehydes and ketones with H_2O_2 [295]. An example and the putative two catalytic cycles proposed are shown in Scheme 7.90.

7.4 Miscellaneous

7.4.1 Telluride Catalyzed Conversion of Thiocarbonyl Compounds to Oxo Analogues

Ley et al. reported a catalytic method for the conversion of thiocarbonyl compounds to their oxo analogues with diaryltellurium oxide. The telluride is recycled with a halogenating agent like 1,2-dibromotetrachloroethane, water and base (Scheme 7.91) [362, 363].

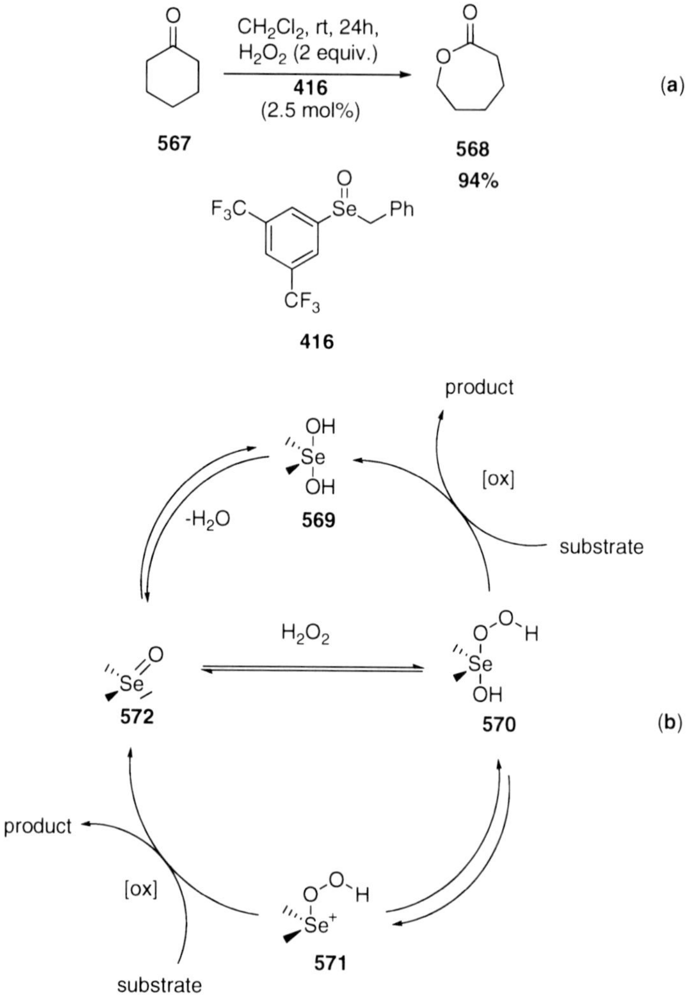

CH2Cl2, rt, 24h,
H2O2 (2 equiv.)
416
(2.5 mol%)
567
568
94%
416
(a)

product
[ox]
-H2O
569
substrate
H2O2
572
570
(b)
product
[ox]
571
substrate

Scheme 7.90 (**a**) Baeyer-Villiger oxidation catalyzed by selenoxide **30**; (**b**) Two putative catalytic cycles proposed for selenoxide organocatalysis of the Baeyer-Villiger reaction

Scheme 7.91 Conversion of thiocarbonyl compounds catalyzed by ditellurium oxide (generated in *situ* from **201**)

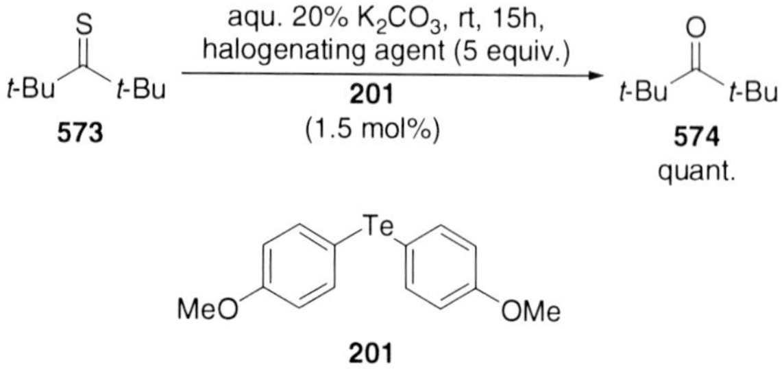

aqu. 20% K_2CO_3, rt, 15h,
halogenating agent (5 equiv.)
201
(1.5 mol%)
573
574
quant.
201

The proposed catalytic cycle of the reaction is shown below (Scheme 7.92).

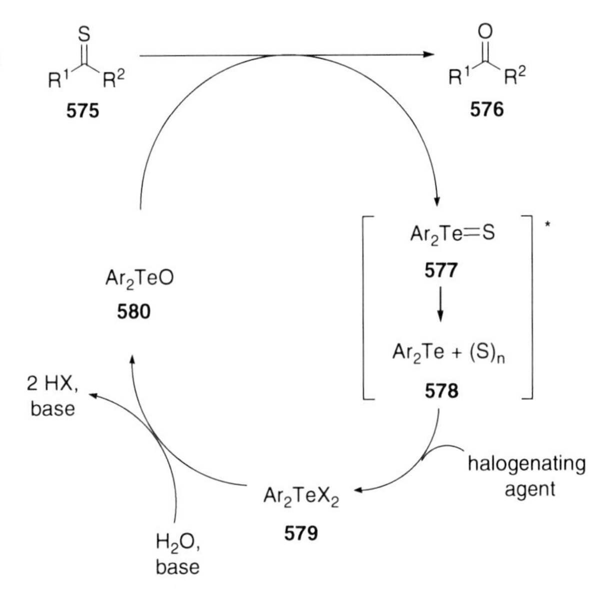

Scheme 7.92 Proposed catalytic cycle (* transformation added by the authors of this article)

7.4.2 *Sulfoxides and Polysulfoxides as Phase Transfer Catalysts in S_N2 Reactions*

Oae et al. reported the use of methyl 2-pyridyl sulfoxide (**582**) as an excellent phase transfer catalyst (PTC) promoting simple S_N2 reactions of alkyl halides. One of the best results is shown in Scheme 7.93 [364].

Also a series of macrocyclic polythiaethers and tetrakis(alkyl-sulfinylmethyl) methanes were tried, the best was catalyst **587** (Scheme 7.94).

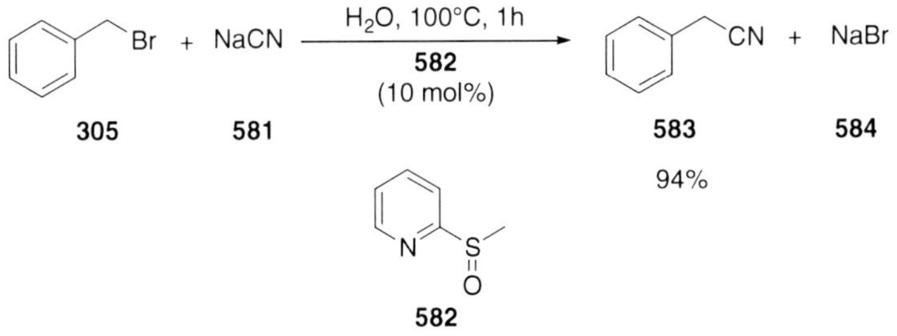

Br + NaCN → (H_2O, 100°C, 1h; **582** (10 mol%)) CN + NaBr

305 **581** **583** (94%) **584**

582

Scheme 7.93 Pyridyl sulfoxide as phase transfer catalyst in the S_N2 reactions of alkyl halides

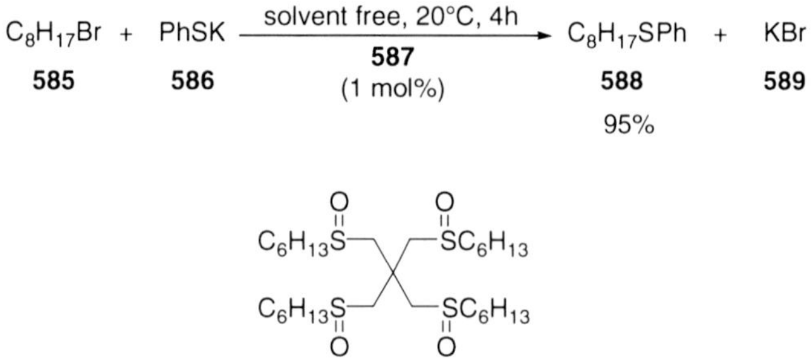

$$C_8H_{17}Br + PhSK \xrightarrow[\textbf{587}\ (1\ mol\%)]{solvent\ free,\ 20°C,\ 4h} C_8H_{17}SPh + KBr$$

585 **586** **588** (95%) **589**

587

Scheme 7.94 S_N2 reactions of alkyl halides catalyzed by tetrakis(alkyl-sulfinylmethyl) methane **587**

For example in the alkylation of phenylacetonitrile and phenylacetone in aqueous NaOH generates the monoalkylated products selectively and in high yields (Table 7.40) [365–368]. In a similar approach, polymeric sulfoxides were used as PTCs in S_N2 reactions by Kondo et al [369, 370].

Table 7.40 Monoalkylation of phenylacetonitrile and phenylacetone catalyzed by sulfoxide **587**

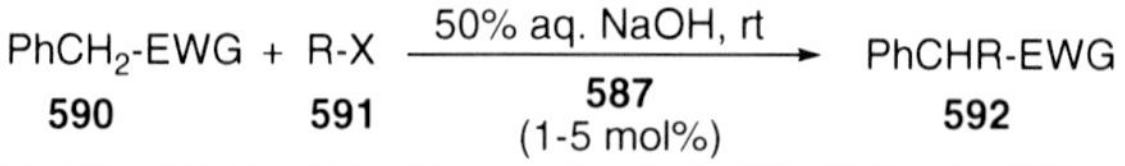

$$PhCH_2\text{-}EWG + R\text{-}X \xrightarrow[\textbf{587}\ (1\text{-}5\ mol\%)]{50\%\ aq.\ NaOH,\ rt} PhCHR\text{-}EWG$$

590 **591** **592**

Entry 1	EWG	R	Time (h)	mol%	Yield (%)
1	CN	Et	13	5	92
2	COMe	Me	0.65	1	96

References

1. Nahmany M, Melman A (2004) Org Biomol Chem 2:1563
2. Otera J (2003) Esterification: methods reactions and applications. Wiley-VCH, Weinheim
3. Green TW, Wuts PGM (1991) In protective groups in organic synthesis. Wiley-VCH, New York
4. Hanson JR (1999) In protective groups in organic synthesis. Blackwell Science, Malden
5. Blow DM (1976) Acc Chem Res 9:145
6. Hedstrom L (2002) Chem Rev 102:4501
7. Breslow R, Dong SD (1998) Chem Rev 98:1997
8. Ghosh M, Conroy JL, Seto CT (1999) Angew Chem 111:575; Angew Chem Int Ed 38:514
9. Barrett AJ, Tolle DP, Rawlings ND (2003) Biol Chem 384:873
10. Di Cera E, Page MJ (2008) Cell Mol Life Sci 65:1220
11. Rawling ND, Barrett AJ, Bateman A (2010) Nucleic Acids Res 38:D227
12. Perona JJ, Craik CS (1995) Protein Sci 4:337
13. Gromer S, Wessjohann LA, Eubel J, Brandt W (2006) Chembiochem 7:1649
14. Siegel B, Breslow R (1975) J Am Chem Soc 97:6869
15. Czarniecki MF, Breslow R (1978) J Am Chem Soc 100:7771

16. Breslow R, Czarniecki MF, Emert J, Hamaguchi H (1980) J Am Chem Soc 102:762
17. Trainor GL, Breslow R (1981) J Am Chem Soc 103:154
18. Breslow R, Trainor G, Ueno A (1983) J Am Chem Soc 105:2739
19. le Noble WJ, Srivastava S, Breslow R, Trainor G (1983) J Am Chem Soc 105:2745
20. Breslow R, Chung S (1990) Tetrahedron Lett 31:631
21. D'Souza TV, Bender ML (1987) Acc Chem Res 20:146
22. Zimmerman SC (1989) Tetrahedron Lett 30:4357
23. Rao KR, Srinivasan TN, Bhanumathi N, Sattur P B (1990) J Chem Soc Chem Commun 10
24. Dong Z, Li X, Liang K, Mao S, Huang X, Yang B, Xu J, Liu J, Luo G, Shen J (2007) J Org Chem 72:606
25. Chao Y, Cram DJ (1976) J Am Chem Soc 98:15
26. Chao Y, Weisman GR, Sogah GDY, Cram DJ (1979) J Am Chem Soc 101:4948
27. Cram DJ, Katz HE (1983) J Am Chem Soc 105:135
28. Cram JD, Katz HE, Dicker IB (1984) J Am Chem Soc 106:4987
29. Cram JD, Dicker IB, Lauer M, Knobler CB, Trueblood KN (1984) J Am Chem Soc 106:7150
30. Cram JD, Lam PY-S, Ho SP (1986) J Am Chem Soc 108:839
31. Cram JD, Lam PY-S (1986) Tetrahedron 42:1607
32. Cram JD (1988) Angew Chem Int Ed 27:1009
33. Delort E, Darbre T, Reymond J-L (2004) J Am Chem Soc 126:15642
34. Delort E, Nguyen-Trung N- Q, Darbre T, Reymond J-L (2006) J Org Chem 71:4468
35. Darbre T, Reymond J-L (2006) Acc Chem Res 39:925
36. Javor S, Delort E, Darbre T, Reymond J-L (2007) J Am Chem Soc 129:13238
37. Jencks WP, Carriuolo J (1960) J Am Chem Soc 82:1778
38. Bruice TC, York JL (1961) J Am Chem Soc 83:1382
39. Page MI, Jenkers WP (1972) J Am Chem Soc 94:8818
40. Werber MM, Shalitin Y (1973) Bioorg Chem 2:202
41. Khan MN (1985) J Org Chem 50:4851
42. Hine J, Khan MN (1977) J Am Chem Soc 99:3847
43. Moss RA, Nahas RC, Ramaswami S (1977) J Am Chem Soc 99:627
44. Tonellato UK (1977) J Chem Soc Perkin Trans 2:821
45. Skorey KI, Somayaji V, Brown RS (1989) J Am Chem Soc 111:1445
46. Gruhn WB, Bender ML (1975) Bioorg Chem 4:219
47. Hoffmann HMR, Schrake O (1998) Tetrahedron Asymmetr 9:1051
48. Schrake O, Franz MH, Wartchow R, Hoffmann HMR (2000) Tetrahedron 56:4453
49. Steglich W., Höfle G (1969) Angew Chem 81:1001. Angew Chem Int Ed 8:981
50. Höfle G, Steglich W (1972) Synthesis 619
51. Waddell TG, Rambalakos T, Christie KR (1990) J Org Chem 55:4765
52. Wessjohann LA, Zhu M (2008) Adv Synth Catal 350:107
53. Sammakia T, Hurley TB (1996) J Am Chem Soc 118:8967
54. Sammakia T, Hurley TB (1999) J Org Chem 64:4652
55. Sammakia T, Hurley TB (2000) J Org Chem 65:974
56. Wayman KA, Sammakia T (2003) Org Lett 5:4105
57. Notte GT, Sammakia T, Steel PJ (2005) J Am Chem Soc 127:13502
58. Notte GT, Sammakia T (2006) J Am Chem Soc 128:4230
59. Ema T, Tanida D, Matsukawa T, Sakai T (2008) Chem Commun 957
60. Berkessel A, Gröger H (2005) Asymmetric organocatalysis. Wiley-VCH, Weinheim
61. Dalko PI (2005) Enantioselective organocatalysis. Wiley-VCH, Weinheim
62. Atodiresei I, Schiffers I, Bolm C (2007) Chem Rev 107:5683
63. Tian SK, Chen Y, Hang J, Tang L, McDaid P, Deng L (2004) Acc Chem Res 37:621
64. Hiratake J, Yamamoto Y, Oda J (1985) J Chem Soc Chem Commun 1717
65. Hiratake J, Inagaki M, Yamamoto Y, Oda J (1987) J Chem Soc, Perkin Trans 1:1053
66. Aitken RA, Gopal J, Hirst JA (1988) J Chem Soc Chem Commun 632
67. Aitken RA, Gopal J (1990) Tetrahedron Asymmetr 1:517

68. Bolm C, Gerlach A, Dinter CL (1999) Synlett 195
69. Bolm C, Schiffers I, Dinter CL, Gerlach A (2000) J Org Chem 65:6984
70. Bolm C, Schiffers I, Atodiresei I, Hackenberger CPR (2003) Tetrahedron Asymmetr 14:3455
71. Bolm C, Atodiresei I, Schiffers I (2005) Org Synth 82:120
72. Rantanen T, Schiffers I, Bolm C (2007) Org Process Res Dev 11:592
73. Chen Y, Tian S-K, Deng L (2000) J Am Chem Soc 122:9542
74. Deng L, Chen Y, Tian S (2001) PCT Int Appl WO 2001074741 A2. Chem Abstr 135:288345
75. Deng L, Chen Y, Tian S (2004) US Patent 20040082809 A1. Chem Abstr 140:374734
76. Kolb HC, Van Nieuwenhze MS, Sharpless KB (1994) Chem Rev 94:2483
77. Kolb HC, Sharpless KB (2004) In transition metals for organic synthesis. Wiley-VCH, Weinheim
78. Choi C, Tian S-K, Deng L (2001) Synthesis 1737
79. Chen Y, Deng L (2001) J Am Chem Soc 123:11302
80. Hang J, Deng L (2009) Bioorg Med Chem Lett 19:3856
81. Ishii Y, Fujimoto R, Mikami M, Murakami S, Miki Y, Furukawa Y (2007) Org Process Res Dev 11:609
82. Ivši T, Hameršak Z (2009) Tetrahedron Asymmetr 20:1095
83. Uozumi Y, Yasoshima K, Miyachi T, Nagai S-I (2001) Tetrahedron Lett 42:411
84. Uozumi Y, Mizutani K, Nagai S-I (2001) Tetrahedron Lett 42:407
85. Schiffers I, Rantanen T, Schmidt F, Bergmans W, Zani L, Bolm C (2006) J Org Chem 71:2320
86. Okamatsu T, Irie R, Katsuki T (2007) Synlett 10:1569
87. Theil F (1995) Chem Rev 95:2203
88. Ghanem A, Aboul-Enein HY (2005) Chirality 17:1
89. Vedejs E, Chen X (1996) J Am Chem Soc 118:1809
90. Kawabata T, Nagato M, Takasu K, Fuji K (1997) J Am Chem Soc 119:3169
91. Ruble JC, Tweddell J, Fu GC (1998) J Org Chem 63:2794
92. Miller SJ, Copeland GT, Papaioannou N, Horstmann TE, Ruel EM (1998) J Am Chem Soc 120:1629
93. Sano T, Imai K, Ohashi K, Oriyama T (1999) Chem Lett 28:265
94. Vedejs E, Daugulis O (1999) J Am Chem Soc 121:5813
95. Naraku G, Himomoto N, Hanamoto T, Inanaga J (2000) Enantiomer 5:135
96. Spivey AC, Fekner T, Spey SE (2000) J Org Chem 65:3154
97. Lin M-H, RajanBabu TV (2002) Org Lett 4:1607
98. Priem G, Pelotier B, Macdonald SJF, Anson MS, Campbell IB (2003) J Org Chem 68:3844
99. Vedejs E, Daugulis O (2003) J Am Chem Soc 125:4166
100. Suzuki Y, Yamauchi K, Muramatsu K, Sato M (2004) Chem Commun 2770
101. Ishihara K, Kosugi Y, Akakura M (2004) J Am Chem Soc 126:12212
102. Birman VB, Uffman EW, Jiang H, Li X, Kilbane CJ (2004) J Am Chem Soc 126:12226
103. Fu GC (2004) Acc Chem Res 37:542
104. Terakado D, Koutaka H, Oriyama T (2005) Tetrahedron Asymmetr 16:1157
105. Yamada S, Misono T, Iwai Y (2005) Tetrahedron Lett 46:2239
106. Poisson T, Penhoat M, Papamicael C, Dupas G, Dalla V, Marsais F, Levacher V (2005) Synlett 2285
107. Kano T, Sasaki K, Maruoka K (2005) Org Lett 7:1347
108. Duhamel L, Herman T (1985) Tetrahedron Lett 26:3099
109. Mizuta S, Sadamori M, Fujimoto T, Yamamoto I (2003) Angew Chem 115:3505
110. Mizuta S, Tsuzuki T, Fujimoto T, Yamamoto I (2005) Org Lett 7:3633
111. Mizuta S, Ohtsubo Y, Tsuzuki T, Fujimoto T, Yamamoto I (2006) Tetrahedron Lett 47:8227
112. Kawabata T, Yamamoto K, Momose Y, Yoshida H, Nagaoka Y, Fuji K (2001) Chem Commun 2700
113. Fernández I, Valdivia V, Leal MP, Khiar N (2007) Org Lett 9:2215
114. Effenberger F (1994) Angew Chem 106:1609

115. Effenberger F (1996) Enantiomer 1:359
116. Effenberger F, Forster S, Wajant H (2000) Curr Opin Biotechnol 11:532
117. Effenberger F, Foerster S, Kobler C (2007) In: Patel RN (ed.) Biocatalysis in the pharmaceutical and biotechnology industries. CRC, Boca Raton
118. Holmes IP, Kagan HB (2000) Tetrahedron Lett 41:7453
119. Holmes IP, Kagan HB (2000) Tetrahedron Lett 41:7457
120. Strecker A (1850) Ann Chem Pharm 75:27
121. Vachal P, Jacobsen EN (2000) Org Lett 2:867
122. Vachal P, Jacobsen EN (2002) J Am Chem Soc 124:10012
123. Kato N, Suzuki M, Kanai M, Shibasaki M (2004) Tetrahedron Lett 45:3147
124. Kanai M, Kato N, Ichikawa E, Shibasaki M (2005) Pure Appl Chem 77:2047
125. Liu B, Feng X, Chen F, Zhang G, Cui X, Jiang Y (2001) Synlett 10:1551
126. Jiao ZG, Feng XM, Liu B, Chen FX, Zhang GL, Jiang YZ (2003) Eur J Org Chem 19:3818
127. Hou Z, Wang J, Liu X, Feng X (2008) Eur J Org Chem 14:4484
128. Rueping M, Sugiono E, Moreth SA (2007) Adv Synth Catal 349:759
129. Pei D, Zhang Y, Wie S, Wang M, Sun J (2008) Adv Synth Catal 350:619
130. Nelson SG (1988) Tetrahedron Asymmetr 9:357
131. Gröger H, Vogl EM, Shibasaki M (1998) J Eur Chem 4:1137
132. Mahrwald R (1999) Chem Rev 99:1095
133. Machajewski TD, Wong C-H (2000) Angew Chem Int Ed 39:1352
134. Palomo C, Oiarbide M, García JM (2004) Chem Soc Rev 33:65
135. Hamada T, Manabe K, Ishikawa S, Nagayama S, Shiro M, Kobayashi S (2003) J Am Chem Soc 125:2989
136. Saito S, Yamamoto H (2004) Acc Chem Res 37:570
137. Akiyama T (2007) Chem Rev 107:5744
138. Yanagisawa A, Arai T (2008) Chem Commun 1165
139. Terada M (2008) Chem Commun 4097
140. Denmark SE, Wynn T, Beutner GL (2002) J Am Chem Soc 124:13405
141. Denmark SE, Beutner GL (2008) Angew Chem Int Ed 47:1560
142. Denmark SE, Heemstra JR (2007) J Org Chem 72:5668
143. Denmark SE, Beutner GL, Wynn T, Eastgate MD (2005) J Am Chem Soc 127:3774
144. Denmark SE, Chung W (2008) Angew Chem Int Ed 47:1867
145. Denmark SE, Lee W (2008) Chem Asian J 3:327
146. McGarrigle EM, Myers EL, Illa O, Shaw MA, Riches SL, Aggarwal VK (2007) Chem Rev 107:5841
147. Kataoka T, Iwama T, Tsujiyama S-I (1998) Chem Commun 197
148. Kataoka T, Iwama T, Tsujiyama S-I, Iwamura T, Watanabe S-i (1998) Tetrahedron 54:11813
149. Kataoka T, Iwama T, Kinoshita H, Tsujiyama S, Tsurukami Y, Watanabe S (1999) Synlett 2:197
150. Kataoka T, Iwama T, Kinoshita H, Tsurukami Y, Fujita M, Honda E, Iwamura T, Watanabe S-i (2000) J Organometal Chem 611:455
151. Kataoka T, Kinoshita H, Iwama T, Tsujiyama S-i, Iwamura T, Watanabe SI, Muraoka O, Tanabe G (2000) Tetrahedron 56:4725
152. Kataoka T, Kinoshita H, Kinoshita S, Iwamura T, Watanabe S-I (2000) Angew Chem Int Ed 39:2358
153. Kinoshita S, Kinoshita H, Iwamura T, Watanabe S-I, Kataoka T (2003) Chem Eur J 9:1496
154. Basavaiah D, Muthukumaran K, Sreenivasulu B (1999) Synlett 8:1249
155. Bauer T, Tarasiuk J (2001) Tetrahedron Asymmetr 12:1741
156. Kataoka T, Iwama T, Tsujiyama S-I, Kanematsu K, Iwamura T, Watanabe S-i (1999) Chem Lett 257
157. Rao JS, Brière J-F, Metzner P, Basavaiah D (2006) Tetrahedron Lett 47:3553
158. Suzuki H, Kondo A, Osuka A (1985) Bull Chem Soc Jpn 58:1335
159. Butcher ST, Detty MR (1998) J Org Chem 63:177
160. Silveira CC, Lenardão EJ, Comasseto JV (1994) Synth Commun 24:575
161. Huang Y-Z, Shi L-L, Li S-W, Wen X-Q (1989) J Chem Soc Perk Trans I 2397

162. Huang Z-Z, Ye S, Xia W, Tang Y (2001) Chem Commun 1384
163. Deng X-M, Cai P, Ye S, Sun X-L, Liao W-W, Li K, Tang Y, Wu Y-D, Dai L-X (2006) J Am Chem Soc 128:9730
164. Huang Y-Z, Tang Y, Zhou Z-L, Xia W, Shi L-P (1994) J Chem Soc Perkin Trans 1:893
165. Liao W-W, Li Kai, Tang Y (2003) J Am Chem Soc 125:13030
166. Ye L-W, Sun X-L, Li C-Y, Tang Y (2007) J Org Chem 72:1335
167. Ye L-W, Sun X-L, Zhu C-Y, Tang Y (2006) Org Lett 8:3853
168. Aggarwal VK, Smith HW, Jones R VH, Fieldhouse R (1997) Chem Commun 1785
169. Aggarwal VK, Smith HW, Hynd G, Jones RVH, Fieldhouse R, Spey SE (2000) J Chem Soc Perkin Trans 1:3267
170. Aggarwal VK, Alonso E, Fang G, Ferrara M, Hynd G, Porcelloni M (2001) Angew Chem Int Ed 40:1433
171. Fulton JR, Aggarwal VK, Vicente J (2005) Eur J Org Chem 1479
172. Saito T, Sakairi M, Akiba D (2001) Tetrahedron Lett 42:5451
173. Li A-H, Dai L-X, Hou X-L (1996) J Chem Soc Perkin Trans 1:867
174. Aggarwal VK, Abdel-Rahman H, Thompson A, Mattison B, Jones RVH (1994) Phosphorus Sulfur Silicon Relat Elem 95 and 96:283
175. Aggarwal VK, Thompson A, Jones RVH, Standen MCH (1996) J Org Chem 61:8368
176. Aggarwal VK, Abdel-Rahman H, Thompson A, Mattison B, Jones RVH (1997) Phosphorus Sulfur Silicon Relat Elem 120 and 121:361
177. Aggarwal VK, Ferrara M, O'Brien CJ, Thompson A, Jones RVH, Fieldhouse R (2001) J Chem Soc Perkin Trans 1:1635
178. Furukawa N, Okano K, Fujihara H (1987) Nippon Kagaku Kaishi 1353
179. Furukawa N, Sugihara Y, Fujihara H (1989) J Org Chem 54:4222
180. Li A-H, Dai L-X, Hou X-L, Huang Y-Z, Li F-W (1996) J Org Chem 61:489
181. Julienne K, Metzner P (1998) J Org Chem 63:4532
182. Julienne K, Metzner P, Henryon V (1999) J Chem Soc Perk Trans 1:731
183. Julienne K, Metzner P, Greiner A, Henryon V (1999) Phosphorus Sulfur Silicon Relat Elem 153–154:341
184. Zanardi J, Leriverend C, Aubert D, Julienne K, Metzner P (2001) J Org Chem 66:5620
185. Minière S, Reboul V, Arrayás RG, Metzner P, Carreto JC (2003) Synthesis 14:2249
186. Minière S, Reboul V, Metzner P (2005) ARKIVOC 11:161
187. Hayakawa R, Shimizu M (1999) Synlett 8:1328
188. Saito T, Akiba D, Sakairi M, Kanazawa S (2001) Tetrahedron Lett 42:57
189. Saito T, Akiba D, Sakairi M, Ishikawa K, Otani T (2004) ARKIVOC 2:152
190. Winn CL, Bellenie BR, Goodman JM (2002) Tetrahedron Lett 43:5427
191. Miyake Y, Oyamada A, Nishibayashi Y, Uemura S (2002) Heteroatm Chem 13:270
192. Zanardi J, Reboul V, Metzner P (2004) Bull Korean Chem Soc 25:1695
193. Minière S, Reboul V, Metzner P, Fochi M, Bonini BF (2004) Tetrahedron Asymm 15:3275
194. Davoust M, Brière J-F, Jaffrès P-A, Metzner P (2005) J Org Chem 70:4166
195. Hansch M, Illa O, McGarrigle EM, Aggarwal VK (2008) Chem Asian J 3:1657
196. Gui Y, Li J, Guo C-S, Li X-L, Lu Z-F, Huang Z-Z (2008) Adv Synth Catal 350:2483
197. Lefranc B, Valla A, Ethiraj K, Jaubert J-N, Metzner P, Brière J-F (2008) Synlett 11:1679
198. Zhou Z-L, Shi L-L, Huang Y-Z (1990) Tetrahedron Lett 31:7657
199. Takada H, Metzner P, Philouze C (2001) Chem Comm 22:2350
200. Brière J-F, Takada H, Metzner P (2005) Phosphorus Sulfur Silicon Relat Elem 180:965
201. Zanardi J, Lamazure D, Minière S, Reboul V, Metzner P (2002) J Org Chem 67:9083
202. Li K, Deng X-M, Tang Y (2003) Chem Comm 2074
203. Li J, Hu F, Xie X-K, Liu F, Huang Z-Z (2009) Cat Commun 11:276
204. Aggarwal VK, Abdel-Rahman H, Jones RVH, Lee HY, Reid BD (1994) J Am Chem Soc 116:5973
205. Aggarwal VK, Thompson A, Jones RVH, Standen M (1995) Tetrahedron Asymm 6:2557
206. Aggarwal VK, Ford JG, Thompson A, Jones RVH, Standen MCH (1996) J Am Chem Soc 118:7004

207. Aggarwal VK, Abdel-Rahman H, Jones RVH, Standen MCH (1995) Tetrahedron Lett 36:1731
208. Aggarwal VK, Abdel-Rahman H, Fan L, Jones RVH, Standen MCH (1996) Chem Eur J 2:1024
209. Aggarwal VK, Ford JG, Fonquerna S, Adams H, Jones RVH, Fieldhouse R (1998) J Am Chem Soc 120:8328
210. Aggarwal VK, Alonso E, Hynd G, Lydon KM, Palmer MJ, Porcelloni M, Studley JR (2001) Angew Chem Int Ed 40:1430
211. Aggarwal VK, Angelaud R, Bihan D, Blackburn P, Fieldhouse R, Fonquerna SJ, Ford GD, Hynd G, Jones E, Jones RVH, Jubault P, Palmer MJ, Ratcliffe PD, Adams H (2001) J Chem Soc Perkin Trans 1:2604
212. Aggarwal VK, Alonso E, Bae I, Hynd G, Lydon KM, Palmer MJ, Patel M, Porcelloni M, Richardson J, Stenson RA, Studley JR, Vasse J-L, Winn CL (2003) J Am Chem Soc 125:10926
213. Marco-Contelles J, Molina MT, Anjum S (2004) Chem Rev 104:2857
214. Xia Q-H, Ge H-Q, Ye C-P, Liu Z-M, Su K-X (2005) Chem Rev 105:1603
215. Barlan AU, Zhang W, Yamamoto H (2007) Tetrahedron 63:6075
216. Katsuky T (2000) Catalytic asymmetric synthesis. Wiley-VCH, New York
217. Denmark S E, Wu Z (1999) Synlett 847
218. Frohn M, Shi Y (2000) Synthesis 1979
219. Shi Y (2002) J Synth Org Chem Jpn 60:342
220. Shi Y (2004) In: Bäckvall JE (ed.) Modern oxidation methods. Wiley-VCH, Weinheim
221. Shi Y (2004) Acc Chem Res 37:488
222. Yang D (2004) Acc Chem Res 37:497
223. Shi Y (2006) In handbook of chiral chemicals. Taylor & Francis, Boca Raton
224. Wong OA, Shi Y (2008) Chem Rev 108:3958
225. Edwards JO, Pater RH, Curci R, Di Furia F (1976) Photochem Photobiol 30:63
226. Curci R, Fiorentino M, Troisi L, Edwards JO, Pater RH (1980) J Org Chem 45:4758
227. Gallopo AR, Edwards JO (1981) J Org Chem 46:1684
228. Cicala G, Curci R, Fiorentino M, Laricchiuta O (1982) J Org Chem 47:2670
229. Corey PF, Ward FE (1986) J Org Chem 51:1925
230. Adam W, Hadjiarapoglou L, Smerz A (1991) Chem Ber 124:227
231. Kurihara M, Ito S, Tsutsumi N, Miyata N (1994) Tetrahedron Lett 35:1577
232. Denmark SE, Forbes DC, Hays DS, DePue JS, Wilde RG (1995) J Org Chem 60:1391
233. Yang D, Wong M-K, Yip Y-C (1995) J Org Chem 60:3887
234. Denmark SE, Wu Z (1997) J Org Chem 62:8964
235. Boehlow TR, Buxton PC, Grocock EL, Marples BA, Waddington VL (1998) Tetrahedron Lett 39:1839
236. Denmark SE, Wu Z (1998) J Org Chem 63:2810
237. Yang D, Yip Y-C, Jiao G-S Wong M-K (1998) J Org Chem 63:8952
238. Yang D, Yip Y-C, Tang M-W, Wong M-K, Cheung K-K (1998) J Org Chem 63:9888
239. Nieto N, Munslow IJ, Barr J, Benet-Buchholz J, Vidal Ferran A (2008) Org Biomol Chem 6:2276
240. Curci R, Fiorentino M, Serio M R (1984) Chem Commun 155
241. Curci R, D'Accolti L, Fiorentino M, Rosa A (1995) Tetrahedron Lett 36:5831
242. Yang D, Yip Y-C, Tang M-W, Wong M-K, Zheng J-H, Cheung K-K (1996) J Am Chem Soc 118:491
243. Yang D, Wang X-C, Wong M-K, Yip Y-C, Tang M-W (1996) J Am Chem Soc 118:11311
244. Yang D, Wong M-K, Yip Y-C, Wang X-C, Tang M-W, Zheng J-H, Cheung K-K (1998) J Am Chem Soc 120:5943
245. Song CE, Kim YH, Lee KC, Lee S-G, Jin BW (1997) Tetrahedron Asymmetr 8:2921
246. Kim YH, Lee KC, Chi DY, Lee S-G, Song CE (1999) Bull Korean Chem Soc 20:831
247. Denmark SE, Matsuhashi H (2002) J Org Chem 67:3479
248. Stearman CJ, Behar V (2002) Tetrahedron Lett 43:1943

249. Matsumoto K, Tomioka K (2001) Heterocycles 54:615
250. Matsumoto K, Tomioka K (2001) Chem Pharm Bull 49:1653
251. Matsumoto K, Tomioka K (2002) Tetrahedron Lett 43:631
252. Armstrong A, Hayter B R (1998) Chem Commun 621
253. Armstrong A, Ahmed G, Dominguez-Fernandez B, Hayter BR, Wailes JS (2002) J Org Chem 67:8610
254. Armstrong A, Ahmed G, Dominguez-Fernandez B, Tsuchiya T (2006) Tetrahedron 62:6614
255. Klein S, Roberts SM (2002) J Chem Soc Perkin Trans 1:2686
256. Tu Y, Wang Z-X, Shi Y (1996) J Am Chem Soc 118:9806
257. Wang Z-X, Tu Y, Frohn M, Zhang J-R, Shi Y (1997) J Am Chem Soc 119:11224
258. Mio S, Kumagawa Y, Sugai S (1991) Tetrahedron 47:2133
259. Tu Y, Frohn M, Wang Z-X, Shi Y (2003) Org Synth 80:1
260. Tu Y, Wang Z-X, Frohn M, He M, Yu H, Tang Y, Shi Y (1998) J Org Chem 63:8475
261. Wang Z-X, Miller SM, Anderson OP, Shi Y (2001) J Org Chem 66:521
262. Tian H, She X, Shi Y (2001) Org Lett 3:715
263. Wu X-Y, She X, Shi Y (2002) J Am Chem Soc 124:8792
264. Wang B, Wu X-Y, Wong OA, Nettles B, Zhao M-X, Chen D, Shi Y (2009) J Org Chem 74:3986
265. Wong OA, Shi Y (2009) J Org Chem 74:8377
266. Tian H, She X, Shu L, Yu H, Shi Y (2000) J Am Chem Soc 122:11551
267. Tian H, She X, Xu J, Shi Y (2001) Org Lett 3:1929
268. Tian H, She X, Shu L, Yu H, Shi Y (2002) J Org Chem 67:2435
269. Hickey M, Goeddel D, Crane Z, Shi Y (2004) Proc Natl Acad Sci U S A 101:5794
270. Crane Z, Goeddel D, Gan Y, Shi Y (2005) Tetrahedron 61:6409
271. Shu L, Wang P, Gan Y, Shi Y (2003) Org Lett 5:293
272. Shu L, Shi Y (2004) Tetrahedron Lett 45:8115
273. Wong OA, Shi Y (2006) J Org Chem 71:3973
274. Burke CP, Shi Y (2007) J Org Chem 72:4093
275. Wang B, Wong OA, Zhao M-X, Shi Y (2008) J Org Chem 73:9539
276. Wong OA, Wang B, Zhao M-X, Shi Y (2009) J Org Chem 74:6335
277. Shing TKM, Leung GYC (2002) Tetrahedron 58:7545
278. Shing TKM, Leung YC, Yeung KW (2003) Tetrahedron 59:2159
279. Bez G, Zhao C-G (2003) Tetrahedron Lett 44:7403
280. Boutureira O, McGouran JF, Stafford RL, Emmerson DPG, Davis BG (2009) Org Biomol Chem 7:4285
281. Wang Z-X, Shi Y (1997) J Org Chem 62:8622
282. Wang Z-X, Miller SM, Anderson OP, Shi Y (1999) J Org Chem 64:6443
283. Adam W, Saha-Möller CR, Zhao C-G (1999) Tetrahedron Asymmetr 10:2749
284. Yang D, Yip Y-C, Chen J, Cheung K-K (1998) J Am Chem Soc 120:7659
285. Solladié-Cavallo A, Bouérat L (2000) Tetrahedron Asymmetr 11:935
286. Solladié-Cavallo A, Bouérat L (2000) Org Lett 2:3531
287. Solladié-Cavallo A, Jierry L, Norouzi-Arasi H, Tahmassebi D (2004) J Fluorine Chem 125:1371
288. Solladié-Cavallo A, Jierry L, Klein A, Schmitt M, Welter R (2004) Tetrahedron Asymmetr 15:3891
289. Bortolini O, Fantin G, Fogagnolo M, Mari L (2004) Tetrahedron Asymmetr 15:3831
290. Bortolini O, Fantin G, Fogagnolo M, Mari L (2006) Tetrahedron 62:4482
291. Armstrong A, Hayter BR (1999) Tetrahedron 1999:11119
292. Chan W-K, Yu W-Y, Che C-M, Wong M-K (2003) J Org Chem 68:6576
293. Rousseau C, Christensen B, Petersen TE, Bols M (2004) Org Biomol Chem 2:3476
294. Betzemeier B, Lhermitte F, Knochel P (1999) Synlett 1999:489
295. Goodman MA, Detty MR (2006) Synlett 2006:1100
296. Weitz E, Scheffer A (1921) Chem Ber 54:2327
297. Bäckvall JE (2004) Modern oxidation methods. Wiley-VCH, Weinheim, 2004

298. Lattanzi A (2005) Org Lett 7:2579
299. Lattanzi A (2006) Adv Synth Catal 348:339
300. Bortolini O, Fogagnolo M, Fantin G, Maietti S, Medici A (2001) Tetrahedron Asymmetr 12:1113
301. Bortolini O, Fantin G, Fogagnolo M, Forlani R, Maietti S, Pedrini P (2002) J Org Chem 67:5802
302. Imashiro R, Seki M (2004) J Org Chem 69:4216
303. Lattanzi A, Russo A (2006) Tetrahedron 62:12264
304. Russo A, Lattanzi A (2008) Eur J Org Chem 2767
305. Lattanzi A (2009) Chem Commun 1452
306. Li Y, Liu X, Yang Y, Zhao G (2007) J Org Chem 72:288
307. Cui H, Li Y, Zheng C, Zhao G, Zhu S (2008) J Fluorine Chem 129:45
308. Lu J, Xu Y-H, Liu F, Loh T-P (2008) Tetrahedron Lett 49:6007
309. Torii S, Uneyama K, Ono M, Bannou T (1981) J Am Chem Soc 103:4606
310. Iwaoka M, Tomoda S (1992) J Chem Soc Chem Commun 1165
311. Tiecco M, Testaferri L, Tingoli M, Bagnoli L, Santi C (1993) J Chem Soc Chem Commun 637
312. Tiecco M, Testaferri L, Tingoli M, Bagnoli L, Santi C (1993) Synlett 798
313. Tiecco M, Testaferri L, Tingoli M, Marini F (1994) Synlett 373
314. Tiecco M, Testaferri L, Santi C (1999) Eur J Org Chem 797
315. Fujita K-I, Iwaoka M, Tomoda S (1994) Chem Lett 923
316. Fukuzawa S-i, Takahashi K, Kato H, Yamazaki H (1997) J Org Chem 62:7711
317. Wirth T, Häuptli S, Leuenberger M (1998) Tetrahedron Asymmetr 9:547
318. Tiecco M, Testaferri L, Santi C, Tomassini C, Marini F, Bagnoli L, Temperini A (2002) Chem Eur J 8:1118
319. Niyomura O, Cox M, Wirth T (2006) Synlett 251
320. Browne DM, Niyomura O, Wirth T (2007) Org Lett 9:3169
321. Shahzad SA, Venin C, Wirth T (2010) Eur J Org Chem 3465
322. Santoro S, Santi C, Sabatini M, Testaferri L, Tiecco M (2008) Adv Synth Catal 350:2881
323. de Nooy AEJ, Besemer AC van Bekkum H (1996) Synthesis 1153
324. Osa T (1998) New challenges in organic electrochemistry. Gordon & Breach, Amsterdam
325. Bobbitt JM, Ma Z, Bolz D, Osa T, Kashiwagi Y, Yanagisawa Y, Kurashima F, Anzai J, Tacorante-Morales JE (1998) Spec Publ R Soc Chem 216:200
326. Adam W, Saha-Möller CR, Pralhad A, Ganeshpure PA (2001) Chem Rev 101:3499
327. Zhao M, Mano J, Li E, Song Z, Tschaen DM, Grabowski EJ, Reider PJ (1999) J Org Chem 64:2564
328. Semmelhack MF, Chou CS, Cortes DA (1983) J Am Chem Soc 105:4492
329. Semmelhack MF, Schmid CR (1983) J Am Chem Soc 105:6732
330. Semmelhack MF, Schmid CR, Corte's DA (1986) Tetrahedron Lett 27:1119
331. Kim SS, Nehru K (2002) Synlett 616
332. Golubev VA, Rosantsev EG, Neiman M (1965) B Bull Acad Sci USSR, Div Chem Sci 1898. (1966) Chem Abstr 64:11164
333. Anelli PL, Montanari F, Quici S (1990) Org Synth 69:212
334. Anelli PL, Banfi S, Montanari F, Quici S (1989) J Org Chem 54:2970
335. Anelli PL, Biffi C, Montanari F, Quici S (1987) J Org Chem 52:2559
336. Miyazawa T, Endo T, Okarawa M (1985) J Polym Sci Polym Chem Ed 23:1527
337. Osa T, Akiba U, Segawa I, Bobbit JM (1988) Chem Lett 1423
338. Dijksman A, Arends I W C E, Sheldon R A (2000) Chem Commun 271
339. Benaglia M, Puglisi A, Holczknecht O, Quici S, Pozzi G (2005) Tetrahedron 61:12058
340. Bolm C, Fey T (1999) Chem Commun 1795
341. Verhoef MJ, Peters JA, van Bekkum H (1999) Stud Surf Sci Catal 125:465
342. Karimi B, Biglari A, Clark JH, Budarin V (2007) Angew Chem Int Ed 46:7210
343. Ciriminna R, Blum J, Avnir D, Pagliaro M (2000) Chem Commun 1441
344. Ciriminna R, Bolm C, Fey T, Pagliaro M (2002) Adv Synth Catal 344:159
345. Schätz A, Grass RN, Stark WJ, Reiser O (2008) Eur J Chem 14:8262

346. Rychnovsky SD, McLernon TL, Rajapakse H (1996) J Org Chem 61:1194
347. Holczknecht O, Cavazzini M, Quici S, Shepperson I, Pozzi G (2005) Adv Synth Catal 347:677
348. Ciriminna R, Magliaro M (2010) Org Process Res Dev 14:245
349. Rychnovsky SD, Vaidyanathan R, Beauchamp T, Lin R, Farmer PJ (1999) J Org Chem 64:6745
350. Shibuya M, Tomizawa M, Suzuki I, Iwabuchi Y (2006) J Am Chem Soc 128:8412
351. Shibuya M, Tomizawa M, Sasano Y, Iwabuchi Y (2009) J Org Chem 74:4619
352. Tomizawa M, Shibuya M, Iwabuchi Y (2009) Org Lett 11:1829
353. Demizu Y, Shiigi H, Oda T, Matsumura Y, Onomura O (2008) Tetrahedron Lett 49:48
354. Shiigi H, Mori H, Tanaka T, Demizu Y, Onomura O (2008) Tetrahedron Lett 49:5247
355. Grieco P A, Yokoyama Y, Gilman S, Ohfune Y (1977) J Chem Soc Chem Commun 870
356. Taylor RT, Flood LA (1983) J Org Chem 48:5160
357. Syper L (1987) Tetrahedron 43:2853
358. Syper L (1989) Synthesis 167
359. ten Brink G-J, Vis JM, Arends IWCE, Sheldon RA (2001) J Org Chem 66:2429
360. ten Brink G-J, Vis JM, Arends IWCE, Sheldon RA (2002) Tetrahedron 58:3977
361. Ichikawa H, Usami Y, Arimoto M (2005) Tetrahedron Lett 46:8665
362. Ley SV, Meerholz CA (1980) Tetrahedron Lett 21:1785
363. Ley SV, Meerholz CA (1981) Tetrahedron 37:213
364. Furukawa N, Kishimoto K, Ogawa S, Kawai T, Fujihara H, Oae S (1981) Tetrahedron Lett 22:4409
365. Fujihara H, Imaoka K, Furukawa N, Oae S (1981) Chem Lett 1293
366. Furukawa N, Imaoka K, Fujihara H, Oae S (1982) Chem Lett 1421
367. Fujihara H, Imaoka K, Furukawa N (1986) J Chem Soc Perk Trans I 333
368. Fujihara H, Imaoka K, Furukawa N (1986) J Chem Soc Perk Trans I 465
369. Kondo S, Ohta K, Ojika R, Yasui H, Tsuda K (1985) Makromol Chem 186:1
370. Kondo S, Yasui H, Tsuda K (1989) Makromol Chem 190:2079
371. Ishii Y, Nakayama K, Takeno M, Sakaguchi S, Iwahama T, Nishiyama Y (1995) J Org Chem 60:3934
372. Santi C, Santoro S, Battistelle B (2010) Curr Org Chem 14:2442

Index

R. Mahrwald (ed.), *Enantioselective Organocatalyzed Reactions I: Enantioselective Oxidation, Reduction, Functionalization and Desymmetrization*, DOI 10.1007/978-90-481-3865-4, © Springer Science+Business Media B.V. 2011